Parallel Computing

Roman Trobec · Marián Vajteršic · Peter Zinterhof
Editors

Parallel Computing

Numerics, Applications, and Trends

 Springer

Editors

Roman Trobec
Dept. of Comm. Systems
Jožef Stefan Institute
Jamova 39
SI-1000 Ljubljana
Slovenia
roman.trobec@ijs.si

Marián Vajteršic
Department of Computer Sciences
University of Salzburg
Jakob–Haringer Str. 2
5020 Salzburg
Austria
marian@cosy.sbg.ac.at

and

Peter Zinterhof
Department of Computer Sciences
University of Salzburg
Jakob–Haringer Str. 2
5020 Salzburg
Austria
peter.zinterhof@sbg.ac.at

Mathematical Institute
Department of Informatics
Slovak Academy of Sciences
Dúbravská 9
840 00 Bratislava
Slovakia
vajtersic@savba.sk

ISBN 978-1-84882-408-9   e-ISBN 978-1-84882-409-6
DOI 10.1007/978-1-84882-409-6
Springer Dordrecht Heidelberg London New York

British Library Cataloguing in Publication Data
A catalogue record for this book is available from the British Library

Library of Congress Control Number: 2009926892

Printed on acid-free paper

Springer is part of Springer Science+Business Media (www.springer.com)

To all who make our lives worthwhile.

Preface

The use of parallel programming and architectures is essential for simulating and solving problems in modern computational practice. There has been rapid progress in microprocessor architecture, interconnection technology and software development, which are influencing directly the rapid growth of parallel and distributed computing. However, in order to make these benefits usable in practice, this development must be accompanied by progress in the design, analysis and application aspects of parallel algorithms. In particular, new approaches from parallel numerics are important for solving complex computational problems on parallel and/or distributed systems.

The contributions to this book are focused on topics most concerned in the trends of today's parallel computing. These range from parallel algorithmics, programming, tools, network computing to future parallel computing. Particular attention is paid to parallel numerics: linear algebra, differential equations, numerical integration, number theory and their applications in computer simulations, which together form the kernel of the monograph. We expect that the book will be of interest to scientists working on parallel computing, doctoral students, teachers, engineers and mathematicians dealing with numerical applications and computer simulations of natural phenomena.

The roots of this book are in *Parallel Numerics*, an initiative that has been active in the Central European Region since 1994, starting with the Central European Initiative (CEI) joint research project *Programming Environments, Algorithms, Applications, Compilers and Tools for Parallel Computation* (PACT). The initial scope was focused on the new results and ideas related to parallel numerics. Later the research and applied interests were broadened to theoretical and practical aspects of parallel and distributed computing, creating a fruitful combination of theoretical and applied research. Besides numerical applications, the parallel solution of financial, medical and other problems from the natural and technical sciences has been incorporated. We are glad to see that the output of this initiative has become useful in everyday computational practice, through adopting new algorithmic solutions and/or progressive programming techniques and architectural improvements.

We are grateful to the authors for their valuable contributions which make this book rich in content. We are grateful also to Springer-Verlag London for their excellent technical and editorial support. We are indebted to the Jožef Stefan Institute and the University of Salzburg for their generous support of our work.

Roman Trobec

Ljubljana, Salzburg   Marián Vajteršic
December, 2008   Peter Zinterhof

List of Contributors

Selim G. Akl
School of Computing, Queen's University, Kingston, Ontario, Canada,
e-mail: akl@cs.queensu.ca

Tore Birkeland
Department of Mathematics, University of Bergen, Norway,
e-mail: Tore.Birkeland@math.uib.no

Urban Borštnik
National Institute of Chemistry, Ljubljana, Slovenia,
e-mail: urban@cmm.ki.si

Rajkumar Buyya
Grid Computing and Distributed Systems Laboratory, University of Melbourne,
Australia, e-mail: raj@csse.unimelb.edu.au

Uroš Čibej
Faculty of Computer and Information Science, University of Ljubljana, Slovenia,
e-mail: uros.cibej@fri.uni-lj.si

Matjaž Depolli
Department of Communication Systems, Jožef Stefan Institute, Ljubljana, Slovenia,
e-mail: matjaz.depolli@ijs.si

Stefan Donath
Chair for System Simulation, Friedrich-Alexander University Erlangen-Nuremberg,
Germany, e-mail: stefan.donath@informatik.uni-erlangen.de

Thomas Fahringer
Institute for Computer Science, University of Innsbruck, Austria,
e-mail: Thomas.Fahringer@uibk.ac.at

Christian Feichtinger
Chair for System Simulation, Friedrich-Alexander University Erlangen-
Nuremberg, Germany, e-mail: `christian.feichtinger@informatik.`
`uni-erlangen.de`

Bogdan Filipič
Department of Intelligent Systems, Jožef Stefan Institute, Ljubljana, Slovenia,
e-mail: `bogdan.filipic@ijs.si`

Jan Götz
Chair for System Simulation, Friedrich-Alexander University Erlangen-Nuremberg,
Germany, e-mail: `jan.goetz@informatik.uni-erlangen.de`

Andreas Grothey
School of Mathematics, University of Edinburgh, England,
e-mail: `A.Grothey@ed.ac.uk`

Klaus Iglberger
Chair for System Simulation, Friedrich-Alexander University Erlangen-Nuremberg,
Germany, e-mail: `klaus.iglberger@informatik.uni-erlangen.de`

Dušanka Janežič
National Institute of Chemistry, Ljubljana, Slovenia,
e-mail: `dusa@cmm.ki.si`

Peter Jez
Department of Computer Sciences, University of Salzburg, Austria,
e-mail: `peter.jez@cosy.sbg.ac.at`

Rade Kutil
Department of Computer Sciences, University of Salzburg, Austria,
e-mail: `rkutil@cosy.sbg.ac.at`

Marius Nagy
School of Computing, Queen's University, Kingston, Ontario, Canada,
e-mail: `marius@cs.queensu.ca`

Gabriel Okša
Mathematical Institute, Department of Informatics, Slovak Academy of Sciences,
Bratislava, Slovakia, e-mail: `Gabriel.Oksa@savba.sk`

Matej Praprotnik
National Institute of Chemistry, Ljubljana, Slovenia,
e-mail: `praprot@cmm.ki.si`

Jarmo Rantakokko
Department of Information Technology, Uppsala University, Sweden,
e-mail: Jarmo.Rantakokko@it.uu.se

Ulrich Rüde
Chair for System Simulation, Friedrich-Alexander University Erlangen-Nuremberg,
Germany, e-mail: ulrich.ruede@informatik.uni-erlangen.de

Tor Sørevik
Department of Mathematics, University of Bergen, Norway,
e-mail: tor.sorevik@math.uib.no

Anthony Sulistio
Grid Computing and Distributed Systems Laboratory, University of Melbourne,
Australia, e-mail: anthony@csse.unimelb.edu.au

Michael Thuné
Department of Information Technology, Uppsala University, Sweden,
e-mail: michael.thune@it.uu.se

Roman Trobec
Department of Communication Systems, Jožef Stefan Institute, Ljubljana, Slovenia,
e-mail: roman.trobec@ijs.si

Andreas Uhl
Department of Computer Sciences, University of Salzburg, Austria,
e-mail: uhl@cosy.sbg.ac.at

Marián Vajteršic
Department of Computer Sciences, University of Salzburg, Austria, and
Mathematical Institute, Department of Informatics, Slovak Academy of Sciences,
Bratislava, Slovakia,
e-mail: marian@cosy.sbg.ac.at

Peter Zinterhof
Department of Computer Sciences, University of Salzburg, Austria,
e-mail: peter.zinterhof@sbg.ac.at

Contents

About the Editors

Roman Trobec received MSc and PhD degrees in electrical engineering and computer science from University of Ljubljana, Slovenia, in 1979 and 1988, respectively. He is with the Department of Communication Systems at the Jožef Stefan Institute since 1976, holding currently the position of principal investigator. Since 2001 he is Associate Professor at the University of Ljubljana, Faculty of Electrical Engineering and Faculty of Computer and Information Science and visiting professor at the University of Salzburg, Department of Scientific Computing.

His research and working experiences are in the area of parallel and distributed computing, interconnection networks, scientific computing, computer simulations, computer-aided applications, wireless sensor networks and advanced biosignal analysis. He has published over seventy scientific papers in international recognized journals and six book chapters. He is a program or editorial board member of several international conferences and journals. He is cooperating on various national and EU scientific and applied projects from the areas relevant to his research.

Marián Vajteršic graduated in Numerical Mathematics from Comenius University, Bratislava (Slovak Republic) in 1974. He received his CSc (candidate of sciences) degree in mathematics from the same university in 1984 and he defended there the DrSc (doctor of sciences) degree in 1997. In 1995, he obtained the habilitation degree in numerical mathematics and parallel processing from the University of Salzburg (Austria).

His research activity is focused on the area of parallel numerical algorithms for high-performance computer systems. He is author of two monographs, co-author of three other books and of more than 100 scientific papers. Since 1974, he is with the Slovak Academy of Sciences in Bratislava, Slovakia. As a visiting professor he has been with the universities of Vienna, Bologna, Milan, Linz, Salzburg, Amiens and Munich. Since 2002 he is a Professor at the Department of Computer Sciences at the University of Salzburg, Austria.

He holds a fellowship of the Alexander von Humboldt-Foundation and the Royal Norwegian Society. He coordinated a number of national and international projects (e.g. EU, NATO). He is a member of the editorial boards of multiple international journals (e.g. PPL, IJPEDS, CAI) and programme committees of conferences in parallel computing and numerical linear algebra.

Peter Zinterhof received his PhD in mathematics after a visiting scholarship at Steklov Institute in Moscow from University of Vienna in 1968. He defended his habilitation in mathematics at Technical University of Vienna in 1971. He is Professor for Mathematics since 1972 and also Professor for Theoretical Informatics since 1989. He served as Dean of the faculty of Natural Sciences and as Head of the Mathematics and computer science departments of the University of Salzburg for many years and is actual Head of the department for Computer Science of the University of Salzburg. He is program or editorial board member of several international conferences and journals. His research interests are parallel computation, number theoretical numerics, uniform distribution of sequences, Monte-Carlo- and Quasi-Monte-Carlo-Methods, stochastics and reliability theory, and Hilbert space methods in numerics and applied mathematics. He is the co-founder of the Austrian Center for Parallel Computation (ACPC) and founder of the Computer Science Department of the Salzburg University. He is author or co-author of more than hundred original papers and co-author of several books and book chapters.

Chapter 1
Overview – Parallel Computing: Numerics, Applications, and Trends

Marián Vajteršic, Peter Zinterhof and Roman Trobec

Abstract

This book is intended for researchers and practitioners as a foundation for modern parallel computing with several of its important parallel applications, and also for students as a basic or supplementary book to accompany advanced courses on parallel computing. Fifteen chapters cover the most important issues in parallel computing, from basic principles through more complex theoretical problems and applications, together with future parallel paradigms including quantum computing. Each chapter is written on two levels: a more general overview and a more specific example of theory or practice. In this introductory chapter some views regarding state-of-the-art and trends in parallelism are given, accompanied by a summary of individual chapters.

1.1 Introduction

The use of parallel processing is today essential for solving practical problems in science and engineering. Parallelism is a way of speeding up computations which make high time and memory demands. Historically [1], parallelism was first aimed at speedup, as its primary objective, which was characterized by the Amdahl's law.

Marián Vajteršic
Department of Computer Sciences, University of Salzburg, J.-Haringer-Strasse 2,
5020 Salzburg, Austria, and Mathematical Institute, Department of Informatics, Slovak Academy
of Sciences, Bratislava, Slovakia, e-mail: marian@cosy.sbg.ac.at

Peter Zinterhof
Department of Computer Sciences, University of Salzburg, J.-Haringer-Strasse 2,
5020 Salzburg, Austria, e-mail: peter.zinterhof@sbg.ac.at

Roman Trobec
Department of Communication Systems, Jožef Stefan Institute, Jamova cesta 39, 1000 Ljubljana,
Slovenia, e-mail: roman.trobec@ijs.si

R. Trobec et al. (eds.), *Parallel Computing*, DOI 10.1007/978-1-84882-409-6_1,
© Springer-Verlag London Limited 2009

The Gustafson–Barsis' law then turned the attention to scalability, the goal being to keep the time constant as the problem size grows. The consequence of this was a construction of systems with increasing numbers of processors, which were mostly special purpose and hence costly not only from the hardware point of view but also because of the accompanying software development. So parallelism approached the cluster era, with cheap systems configured as clusters of desktop computers, which were interconnected by special or commodity networks. And the progress continues: at the microprocessor level parallelism is encountered in multicores and manycores. The ambition of the Grid and Cloud ideas is to elevate parallelism to the highest level, making the computational power as widely available to that extent as we know it for electricity. Further, on the horizon there are radically new solutions such as quantum computing, optical computing, and chips working on biological structures, which all possess a potential for future parallel computing.

All this progress is tightly connected with development of methods and algorithms for these systems. Without effort on this field, the computational power of supermodern computers, which are today available in a very rich variety of products, cannot be exploited. Fast and robust algorithms, programmed with the necessary skill, are essential for efficiently solving urgent scientific problems in physics, medicine, biology, pharmacy, meteorology, information retrieval, and other areas, where parallelism is already a domestic principle.

This book deals with a broad spectrum of issues related to parallelism. It is a reflection of new opportunities and challenges which require major progress in architecture, system scalability, and interconnect technologies for algorithm designers, system architects, and software developers when aiming to master solutions of larger and more complex problems in parallel.

1.1.1 Parallel Numerics

Primary attention is paid to parallel numerics, since the kernel group of the authors of this book were participating at the *ParNum* initiative [2], focused on design and implementation of parallel numerical methods and their applications. In order to exploit all the power that modern parallel systems offer, the algorithm design for solving numerical problems is of a crucial importance. Numerical algorithms are widely used by solving intensive problems in a great variety of application areas in natural and technical sciences. Problems related to the algorithm area include parallelization paradigms, optimal data structures, analysis of rounding errors of parallel numerical computations, optimalization techniques for various parallel platforms, parallel numerical libraries, testing and benchmarking, performance tuning, etc. Without considering the intimate relationship between algorithms on the one hand and languages, tools, and environments for their programming on contemporary computer architectures on the other, parallel numerical computing could never be successful.

The topical issue in parallel numerics is the parallel linear algebra [3]. It is because, in our opinion, the work of parallel linear algebra solvers is fundamental to solving scientific computing applications. Over the past two decades the development of efficient linear algebraic techniques has become an important feature of research in many theoretical and application areas of parallel computing. Novel, highly efficient methods, algorithms, and implementations have been developed and are now in widespread use.

Some of the relevant themes in this problem area include:

- parallel methods for solving large dense, sparse, and structured linear systems of equations: direct, iterative, domain decomposition
- parallel eigenvalue and singular value computations
- parallel matrix multiplication and combinatorial scientific computing
- parallel linear algebra methods in data mining
- parallel linear algebra solvers for ordinary and partial differential equations
- parallel least squares and optimization.

The libraries LAPACK (Linear Algebra PACKage) [4] and ScaLAPACK (Scalable LAPACK) [5] are still main sources of software for linear algebra. In LAPACK algorithms, parallelism is exploited at the level of the BLAS [6] operations. ScaLAPACK provides optimized high performance scalable algorithms, based on the distributed memory versions (PBLAS) of BLAS and on communication subprograms (BLACS), which makes it suitable for parallel computers with distributed memory.

The project FLAME (Formal Linear Algebra Methods Environment) [7] provides a new way toward the systematic and mechanical development of libraries for this field. Many obstacles in library production (e.g., effort required to identify candidate algorithms, formal correctness approval through extensive testing, numerical stability analysis, and the time for empirical tuning) may be overcome by applying modern software engineering approaches and tools. The effort and expense of upgrading for new architectures can thus be reduced. Within this, a new notation for expressing dense linear algorithms is being formulated, raising the level of abstraction at which the algorithm is represented.

For problems involving very large matrices, the so-called OOC (out-of-core) algorithms have been developed [8]. It is often the case that the data structures of the problem are too large to fit in the memory; therefore, it is reasonable to use disk storage which is cheap and available in larger quantity. OOC LU decomposition of a nonsymmetric dense matrix with pivoting [9], sparse symmetric–indefinite factorization [10], and QR and SVD decompositions [11] are examples of such algorithms.

New trends in this area are influenced by the existence of multicore systems, for which the library routines have to be reformulated or new algorithms have to be developed. The exploitation of fine grain parallelism becomes a major requirement and introduces the necessity of loose synchronization in the parallel execution of an operation. Algorithms for the Cholesky, LU, and QR factorization, where the operations can be represented as a sequence of small tasks that operate on square blocks of data, are already available [12]. These tasks can be scheduled dynamically for execution based on the dependences between them and on the availability of computational resources. This may result in an out-of-order execution of the tasks which will completely hide the presence of intrinsically sequential tasks in the factorization.

1.1.2 Parallel Architectures

It is out of the scope of this book to cover the huge progress in computer architecture, but we have to at least mention the revolutionary multicore idea in processor design. What is the motivation for it? One could name power consumption, heat dissipation, minimization of communication distance, and other physical limitations which are factors pushing the microprocessor industry toward multicores.

Multicore architectures have started a new era of computing and boost performance and efficiency of parallel programs. The problem is, however, how to execute sequential programs efficiently and reliably on multicores. Multicore architectures usually have many cores that are weaker than those in a monolithic out-of-order core; however, the aggregation of many small cores provides better computing and power efficiency than a monolithic one. It is expected that with multi- and manycore-based systems, performance increase on the microprocessor side will continue according to Moore's law, at least in the near future. However, the already limited memory access is expected to become more of a problem with multiple cores on a chip, and gets more complex because of the required hierarchies of cache memory. In addition, the increasingly hybrid and hierarchical design of high-end systems, including accelerator components, is another obstacle for programming. These issues will have a major impact on overall performance in future systems. Nanoscale manufacturing processes increase the relevance of reliability and availability while the future appearance of tens or hundreds of cores on a single chip moves the programmability to a center stage.

This confirms our earlier statement that computation- and data-intensive tasks can benefit from the hardware's full potential only if both processor and architecture features are taken into account at all stages from the early algorithm design to the final implementation.

Problem areas related to efficient use of multicore systems include

- design of multicore-aware approaches for large-scale parallel simulations in both implementation and algorithm design, including scalability studies
- compiler techniques and code optimization strategies for parallel systems with multicore processors
- tools for performance and cache behavior analysis (including cache simulation) for parallel systems with multicore processors
- parallelization with appropriate programming models and tool support for multicore and hybrid platforms.

One example of a novel multicore architecture is the Cell Broadband Engine (Cell BE) developed jointly by IBM, Sony, and Toshiba [13]. It is an innovative heterogeneous multicore chip that differs significantly from conventional multiprocessor or multicore architectures. It represents a new execution platform for computation-intensive applications that reflect both the requirements of future computational workloads and manufacturing constraints. The Cell BE is a heterogeneous chip achieving more than 200 GFlop per chip. The initial target of the Cell BE was the Sony's PlayStation 3 game console, which boasts a chip with nine CPUs for faster and more realistic video gaming. Soon it became evident that the impressive computational power of Cell BE could make it potentially well suited for other applications such as visualization, multimedia processing, and various scientific and technical workloads [14].

The number of cores on a single chip continues to increase: there are actual reports of 64 [15] and even 80 [16] cores accommodated on a single chip. All these new designs confirm the evidence that the exploitation of their spectacular capabilities will require new tools, new algorithms, and a new way of looking at the programming.

Network-on-Chip (NoC) constitutes a practical approach in the architecture of communication networks [17]. It is driven by the advanced CMOS nanotechnology and its function is to provide delivery between the source node and the destination within large VLSI (Very Large Scale Integration) systems implemented on a chip. An NoC can provide separation between computation and communication. The fact is that while the computation logic grows, the performance of on-chip interconnections does not scale as well. Wire delay dominates gate delay and the gap between wire delay and gate delay becomes wider as process technology improves. In addition, human design productivity cannot keep up with the growth rate of available circuits on a single chip. NoC enables wire segmentation and wire sharing design techniques to be applied to resolve the performance bottleneck due to wire delay [18].

In particular, NoC switches must be small, energy-efficient, and fast. The routing algorithms should be implemented by simple logic, and the number of data buffers should be minimal. Most NoCs are based on a regular mesh-like pattern, but there are already thoughts about using heterogeneous and less regular topologies, enabling application-specific designs [19].

Field Programmable Gate Arrays (FPGAs) [20] are a novel series of building blocks for massively parallel systems. They are programmable computing devices that are given their final functionality shape by the customer or designer following manufacture. FPGAs are wired together from programmable logic components, using a hierarchy of reconfigurable interconnections. Due to their flexibility, FPGAs are well suited for constructing special-purpose hardware. Moreover, they may be used to design reconfigurable massively parallel architectures [21]. Because of the many ways in which FPGAs can be arranged into massively parallel systems, they have achieved a high profile in modern High-Performance Computing (HPC) environments [22].

These programmable chips are useful for applications where parallelism can be applied massively, e.g., in signal processing [23] or cryptography [24]. In particular, both FFT-based and convolution-based numerical computations can be executed efficiently on FPGAs instead of general-purpose microprocessors [25]. An iterative image restoration algorithm has been developed to suit FPGAs implementation in [26]. Performance results, obtained from an actual implementation of this approach on a Xilinx FPGA, demonstrate the advantage of this attractive hardware concept also for solving more demanding signal processing tasks.

Parallel linear algebra solvers have also been proposed for FPGAs with promising speedups achieved over CPUs. Matrix inversion and LU decomposition with pivoting and the so-called mixed-precision direct linear system solvers have been designed and examined from this perspective in [27]. A family of fixed-point linear algebra intellectual property (IP) cores for Xilinx FPGA devices is already on offer commercially [28]. These cores constitute direct implementations of Matlab procedures in silicon.

With each newcomer in architecture there is a need for researchers from academia and industry to share and exchange their experience, discuss challenges, and report state-of-the-art and in-progress research of all features of a new technology. In particular, experience from application developers with its use, performance of real applications, results from the implementation of tools supporting the development, and final execution are required in order to provide a final answer about its real potential.

1.1.3 Scalability

The most used choice for large-scale computing is the parallelism for supercomputer clusters with distributed memory. This parallelization is of coarse-grain type, where large portions of programs should run on processors independently, without mutual communication. For communication, the message passing standard is adopted. Also a proper load balancing of tasks among the processors has to be optimized in order to achieve reasonable performance figures. Thus, the problems which are going to be parallelized have to be analyzed and solved in such a way that these requirements are met.

Other option for parallel solution of challenging applications and for processing huge and complex information sets of all kinds are the heterogenous distributed systems. These range from simple networks of desktop computers to highly complex Grid computing environments. Such computational systems have been preferred due to their reduced costs and inherent scalability, but their efficient use poses many challenges in terms of information access, storage and retrieval, as well as in terms of algorithms for efficient management, transfer, and analysis of huge volumes of data and knowledge.

Grid computing, peer-to-peer technology, data and knowledge bases, distributed information retrieval technology, and networking technology all these converge by addressing the scalability problem [29]. Furthermore, with the advent of emerging computing architectures, e.g., SMTs (Simultaneous Multithreading) [30], GPUs

(Graphics Processing Unit) [31], multicores, etc. The importance of designing techniques explicitly targeting these systems is becoming more and more important. Some areas where scalability plays a role are:

- parallel and distributed information retrieval
- Grid information systems
- peer-to-peer systems
- mobile systems
- Web services
- multimedia information systems
- emerging computing architectures (SMTs, GPUs, multicores)
- data mining
- information security
- very large databases.

1.1.4 Supercomputers

In 2008, supercomputing entered the petaflop/s era. In the TOP500 list of the most powerful computers is a system which for the first time in history delivered a peak performance of 1.026 petaflop/s [32]. The system is named Roadrunner and was built by IBM [33] for the Los Alamos National Laboratory in New Mexico. It is based on the IBM QS22 [34] blades which are built with advanced versions of the processor in the Sony PlayStation 3.

Rank 2 on the actual TOP500 list, which has been produced since 1993 twice a year and brings a survey of trends and changes in the global supercomputer area, is now reserved for IBM Blue Gene/L with 478.2 teraflop/s. This computer is installed at the Lawrence Livermore National Laboratory and it has been the world's number one from 2004.

The first five positions occupy systems in the U.S. Number 3 is the new IBM Blue Gene/P (450.3 teraflop/s) at the Argonne National Laboratory, followed by the new Sun SunBlade x6420 Ranger system (326 teraflop/s) at the Texas Advanced Computing Center at the University of Texas in Austin, and the upgraded Cray XT4 Jaguar (205 teraflop/s) at the Oak Ridge National Laboratory. The first ranked site in Europe is Forschungszentrum Jülich on rank 6 with its BlueGene/P of 180 teraflop/s. Among the first ten systems, there are also two installations in France.

The advent of multicores is also apparent in these systems. Quad-core processors are used in 283 systems and 203 systems are using dual-core processors. Only 11 systems still use single-core processors, and three systems use IBM's advanced Sony PlayStation 3 processor with 9 cores.

IBM is the leader in the list for the total number of systems and also for the performance. It leads in systems with 210 systems (42 percent) over Hewlett-Packard with 183 systems (36.6 percent) and in performance with 48 percent of installed total performance, compared to Hewlett-Packard with 22.4 percent. The rapid progress in supercomputers is illustrated also by the fact that the last system on the current list would have been listed at position 200 just six months ago.

For the first time, the TOP500 list also provides energy efficiency calculations (in Mflop/s/Watt). The Roadrunner is not only the most powerful but also the most energy-efficient supercomputer at the moment. It uses the IBM QS22 Cell processor, which has an efficiency up to 488 Mflop/s/Watt, while in second position, also held by IBM with BlueGene/P systems, is this value up to 376 Mflop/s/Watt.

1.1.5 Grid Computing

Grid computing occupies the other side of the parallel computational spectrum. It tries to fulfill a long-term goal of the user community to have computing power available anytime and at any place just by simply plugging a terminal into the net. However, we are still far from this goal, but the idea is still challenging. In such areas as scientific and enterprise computing there are already results from utilization of geographically dispersed computer facilities. The main problem is that the systems included in Grid are heterogeneous. It is assumed that Grid comprises a large, collaborative system consisting of different hardware platforms, operating systems, software libraries, and applications providing the ability to perform higher throughput computing. While some differences may be relatively easy to overcome, such as different processor architectures or operating systems, other differences are more subtle, such as binary patches or simply placement within the file system, and carry a much greater potential for failure.

A broad range of both commercial and scientific toolkits for Grid computing exists. The Open Grid Forum (OGF) [35] holds the standardization position and the result of this effort is the reference implementation Globus Toolkit [36]. The design of the common Grid architecture of the OGF is referred to as the Open Grid Services Architecture. The approach taken by the OGF relies fundamentally on a web-based service-oriented architecture, as embodied by the World Wide Web consortium's web service standards.

More problems already examined for Grid implementation come from parallel numerics. Monte Carlo methods appear to be good examples of this effort [37]. Monte Carlo for matrix operations is applied in Grid environment in [38], and Quasi Monte Carlo for high-dimensional integration is published in [39]. Also a generalized eigenvalue problem was solved in Grid environment [40]. For linear algebra, a scientific computing environment based on web services is presented in [41]. It allows users to perform their linear algebra tasks without explicitly calling the libraries like BLAS, LAPACK or ScaLAPACK, and software tools, as well as without installing related software on local computers. A user enters an algebraic formula, as, e.g., in Matlab, which is evaluated for determining the combinations of services answering the user request. Services are then executed over the Grid using the DIET (Distributed Interactive Engineering Toolbox) [42] middleware.

Information retrieval is an application that really can take advantage from Grid computing. Because of the large amount of distributed data, well-known online information retrieval engines are already powered by proprietary Grid-like technologies. In [43], the parallel SVD (Singular Value Decomposition) was the kernel operation in the application of the LSI (Latent Semantic Indexing) principle for algebraic data retrieval approach on a Grid.

Parallel numerical methods are also included in more complex applications, e.g., from medicine, multimedia, meteorology, and bioinformatics, which are solved on a Grid within the national project AGRID2 (Austrian Grid) [44], which includes leading Austrian research institutions in advanced computing technologies with partners working in Grid-relevant application areas. One of the applications is the image retrieval using the widespread vector-space model with the help of Grid middleware, where Householder orthogonalizations were used [45].

In this context, the rapidly growing area of the Cloud computing has to be noted [46]. This approach is based on the Internet as a vehicle to satisfy the computing needs of the consumers. Such technologies developed for Internet, such as SaaS (Software as a Service) [47], Web2.0, and GoogleApps [48], enable the use of remote computing capacities of a huge numbers of servers. Without having knowledge about the organization of the services behind it, the application is put into the cloud of elastic computing power, where it will be processed on the usual payment basis. There are already scientific clouds available, such as Nimbus (University of Chicago), Stratus (University of Florida), Wispy (Purdue University), and Kupa (Masaryk University, Czech Republic) [49], as well as those for commercial purposes, offered by vendors such as Amazon, Google, and Salesforcee [50]. However, this challenging idea needs a long period of evolution before it becomes a widely adopted technology. The first obstacle is that not all applications are naturally suited to this concept. Generally, the applications should be split into a relatively large number of subtasks, all of which can be executed concurrently. The second problem lies in the software. Only a few applications are coded in such a manner that they can run on a large number of servers.

1.1.6 Parallel Programming Languages

For programming of parallel computers, MPI (Message Passing Interface) [51], which conforms to the message-passing model, is the most widely considered tool at present. It is an example of a successful standardization process, which has been followed by a number of realizations such as MPICH2 [52], LAM/MPI [53], or implementations by principal vendors (IBM, Sun, and Intel). The implementations contain parallelization functions with language interface to Fortran and C/C++. The state-of-the-art of the MPI standard is called MPI-2, which adds one-sided communication, dynamic processes, and parallel I/O.

PVM (Parallel Virtual Machine) [54] is another example of a message-passing library. A pool of workstations was configured by PVM to a virtual parallel machine capable of performing concurrent computations, like a high-performance computer, but with significantly lower cost.

The so-called PGAS (Partitioned Global Address Space) languages [55] constitute an improvement of message-passing libraries that are designed for a memory model where the global address space is logically partitioned in such a way that a portion of it is local to each processor. The main representatives of this group are Co-array Fortran [56], UPC (Unified Parallel C) [57], and Titanium [58].

Since most of scientific computing applications are written in Fortran, a parallel version of this language was expected to be echoed by parallel programmers. With this objective, parallel versions of this language Vienna Fortran [59] and HPF (High Performance Fortran) [60], based respectively on Fortran77 and Fortran90, were developed. HPF is a representative data-parallel language in which the basic idea is to implement vector and matrix operations in parallel, whereby operations on array elements are executed simultaneously across available processors. Moreover, the responsibility for low-level details of the implementation is transferred from the programmer to the compiler. This makes data-parallel programming attractive for use. Despite this fact, neither the first version of this language nor its successor HPF-2 has gained the acceptance as expected. The best-known commercial HPF compiler is provided by Portland Group [61] and is a good tool for those who want to parallelize their Fortran codes with operations on large arrays.

In the shared-memory world, the programming approaches are based on threads. There are more options for employment of threads. The first is to use the thread library, from which the standard represents POSIX (Pthreads) [62]. Its weakness is that it has marginal support for data-parallel operations that frequently occur in scientific computing.

The most frequently used approach for shared-memory parallelization is still OpenMP [63]. Targeting Fortran or C/C++, it enriches the programming language by a set of compiler directives, used to describe the parallelism in the source code, and a small library of supporting routines. It is the compiler that arranges for the low-level work and creates, when necessary, a number of threads to process simultaneously parallel regions in the code. OpenMP programming is simpler and on a higher level than programming with Pthreads or MPI. Because of its success, there is a development attempting to bring OpenMP also onto the distributed-memory territory. Initiated by Intel, its name is Cluster OpenMP [64] and is related to the distributed shared-memory concept. On a software basis, it provides cluster nodes with access to a large shared memory, in addition to their non-shared private memories.

The popular Matlab [65] environment also provides parallel programming support for both task- and data-parallelism. The Matlab producer MathWorks developed the parallelized version Parallel Computing Toolbox [66] which enables to solve computation- and data-intensive problems using Matlab on multicore and multiprocessor computers. Parallel processing constructs such as parallel for–loops and code blocks, distributed arrays, parallel numerical algorithms, and message-passing functions enable task- and data-parallel algorithms to be implemented in Matlab at a high level without programming for specific hardware and network architectures. Thus, converting serial Matlab applications to parallel Matlab applications is done easily by few code modifications and without programming in a low-level language.

It is obvious that, with the advent of multi- and manycore processors and heterogeneous architectures, hardware deep memory hierarchies and exponentially increasing numbers of processors, new programming models, and support have to be proposed. These factors dictate the use of language features that provide higher level abstractions than do C or older Fortran standards. Also, object-oriented programming, represented by C++ and to an increasing extent by Java and Python [67], is not able to meet these demands.

A novel and promising proposal in the area of parallel programming languages is going to be Chapel [68]. This language is under development by Cray within the Cascade Program. This project is part of DARPA's High Productivity Computing Systems (HPCS) initiative [69], whose main objective is to provide a new generation of economically viable HPCS. HPCS program focuses on productivity in the HPC domain, which is defined as a combination of performance, programmability, portability, and robustness. The ambition of the initiative is to increase the productivity by a factor of 10 by the end of the period 2007–2010.

Chapel uses concepts from already developed languages, mainly from the HPF. It is based on the multithread execution model where the parallelism is expressed in terms of independent computations using threads. It supports high-level abstractions for data parallelism, task parallelism, and nested parallelism. It enables optimization for the data locality. Object-oriented features are also included. Interoperability with other languages (C, Fortran, C++, Python, and Java) is also foreseen. On November 15, 2008, the first public release of Chapel became available for download.

Results from the study [70] are interesting, from the parallel programming reality point of view. A pool of 250 participants answered a question about parallel programming systems they use in practice. It was not surprising that MPI was the clear winner, followed by POSIX Threads and OpenMP. Together with Java Threads [71], these three systems are widely known and used tools for parallel programming by the practitioners who participated in the above worldwide inquiry.

1.1.7 Parallel Compilers

One of important factors contributing to the performance of parallel applications is the quality of the generated code, which depends heavily on the compiler. A program written in a high-level language, such as C/C++ and Fortran, has to be converted into the machine instruction set of the architecture being used. This kind of translation is done by compilers and interpreters. Parallel compilers are programs that try to parallelize the process of program compilation.

The appearance of processors with simultaneous multithreading, and currently with the advent of chip multiprocessors with an increasing number of cores, stimulates the interest of both industry and academia to improve compilers to be able to exploit this type of parallelism. In these systems, the already traditional gap between processor speed and memory speed is now overbridged by the increase in the number of computing cores that demand data from the same memory. As a result, understanding the complex interaction between software and memory hierarchy hardware, and the implementation of automatic compiler techniques that lead to optimal exploitation of this hierarchy are, and will continue to be, hot research topics.

Interest in shared-memory compilers, motivated particularly by the advent of multicores, has also stimulated new developments, of which OpenUH [72] is one. It is a portable OpenMP compiler which fully supports the latest OPenMP version and contains, in addition, many analysis and optimization passes. Another example is the adaption of OpenMP to Java, called JaMP, described in [73]. This system proposes new extensions to OpenMP which are better suited to the object-oriented programming by Java.

Another current trend is due to the growing availability and complexity of embedded processors. Here, much more than in the HPC area, we may observe a development of new processors with novel features that can be very compiler-dependent such as VLIW (Very Large Instruction Word) and complex register file architectures. The difficulty in programming these architectures and the strong demands for real-time response drive the developers mostly to low-level manual programming. Hence, the understanding and design of powerful analysis and heuristics for compilers oriented to these architectures is an attractive research topic.

It is recommended that, instead of completely re-engineering compilers for parallelism, it is more efficient to use autotuners that search to yield efficient parallel code [74]. The arguments for looking in new direction are obvious: new functions added to compilers need a large programming investment and testing and, due to the limitations of exiting compilers, peak performance may still require labor-intensive manual programming of parallel issues, such as data layout, data movement, and processor synchronization.

Despite the fact that autotuners for parallel codes do not exit so far, their use would be a promising way to overcome these obstacles. Autotuners [75] work on a search principle and optimize a set of library kernels by generating many variants of a given kernel and benchmarking each variant by running on the target platform. The search process effectively tries many or all optimization switches. In many cases, the autotuned code is faster than vendor libraries that were specifically hand-tuned for the target machine. Autotuners for parallel codes have to take into account the fact that, for a given problem to be solved, there may be several parallel algorithms, each with alternative parallel data layouts. The optimal choice may depend not only on the processor architecture, but also on the parallelism of the computer, as well as the network bandwidth and latency.

It is expected that compiler development will continue to be driven by the appearance of new needs, novel hardware architectures, and programming languages and paradigms. The more that parallelism will be offered at different levels of future computer systems, the greater will be the motivation to develop programs for them in an easy and user-friendly manner, for which new compiler solutions are inevitable.

1.2 Book Chapters

In this part of the book's introductory chapter, a brief annotation of subsequent chapters, with some information on the authors and their institutions, will be given.

1.2.1 Introduction to Parallel Computation

Selim G. Akl and Marius Nagy, Kingston, Ontario, Canada

The first author of this chapter is a well-known expert in the area of algorithm design and analysis, in particular for problems in parallel computing. His recent research interests are focused on unconventional computing [76]. He published fundamental books related to parallel models and algorithms, of which [77, 78] could be counted as classics among textbooks in this area. He is editor in chief of Parallel Processing Letters (World Scientific), one of the most prestigious international journals devoted to parallelism.

His chapter, co-authored by M. Nagy, provides an overview of the fundamental concepts and ideas in the field of parallel computation. The first part gives the reader useful introductory information about possible computational models, algorithm design methods, and specific analysis techniques, which are helpful to understand the themes discussed in further chapters of the book, especially for a non-expert audience.

Fundamental parallel classical models, ranging from the shared-memory model, through combinatorial circuits, clusters, up to the Grid are included and explained. The section devoted to the design methods for parallel algorithms is written from the perspective that the algorithm designer has to consider necessarily the characteristic features of the model being used. The design methodology consists of four basic steps: partitioning (decomposition to a maximum number of concurrently executable tasks), communication analysis (evaluating the amount of communication among tasks), granularity control (reduction of communication requirements by agglomeration), and mapping (assigning tasks to processors of the model). The process is illustrated for the Gaussian elimination on the mesh of trees topology.

Parallel algorithms for conventional computations, i.e., for computations which are frequent components of various applications, are illustrated by designs for prefix-sum and suffix-sum computations on a linked list and by sorting on a model with three configured buses. Following the authors' view on classification of computations, for the so-called unconventional computations, they present interesting algorithmic solutions which confirm that parallelism is beneficial from the point of view of speedup and quality-up.

The second part of the chapter concludes with a surprising and maybe, at the same time, provocative claim that the notion of universality in computation is false. It is based on the observation that no machine can claim universality since there will always be a larger set of problems that such a machine cannot solve. Parallelism should deliver a proof of the nonuniversality claim: there is no finite computational device, sequential or parallel, conventional or unconventional, which is able to simulate all others [79].

1.2.2 Tools for Parallel and Distributed Computing

Thomas Fahringer, Innsbruck, Austria

The author is the head of the Institute of Computer Science, University of Innsbruck, Austria, where he also leads the Distributed and Parallel Systems Group. Before joining the University of Innsbruck, he worked at the University of Vienna. T. Fahringer was involved in numerous national and international research projects including AURORA [80] and the Austrian Grid [81]. His group currently coordinates two EU projects (edutain@grid and EC-GIN). He has published three books, the most recent one by Springer [82]. His current research contributions are related to the area of compilers and programming languages and tools for distributed and parallel systems.

In our book, the author presents the ASKALON system [83], which comprises a tool set to support performance-oriented development of parallel and distributed applications. It was originally designed for clusters of multiprocessors and has later on been extended to a powerful environment for application development and optimization for Grid and Cloud applications. An important aspect of ASKALON deals with the problem to port existing software tools to Grid and Cloud infrastructures which turned into an increasingly serious problem over the last couple of years and has not been thoroughly addressed by the scientific community. Portability and interoperability problems of software tools on the Grid and on Clouds are caused mostly by the heterogeneous and often machine-dependent nature of tools, by complex operating system and compiler dependencies as well as by differences and incompatibilities in tool functionality, interfaces, and other proprietary solutions.

The chapter starts with a detailed description of related work that reflects previous and ongoing work in this area. The overall architecture of the ASKALON development and runtime environment and its basic functionality is outlined in the first section devoted to this system. A separate section is reserved for the detailed introduction of each of the four constituent tools, the structure of which is based on the composition and sharing of remote Web services.

ASKALON's four tools are coherently integrated into a Web service-based distributed architecture which enables their interoperability. SCALEA is a performance instrumentation, measurement, and analysis tool for parallel programs that supports automatic generation of instrumentation code and delivers required performance parameters during or after program execution. ZENTURIO is a general purpose experiment management tool for automatic generation and conduction of a large number of experiments in the context of large-scale performance and parameter studies for clusters and Grid and Cloud architectures. AKSUM provides the user with an interface for analysis of performance data of several experiments and defines how experiments are generated and executed. After the necessary information has been provided, AKSUM automatically conducts performance analysis without any user interference. The Grid-Prophet can predict the performance of parallel and distributed applications based on machine learning techniques.

The chapter is concluded by numerous experiments which demonstrate the usefulness and effectiveness of the ASKALON development and runtime environment for a variety of real-world applications.

1.2.3 Grid Computing

Uroš Čibej, Ljubljana, Slovenia
Anthony Sulistio and Rajkumar Buyya, Melbourne, Australia

Grid computing occupies one of the most trendy positions in contemporary computer science. It offers obvious benefits and has a big potential to be accepted as a new computing technology for and beyond parallel computing. Clearly these are good arguments to include into our publication also a chapter devoted to this evolving topic. Grid computing has been recognized as a new strategy in computing; almost each IT-developed country has started its national Grid initiative. As an example of integration on international level, the EU Project EGEE (Enabling Grids for E-SciencE) [84] could be named, which is the world's largest scientific Grid and achieves a scale comparable to today's largest supercomputers. It encompasses more than 37,000 CPUs and 15 Petabytes of memory with servers at 240 sites in 40 countries. The latest efforts of the European Grid Initiative (EGI) [85] are trying to establish a permanent Grid infrastructure for Europe.

The fact that Grid is a worldwide echoed paradigm for modern computing is also confirmed by the occupation of authors of this chapter; they work at research sites on two geographically distanced continents. The first author is working on more topics of Grid computing: e.g., tool development for visualizing data Grid simulations, development of data Grid simulators, and on writing Grid-aware scientific applications.

His coauthors are with the Grid Computing and Distributed Systems (GRIDS) Laboratory at the University of Melbourne, the research of which is centered around the Gridbus project [86]. From most recent publications related to Grid, the upcoming monograph [87] will bring an overview of the Grid economy research and technologies that will facilitate a global commercial Grid system. In [88], a common work of all three authors concerning data Grids is presented.

Grid computing can be characterized as the utilization of widely distributed resources within virtual organizations for the solution of large-scale problems from science and, most recently, also from business. The authors write about the motivation for developing and using the Grid; they bring an useful overview of technologies behind it and also describe practical applications and new promising future trends.

The chapter starts with the categorization of Grids. In the second section, actual problems related to Grid computing, such as quality-guarantee of services, resource regulation, and data management are discussed. In order to make Grid available to users, the middleware plays a key role. The third section concentrates on description of three toolkits: Globus, gLite, and Gridbus. From tools for computationally intensive applications, MPICH-G2, Cactus, Condor-G, and GridSolve are put into context. Also application-specific Grid projects from medicine, astronomy, finance, and digital libraries are mentioned, which document that this new paradigm is not only a domain of IT researchers, but it is convincing now also to practitioners.

The efficiency of a Grid depends heavily on the way how the management of jobs and resources is organized. Authors survey advance reservation strategies. Also new trends are sketched, pointing out that Cloud computing is the most promising new paradigm. It is interesting to note that this idea was initiated by the industrial enterprises and afterward it became a hot topic for the research community. It is in contrary to the Grid, which was formulated as a concept first by the academicians and later on it found acceptance by users from a practice.

1.2.4 Parallel Structured Adaptive Mesh Refinement

Jarmo Rantakokko and Michael Thuné, Uppsala, Sweden

The techniques for mesh construction and mesh refinement are of crucial importance in many areas of large-scale computing, where PDEs (Partial Differential Equations) are to be solved numerically. The topic of this chapter is Structured Adaptive Mesh Refinement (SAMR), which is used in simulations where the problem domain is divided into rectangular subdomains, upon each of them a structured mesh is superimposed. SAMR enables to automatically adapt the mesh granularity in subdomains, according to the resolution quality required to represent important features contained in them. Parallelism for this type of problems can be exploited naturally. However, in simulations the refinement on subdomains has to be performed dynamically, in order to reply to desired changes in the resolution. For parallel implementation it means to take into account dynamically changing work load, data volume, and communication pattern at runtime. As a consequence, dynamic load balancing techniques have to be applied, which have implications for the parallelization strategy.

The authors are primarily oriented to numerical scientific computing in two of its key facets: methods [89, 90] and software [91, 92]. Their chapter gives an overview of SAMR approaches and reports on new results obtained in framework of the ongoing research project Dynamic Load Balancing of Parallel SAMR Applications, which is under development by the research group Software Aspects of High-Performance Computing of the Scientific Computing Division at Uppsala University. This university has a long tradition and high reputation in scientific computing. This is concentrated around the Department of Information Technology, which is linked to more research centers, among them also the Uppsala Multidisciplinary Center for Advanced Computational Science (UPPMAX), which provides high-performance computers and know-how of HPC.

The chapter starts with a brief introductory survey of SAMR techniques and software packages. The main part of the chapter deals with various issues related to implementation of SAMR on parallel computers. From a computer science point of view, such aspects as data dependencies, parallel programming models, data placement strategy, and appropriate choice of the dynamic load balancing are discussed for systems based on the shared memory as well as the distributed memory. Various approaches and algorithms are presented in the fifth section, where a place is also given to own results achieved by the authors. The conclusion of their comparison is that there is no single best alternative under all circumstances because the appropriate choice of all the above-mentioned aspects to be considered by the parallelization depends on both the application state and the computer platform.

The chapter ends with a section about the future research plans, where the objective is to equip SAMR-based simulation software with additional features of self-configuration and self-optimization. Two self-optimization functions, which the software should be able to do automatically, will be focused: selection of the amount of parallel resources and selection of the details of the model. This progress toward autonomicity will make the SAMR software more attractive to a broad community of scientists and engineers who are working on simulation of large-scale applications, without being primarily parallel computing experts.

1.2.5 Applications and Parallel Implementation of QMC Integration

Peter Jez, Andreas Uhl, and Peter Zinterhof, Salzburg, Austria

The team of authors, affiliated to the Department of Computer Sciences of the University of Salzburg, combine expertise in advanced numerical algorithms with experience in parallel computing. This twofold competence enables synergies, which profile the department in the area of Scientific Computing. The department is led by P. Zinterhof, whose research roots are in high-dimensional number theoretic numerics [93] and signal and image processing [94]. He has been coordinator of the *ParNum* initiative, which was the actual background on which the cooperation in parallel numerics for a majority of authors of this book started. The intensity and productivity of A. Uhl's research is evident from his wide–scope publications and projects in the area of image and video processing, multimedia security [95], biometrics [96], medical imaging [97], and parallel numerical mathematics [98]. P. Jez is specialized in Quasi-Monte Carlo (QMC) methods [99]. In particular, high-dimensional numerical integration and QMC methods belong to the topics making Salzburg visible in the Scientific Computing community. Methods from both the areas offer a potential for parallelization. It is a motivation for this chapter, where the authors deal with numerical computation of integrals over a high-dimensional domain. The approach applied for it is the QMC, due to the special choice of de-terministic point-sequences for the integration, the use of which allows to increase significantly the convergence rate against randomly chosen sequences, as it is the case by the Monte Carlo (MC) methods.

In the first part of the chapter, some theoretical results about QMC integration over a high-dimensional domain are presented. In this context, the application of the reproducing kernel Hilbert spaces within the QMC integration is discussed. Im-portance of integrals of the considered type is illustrated for two application areas: solving PDEs and finance engineering.

The second part of the chapter is devoted to the parallelization of the integration computations and their implementations. It is introduced by a survey of parallel and distributed QMC approaches. Due to the convergence rate of QMC integration, the integrand must be evaluated on a huge amount of integration nodes. Here is the main space for parallelism because these evaluations can be splitted into independent sub-tasks, which can be assigned to different processors for a simultaneous execution. Three different concepts of parallelization have been under consideration: block-ing, leaping, and parametrization. The experiments were performed on the cluster *Gaisberg*, installed at the Department, which is configured as a two-dimensional (2D) 6×6 Scalable Coherent Interface (SCI) torus with dual-core AMD Athlon MP2800+ processors.

Different low discrepancy sequences, among them Good Lattice Points and Zin-terhof sequences, were examined in parallel. The integration results for sample integrals are compared and illustrated by a number of figures accompanying the chapter.

1.2.6 Parallel Evolutionary Computation Framework for Single- and Multiobjective Optimization

Bogdan Filipič and Matjaž Depolli, Ljubljana, Slovenia

Both authors are working at the Jožef Stefan Institute in Ljubljana, which is the leading institution in progressive IT-related research in Slovenia. The first author is with the Department of Intelligent Systems, which is active in areas of artificial intelligence, intelligent systems, information systems, information society, medical informatics, natural language processing, and cognitive sciences. His research is primarily oriented to evolutionary computations, the scope of which ranges from the algorithm design to practical applications in dynamic system control and industrial process optimization. Optimization for industry-relevant applications is presented in two recent papers [100, 101]. The second author is the youngest contributor to this book. Since 2005, he is with the Department of Communication Systems, which is specialized also at the development of algorithms for digital signal processing and parallel computing. His research interests include evolutionary computation, artificial intelligence, and knowledge discovery applications in communication systems [102].

Evolutionary computing is a name used from early 1990s to embrace subareas such as evolutionary programming, genetic algorithms, evolution strategies, and genetic programming. To the area of evolutionary computing belong several computational techniques, which reflect to some extent the evolution of biological life in the natural world. It became a popular topic in modern computer science, because the evolutionary algorithms enable to search, applying the adaptation principle, for optimal solutions of problems related to a given environment. Components of an evolutionary algorithm, which selects for the survival the fittest individual, are definition of individuals, fitness function, population, parent selection mechanism, mutation, and survivor selection. Many problems from science, technics, economy, and production were solved effectively using these algorithms. A weakness of evolutionary algorithms is their computational complexity, which is due to the iterative population-based search of the solution space. From the point of view of parallelism, these algorithms are naturally parallelizable since processing a population of candidate solutions can be done concurrently.

The evolutionary computing principle is applied in this chapter to solving numerical optimization problems with one ore more objectives. A framework for the selection process is developed and it is shown how to solve within it a high-dimensional optimization task from industrial practice. At the beginning of the chapter, a formal definition of optimization problems is given. A difference between single- and multiobjective optimization is explained and a review of the concepts needed to deal with multiobjective optimization problems, such as the dominance relation and Pareto optimality, is presented. The framework uses differential evolution, which is a particular kind of evolutionary algorithms. After a description of its original single-objective optimization form, its extension for dealing with multiple objectives is formulated. The final section of the chapter reports on a parallelization of the proposed evolution algorithm on a computer cluster.

The performance of the framework is evaluated by solving an industrial optimization problem. The task comes from metallurgy, where tuning of coolant flows in the steel casting process helps to increase the steel quality. The optimization problem is solved for both the single- and multiobjective variants in parallel and analysis of achieved results is given.

1.2.7 WaLBerla: Exploiting Massively Parallel Systems for Lattice Boltzmann Simulations

Christian Feichtinger, Jan Götz, Stefan Donath, Klaus Iglberger, and Ulrich Rüde, Erlangen, Germany

Erlangen belongs to mostly known and competent sites for Computer Science research and education in Germany. Historically, it is connected to such names as W. Händler, F. Hofmann, H. J. Schneider, und U. Herzog, who pioneered parallel computing from the mid of 1970s [103]. Experimental parallel multiprocessors EGPA [104, 105] and DIRMU [106] were constructed and operated in Erlangen in the first half of 1980s and at that time they were way-giving pointers to further progress in parallel architectures, operating systems, programming, and algorithmics. The Erlangen school in parallel systems found its continuation in later works of A. Bode in Munich, E. Mehle in Lübeck, and J. Volkert in Linz. In Erlangen started in 1981 also the CONPAR conference series which continues today under the name EuroPar and evolved to be the first forum for parallel processing presentations in Europe.

The authors of this chapter are from the Chair 10 of the Department of Computer Science, University Erlangen–Nuremberg. This Chair widens the profile of the Erlangen's Computer Science through the engagement in the area of system simulation. A research at the Chair 10 is performed by four working groups: High-Performance Computing, Algorithms for Simulation, Complex Flows, and Laser Simulation. There is a noteworthy number of ongoing research projects, in which the Chair is engaged. Their topics are numerical simulation, multigrid methods, massively parallel solvers, and solving PDEs [107–109]. One of the projects is WaLBerla [110], which is the theme of authors' contribution to this publication (the acronym resembles the hill Walberla in Franconia, the region of northern Bavaria, where Erlangen is located).

A motivation for the project is a development of a software tool for solving a broad class of computational fluid dynamics applications by the lattice Boltzmann methods, which promise to be a modern alternative to the classical solvers based on the Navier–Stokes equations. The goals of WaLBerla are easy adaptivity and extensibility for new fluid problems, physical correctness, high performance, and easy to be used. The WaLBerla is planned to be a comprehensive tool rich in features as well as a library for efficient development of new applications based on fluid simulation. This ambition seems to be realistic, taking into account experience in parallel numerics and software abilities of the developers' team.

In this chapter, the authors focus on the parallelization of the framework, which is based on a domain partitioning scheme named *patch concept*. A parallelization is the only possible alternative for obtaining time-acceptable simulations for Boltzmann solvers: as authors say, for a typical application one time step would run 3 hours on a serial single-core computer, while a supercomputer with 4864 dual-core processors will reduce this time to 1.5 seconds. Totally, for 50,000 time steps the simulation time would be 20 hours compared to 17 years for the nonparallel realization.

The text starts with a brief overview of the lattice Boltzmann method in 3D, followed by the introduction of the waLBerla parallelization concept. This is based on patches, which characterize partitioning of the fluid domain and represent the basic components for the parallelization, the optimization strategies, and the flexibility, which are needed for the integration of further applications. When patches belong to the same process, the so-called local communication is applied for data exchange among them, whereas interprocess communication is done via MPI. Serial tests are described and a detailed and comprehensive report on parallel studies is given for five parallel systems, among them for the HLRB II at the Leibniz Rechenzentrum in Munich and for the JUICE Cell Cluster at the Forschungszentrum in Jülich, which belong to ten most powerful systems in Europe. These tests used up to 810 cores and a domain containing up to 1530^3 lattice cells. Also the advanced multicore processor Cell BE of IBM was examined by computing the blood flow simulation. The fact was confirmed that a good performance is possible for these new systems only when the implementations are performed carefully, using architecture-related optimizations. In the concluding section, an outlook for future work is given: tests on systems beyond 1000 cores and a development of new fluid dynamics applications with free surfaces and moving objects.

1.2.8 Parallel Pseudo-Spectral Methods for the Solution of the Time Dependent Schrödinger Equation

Tore Birkeland and Tor Sørevik, Bergen, Norway

Both authors are affiliated with the University of Bergen, which has a strong tradition in computational sciences. In particular, this university was the early starter with parallel computers. Since 1985, when the first commercially available parallel system Intel Hypercube iPSC-1 was installed, in Bergen were operated more than a dozen different parallel HPC-platforms, and for more than two decades it has been visible as a top European supercomputing center. On the November 2008 TOP500 list, their Cray XT4 QuadCore system is ranked as no. 65. Today the focus among the computational scientists at the University of Bergen has shifted from basic research on core parallel algorithm and their implementation to solving large-scale computational problems in applied sciences as exemplified by their contribution in this book. Their chapter fits perfectly with the main objective of our book: it deals with a computationally difficult and relevant problem from numerics, which is out of limit to be solvable serially and for which parallelism is the only way for obtaining acceptable and usable results. Moreover, the parallelism employed is not straightforward; there is an intricacy in the formulation of the parallelization strategy and in the implementation as well. Thus, the know-how from numerics as well as computer science has to be combined in order to reach a satisfactory result.

T. Sørevik's research backgrounds do combine these qualifications. He was formerly with the Department of Informatics, focusing on parallel computing.

Currently he is affiliated with the Department of Mathematics, focusing on numerical solution of high-dimensional problems. He has a profound background in numerical mathematics and long-term experience with programming high-performance parallel computers. In particular, constructing lattice rules for high-dimensional integration [111] and numerical solution of PDEs [112] belong to his favorite numerical research topics. Topics such as data partitioning [113], load balancing [114], programming models, and parallel applications are problem areas from parallelism he has been involved in.

T. Birkeland is his Ph.D. student developing new numerical techniques and software for solving the time-dependent Schrödinger equation, which is the theme of their chapter. In addition to physics and numerical mathematics, he also has a strong interest in modern software engineering.

The problem under discussion is from quantum mechanics. Due to complicated interactions in few electron systems, approximation techniques from computational chemistry, simplifying the coupling between particles, cannot be used. The full quantum mechanical solution is required and it can easily become extremely computationally demanding, making parallel computing a necessity. The split-step propagator method was chosen for the solution, because each fractional step corresponds to simpler differential operators, which can easily and accurately be approximated by spectral methods. Furthermore, the technique can easily be parallelized provided the data is properly distributed. However, the standard data distribution has limited scalability; thus an improved data distribution scheme is devised and implemented.

The text starts with the problem formulation and the presentation of the split-operator technique for time-stepping. It follows with the discussion on choice of the coordinate systems. The next section deals with the parallelization of the multidimensional FFT, since each partial operator typically requires a different spectral representation. Thus between each partial step there is a need for transforming the data. Parallelizing such transformations on huge data sets on distributed memory computers usually leads to a communication bottleneck. The authors propose a data redistribution algorithm which helps to overcome this difficulty.

In order to minimize the discretization error and to maximize the performance, one has to make problem-dependent choice of appropriate coordinate system, corresponding discrete representation, and a time propagation scheme. A framework called PyProp is described, which makes it possible to make choices between different methods at runtime. This enables researchers to choose methods based on testing on the given problem instead of relying on a priori estimates. PyProp tries to combine the flexibility of object-oriented programming (C++) with the convenience of high-level scripting language (Python) and high-performance computational libraries (blitz++, FFTW, and LAPACK). The ionization of the hydrogen molecular ion was taken as the computational example to demonstrate some of the features of PyProp. Experiments on the Cray XT4 show good speedup (over 1000 against run on a single processor) and scalability behavior of the produced solution strategy.

In the outlooks, the authors proclaim that parallel computing will be an indispensable tool for their further research on this topic, which will be conducted toward higher dimensions, new algorithms, and implementation improvements.

1.2.9 Parallel Approaches in Molecular Dynamics Simulations

Dušanka Janežič, Urban Borštnik, and Matej Praprotnik, Ljubljana, Slovenia

Another area where parallel numerics has been adopted successfully is molecular dynamics. This contribution brings a view on parallel approaches in molecular dynamics simulation.

It is written by a team of authors from the Center for Molecular Modeling at the National Institute of Chemistry, Slovenia. The research activities of the Center tackle the molecular modeling in more of its aspects: in theory, hardware, software, applications, and also in education. A project related to parallel computing area is named VRANA, which aims to build low-cost parallel computers for computations needed for the Center-specific research. The systems are based on Beowulf clusters connected by torus or other point-to-point based parallel topologies, avoiding expensive network switching technology. Currently, their VRANA clusters under operation contain altogether over 700 cores.

D. Janežič is the head of the Center. She is experienced in application of numerical methods for solving simulation problems in molecular dynamics. Her current research focuses on the development of symplectic integration algorithms for molecular dynamics simulations and their application to simulations of macromolecules to provide better understanding of biological systems [115]. U. Borštnik contributed through parallel implementations to the simulation research [116]. M. Praprotnik is working on computer simulation of soft matter [117]. He is focused on both development of new simulation techniques and their application to realistic systems.

A general problem with parallel simulations in molecular dynamics represent the interactions among all atoms of the simulated system, which is the most computationally demanding part of the whole simulation. Parallel methods differ in their distribution of these calculations among the processors.

Concretely, the chapter deals with parallelization of the split integration symplectic method for the numerical solution of molecular dynamics equations. A computer program for this method, designed to run on specialized parallel computers, is presented. The molecular dynamics integration is performed by a new integration method, which analytically treats high-frequency vibrational motion and thus enables the use of longer simulation time steps. The low-frequency motion is computed in parallel for each simulation time step. The performance of simulation on specialized computers is analyzed and a comparison to the implementations on standard personal computers is given. It is shown that the combination of the new integration method with two specialized parallel computers is an effective way to significantly increase the speed of molecular dynamics simulations. Also a parallellized version for the distributed diagonal force decomposition method has been developed. It is shown that compared to other methods in molecular dynamics simulation, its communication requirements are lower and it features dynamic load balancing.

This chapter clearly illustrates that parallel processing for this type of applications has to consider special granularity and interaction of the computational patterns, which have implications for both the algorithmic design and the parallel computer topology.

1.2.10 Parallel Computer Simulation of Heat Transfer in Bio-Tissue

Roman Trobec, Ljubljana, Slovenia

This chapter deals with the application of parallel numerical methods in medicine. The concrete subject is the parallel computer simulation of heat transfer in parts of a human body. This application is of great practical importance, because the temperature in human tissues, which is an important factor in surgery, physiology, and other medical domains, can be estimated by computer simulations. The obtained results, based on numerical solutions of partial differential equations, have practical advantages for patients, because they bring some additional information, which could improve chances for positive outcomes of many medical treatments.

The author is with the Department of Communication Systems at the Jožef Stefan Institute. From the beginning in 1993, R. Trobec and his team were with the *ParNum* project, and thanks to him Ljubljana, together with Salzburg and Bratislava, became one of the pillar-centers of this research initiative. He was the main organizer of the three successful ParNum workshops in Slovenia: Gozd Martuljek (1996), Bled (2002), and Portorož (2005) [118], which significantly contributed to the scientific synergy and to the establishing of the initiative on the international scene.

Having a strong background in numerical mathematics, his research and working experience are mainly in the area of parallel and distributed computing, interconnection networks, computer simulations, and advanced signal analysis. In particular, scientific computing applications in medicine [119, 120], where numerical and parallel computing meet together, is his most recent publication field. He has been working on parallel network topologies and their performances [121, 122]. He also contributes into the research of parallel numerics, as documented, e.g., by his recent paper [123] about meshless methods for solving the diffusion equation.

After a clearly written introduction to the chapter's topic, a description of principal steps of computer simulations is presented, including mathematical modeling and geometric modeling, domain discretization, numerical solution, and validation of simulated results and their visualization. Numerical solution of partial differential equations is discussed in the third section, with particular attention paid to a solution of time-dependent problems. For these problems, explicit and implicit numerical schemes are presented. The computational and memory complexity for generating the linear algebraic system, related to the finite difference approximation and for its solution by linear algebra procedures, are also mentioned in this relationship. The diffusion equation, which represents the basic equation that describes the heat transfer, is a subject of the subsequent section. Analytical as well as numerical solution approaches are presented for this equation, whereby the formulas for explicit difference scheme are given for both 2D and 3D problems. The numerical finite difference solution of a simple test case is compared to its analytical solution, for inhomogeneous domain, confirming that they match even for the case of very different diffusivities.

The part of a human body being simulated was the knee joint. For temperature modeling, the so-called bio-heat equation was used, which incorporates heat conduction, heat transfer between blood and tissues, and heat production by metabolism. A model of the knee has been developed from digital images coming from the VHD (Visible Human Dataset) project. All major tissues were incorporated in the model. The simulation program computed temperature evolution in the knee during a 2-hour surgery and also during the postoperative 2-hour cooling phase.

The parallel implementation is based on the domain decomposition technique and the message-passing communication. The numerical approach selected was the explicit finite difference method. Mapping of computational subdomains to the processing nodes of parallel computer is discussed and the theoretical performance of the proposed parallel algorithm is analyzed. The implementation of all simulation steps is described in detail. Execution time is measured on a computer cluster with different number of processors and compared with theoretical predictions. Results obtained by the simulation have been verified by measurements.

The work and results presented in this chapter are convincing enough for a conclusion that parallel computer simulations can be of great use in medicine either for planning a surgery or for evaluating different doctrines of medical treatment or for testing of new medical methods.

1.2.11 SVD Computing in LSI Applications for Data Retrieval

Gabriel Okša, Bratislava, Slovakia and Marián Vajteršic, Salzburg, Austria

Research activities of both authors are tightly coupled to the Department of Informatics at the Institute of Mathematics of the Slovak Academy of Sciences. Due to its renowned position in the area of parallel numerics, this site has been one of the key partners of the *ParNum* initiative. Particularly, Bratislava has a long research tradition in the field of parallel numerical algorithms. The fact that parallelism represents a revolutionary trend in computing was recognized there from its beginning: already in the early 1970s a young team of mathematicians working with J. Mikloško published first works related to this topic. At the Slovak Academy of Sciences, a first prototype of a massively parallel computer in the former Eastern block was built in the early 1980s, under the coordination of I. Plander. Acquired know-how in parallel algorithmics was exploited for development of the application software for this system. In the scope of this project, a library with more than 100 assembler-written parallel routines from various application areas of numerical and discrete mathematics was produced [124].

The first author is an internationally acknowledged expert on numerical linear algebra. His research career started at Nuclear Power Plant Research Institute in Trnava, Slovakia, where he was engaged in solving theoretical and implementation problems of stationary diagnostic systems for the monitoring of technical status of nuclear power plants. He has also been interested in work on parallel algorithms, particularly for systolic arrays. After his move to the Academy of Sciences, his research has become more intensively oriented to parallel linear algebra, especially to fast and reliable algorithms for structured matrices (e.g., Toeplitz, Vandermonde, Cauchy, etc.) [125] and eigenvalue and singular value problems [126,127]. Recently, his research scope has been broadened through parallel PDE solvers and algebraic methods for data retrieval.

The second author is one of the founders of *ParNum*. He published and coedited more monographs on parallel algorithms and scientific computing, e.g., [128, 129]. He has coordinated a number of national and international projects in parallel computing (funded, e.g., by EU and NATO) and has been on boards of important conferences (e.g., EuroPar) and journals (e.g., PPL and IJPEDS) in this area. His recent research is focused on parallelization of the Singular Value Decomposition (SVD) and its applications for multiprocessors and Grids [130].

Their chapter illustrates a situation when a method from the numerical linear algebra can be efficiently used in an application, which is nonnumerical in its nature. Moreover, the role of parallelism is shown to be indispensable for getting solutions of problems which are practically relevant. The authors write how the problem of text data retrieval in huge documents sets is solved in parallel by the reduced SVD approach. Easy and reliable manipulation with text data by computers requires an efficient encoding. In a mathematical model adopted by the authors, the Latent Semantic Indexing (LSI) is used, where the occurence of text terms in documents is represented by the so-called term–document matrix. In real-life applications, the number of retrieved documents is large and, hence, computations with large-sized matrices are a consequence in such cases. Therefore, a decomposition technique is used to reduce the dimension of the searching space. The truncated SVD not only enables to reduce the space complexity, but also decreases the storage requirements and, what is even more important, it decreases the time for real-time query analysis and data retrieval. Thus, SVD becomes the kernel computational operation of the whole LSI-based retrieval process. Due to their long-term expertise with parallel SVD algorithms, the authors show how to perform these computations efficiently.

The chapter is introduced by a short description of the LSI model and the truncated SVD. The next two sections deal with adaptation of the LSI model to the text-retrieval problem in changing environments, like the World Wide Web (WWW). The former of the sections is devoted to the updating problem. Both cases are considered: i.e., when the number of terms as well as documents have increased against the original problem. It is shown, how the decomposition of the original term–document matrix can be exploited advantageously for solving the enlarged problem. In the latter section, the downdating problem, also for both cases, is formulated mathematically. The computational kernel for all these cases is similar: it requires to solve SVD of upper triangular matrices.

It is a reason for a presentation of the Kogbetliantz method, which preserves the triangular structure of the matrix during the elimination process. Newly developed parallel algorithms, aiming to use level-3 routines of the BLAS library, are presented for the implementation of this method. Another alternative to solve the kernel problem represents a two-sided block-Jacobi SVD method. The authors make use of an original idea of M. Bečka, who proposed a dynamic ordering for the annihilation of the off-diagonal matrix blocks. In this approach, the eliminations in each sweep do not proceed according to a statically prescribed order, but that is generated during the runtime. It is a highly efficient technique, because it picks up for elimination always as first those blocks, which contribute mostly to the non-diagonal structure of the matrix to be diagonalized. Parallel numerical experiments with this method confirm savings in number of sweeps against the statical orderings.

In the following section, the LSI model is parallelized for a Grid environment, consisting of a number of individual computing nodes, where only a restricted intercommunication is possible. Both updating and downdating are analyzed and the organization of the global retrieval process by a comparison of retrieval results between individual nodes is proposed. Finally, another strategy of storage of documents and their retrieval is discussed. A situation is considered when the individual nodes of a distributed system can communicate and mutually exchange data. In the first phase, each processor generates its own reduced SVD for a data block assigned to it. An algorithm is proposed on how a global approximation of the original term–document matrix can be built upon the partial approximations produced by individual processors. Thus, a completely different retrieval strategy, compared to the Grid system, can be formulated. A discussion of some interesting scenarios for the final retrieval, that can arise in real life, concludes the chapter.

1.2.12 Short-Vector SIMD Parallelization in Signal Processing

Rade Kutil, Salzburg, Austria

The working field of R. Kutil is concentrated on parallelism of signal and image processing applications. In particular, he achieved original results concerning parallelization of wavelets, e.g., in filtering by using Single Instruction Multiple Data (SIMD) extensions [131], in 2D lifting [132], and in encryption [133]. Recently, he is also involved in a project dealing with algebraic methods for retrieval of images in distributed Grid environment [43]. His another research theme is computer arithmetics, where he proposed a language called Alluvion for algorithm representation [134]. Through these topics, the research landscape of the Scientific Computing working group of the Department of Computational Sciences at the University of Salzbug has been enriched. He also has insight into architectures of modern processors and supercomputers. He is experienced in all facets of modern programming, ranging from microprogramming through high-level languages up to the programming tools for distributed and shared-memory supercomputers. This know-how is reflected in this chapter, where he writes about application of the SIMD parallelism in digital signal processing.

Parallelism in various forms influences also the architecture of CPU cores of modern processors. Particularly, for digital signal processors, the SIMD principle can be exploited efficiently. It is because in signal processing a large amount of data is processed continuously under the same instruction stream. For this type of data processing, the so-called short-vector SIMD units represent cost-efficient extensions for performance enhancement on the intraprocessor level. SIMD extensions are not restricted only to the domain of specialized processors. They are included in almost all general-purpose processors, which makes SIMD still present also in high-performance computing.

Parallelism in this chapter concerns signal processing algorithms, which are characterized by almost regular computational and communication patterns. These properties make them, from early parallel era already, exemplary candidates for execution on almost all parallel architectures (e.g., systolic arrays, hypercubes, pyramids, and toroidal massively parallel systems). The availability of processors with inherently implemented SIMD parallelism on short vectors brings a new challenge to examine the signal processing algorithms from this perspective.

The introduction brings a description of basic operations for both classes of signal processing algorithms: global and local. Global work on data blocks are based on global operators (mostly orthogonal transforms like the discrete Fourier transform), while the local ones are convolution-based. Afterward, the SIMD short-vector parallelization principle is explained and an overview of general vectorization approaches is presented. In the next section, the loop transformations as well as novel vectorization approaches are combined and evaluated for some representative local convolution-based operators. The operators considered are related to simple filters as well as to more complex filter banks from the field of wavelet transforms. Experimental results on an Intel Pentium, with SIMD extensions capable to process vectors of four single-precision numbers, are shown and thoroughly discussed. The subsequent section deals with vectorization of the so-called recursive algorithms, where the output data are reused as inputs for the next computational step, as it is a case, e.g., in the Infinite Impulse Filters (IIR). Two algorithmic approaches have been developed: one is based on a straightforward loop rescheduling, while the other uses algebraic transformations to resolve occurring data dependencies. It is shown that for certain values of filter parameters, the proposed algorithm is faster than the professional IPP library implementation.

For a presentation of vectorization strategies in the class of global operators, the Fast Fourier Transform (FFT) has been chosen. It is because the other fast orthogonal transforms are of similar structure and hence the vectorization strategies for them would be basically the same. Since the input data are complex, the data layout for their real and imaginary parts has to be done carefully in order to suit to the basic operations involved in the transform. Then the vectorization operations for the basic FFT block are formulated. Also the principle of automatic tuning is discussed in this context. The final technical section of the chapter deals with the so-called mixed algorithms. These are approaches which cannot be identified fully with one of the two classes mentioned above. The wavelet transform was chosen as a representative of convolution where recursion is involved. Also vectorized versions for the multidimensional transforms are described and corresponding speedup figures are depicted.

The author's conclusion is that the SIMD short-vector vectorization brings performance improvement for digital signal computations, but a creation of the best technique is not a simple task and it is to be done for each application individually. A matter of a future research is a development of languages with automatic vectorization features.

1.2.13 Financial Applications: Parallel Portfolio Optimization

Andreas Grothey, Edinburgh, United Kingdom

The application area of this chapter is nowadays a hot actual theme for almost everyone. The worldwide finance crisis has not suppressed the role of computers in the finance sector – in contrary, we learned that it became a consequence of questionable and unqualified handling caused by the human factor. Hence, the necessity of qualified computer-supported decisions is expected to increase, in order to avoid steps which could lead to such critical and unpredictable situations as we are witnessing on these days.

Finance applications such as portfolio management, risk estimations, and design of new credit and saving products call for employment of computers which could solve related mathematical problems. Because these are mostly large-scale problems, high-performance computing is adopted to increase the speedup and reliability of such computations. This chapter is focused on the portfolio managment area and outlines the position of parallelism in solving multistage stochastic optimization problems.

It is contributed by A. Grothey, a mathematician, with research interests centered on mathematical optimization. In particular, he is specialized in Interior Point Methods (IPM) [135], development of object-oriented parallel solvers, decomposition methods for large-scale nonlinear nonconvex constrained optimization, and modeling approaches for large-structured problems.

The practical result of this research is the Object-Oriented Parallel Solver (OOPS) [136, 137]. It is a parallel interior point code that exploits any special structure in the Hessian and Jacobian matrices. The solver is implemented using object-oriented programming techniques. It solves very large linear (LP), quadratic (QP), and nonlinear (NLP) problems. The code is developed for massively parallel platforms and capable to solve problems of sizes up to 10^9 variables [138]. Indeed, OOPS has been used to solve a QP problem with $1,010,507,968$ variables and $352,875,799$ constraints.

The chapter is introduced by explaining the position and objective of portfolio management in finance. The objective could be characterized as trade-off between investment return and associated risk. Different formulations of the problem are mentioned, including the classical mean-variance model and models, where a number of additional classes of constraints are incorporated. The model discussed in detail is the so-called Asset and Liability Management (ALM) which plays an important role for long-term investments such as issued by insurances or pension funds. Portfolio optimization problems are usually treated as stochastic programming problems. In the second section, this framework for the mathematical description of the ALM model is reviewed. Various aspects of the model, such as variables, constraints, objective, and the structure, are formulated in the subsequent text. Particularly, the constraint matrix with its nested form is presented and its suitability for parallel computations is examined. The review part of the models is completed with the extensions to the prototype ALM model and their consequences for development of parallel solution approaches.

The model and its extensions lead to large-sized problems with millions of unknowns and more, which cannot be solvable without parallelism. Therefore, parallel solution approaches for stochastic programming problems are treated in subsequent sections. Two main directions represent here: decomposition and IPM. Particular attention is paid to the application of the IPM to the multistage stochastic programming problem. The presentation of resulting matrices is nicely structured and all linear algebra operations clearly stated. An overview of different implementations on a variety of computing platforms ranging from dedicated parallel machines to PC clusters up to Grid environments is given. In the final section, three parallel IPM approaches are compared, among them also the OOPS codeveloped by the author at the School of Mathematics in Edinburgh. A general conclusion for parallelization is that the method and parallelization techniques have to be adapted to both the specific portfolio management model formulation and the available computing platform.

The author is convinced that new developments in parallel architecture will pose new challenges for further stimulation of research in parallel finance applications.

1.2.14 Future of Parallel Computing

Selim G. Akl and Marius Nagy, Kingston, Ontario, Canada

Both authors are members of the Parallel and Unconventional Computation Group in the School of Computing at Queen's University in Kingston. The research scope of the Group is wide: it is tackling not only classical topics in parallelism, such as, e.g., implementations and applications of shared-memory models, properties of interconnection networks and their algorithms, and fundamental theorems in parallel computation, but also themes related to new trends, such as, e.g., new paradigms in parallel computation, energy-aware computing, quantum computing [139], parallel biomolecular computing [140], scheduling algorithms for wireless mobile ad hoc networks [141], sensor networks [142], and unconventional computing [143].

Due to the topics on this list, each of which is supported sufficiently by relevant publications, it is apparent that with such scientific background the authors are capable enough to say a competent word about outlooks and future trends in parallel computation.

This chapter closes the arc of themes spanned by our publication. It brings a view on innovative trends and outlines perspectives in parallel computing. The progress in computing is generally driven on one side by investigations in theoretical computer science, which provides studies on models inspired by physical, chemical, and biological phenomena coming from nature. The second source for this progress are revolutionary technological and software achievements, which give us opportunities to build and program ever more powerful systems capable of approaching solutions of grand challenge problems. The third driving force is the never declining need to employ computers in the increasingly wider and more complex spectrum of tasks tackled by modern society.

The chapter is introduced by arguments for the fundamental role parallel processing continues to occupy in the theory of computing. It is shown that the idea of massive parallelism is reflected in unconventional models of computation such as DNA computing, membrane-computing systems, quantum computing, and reaction–diffusion computers. Also, the accelerating machine model can be thought of as deriving its power from doubling the number of processing units (operating in parallel) at each step.

The text is primarily focused on parallelism in quantum computing. This computing paradigm, impacted by quantum mechanics, is foreseen as one of the most promising candidates for future computers. This expectation is supported by a presentation of five problems in quantum information processing, for which only a parallel approach can guarantee a reliable solution.

After reviewing the basic notion from quantum mechanics, such as qubit, superposition, measurement, and interference, the mathematical framework for this type of information processing is presented. It is explained how the quantum states and the transformations acting on them can be expressed in terms of the linear algebra apparatus and Hilbert spaces. In the next section, the mechanism of quantum parallelism is described and it is shown that, through its application, exponential speedup over a conventional computer is achieved for some applications. The above-mentioned five examples are then treated in detail. Their common feature is their evolving nature, i.e., their characteristics vary during the computational process. Because of their dynamic nature, these computations may be labeled as unconventional as opposed to the conventional computations characterized, e.g., by simulations on the Turing machine.

The Quantum Fourier Transform (QFT) is the first of these examples: it is shown that the quantum mechanical principle of superposition of states allows all possible inputs to be processed at the same time which brings a distinctive advantage over the classical FFT algorithm. In this approach, parallelism can be exploited and the analysis confirms the complexity savings. The second treated problem concerns quantum decoherence, which is due to the sensitivity of qubits, when interactions are taking place between the quantum computing system and the environment. This affects the values stored in quantum registers because these may significantly differ before and after the decoherence threshold. The paradigm here is that of time-varying variables, and the use of parallelism overcomes the decoherence effect. Parallelism also helps to correct other errors occurring in quantum computations. A known fact is that the complexity of errors in quantum computing increases with time. Parallelism splits the computation into smaller independent tasks and thus, through shortening the total execution time, the probability of error is reduced and eventual register recovery is much easier. One way for corrections is the construction of quantum error–correcting codes. These are based on the idea of digitizing the errors, where the error is expressed as a linear combination of the so-called Pauli error operators. Another technique presented here is the so-called error correction via symmetrization. The next problem relates to quantum states and their distinguishability. It is shown that this problem can be solved successfully only in parallel, enabling a simultaneous measurement of all qubits. A global condition is a condition on variables of a problem describing the input state, which has to be obeyed at every step of the computational process. In quantum computing, there exist transformations with such constraints. When some qubits fail during the quantum computational process, the constraint is not satisfied. Parallelism makes it possible to identify the difference in qubits simultaneously and, thus, a guarantee is given that the initial state and the final one coincide.

The looking ahead section of the chapter ends with an optimistic perspective for parallelism. Quantum computing as well as other new trends in biology-, chemistry-, and physics-motivated computing are all testimony to the fact that parallelism is universally applicable and that the future of computing cannot be conceived without parallel processing.

1.3 Conclusions

The book brings together several aspects of parallel processing. The first is parallel numerics, which is fundamental for the design of parallel algorithms used in high-performance scientific computing. Parallel computing applications comprise the second aspect. The book shows, on a representative palette of application domains, that parallelism is the only possible vehicle for solving large and time-demanding problems arising in the computational practice. The final aspect relates to modern trends and outlooks in this area. In the first part of this introduction and in the texts of individual chapters itself, sufficient arguments and working examples are given to testify the fact that parallelism is still vital, delivering new impulses not only for theory but also for practice and industry. All authors assert in their chapters that parallelism is inevitable for a successful continuation of their further research on open problems related to their problem area. In particular, the final chapter shows that all promising future computing paradigms will also possess the idea of parallelism in their substance.

The first part of this chapter has been devoted to a brief sketch of some actual themes and trends in parallel computing. We start with parallel numerics, which is reflected directly or indirectly in almost all the chapters. A particular focus is on achievements in numerical linear algebra, which is a basis for solving matrix-formulated problems. In this context, a new library concept, OOC algorithmic proposals, and solutions motivated by the multicore architecture are presented.

Developments in parallel numerical algorithms go hand in hand with progress in parallel architectures. New multicore architectures and networks on chip are at the center of our attention, when presenting advances in this area.

The important role, played by scalability in modern supercomputing, is outlined in the next subsection. In order to provide hard evidence of the huge progress in supercomputing, an excerpt from the recent list of the most powerful supercomputers in the world is also presented, together with some of their parameters.

Grid and Cloud computing can be viewed as new paradigms in parallel distributed computing, and some numerical solutions already available for these environments have been surveyed.

The battle for performance of a parallel numerical algorithm is usually won or lost in the way it is programmed in parallel. For this reason, we present not only existing parallel programming approaches, but also a new project aiming at the development of a programming friendly and function-rich high-level language for emerging, economically viable HPCS.

The introductory part of this chapter is concluded by a discussion on parallel compilers. Developments in this, probably the hardest field in the parallel area, are decisive in whether parallel computing will gain acceptance not only by experts but also by a broad clientele of computer users. We provide information about work on new compilers, motivated primarily by the ever increasing employment of multicore processors, together with an interesting view on using the autotuners in engineering of parallel compilers.

The contents of the chapters are described in the second part of this introductory chapter. The selection of individual chapters and their ordering reflect the main intention of the book: to present modern parallel numerical computing in all the richness of its facets through real parallel applications. We have tried to mediate there the whole spectrum of themes of the book and to point out crossovers of individual texts with the topics treated substantially in the first part of the introduction.

In order to make key notions and terminology from parallel computing understandable to nonspecialized readers, S. Akl and M. Nagy describe in the first part of their chapter existing relevant parallel models and show illustrative examples of parallel algorithms for some selected topologies. In the second part, unconventional parallel computer models are formulated and the problem of universality in computation is discussed.

The evolving Grid and Cloud computing paradigms are covered in the book through the next two chapters. The chapter contributed by T. Fahringer tackles this topic from the perspective of software tool development. After presenting the state-of-the-art in this area, he describes in detail an actual system, consisting of four integrated tools supporting the design of applications for these environments. The results of tests confirm that it is a practical and robust software system for automatic generation, management, and performance prediction of applications for Web-driven parallel applications. U. Čibej and his coauthors reserve more place in their contribution to a systematic overview of the middleware toolkits and to Grid projects from various application domains, such as medicine, astronomy, and finance, where Grid computing has already demonstrated its usefulness.

As noted above, the main aim reflected in this book is to demonstrate the relevance of parallel numerics within parallel computing. The first representative of this research line is the paper by J. Rantakokko and M. Thuné, who identify and analyze the role of parallelism for the construction and refinement of meshes, which is a crucial part of many simulations in large-scale scientific computing. Their study on parallelism for this problem establishes the relations between the dynamically changing work load, data volume, and communications. They have proposed an automatic system for mesh refinement in runtime, which will be extended through the self-configuration and self-optimization features.

Another theme from the parallel numerics is treated in the chapter by P. Jez et al. They propose a parallel implementation for multidimensional integration by a QMC method. This problem is relevant, e.g., to finance models, and the parallelism can be exploited there beneficially because the integrand has to be evaluated in a huge number of points. Various parallelizing strategies have been developed and examined experimentally on a supercomputer cluster.

Attention is given in the subsequent chapter to single- and multiobjective optimizations. B. Filipič and M. Depolli present a parallel framework for numerical optimization of practical problems from industry. Their approach is based on evolutionary computing. Because of its high complexity, this type of algorithm is critical for realization on serial computers. The authors explain the principle on which the evolutionary algorithms work, pointing out their potential for parallel exploitation. They illustrate their approach with a problem from metallurgy and demonstrate the advantages gained by the parallel solution.

The team led by U. Rüde concentrates on a parallel software tool, mathematically based on the Lattice Boltzmann simulators. These methods offer a new computational vehicle for solving numerical problems in fluid dynamics. The authors report the development, features, and optimization strategies of their parallelization concept. For computational experiments, which are highly data- and memory-intensive, some of the most powerful parallel systems in Germany were used, among them those based on the well-known Cell BE multicore processors.

Another kernel chapter of the book also deals with a large-scale numerical application arising by solving PDEs. T. Sørevik and T. Birkeland treat the time-dependent Schrödinger equation, which describes processes in quantum mechanics. The approximate solution for the full problem is very time-consuming and not achievable in practice without parallelism. The contribution provides an in-depth analysis of the parallelizable blocks, among them the multidimensional FFT. A parallel software system for testing and simulating various methods is described. It features different languages and libraries, among them the Python. Satisfactory scalability behavior of the parallel solution has been achieved for a typical example, the hydrogen molecular ion.

The chapter of D. Janežič et al. concentrates on numerics in another computationally demanding problem area: molecular dynamics. A critical issue here is to evaluate interactions between all atoms in the simulated system. A novel contribution presented here is an efficient parallel algorithm for molecular dynamics integration. This is designed for a specialized parallel system, which has been constructed at the authors' home institution.

In medicine, the role of parallelism has been recognized already in both research and practical treatment. This is further confirmed in the chapter by R. Trobec, who writes about a parallel computer simulation of heat transfer in parts of the human body. The mathematical background constitutes the numerical solution of the heat equation by finite-diference methods. The simulation was performed for heat propagation in tissues of the human knee. The simulation results obtained for the time during and after surgery have been promising enough to justify the author's parallel model and its numerical implementation on a parallel cluster.

The book also illustrates the fact that parallel numerical methods can find a use in applications that are, at first sight, non-numerical, such as data retrieval in large document sets. Using the Latent Semantic Indexing (LSI) model for data encryption, the so-called term–document matrix is generated and thus the retrieval problem can be transferred onto the linear algebra platform. G. Okša and M.Vajteršic show how the full searching space can be reduced, applying a truncated Singular Value Decomposition (SVD) method. The focus is on solving retrieval problems in changing environments, where update and downdate of document sets occur dynamically. Original parallel implementations of the SVD are presented and retrieval strategies for various real-life scenarios are discussed.

Signal processing is, from the beginning of the parallel era, one of the most attractive domains for the employment of parallelism. R. Kutil tackles the new opportunity that parallelism offers for this application domain – short-vector SIMD extensions. Here, parallelism is applied on the deepest level of a computational platform – in the CPU cores. Both transform-based and convolution-based operations are parallelized and implemented in this model. Experimental results are obtained for SIMD functions applied to vectors with four single-precision numbers. They indicate that this type of parallelism, when applied carefully, can bring performance enhancement for signal-processing computations.

The next chapter shows that parallel computing is being positioned successfully in the finance sector. Large and complex stochastic financial models call for a treatment by parallel optimization techniques, where the know-how of numerical linear algebra can be exploited. Concretely, A. Grothey presents parallel methods for portfolio management that are based on the Interior Point Method (IPM). They have been implemented on different parallel computing platforms, including the Grid. He asserts the conclusion that successful parallelization of this application has to take into account both the portfolio management model and the computer platform. This is not specific for this case alone, but is a confirmation of a generally known fact in parallel computing.

The last chapter in our collection is devoted to future trends and outlooks in parallel computing. It is written by the same authors who wrote the first chapter. This coincidence is not only symbolic, i.e., that the arc spanning the variety of this book's topics begins and terminates in the same hands, but it is more a consequence of the research competence of the team around S. Akl, which has been deeply rooted in parallel computing since its early beginning. The chapter brings convincing arguments for the importance of parallelism in future computing paradigms, inspired by physics, chemistry, and biology. In particular, the role of parallelism is illustrated by solving examples of problems arising in quantum computing. The chapter sends an optimistic message concerning parallel computing: parallelism is universally applicable, and not only existing advanced paradigms but also the future trends in computing cannot but obey this principle.

We believe that the selection of chapters covered in this monograph provides a representative sample of current achievements and challenges in parallel numerical computing. We anticipate that developments in this area will continue, supported by inspiring inputs coming from theory, applications, and industry.

We hope that the book will be positively echoed not only by specialists from the parallel and HPC communities, but also by a broader audience of readers, interested in computer solutions for their applications.

Acknowledgments The editors are grateful to the authors for all their efforts in contributing to this book and for the smooth and friendly cooperation.

References

1. D. Keyes, Parallel numerical algorithms: An introduction, in: Parallel Numerical Algorithms, D. E. Keyes, A. Sameh, V. Venkatakrishnan (Eds.), Kluwer Academic Publisher, Norwell, MA (1997).
2. M. Vajteršic, R. Trobec, P. Zinterhof, A. Uhl (Eds.), Parallel Numerics 2005, JSI Publisher, Ljubljana (2005).
3. J. Dongarra, I. S. Duff, D. C. Sorensen, H. A. van der Vorst, Numerical Linear Algebra for High-Performance Computers, SIAM, Philadelphia (1998).
4. E. Anderson, Z. Bai, C. Bischof, L. S. Blackford, J. Demmel, J. Dongarra, J. Du Croz, A. Greenbaum, S. Hammarling, A. McKenney, D. Sorensen, LAPACK Users' Guide, SIAM, Philadelphia (1999).
5. L. S. Blackford, J. Choi, A. Cleary, E. D'Azevedo, J. Demmel, I. Dhillon, J. Dongarra, S. Hammarling, G. Henry, A. Petitet, K. Stanley, D. Walker, R. C. Whaley, ScaLAPACK Users' Guide, SIAM, Philadelphia (1997).
6. L. S. Blackford, J. Demmel, J. Dongarra, I. Duff, S. Hammarling, G. Henry, M. Heroux, L. Kaufman, A. Lumsdaine, A. Petitet, R. Pozo, K. Remington, R. C. Whaley, An updated set of basic linear algebra subprograms (BLAS), ACM Transactions on Mathemetical Software 28 (2002), 135–151.
7. R. A. van de Geijn, E. S. Quintana-Orti, The Science of Programming Matrix Computations (2008).
8. S. Toledo, A survey of out-of-core algorithms in numerical linear algebra, in: External Memory Algorithms, J. M. Abello, J. S. Vitter (Eds.), DIMACS Series in Discrete Mathematics

and Theoretical Computer Science, American Mathematical Society Boston, MA (1999) 161–179.

9. G. Wang, A. Monti, G. Quan, Out-of-Core LU decomposition on a multiple-DSP platform, In: Electric Ship Technologies Symposium, IEEE (2007) 275–280.

10. O. Meshar, D. Irony, S. Toledo, An out-of-core sparse symmetric-indefinite factorization method, ACM Transactions on Mathematical Software 32 (2006) 445–471.

11. E. Rabani, S. Toledo, Out-of core SVD and QR decompositions, in: Proceedings of the 10th SIAM Conference on Parallel Processing for Scientific Computing, Norfolk (2001).

12. A. Butari, J. Langou, J. Kurzak, J. Dongarra, A class of parallel tiled linear algebra algorithms for multicore architectures, Technical Report 2007.122, University of Manchester (2007).

13. Cell Broadband Engine,
http://www-01.ibm.com/chips/techlib/techlib.nsf/products/CellBroadbandEngine/.

14. Sony PS3 Cluster (IBM Cell BE), http://moss.csc.ncsu.edu/~mueller/cluster/ps3/.

15. S. Bell et al., TILE64 Processor: A 64-Core SoC with Mesh Interconnect, Digest of Technical Papers, IEEE International (2008).

16. S. Vangal et al., An 80-Tile 1.28TFLOPS Network-on-Chip in 65nm CMOS, Digest of Technical Papers, IEEE International (2007).

17. A. Jantsch, H. Tenhunen (Eds.), Networks on Chip, Kluwer Academic Publishers, Dortdrecht (2003).

18. T. Bjerregaard, S. Mahadevan, A survey of research and practices of Network-on-chip, ACM Computing Surveys 38 (2006).

19. Z. Song, G. Ma , Heterogeneous NoC physical-level low-power implementation, in: Internet Computing in Science and Engineering, IEEE (2008) 379–382.

20. Revolutionary architecture for the next generation platform FPGAs,
http://www.xilinx.com/company/press/kits/asmbl/asmbl_arch_pres.pdf/.

21. S. Kumar, C. Paar, J. Pelzl, G. Pfeiffer, M. Schimmler, A configuration concept for a massive parallel FPGA architecture, in: International Conference on Computer Design (CDES'06) (2006).

22. E. H. D'Hollander, D. Stroobandt, A. Touhafi, Parallel computing with FPGAs – Concepts and applications, in: Parallel Computing and Applications, C. Bishof et al. (Eds.), NIC Series 38, Jülich (2007) 739–740.

23. DSP with FGPAs, http://www.andraka.com/dsp.htm/.

24. T. Wollinger, Ch. Paar, How secure are FPGAs in cryptographic applications, in: Proceedings of International Conference on Field Programmable Logic and Applications (FPL 2003), LNCS 2778, Springer-Verlag, Berlin (2003) 91–100.

25. T. Sansaloni, A. Perez-Pascual, J. Valls, Area-efficient FPGA-based FFT processor, Electronics Letters 39 (2003) 1369–1370.

26. S. O. Memik, A. K. Katsaggelos, M. Sarrafzadeh, Analysis and FPGA implementation of image restoration under resource constraints, IEEE Transactions on Computers 52 (2003) 390–399.

27. J. Sun, High Performance Reconfigurable Computing for Linear Algebra: Design and Performance Analysis, PhD dissertation, University of Tennessee, Knoxville (2008).

28. AccelChip Develops Linear Algebra Cores for Xilinx's FPGA Devices,
http://www.embeddedstar.com/press/content/2005/7/embedded18667.html/.

29. Scalable Definition, http://www.linfo.org/scalable.html/.

30. S. Swanson, L. McDowell, M. Swift, S. Eggers, H. Levy, An evaluation of speculative instruction execution on simultaneous multithreaded processors, Transactions on Computer Systems 21 (2003) 314–340.

31. GPU, http://www.webopedia.com/TERM/G/GPU.html.

32. TOP500 Supercomputer Sites at Mannheim University and University of Tennessee, `http://www.netlib.org/benchmark/top500.html/`.

33. Computer science reaches historic breakthrough, `http://www.ibm.com/ibm/ideasfromibm/us/roadrunner/20080609/index.shtml/`.

34. IBM BladeCenter QS22, `http://www-03.ibm.com/systems/bladecenter/hardware/servers/qs22/index.html`.

35. Open Grid Forum, `http://www.ogf.org/`.

36. Globus Toolkit, `http://www.globus.org/toolkit/`.

37. M. Mascagni, Grid-Based Monte Carlo Application, GRID (2002) 13–24.

38. S. Branford, C. Sahin A. Thandavan, C. Weihrauch, V. N. Alexandrov, I. T. Dimov, Monte Carlo methods for matrix computations on the grid, Future Generation Computer Systems 24 (2008) 605–612.

39. H. Hofbauer, A. Uhl, P. Zinterhof, Quasi Monte Carlo integration in Grid enviroments, Parallel Processing Letters 16 (2006) 285–312.

40. T. Sakurai, Y. Kodaki, H. Tadano, D. Takahashi, M. Sato, U. Nagashima, A parallel method for large sparse generalized eigenvalue problems using a grid RPC system, Future Generation Computer Systems 24 (2008) 613–619.

41. H. Astsatryan, V. Sahakyan, Y. Shoukouryan, M. Daydé, A. Hurault, M. Pantel, E. Caron, A Grid-aware Web interface with advanced service trading for linear algebra calculations, in: International Meeting High Performance Computing for Computational Science (VECPAR 2008) (2008).

42. E. Caron, F. Desprez, DIET: A scalable toolbox to build network enabled servers on the Grid, International Journal of High Performance Computing Applications 20 (2006) 335–352.

43. M. Watzl, R. Kutil, Distributed information retrieval using LSI, in: Proceedings of the 1st Austrian Grid Symposium, J. Volkert, T. Fahringer, D. Kranzlmüller, W. Schreiner (Eds.), Austrian Computer Society, Schloss Hagenberg (2006), 90–101.

44. Austrian Grid 2: Distributed Supercomputing in the Grid, `http://www.risc.uni-linz.ac.at/projects/agrid2/`.

45. T. Berka, Parallel image retrieval on the grid using the vector space model, Master Thesis, University of Salzburg (2008).

46. W. P. Carey, Cloud computing: The evolution of Software-as-a-Service, Knowledge@W.P.Carey, 2008.

47. K. Bennett, P. Layzell, D. Budgen, P. Brereton, L. Macaulay, M. Munro, Service-based software: The future for flexible software, in: Proceedings of the Seventh Asia-Pacific Software Engineering Conference (2000) 214.

48. Google Apps: Web Applications for Communication and Collaboration, `http://www.google.com/apps/`.

49. Science Clouds, `http://workspace.globus.org/`.

50. R. Martin J., N. Hoover, Guide to cloud computing, InformationWeek, 2008.

51. Message Passing Interface Forum, `http://www.mpi-forum.org/`.

52. MPICH2, `http://www.mcs.anl.gov/research/projects/mpich2/`.

53. LAM/MPI Parallel Computing, `http://www.lam-mpi.org/`.

54. PVM – Parallel Virtual Machine, `http://www.csm.ornl.gov/pvm/`.

55. B. Carlson, T. El-Ghazawi, R. Numerich, K. Yelick, Programming in the partitioned global address space model, Tutorial at Supercomputing 2003, `http://upc.gwu.edu/`.

56. Co-Array Fortran, `http://www.co-array.org/`.

57. Unified Parallel C, `http://upc.gwu.edu/`.

58. Titanium, `http://titanium.cs.berkeley.edu/`.

59. B. Chapman, P. Mehrotra, H. Zima, Programming in Vienna Fortran, Scientific Programming 1 (1992), 31–50

60. High Performance Fortran, `http://www.netlib.org/hpf/`.

61. The Portland Group, `http://www.pgroup.com/`.

62. POSIX Threads Programming, `https://computing.llnl.gov/tutorials/pthreads/`.
63. OpenMP, `http://www.openmp.org/`.
64. Cluster OpenMP, User's Guide, Version 9.1, INTEL, 2006.
65. The MathWorks, `http://www.mathworks.com/`.
66. Parallel Programming in MATLAB, `http://www.mathworks.com/products/parallel-computing/parallel/`.
67. Python Programming Language, `http://www.python.org/`.
68. B. L. Chamberlain, D. Callahan, H. P. Zima, Parallel programmability and the Chapel language, International Journal of High Performance Computing Applications 21 (2007) 291–312.
69. High Productivity Computer Systems, `http://www.highproductivity.org/`.
70. M. Süss, C. Leopold, Evaluating the state of the art of parallel programming systems, Technical Report KIS 1/2005, University of Kassel (2005).
71. Introduction to Java Threads, `http://www.javaworld.com/javaworld/jw-04-1996/jw-04-threads.html/`.
72. Ch. Liao, O. Hernandez, B. Chapman, W. Chen, W. Zheng, OpenUH: An optimizing, portable OpenMP compiler, Concurrency and Computation: Practice and Experience 19 (2007) 2317–2332.
73. M. Klemm, M. Bezold, R. Veldema, M. Philippsen, JaMP: An implementation of OpenMP for a Java DSM, Concurrency and Computation: Practice and Experience 19 (2007) 2333–2352.
74. K. Asanovic et al., The landscape of parallel computing research: A view from Berkley, Technical Report UCB/EECS-2006-183, University of California, Berkley (2006).
75. K. Yelick, Automatic Performance Tuning Workshop, `http://cscads.rice.edu/`.
76. S. G. Akl, Unconventional computational problems with consequences to universality, International Journal of Unconventional Computing 4 (2008) 89–98.
77. S. G. Akl, The Design and Analysis of Parallel Algorithms, Prentice Hall, Upper Saddle River, NJ (1989).
78. S. G. Akl, Parallel Computation: Models and Methods, Prentice Hall, Upper Saddle River, NJ (1997).
79. M. Nagy, S. G. Akl, Parallelism in quantum information processing defeats the Universal Computer, Parallel Processing Letters 17 (2007) 233–262.
80. AURORA 2007 Conference on Scientific Computing, `http://www.vcpc.univie.ac.at/aurora/`.
81. The Austrian Grid Consortium, `http://www.austriangrid.at/`.
82. R. Prodan, T. Fahringer, Grid Computing. Experiment Management, Tool Integration and Scientific Workflows, LNCS 4340, Springer-Verlag, Berlin, (2007).
83. T. Fahringer, R. Prodan, R. Duan, J. Hofer, F. Nadeem, F. Nerieri, S. Podlipnig, J. Qin, M. Siddiqui, H.-L. Truong, A. Villazon, M. Wieczorek, ASKALON: A development and Grid computing environment for scientific workflows, in: Workflows for eScience: Scientific Workflows for Grids, I. J. Taylor, E. Deelman, D. B. Gannon, M. Shields (Eds.), Springer-Verlag, Berlin, (2007) 450–474.
84. Enabling Grids for E-sciencE (EGEE), `http://www.eu-egee.org/`.
85. The European Grid Initiative (EGI), `http://web.eu-egi.eu/`.
86. Gridbus News Blog, `http://www.gridbus.org/`.
87. R. Buyya, K. Bubendorfer (Eds.), Market Oriented Grid and Utility Computing, Wiley Press, New York (2009).
88. A. Sulistio, U. Čibej, S. Venugopal, B. Robič, R. Buyya, A toolkit for modelling and simulating data Grids: an extension to GridSim, Concurrency and Computation: Practice and Experience 20 (2008) 1591–1609.
89. M. Thuné, Straightforward partitioning of composite grids for explicit difference methods, Parallel Computing 17 (1991) 665–672.

90. M. Ljungberg, K. Otto, M. Thuné, Design and usability of a PDE solver framework for curvilinear coordinates, Advances in Engineering Software, 37 (2006) 814–825.

91. J. Rantakokko, A dynamic MPI-OpenMP model for structured adaptive mesh refinement, Parallel Processing Letters 15 (2005) 37–47.

92. M. Nordén, H. Löf, J. Rantakokko, S. Holmgren, Geographical locality and dynamic data migration for OpenMP implementations of adaptive PDE solvers, in: OpenMP Shared Memory Parallel Programming, M. S. Müller, B. M. Chapman, B. R. de Supinski, A. D. Malony, M. Voss (Eds.), LNCS 4315, Springer-Verlag, Berlin, (2008) 382–393.

93. C. Amstler, P. Zinterhof, Uniform distribution, discrepancy, and reproducing kernel Hilbert spaces, Journal of Complexity 17 (2001) 497–515.

94. T. Schell, A. Uhl, P. Zinterhof, Measures of uniform distribution in wavelet based image compression, Monte Carlo Methods and Applications 10 (2004) 587–598.

95. A. Uhl, A. Pommer, Image and Video Encryption: From Digital Rights Management to Secured Personal Communication, Springer-Verlag, 2004.

96. A. Uhl, P. Wild, Footprint-based biometric verification, Journal of Electronic Imaging 17 (2008).

97. M. Haefner, R. Kwitt, A. Uhl, A. Gangl, F. Wrba, A. Vecsei, Computer-assisted pit-pattern classification in different wavelet domains for supporting dignity assessment of colonic polyps, Pattern Recognition, doi:10.1016/j.patcog.2008.07.012, Elsevier Ltd. Amsterdam (2008).

98. H. Hofbauer, A. Uhl, P. Zinterhof, Zinterhof sequences in Grid-based numerical integration, in: Monte Carlo and Quasi-Monte Carlo Methods 2006, A. Keller, S. Heinrich, H. Niederreiter (Eds.), Springer-Verlag, Berlin (2008), 495–510.

99. P. Jez, Approximative solution of the Cauchy problem of the homogenous heat conduction equation by Quasi Monte Carlo methods, Technical Report 2007-03, Department of Computer Sciences, University of Salzburg, 2007.

100. T. Tusar, P. Korošec, G. Papa, B. Filipič, J. Šilc, A comparative study of stochastic optimization methods in electric motor design, Applied Intelligence 27 (2007) 101–111.

101. I. Fister, M. Mernik, B. Filipič, Optimization of markers in clothing industry, Engineering Applications of Artificial Intelligence 21 (2008) 669–678.

102. M. Depolli, V. Avbelj, R. Trobec, Computer-simulated alternative modes of U-wave genesis, Journal of Cardiovascular Electrophysiology 19 (2008) 84–89.

103. Wolfgang Händler, Fridolin Hofmann, Hans Jürgen Schneider, A general purpose array with a broad spectrum of applications, Computer Architecture (1975) 311–335.

104. W. Händler, U. Herzog, F. Hofmann, H. J. Schneider, Multiprozessoren für breite Anwendungsbereiche: Erlangen General Purpose Array, ARCS (1984) 195–208.

105. A. Bode, G. Fritsch, W. Händler, W. Henning, F. Hofmann, J. Volkert, Multi-Grid oriented computer architecture, International Conference on Parallel Processing (1985) 89–95.

106. W. Händler, E. Maehle, K. Wirl, Dirmu multiprocessor configurations, International Conference on Parallel Processing (1985) 652–656.

107. M. Stürmer, H. Köstler, U. Rüde, A fast full multigrid solver for applications in image processing, Numerical Linear Algebra with Applications 15 (2008) 187–200.

108. H. Köstler, R. Schmid, U. Rüde, Ch. Scheit, A parallel multigrid accelerated Poisson solver for ab initio molecular dynamics application, Computing and Visualization in Science 11 (2008) 115–122.

109. Ch. Freundl, T. Gradl, U. Rüde, Towards Petascale multilevel finite-element solvers, in: Petascale Computing: Algorithms and Applications, D. A. Bader (Ed.), Chapman and Hall/CRC, Boca Raton, FL (2008) 375–389.

110. S. Donath, J. Götz, S. Bergler, Ch. Feichtinger, K. Iglberger, U. Rüde, waLBerla: The need for large-scale supercomputers, in: High Performance Computing in Science and Engineering, S. Wagner, M. Steinmetz, A. Bode, M. Brehm (Eds.) Springer-Verlag, Berlin (2008) 459–473.

111. J. N. Lyness, T. Sørevik, Five dimensional K-optimal lattice rules, Mathematics of Computation 75 (2006) 1467–1480.

112. T. Sørevik, L. B. Madsen, J. P. Hansen, A spectral method for integration of the time-dependent Schrödinger equation in hyperspherical coordinates, Journal of Physics A: Mathematical and General 38 (2005) 6977–6985.

113. T. Birkeland, T. Sørevik, Parallel redistribution of multidimensional data, in: Parallel Computing: Architecture, Algorithms and Applications, C. Bishop et al. (Eds.), John von Neumann Institute for Computing, NIC Series 38 (2007) 433–440, 2007.

114. R. Blikberg, T. Sørevik, Load balancing and OpenMP implementation of nested parallelism, Parallel Computing 31 (2005) 984–998.

115. U. Borštnik, D. Janežič, Symplectic molecular dynamics simulations on specially designed parallel computers, Journal of Chemical Information and Modeling 45 (2005) 1600–1604.

116. U. Borštnik, M. Hodoscek, D. Janežič, Improving the performance of molecular dynamics simulations on parallel clusters, Journal of Chemical Information and Modeling 44 (2004) 359–364.

117. M. Praprotnik, S. Hocevar, M. Hodoscek, M. Penca, D. Janežič, New all-atom force field for molecular dynamics simulation of an AlPO4-34 molecular sieve, Journal of Computational Chemistry 29 (2008) 122–129.

118. ParNum 2005, http://www.cosy.sbg.ac.at/events/parnum05/.

119. P. Trunk, J. Močnik, R. Trobec, B. Geršak, 3D heart model for computer simulations in cardiac surgery, Computers in Biology and Medicine 37 (2007) 1398–1403.

120. R. Trobec, M. Šterk, S. Almawed, M. Veselko, Computer simulation of topical knee cooling, Computers in Biology and Medicine 38 (2008) 1076–1083.

121. I. Rozman, M. Šterk, J. Močnik, B. Robič, R. Trobec, Performance measurements of computing networks, Parallel and Distributed Computing Practices 9 (2008) 143–150.

122. R. Trobec, M. Šterk, B. Robič, Computational complexity and parallelization of the meshless local Petrov-Galerkin method, Computers and Structures 87 (2009) 81–90.

123. M. Šterk, R. Trobec, Meshless solution of a diffusion equation with parameter optimization and error analysis, Engineering Analysis with Boundary Elements 32 (2008) 567–577.

124. J. Mikloško, R. Klette, M. Vajteršic, I. Vrťo, Fast Algorithms and their Implementation on Specialized Computers, North-Holland, Amsterdam (1989).

125. G. Okša, M. Bečka, M. Vajteršic, Parallel algorithm for matrix multiplication by Gramian of Toeplitz-block matrix, Proceedings of 20th IASTED Conference Applied Informatics 2002, ACTA Press (2002) 53–58.

126. G. Okša, M. Vajteršic, Efficient pre-processing in the parallel block-Jacobi SVD algorithm, Parallel Computing 32 (2006) 166–176.

127. G. Okša, M. Vajteršic, Preconditioned parallel block-Jacobi SVD algorithm, Parallel Processing Letters 16 (2006) 371–380.

128. M. Vajteršic, Algorithms for Elliptic Problems: Efficient Sequential and Parallel Solvers, Kluwer Academic Publisher, Dordrecht-Boston (1993).

129. L. Grandinetti, J. Kowalik, M. Vajteršic (Eds.), High Performance Computing: Technology and Applications, Kluwer Academic Publisher, Dordrecht (1997).

130. M. Bečka, G. Okša, M. Vajteršic, Dynamic ordering for a parallel block-Jacobi SVD method, Parallel Computing 28 (2002) 243–262.

131. R. Kutil, P. Eder, Parallelization of wavelet filters using SIMD extensions, Parallel Processing Letters 16 (2006) 335–349.

132. R. Kutil, A single-loop approach to SIMD parallelization of 2-D wavelet lifting, in: Proceedings of the 14th Euromicro Conference on Parallel, Distributed and Network-based Processing (2006) 413–420.

133. D. Engel, R. Kutil, A. Uhl, A symbolic transform attack on lightweight encryption based on wavelet filter parameterization, in: Proceedings of ACM Multimedia and Security Workshop (2006) 202–207.

134. R. Kutil, Alluvion – A language for computer arithmetic algorithms, in: Scientific Computing in Salzburg, Austrian Computer Society, Vienna (2005) 113–120.

135. J. Gondzio, A. Grothey, A new unblocking technique to warmstart interior point methods based on sensitivity analysis, SIAM Journal on Optimization 19 (2008) 1184–1210.

136. J. Gondzio, A. Grothey, Solving nonlinear portfolio optimization problems with the primal-dual interior point method, European Journal of Operational Research 181 (2007) 1019–1029.

137. J. Gondzio, A. Grothey, Parallel interior point solver for structured quadratic programs: Application to financial planning problems, Annals of Operations Research 152 (2007) 319–339.

138. J. Gondzio, A. Grothey, Direct solution of linear systems of size 10^9 arising in optimization with interior point methods, in: Parallel Processing and Applied Mathematics 2005, R. Wyrzykowski, J. Dongarra, N. Meyer, J. Wasniewski (Eds.), LNCS 3911, Springer-Verlag, Berlin (2006) 513–525.

139. M. Nagy, S. G. Akl, Coping with decoherence: Parallelizing the quantum Fourier transform, Proceedings of the Nineteenth International Conference on Parallel and Distributed Computing Systems, San Francisco, (2006) 108–113.

140. N. Nagy, S. G. Akl, Aspects of biomolecular computing, Parallel Processing Letters 17 (2007) 185–211.

141. W. M. Alsalih, S. G. Akl, H. Hassanein, Cooperative ad hoc computing: towards enabling cooperative processing in wireless environments, International Journal of Parallel, Emergent and Distributed Systems 23 (2008) 59–79.

142. N. Nagy, M. Nagy, S. G. Akl, Quantum wireless sensor networks, in: Proceedings of the Seventh International Conference on Unconventional Computation, C. S. Calude, et al. (Eds.), LNCS 5204, Springer-Verlag, Berlin (2008) 177–188.

143. M. Nagy, S. G. Akl, Quantum computing: beyond the limits of conventional computation, International Journal of Parallel, Emergent and Distributed Systems 22 (2007) 123–135.

Chapter 2
Introduction to Parallel Computation

Selim G. Akl and Marius Nagy

Abstract

This chapter is intended to provide an overview of the fundamental concepts and ideas shaping the field of parallel computation. If serial (or sequential) algorithms are designed for the generic uni-processor architecture of the Random Access Machine (RAM), in the case of parallel algorithms there are a variety of models and architectures supporting the parallel mode of operation: shared-memory models, interconnection networks, combinational circuits, clusters and grids.

Sometimes, the methods used in designing sequential algorithms can also lead to efficient parallel algorithms, as it is the case with divide and conquer techniques. In other cases, the particularities of a certain model or architecture impose specific tools and methods that need to be used in order to fully exploit the potential offered by that model. In all situations, however, we seek an improvement either in the running time of the parallel algorithm or in the quality of the solution produced by the parallel algorithm with respect to the best sequential algorithm dealing with the same problem.

The improvement in performance can even become superlinear with respect to the number of processors employed by the parallel model under consideration. This is the case, for example, of computations performed under real-time constraints, when the deadlines imposed on the availability of the input and/or output data leave little room for sequentially simulating the parallel approach. Furthermore, in the examples presented at the end of the chapter, the impossibility to simulate a parallel solution on a sequential machine is due to the intrinsically parallel nature of the computation, rather than being an artifact of externally imposed time constraints.

Selim G. Akl
School of Computing, Queen's University, Kingston, Ontario, Canada,
e-mail: akl@cs.queensu.ca

Marius Nagy
School of Computing, Queen's University, Kingston, Ontario, Canada,
e-mail: marius@cs.queensu.ca

R. Trobec et al. (eds.), *Parallel Computing*, DOI 10.1007/978-1-84882-409-6_2,
© Springer-Verlag London Limited 2009

In this respect, parallelism proves to be the vehicle leading to a Non-Universality result in computing: there is no finite computational device, sequential or parallel, conventional or unconventional, that is able to simulate all others.

2.1 Introduction

In our sophisticated modern world, time is perhaps the most precious commodity. We live our lives in the fast lane, always trying to buy more time. In this world, speed is of the essence and efficiency translates naturally into how fast (and sometimes how well) we can solve the problems we face. To this end, parallel computing, the central theme of this book, is perhaps our greatest ally.

Indeed, the main motivation for parallel computing is to speed up computation. The pervasive nature of computers nowadays makes it possible for huge amounts of data to be acquired and stored in large databases for future analysis, data mining, referencing, etc. In some cases, the amount of information that needs to be processed is so huge, that the time required to complete the job becomes prohibitively long. As an illustrative example, imagine you are charged with the following task: given a phone number, you are required to look in the phone book for the name and address of the person whose phone number you were given. If you live in a big city, that is, if the phone book is big, then this task is a tedious one if you are to perform it all by yourself. But if you decide to call your friends and each one agrees to look only at the names beginning with a certain letter, for example, then the task is completed much faster.

The simplicity of the example above is intentional, so that the main message is not obstructed by unnecessary details. Often, and this will become apparent from the applications addressed throughout the book, splitting a job among the available processors is not a trivial task and the overhead incurred by parallelization may become significant. Regardless, the message conveyed by the parallel computing paradigm remains the same: If several processors work together (cooperate) to solve a given computational problem, then the time required to complete the task may be greatly reduced.

But time is not the only measure for the advantage gained by using a parallel approach. Sometimes, it is the quality of the solution computed that is greatly improved if more processors are available, in a fixed amount of time. Furthermore, computational scenarios have been identified, in which the only chance to terminate a computation and reach a solution is to have the required number of processors working simultaneously on that respective task. We call such problems *inherently parallel*, because the ability of a parallel computer to be "in more than one place at a time" through its multiple processing elements is a necessary condition to successfully tackle these problems.

The renewed interest in various forms of parallel computing that we are witnessing today can be largely explained by the availability and affordability of computing power. When it becomes increasingly difficult and costly to build faster processors, it seems that the best idea to improve performance is to design architectures and methods that allow several relatively cheap processors to combine their capabilities in order to achieve performances that none of the processors, taken individually, would have been capable of.

There are many different ways in which multiple processors can work together, and parallel algorithms designed to solve a certain computational problem necessarily depend on the underlying architecture. It is the intention of this early chapter to try to guide the reader through the main concepts governing the study of parallel computing: possible computational models, algorithm design methods and specific analysis techniques. The later part of the chapter proves that the field of parallel computing is worth investigating in its own respect, providing useful and sometimes unexpected insights into the theory of computing. Specifically, we discuss the importance and implications of the parallel computing paradigm for various computational environments and formulate a perhaps surprising Non-Universality result in computing.

2.2 Parallel Versus Sequential Computation

Whenever we are in the position of analyzing the performance of a parallel algorithm, we compare it with the best possible sequential approach dealing with the same problem. The sequential scenario represents the reference for any parallelization attempt. Therefore, it is only natural to begin our review of parallel models with a short description of the sequential model of computation.

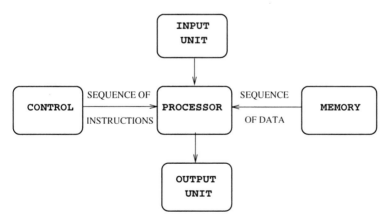

Fig. 2.1 Architecture of a sequential computer.

The main components of a sequential computer are depicted in Fig. 2.1. This design, which coincides with the inception of computer science, proved to be so successful that even today single-processor computers follow the same architecture.

Relevant for our discussion is the observation that a *single* sequence of instructions (the program) is executed, one at a time, by the *single* processing unit on a *single* sequence of data (the problem instance). At each step, the control unit provides the processor with the type of operation to be performed (such as addition, comparison, etc.) and the operands (data) on which the operation is to be carried out. Usually, the data has to be loaded from memory into the processor's internal registers. For this purpose, a memory access unit (not shown in Fig. 2.1 for simplicity) is responsible to create a path from the processor to any desired memory location. Owing its name to this way of working with a memory, the sequential (or serial, or conventional) model of computation sketched above is known as the Random Access Machine (RAM) [1].

When two or more processors must work together to solve the problem at hand, the mode of operation of the parallel machine represents a radical departure from the way the RAM operates. From an algorithmic viewpoint, the problem to be solved is broken into subproblems, which are solved simultaneously, each on a different processor. During this phase, the processors may communicate with each other to exchange partial results. In the final step, the results must be combined to form the solution to the original problem.

In order to support the execution of parallel algorithms, any design for a parallel architecture must address several key issues. What arrangement or topology should be chosen to best organize the processors? How can they be made to communicate efficiently? Should they all execute the same code (program) on different subsets of the input data or each processor is allowed to execute its own particular program? Should the processors operate synchronously or asynchronously? Depending on how these and other related questions are answered, we can have a plethora of different models of parallel computing. The next section reviews some of the most important choices for a parallel model of computation.

2.3 Parallel Computational Models

Parallel and distributed computing models share many important characteristics, the difference being made by how far apart are the processing units that need to collaborate in order to solve a computational problem. Consequently, along with shared-memory models, interconnection networks and combinational circuits (representing three large classes of parallel computers), we also include clusters and grids in our discussion, as possible models of distributed computing.

2.3.1 Shared-Memory Models

A direct extension of the RAM, in which several identical processors are connected to a single shared memory is the Parallel Random Access Machine (PRAM), shown in Fig. 2.2. The Memory Access Unit (MAU) allows any of the N processors to gain access to any particular memory location, whether for the purpose of reading or writing. Thus, the shared memory can be viewed as a communication medium among the processors. Any datum that processor P_i needs to communicate to processor P_j is first written into the memory by P_i, from where it is subsequently read by P_j.

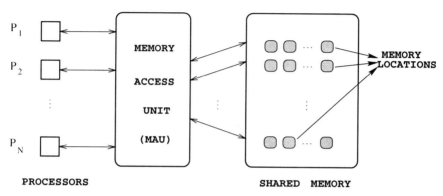

Fig. 2.2 The Parallel Random Access Machine.

Although the model is general enough to allow each processor to execute its own algorithm in a totally asynchronous fashion, many useful applications assume that the processors perform the same operation in a synchronous fashion on different data. In this latter mode of operation, each step of the parallel algorithm can be broken into three phases: a *read* (from memory) phase, a *compute* phase and a *write* (into memory) phase. During the read and write phases, it is possible for more than one processor to simultaneously read from the same memory location (Concurrent Read) or for several processors to write into the same memory location at the same time (Concurrent Write). Obviously, conflicting situations may arise in which different processors attempt to write different values into a single memory location, during the same write phase. Depending on how such conflicts are resolved, several variants of the Concurrent Write (CW) instruction are possible [2]. Also, not all processors must be active in every single step. The algorithm can specify which processors are active in each step, usually by making use of their indices.

The PRAM is a powerful model of parallel computing, especially due to the concurrent forms of memory access available in its repertoire of instructions. But, if the size of the shared memory is big and/or the number of processors N is large, then the MAU endowing the PRAM with its power may become too complex for practical purposes. This puts a limit on the scalability of the PRAM model.

2.3.2 Interconnection Network Models

Interconnection network models try to avoid the bottleneck caused by a MAU having to connect many processors to many memory locations and they do this by abandoning the concept of a shared memory altogether. Instead, each processor has its own local memory and all communications among processors take place via direct links connecting them. Thus, a "network of processors" is formed, with many different topologies possible. Popular topologies include the following: linear array, mesh, tree, hybrid interconnections such as the mesh of trees, hypercube, star, and the enumeration could continue. Figure 2.3 shows, for example, how processors are connected in a hypercube with four dimensions.

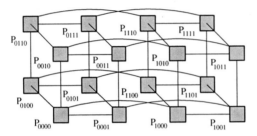

Fig. 2.3 A hypercube interconnection network with four dimensions.

Since we no longer have a shared memory to ensure a fast two-step communication between any pair of processors, sometimes a datum has to travel several links in order to reach its destination. This is because a fully connected network is impractical if the number of processors is not small. Therefore, a good design for an interconnection network is one that aims to keep the *diameter* of the network small. The diameter is defined as the length of the longest distance between pairs of processors in the network. In this context, the distance between two processors in a given topology is taken as the smallest number of links connecting those processors in that particular topology.

An important challenge in achieving a small diameter is to also maintain the number of neighbors (or degree) of each processor to a small value. The obvious reason, as mentioned above for a complete network, is the high cost associated with having many neighbors or even the infeasibility of such a design. Furthermore, there are other criteria that can also be used to compare different topologies. For example, the length of the links between processors and the regularity showed by a specific arrangement are important elements to take into consideration for the extendibility of a network.

In conclusion, it is difficult (if not impossible) to find a single topology that is desirable from all points of view. Therefore, when choosing an interconnection network, one should pay attention to those features that are best suited for the particular problem to be solved.

2.3.3 Circuit Models

Another model of parallel computation is the family of combinational circuits. These circuits consist of very simple processors (such as comparators, for instance) arranged in columns (or stages). The circuit in Fig. 2.4 uses nine comparators arranged in three stages to merge two sequences of four numbers each. The processors in the same column can all operate in parallel.

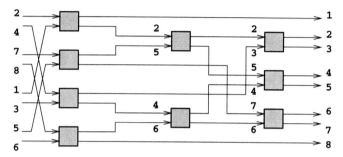

Fig. 2.4 A merging circuit.

The distinctive characteristic of combinational circuits is the fact that they do not have feedback. Data flow through a circuit from input to output, with each processor being used only once during the computation. Assuming that each processor takes one time unit to produce its output(s) from the moment all of its inputs are available, we can assimilate the worst-case running time of a combinational circuit with its *depth* or number of stages. The *width*, on the other hand, or the maximum number of processors in a stage represents the maximum degree of parallelism offered by a combinational circuit. Finally, the total number of processors used by a combinational circuit is known as its *size*.

Naturally, when designing a combinational circuit to solve a certain problem whose input is specified by n input lines, the goal is that the depth and size of the circuit be functions that grow slowly with n. Also, depending on the application, the operation effected by a processor can be a comparison (like in sorting and merging circuits) or any other simple arithmetic or logic operation, such as addition or logical *and*. As in the circuit in Fig. 2.4, many combinational circuits are targeted to address specific problems, but multipurpose circuits exist that may be used for many different kinds of applications.

2.3.4 Clusters

The parallel models discussed so far are characterized by a spatial proximity of the processing elements with which a parallel computer is endowed. With clusters and grids we move to another class of parallel computing models, which can rightfully be labeled as *distributed* computing models, since their computing elements can be distributed over a large area.

A cluster is generally defined as a collection of interconnected stand-alone computers working together as a single, integrated computing resource. The organization in a network is specific to both interconnection network models and clusters, but a node in an interconnection network parallel computer is just a processor, while in a cluster a node is a computing system in its own right, endowed with memory, I/O facilities and an operating system, among other things. In particular, a node could even be a multiprocessor system, such as a shared memory parallel computer. The components of a cluster are usually connected to each other through dedicated, fast links (like in a local area network, for instance) and in many respects may appear as a single system to users and applications. Such a system can provide a cost-effective way to improve performance (speed, reliability, availability, throughput, etc.) compared with supercomputers of similar characteristics.

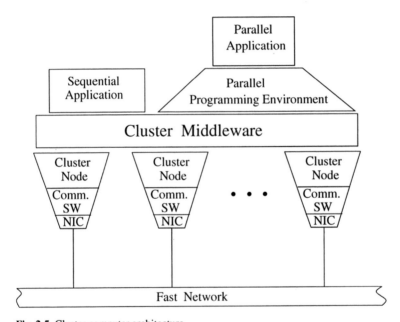

Fig. 2.5 Cluster computer architecture.

The typical architecture of a cluster is depicted in Fig. 2.5. It consists of multiple computing nodes (PCs, workstations, multiprocessor systems) connected to a high speed communication medium (network infrastructure) through Network Interface Cards (NICs). These are responsible for transmitting and receiving packets of data between cluster nodes. The communications software layer endows the node with fast communication protocols and services, offering a means of fast and reliable data communication among cluster nodes. The cluster middleware is an interface between applications and the individual computers composing the cluster and has the important role of making the latter appear as a unified system to the former. Parallel applications run on the cluster with the support of specific parallel programming environments and tools, such as compilers, parallel virtual machines and message passing libraries.

Clusters can have different purposes, according to the particular performance metric they have been deployed to improve. High availability clusters, for example, seek to improve the availability of services offered by the cluster. They achieve this by having redundant nodes which can keep providing services even if some system components fail. Alternatively, load balancing clusters are designed with the main purpose of distributing the workload as evenly as possible among the cluster's nodes. Other possible factors for cluster classification are node ownership, node hardware, node operating system, node configuration and the level of clustering.

Besides offering high performance at a relatively low cost, clusters are easily expandable, thus making cluster computing an increasingly popular form of parallel/distributed computing.

2.3.5 Grids

In some sense, grid computing can be seen as an extreme case of cluster computing. The following properties may help draw a separation line between clusters and grids, although this line is not always very visible. Grids are usually geographically dispersed and consequently more loosely coupled than clusters. The network infrastructure allowing nodes in a grid to communicate is typically the Internet. This may create important additional problems for the grid middleware in charge of managing the access to the grid's resources, as the nodes can be heterogeneous (different operating systems and hardware architectures), having different owners and not fully trusting each other.

One way to deal with faulty or malicious nodes producing erroneous results is to assign a chunk of work randomly to several nodes (presumably with different owners) and check that at least two different nodes report the same answer. Nodes such as laptops or dial-up Internet computers are very unreliable in terms of maintaining connectivity, so one strategy to address this problem is to assign large work chunks in order to avoid the need for a continuous connection. If a node fails to report its results in a reasonable amount of time, that particular job would then have to be reassigned. Of course, the workloads assigned to the nodes must consist of many independent packets of work, such that each computer will perform its task independently of the rest of the grid. In this way, intermediate results obtained on one node do not affect other jobs in progress on some other nodes of the grid. Resources such as storage, on the other hand, may be shared by all the nodes in the grid.

In essence, grid computing aims at making computer power as easy to access as an electric power grid and for this reason grid computing is also known as utility computing, since utilities such as processor time, data and storage space are offered as a service supported by a pool of distributed computing resources. In principle, any application can be "plugged" into the grid in order to take advantage of its resources, but it is the very large tasks that will benefit most from the grid's ability to aggregate the computing power of many different and distant machines into some sort of "virtual supercomputer." Thus, unlike clusters (which usually number hundreds of nodes), a grid may harness the resources of even millions of computers, coordinating their efforts to accomplish tasks that would otherwise be impossible to complete on a single machine due to cost and/or time constraints.

An example of such very large grid is the *SETI@home* project, which uses about three million computers all over the world to analyze the colossal amount of data produced by the Arecibo Observatory radiotelescope in its search for evidence of extra-terrestrial intelligence.

2.4 Parallel Algorithm Design Methods

Designing an algorithm for a particular model of parallel computation is a process that necessarily has to take into consideration the characteristic features, strengths and weaknesses of that model. Therefore, the design of a parallel algorithm is clearly model-dependent. Nevertheless, at a high level, the design methodology for a parallel algorithm consists, in general, of the following four steps:

1. *Partitioning*: The problem is decomposed into fine-grain tasks, maximizing the number of tasks that can be executed simultaneously.
2. *Communication analysis*: Determine what communications are required among the tasks. Sometimes, the output of this step takes the form of a *task graph* with fine-grain tasks as nodes and communication channels as edges.
3. *Granularity control* (or **Agglomeration**): Aims to reduce communication requirements by combining groups of fine-grain tasks into fewer, but larger coarse-grain tasks.
4. *Mapping*: Assign coarse-grain tasks to processors, trying to achieve an optimal tradeoff between communication costs and degree of parallelism.

With respect to the design methodologies we are familiar with from sequential algorithms, the approach sketched above reminds us of the "divide and conquer" technique. Certainly, this is not at all surprising, since by definition, a parallel algorithm has to break the original problem into subproblems and assign them to different processors that may need to communicate in order to exchange partial results.

To exemplify the process of developing a parallel algorithm for a particular model, we show how Gaussian elimination can be implemented on a mesh of trees in order to solve a system of linear equations. The method solves the equation $A \cdot x = b$ (in matrix form) by applying a sequence of transformations to both A and b such that in the end, A is reduced to the identity matrix I. If the original system contains n equations in n unknowns, then the Gaussian elimination algorithm is made up of n iterations. The aim of the ith iteration is to make all elements in column i of A

equal to 0 except for a_{ii}, which becomes 1. This can be accomplished by subtracting multiples of row i from all other rows, according to the transformation:

$$a_{rj}^{(i)} \longleftarrow a_{rj}^{(i-1)} - \frac{a_{ri}^{(i-1)} \cdot a_{ij}^{(i-1)}}{a_{ii}^{(i-1)}}, \text{ for } r \neq i \text{ and } i \leq j \leq n \qquad (2.1)$$

and dividing the elements of row i by the pivot a_{ii}:

$$a_{ij}^{(i)} \longleftarrow \frac{a_{ij}^{(i-1)}}{a_{ii}^{(i-1)}}, i \leq j \leq n. \qquad (2.2)$$

The same transformations are effected on the elements of vector b:

$$b_r^{(i)} \longleftarrow b_r^{(i-1)} - \frac{a_{ri}^{(i-1)} \cdot b_i^{(i-1)}}{a_{ii}^{(i-1)}}, \text{ for } r \neq i, \qquad (2.3)$$

$$b_i^{(i)} \longleftarrow \frac{b_i^{(i-1)}}{a_{ii}^{(i-1)}}. \qquad (2.4)$$

For numerical stability considerations, the pivot is always chosen as the largest element (in absolute value) from the submatrix of A whose upper left corner element is a_{ii}. Therefore, before the new values for the elements of A and b are computed, a search for the pivot is first conducted. If this element is not already on row i and column i, then row i has to be switched with the row containing the pivot and/or column i is interchanged with the column on which the pivot was found. When two rows of A are interchanged, the corresponding elements of b have to be switched as well and similarly, when two columns change places, the corresponding unknowns also have to be interchanged.

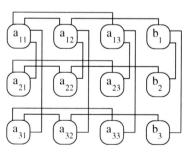

Fig. 2.6 Gaussian elimination on a mesh of trees.

How can the Gaussian elimination algorithm be implemented on a mesh of trees? First of all, note how the processors are connected in this particular topology (Fig. 2.6). The processors are indeed arranged in a mesh-like structure, but they are not connected using regular mesh links. Instead, the processors in each row are connected together to form a binary tree of processors, with the root of the tree being the first processor in that row. Similarly, there is a binary tree of processors for each column of the mesh. This particular way of connecting processors determines the communication patterns. A datum can propagate to all processors on a certain row (or column) in time logarithmic with respect to the number of processors on that row (or column).

We can identify a fine-grain task with the process of computing the new value for an element of A or b. The maximum degree of parallelism is achieved when each such task is mapped onto one processor in the mesh, as illustrated in Fig. 2.6. The processors in the last column also keep track of the unknowns $x_1, x_2, \ldots, x_n$ and are responsible for swapping x_i and x_j whenever columns i and j in A have to be swapped. A description of this is given in Algorithm 2.1.

Algorithm 2.1 Mesh_of_Trees_Gaussian_Elimination

1: **for** $i = 1$ to n **do**
2:     Find the pivot as the largest element $a_{kl}^{(i-1)}$ in the submatrix of A delimited by rows $i, i+1, \ldots, n$ and columns $i, i+1, \ldots, n$.

    - the binary tree over each row i is in charge of finding the maximum element (and its column index) on that row. For this purpose, each node in the tree compares its own value against the values received from its two children and sends to its parent the larger of these three and the column index corresponding to this value. Eventually, the processors in column 1 will store the maximum value (and its column index) in each row.
    - the binary tree over the first column of the mesh is used to find the overall largest element $a_{kl}^{(i-1)}$ together with its row and column numbers, k and l, respectively.

3:     **if** $k \neq i$ **then**
4:         exchange rows i and k using the column trees (b_i and b_k are also swapped).
5:     **end if**
6:     **if** $l \neq i$ **then**
7:         exchange columns i and l using the row trees (x_i and x_l also change places in the last column of the mesh).
8:     **end if**
9:     Distribute $a_{ii}^{(i-1)}$ to all processors in row i using the binary tree over row i.
10:     Compute $a_{ij}^{(i)}$ and $b_i^{(i)}$ according to Equations (2.2) and (2.4).
11:     **for** each row $r \neq i$ in parallel **do**
12:         distribute $a_{ri}^{(i-1)}$ to all processors in row r (using the binary tree over row r).
13:     **end for**
14:     **for** each column j, $1 \leq j \leq n$, in parallel **do**
15:         distribute $a_{ij}^{(i-1)}/a_{ii}^{(i-1)}$ to all processors in column j (using the binary tree over column j).
16:     **end for**
17:     Distribute $b_i^{(i-1)}/a_{ii}^{(i-1)}$ to all processors in the last column using the binary tree over column $n+1$.
18:     Compute in parallel the new values of $a_{rj}^{(i)}$ and $b_r^{(i)}$, $r \neq i$, $i \leq j \leq n$, according to Equations (2.1) and (2.3).
19: **end for**

The most important observation that can be formulated by analyzing Algorithm 2.1 is that by choosing the finest granularity we indeed ensure maximum parallelism, but we must pay a certain communication cost among so many small tasks, each assigned to a different processor. Once the processors have the required data, each computes a single value of the A or b matrix. This allows all processors to compute the new values for the elements of A and b simultaneously, in each iteration. But before the actual computation can take place, we must spend $O(\log n)$ time in each iteration choosing the pivot and distributing (communicating) the necessary data to the processors (we note in passing that the distribution steps 12, 15, and 17 can all be performed simultaneously).

Consequently, we can aggregate $\log n$ fine-grain tasks into one coarse-grain task, which is now charged to compute the new values for $\log n$ elements, sequentially. The overall $O(n \log n)$ running time of the algorithm remains the same, because the time spent on sequential computation is not asymptotically higher than the communication time. Mapping one coarse-grain task to one processor, the problem can still be solved in the same amount of time (asymptotically), but now with fewer processors.

2.5 Theoretical Underpinnings

Similar to the design process, analyzing a parallel algorithm requires specific tools that take into consideration all aspects influencing the performance of a parallel computer. In this section, we discuss the most important performance measures used to evaluate the "goodness" of a parallel algorithm as well as what can and cannot be expected from the parallel computing paradigm vis-à-vis the sequential mode of computation.

2.5.1 Speedup

Historically, the main motivation behind using more than one processor to perform a certain task is to complete the task as quickly as possible, in other words, to speed up computation. It is only natural, therefore, to use the *speedup* achieved by a parallel algorithm designed for a certain problem relative to the best available sequential algorithm for the same problem as a primary means of evaluating parallel algorithms. Formally, the speedup provided by a parallel algorithm is defined as the ratio:

$$S(1,p) = \frac{t_1}{t_p}, \tag{2.5}$$

where t_1 denotes the worst-case running time of the fastest available sequential algorithm for the problem at hand, while t_p is the worst-case running time of the parallel algorithm running on p processors. In the case of Algorithm 2.1, the speedup achieved on a mesh with $n \times (n+1)$ processors is

$$S(1, n \times (n+1)) = \frac{O(n^3)}{O(n \log n)} = O\left(\frac{n^2}{\log n}\right). \tag{2.6}$$

Obviously, when designing a parallel algorithm, the goal is to make this ratio as large as possible. But we should always take into consideration the number of processors that contribute to a certain speedup. In Eq. (2.6), a speedup of $O(n^2/\log n)$ is achieved by $O(n^2)$ processors. As already shown, the same speedup can be obtained using only $O(n^2/\log n)$ processors, by having each processor compute the values of $\log n$ elements, sequentially. In this way, the speedup becomes equal (up to a constant factor) to the number of processors used.

For most computational problems, and especially those we are familiar with, the maximum possible speedup equals the number of processors employed in the parallel computer. As a simpler example, think of an unstructured search in a list of n items (an instance of which is looking for a number in the phone book example, given in the Introduction). If we can partition the list into 10 equal parts and have 10 processors looking simultaneously for the target element, each in a different sublist, then the most we can hope for is to reduce the search time by a factor of 10, in the worst case of the search. This limitation on the largest speedup possible when p processors are employed in the parallel model is expressed in general through the following inequality:

$$S(1, p) \leq p, \tag{2.7}$$

which is usually given as the status of a theorem. The justification is simple. If the speedup is greater than the number of processors, it follows that we can obtain a better sequential algorithm than the fastest possible by simulating the parallel algorithm on a sequential machine (executing, in sequence, the work of each of the p processors on the sequential computer).

The key observation to formulate here is that this simulation is not always possible or it simply does not make sense in some cases. It is certainly possible for the majority of the problems encountered in computer science, problems that we can term as *traditional* or *conventional* (adding a set of numbers stored in memory, searching, sorting, etc.). But as we will show later in this chapter, there are particular computational environments for which the speedup limit stated above (Eq. (2.7)) does not apply. Coming back to standard computations, in many applications even that limit cannot be reached, either because there is no way to decompose the problem such that all processors are busy at all times or because the overhead caused by inter-processor communications is too high.

2.5.2 Slowdown

Similar to how speedup is defined, we can use *slowdown* as a measure of how the performance of a parallel algorithm degrades when the number of processors decreases. More precisely, if a certain computation is completed in time t_p by p processors and in time t_q by q processors, where $q < p$, then the slowdown incurred by reducing the number of processors from p to q is

$$s(p,q) = \frac{t_q}{t_p} \leq \frac{p}{q}. \tag{2.8}$$

The above inequality imposes an upper bound to how much the running time of a parallel algorithm can increase when fewer processors are available. Although it is sometimes referred to as "Brent's theorem" (or Folk Slowdown theorem), Eq. (2.8) has the same scope as the "speedup theorem." It holds only when the work performed by p processors can be simulated on a parallel model endowed with only q processors.

Most conventional problems, however, exhibit this property and, consequently, they fall under the scope of Brent's theorem. For example, the running time of Algorithm 2.1 does not increase asymptotically when we reduce the number of processors in the mesh from $O(n^2)$ to $O(n^2/\log n)$. In other cases, like the searching problem, having fewer processors to perform the search does increase the duration of the procedure, but the slowdown incurred still obeys Eq. (2.8).

On the other hand, there are situations where the inherently parallel nature of the problem to be solved requires a certain number of processors in the model. Anything below the required degree of parallelism can lead to arbitrarily bad running times. Examples of such computing paradigms can be found in Sect. 2.7.

2.5.3 Quality-Up

In some computational environments, having more processors available to work on a certain problem may account for obtaining a *better* solution rather than a faster one. This may look quite surprising at a first glance, since the same solution obtained by a parallel computer can also be arrived at by a sequential machine, if the latter simulates the work of each processor of the former, in sequence. However, we must draw attention again on the fact that such a simulation is not always possible and a typical example is a real-time computational environment where firm deadlines are imposed on when the solution is to be produced.

Also, what constitutes a better solution depends on the problem under consideration. If, for instance, the problem to be solved is an optimization one, then several cooperating processors may compute a solution closer to optimal than a single processor is able to, before the deadline. Alternatively, "better" might mean more accurate for numerical problems, more secure for cryptographic applications, providing superior error correction capabilities when applying error-correcting schemes or higher compression rates for source coding algorithms and so on. But, regardless of the particular context, we need a way to quantify the improvement in quality of the solution computed in parallel with respect to the best one that can be obtained sequentially.

In analogy with speedup (which is an improvement in speed), we can term an improvement in quality as *quality-up* and define it formally as the ratio:

$$\text{quality-up} = \frac{V_n}{V_1}, \tag{2.9}$$

where V_n is the value of the solution obtained in parallel and V_1 is the value of the solution derived sequentially. How V_1 and V_n are defined depends again on the type of application. When the purpose is to maximize a quantity (such as the profit in an optimization problem or the level of security in a cryptographic application), the choice of V_1 and V_n is straightforward. But in those cases where the goal is to minimize a quantity, like the amount of error in the solution to a numerical computation, the value of a numerical solution should be defined as the inverse of the error it contains.

Although for most traditional ways of computing, quality-up is unitary (no gain in quality whatsoever by using parallelism), there are modes of computation for which parallelism can translate directly into an impressive quality-up, even super-linear in the number of processors used (see Sect. 2.7.1.2).

2.5.4 Computations that Seem Inherently Sequential

Striving to design efficient parallel algorithms can sometimes be frustrating. There are problems that seem to resist all our attempts to speed up the computation through parallel processing. In this section we review what is known and what is not known about problems that are not efficiently parallelizable, from the viewpoint of computational complexity theory.

2.5.4.1 Class NC

Intuitively, computations that do not admit an efficient parallel solution are seen as inherently sequential. Therefore, we begin our discussion with a precise definition for the class of problems that can be efficiently solved on a parallel model of computation and then look for inherently sequential problems outside that class.

The class **NC** is the set of decision problems decidable in poly-logarithmic time on a parallel computer with a polynomial number of processors. More formally, a problem belongs to **NC** if it can be solved in $O(\log^c n)$ time using $O(n^k)$ processors, where c and k are constants and n is the size of the problem. For definiteness, the parallel computer can be assumed to be a PRAM. Equivalently, we may think of **NC** as the set of decision problems decidable by uniform Boolean circuits with poly-logarithmic depth (running time) and a polynomial number of gates (processors).

Examples of problems belonging to **NC** are numerous. In our unstructured search problem, if we have as many processors as there are elements in the list, then the search can be performed in just one time unit. An odd-even-merge sorting circuit has a depth of $O(\log^2 n)$ and can sort n numbers using $O(n \log^2 n)$ comparators [1]. Any textbook on parallel algorithms may provide many other examples.

We note in passing that one of the drawbacks of class **NC** is the fact that any sequential algorithm with logarithmic time is in **NC** regardless of its parallel feasibility. A typical example is parallel binary search, which yields a modest speedup and is therefore far from being considered efficiently parallelizable. But since its parallel running time is logarithmic, we do consider binary search as having an efficient parallel solution and therefore, parallelizable.

In order to gain a better understanding about the relationship between parallelizable and non-parallelizable problems, it may be useful to draw an analogy with the relationship between tractable and intractable problems.

2.5.4.2 P-completeness

The complexity class **P** is the set of decision problems that can be solved on a deterministic sequential machine in an amount of time that is polynomial in the size of the input. Just as the problems in **P** can be considered as tractable, so **NC** can be thought of as the class of problems that can be efficiently solved on a parallel computer. Because complexity classes are defined in terms of languages accepted by Turing machines and Turing machines can simulate each other efficiently, it follows that **NC** is a subset of **P**, since the operation of a parallel machine can be simulated on a sequential one with a polynomially bounded overhead.

However, it is not known whether **NC = P**, although most researchers suspect that this is not the case, just as the equality **P = NP** (where **NP** is the set of all decision problems whose solution can be found in polynomial time on a non-deterministic machine) is also suspected to be false. This means that there are probably some tractable problems which are *inherently sequential* and cannot be significantly sped up through a parallel approach. The most difficult problems in **NP** are labeled as **NP-complete**, owing their name to the property that *any* problem in **NP** can be reduced to an **NP-complete** problem in polynomial time. In the same way class **NP-complete** can be thought of as identifying "probably intractable" problems; we can define class **P-complete** as the set of problems seen as "probably not parallelizable" or "probably inherently sequential." Formally, a decision problem is in **P-complete** if it is in **P** and every problem in **P** can be reduced to it using **NC** reductions (reductions that can operate in poly-logarithmic time on a parallel computer with a polynomial number of processors).

Many problems have been proved to belong to **P-complete** and so they are widely believed to be inherently sequential. Here are a few:

- *Circuit value problem*: Given a circuit, the inputs to the circuit and one gate in the circuit, calculate the output of that gate.
- *Linear programming*: Maximize a linear function subject to linear inequality constraints.
- *Horn-satisfiability*: Given a set of Horn clauses, is there a variable assignment which satisfies them?
- *Game of life*: Given an initial configuration of Conway's Game of Life, a particular cell, and a time T (in unary), is that cell alive after T steps?

Finally, analogous to problems that are not proven to be either **NP-complete** or **P-complete** (such as factoring, for example), we have problems that are not known to be either **P-complete** or **NC**, but are still thought to be difficult to parallelize. One example is finding the greatest common divisor of two binary numbers (decision problem form).

2.6 Parallel Algorithms for Conventional Computations

The vast majority of computations carried out every day as part of various applications can be labeled as "conventional." Unless the computational environment exhibits some special properties, for example, imposing restrictions on the availability or reliability of data to be processed, any computation is a conventional or traditional one. Immediate examples that come to mind are sorting, searching, operations on matrices and so on. In Sect. 2.4 we have already seen, in detail, how a conventional problem, namely, solving a system of linear equations using Gaussian elimination, can be solved by a parallel algorithm on a mesh of trees. In what follows, we develop parallel algorithms for two other conventional problems, quite different in nature from Gaussian elimination. The parallel models for which the algorithms are intended are also of a different kind.

2.6.1 Parallel Prefix and Suffix Computations on a Linked List

In this section, we show that a parallel approach can also be useful when working with pointer-based data structures, such as a linked list. A singly linked list L consists of a number of nodes where each node (except the last one) holds a pointer to the next node in the list (Fig. 2.7(a)). Besides this pointer, a node also usually holds a value (that depends on the application) and other necessary information.

For our problem, we assume that a linked list L has been stored in the shared memory of a PRAM. Each node was added to the list at a different time, without global knowledge of the data structure constructed so far or the positions where future nodes will be added. The *list-sequencing* problem asks for the sequence number of each node to be computed. The head of the list has sequence number 1, the node pointed at by the head bears sequence number 2 and so on. If we are given a pointer

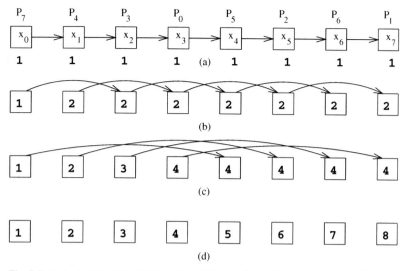

Fig. 2.7 A singly linked list L **(a)** and how it is transformed by Algorithm 2.2 after one iteration of the *while* loop **(b)**; after two iterations **(c)**; after three iterations **(d)**.

to the head of L, we can easily compute the sequence numbers for all nodes in time linear in the size of L, just by following the chain of pointers and adding 1 to the current sequence number at each step. In order to execute this algorithm, a single processor suffices. The question is whether we can do better than a linear running time, if we have more processors available.

For simplicity, assume that the PRAM model in charge of solving the problem has as many processors as there are nodes in L. Thus, each processor knows the location of one distinct node in the list and can therefore access any information stored in that node in constant time (see Fig. 2.7(a)). This may well be the case in practical applications, as each processor may have contributed to the construction of L with the node they are in charge of.

For the list-sequencing problem, the values x_j, $1 \le j \le n$, stored in each node are initially all equal with 1. The parallel algorithm then consists of repeatedly performing the following step: each processor adds its own value to the value stored in its successor node and then updates its pointer to the node following its successor. A graphical illustration of the algorithm, for the particular case of a list with 8 elements is given in Fig. 2.7. Note that the number of final sequence numbers computed at each step doubles. Therefore, it is easy to see that for a list L having n nodes, $O(\log n)$ steps are required to complete the computation and each step takes constant time. In the end, each node will have its successor pointer set to *nil* (Fig. 2.7(d)). A formal description of the algorithm is given as Algorithm 2.2.

The algorithm works with copies of the original *succ* pointers in order to protect them and preserve the initial structure of the list, since all *next* pointers become

---

Algorithm 2.2 PRAM_List_Sequencing

1: **for** all i in parallel **do**
2:     $next(i) \longleftarrow succ(i)$;
3:     $val(i) \longleftarrow 1$.
4: **end for**
5: $finished \longleftarrow false$;
6: **while** (not $finished$) **do**
7:     $finished \longleftarrow true$;
8:     **for** all i in parallel **do**
9:         **if** $next(i) \neq nil$ **then**
10:             $val(next(i)) \longleftarrow val(i) + val(next(i))$;
11:             $next(i) \longleftarrow next(next(i))$;
12:             **if** $next(i) \neq nil$ **then**
13:                 $finished \longleftarrow false$.
14:             **end if**
15:         **end if**
16:     **end for**
17: **end while**

---

nil at the end of the algorithm. In fact, this is the condition for the termination of the algorithm: each node has its *next* pointer set to *nil*. As soon as this pointer becomes *nil* for some node, the processor in charge of that node will remain idle for the remaining of the algorithm. But since a processor has no knowledge about when other processors have completed their work, we use the variable *finished* to detect when all processors have become idle and the algorithm can terminate. In this respect, note that the variable *finished* can be set to *false* at the end of an iteration of the while loop by several processors using a CW instruction from the PRAM repertoire.

The algorithmic technique used to solve the list-sequencing problem has a far broader applicability than just computing the sequence numbers of the elements composing a linked list. If we modify Algorithm 2.2 such that we replace addition by a generic operator "$\circ$," transforming two objects x_i and x_j into an object of the same type $x_i \circ x_j$, then what we obtain is an algorithm performing a *prefix* computation. It is called so because the values stored in each node at the end of the algorithm are $x_1, x_1 \circ x_2, x_1 \circ x_2 \circ x_3, \ldots, x_1 \circ x_2 \circ \cdots \circ x_n$, from the head of the list down to the last element.

On the other hand, in a *suffix* computation, the resulting values would be $x_1 \circ x_2 \circ \cdots \circ x_n, x_2 \circ x_3 \circ \cdots \circ x_n, \ldots, x_n$. Transforming a prefix computation algorithm into a suffix one is easily achieved by replacing the line

$$val(next(i)) \longleftarrow val(i) \circ val(next(i)) \qquad (2.10)$$

in the modified version of Algorithm 2.2 with

$$val(i) \longleftarrow val(i) \circ val(next(i)). \qquad (2.11)$$

If we now instantiate "∘" back to "+" and initialize all values with 1, then instead of the list-sequencing problem we are now solving the list-ranking problem, where we have to compute the *rank* of each node as its distance from the end of the list. Many problems of practical interest can be reduced to a form of prefix or suffix computation; so having an efficient parallel algorithm to deal with these two generic computations is of great importance.

2.6.2 Sorting on a Model with Buses

In Sect. 2.4 we have exemplified the design methodology for developing parallel algorithms on a model where the processors are arranged in a mesh-like structure, but with the regular mesh links replaced by row and column binary trees. The reason behind choosing this topology is to reduce the diameter of a regular mesh of processors, thereby reducing the communication cost among the processors in the model. In doing so, however, we also introduce some disadvantages. The new topology does not possess the regularity and modularity that make a mesh of processors so easy to extend. In addition, the length of the links in a mesh of trees increases with the number of processors in the model, affecting its scalability.

A popular way to make communications inside a mesh of processors easier, while still retaining the attractive features of the model, is to augment them with *buses*. In the context of our discussion, a bus is simply a communication link to which a number of processors are attached in order to transmit and receive data. Depending on the particular technology used to implement them, these buses can be fixed, reconfigurable or optical. In order to present algorithmic techniques that are specific to models enhanced with buses and also prove their usefulness, we develop a parallel algorithm for sorting a sequence of numbers on a mesh with reconfigurable buses.

We begin by describing the model that will allow us to sort an arbitrary sequence of numbers in constant time. Any processor that is not on the border of a regular mesh has four neighbors. In order to communicate with its neighbors, a processor uses direct links. The interface through which a link connects to a processor is called a *port*. Consequently, a processor included in a mesh structure is equipped with four ports: north (N), south (S), west (W) and east (E). In a mesh with reconfigurable buses, any processor is capable of connecting its ports internally, in arbitrary pairs. When combined with the standard (external) mesh links, these internal connections can be used to create paths (buses) of the desired length and shape. Figure 2.8 depicts a mesh in which three buses have been formed by its processors. In particular, if every processor in a row connects its W and E ports together, then we obtain a row bus. Similarly, by connecting the N and S ports for each processor in a column, we create a column bus.

As their name says, the most important property of reconfigurable buses is that they are dynamic, that is, they can change according to the needs of the algorithm, as many times as required. This distinguishes them from fixed buses, which are "hard-wired" into the model and remain unchanged throughout the computation. Any path

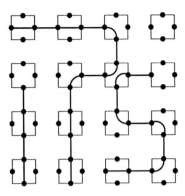

Fig. 2.8 A mesh with three
configured buses.

can be set up in constant time, as we take the number of time units for a processor to connect its ports internally to be a constant. Otherwise, reconfigurable buses behave just like fixed ones: only one processor is allowed to write a datum on a bus at any given time and all processors connected to the bus can read that datum simultaneously.

In order to sort n numbers on a mesh with reconfigurable buses we resort to a common technique called *sorting by enumeration*. For each number in the sequence, we compute its *rank* (position in the sorted sequence) as the number of elements that are smaller than it. We choose to break ties using the index of each element in the original sequence. Consequently, for the purpose of our algorithm, whenever $x_i = x_j$, x_i is considered "smaller" than x_j if and only if $i < j$. Once all ranks are computed, we just need to permute the elements such that in the end each occupies the position indicated by its rank.

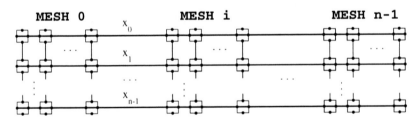

Fig. 2.9 Distribution along rows in sorting by enumeration on a mesh with reconfigurable buses.

For each element x_i, $0 \leq i \leq n - 1$, we assign a distinct mesh (MESH i) to the task of computing its rank (see Fig. 2.9). Therefore, to sort n numbers in constant time we need a mesh with n rows and n^2 columns. At the outset, the sequence to be sorted is stored by the processors in the first column of the mesh: x_0 by $P(0,0)$, x_1 by $P(1,0),\ldots,x_{n-1}$ by $P(n-1,0)$. When the algorithm terminates, the same first column holds the sequence sorted in non-decreasing order from top to bottom. The algorithm, broken into four main steps, is given as Algorithm 2.3.

Algorithm 2.3 Reconfigurable_Buses_Mesh_Sort

1: **I. Distribution:**
2: **for** all processors in parallel **do**
3:     connect W and E ports (creates row buses across the entire mesh).
4: **end for**
5: **for** $i = 0$ to $n - 1$ in parallel **do**
6:     distribute x_i to all processors in row i using the bus on row i (see Fig. 2.9).
7: **end for**
8: **for** $i = 0$ to $n - 1$ in parallel **do**
9:     processors in column 0 of MESH i connect N and S ports (creates a column bus in the first column of each $n \times n$ mesh);
10:     processor $P(i,0)$ distributes x_i to all processors in its column using the column bus.
11: **end for**
12:
13: **II. Comparison:**
14: **for** $i = 0$ to $n - 1$ in parallel **do**
15:     **for** $j = 0$ to $n - 1$ in parallel **do**
16:         $P(j,0)$ from MESH i compares x_j to x_i:
17:         **if** $x_j < x_i$ **then**
18:             $R = 1$
19:         **else**
20:             $R = 0$ (R is a local register in each processor).
21:         **end if**
22:     **end for**
23: **end for**
24:
25: **III. Rank computation:**
26: **for** $i = 0$ to $n - 1$ in parallel **do**
27:     **for** $j = 1$ to $n - 2$ in parallel **do**
28:         all processors in column j of MESH i connect W and E ports (row buses are created for each row of each $n \times n$ mesh).
29:     **end for**
30:     **for** $j = 0$ to $n - 1$ in parallel **do**
31:         $P(j,0)$ in MESH i broadcasts the content of its R register to all processors attached to its row bus (see Fig. 2.10(a)).
32:     **end for**
33: **end for**
34: **for** all processors in parallel **do**
35:     **if** $R = 0$ **then**
36:         connect N and S ports
37:     **else**
38:         connect W with N and S with E (see Fig. 2.10(b)).
39:     **end if**
40: **end for**
41: **for** $i = 0$ to $n - 1$ in parallel **do**
42:     $P(n - 1, 0)$ of MESH i writes a marker symbol on the bus to which its S port is connected (rank of x_i = column index of processor receiving the marker symbol, see Fig. 2.10(c)).
43: **end for**

Algorithm 2.3 Reconfigurable_Buses_Mesh_Sort (Continued)

44: **IV. Permutation:**
45: **for** $i = 0$ to $n - 1$ in parallel **do**
46:    all processors in MESH i connect their N and S ports (creates column buses);
47:    processor $P(0, j)$ in MESH i containing the marker symbol broadcasts j (x_i's rank) to all processors attached to its column bus;
48:    processor $P(i, j)$ in MESH i broadcasts x_i to all processors in column j using the column bus.
49: **end for**
50: **for** all processors in parallel **do**
51:    connect W and E ports (creates row buses across the entire mesh).
52: **end for**
53: **for** $i = 0$ to $n - 1$ in parallel **do**
54:    $P(j, j)$ of MESH i (j is x_i's rank) broadcasts x_i along its row bus to be read (received) by processor on row j in the first column of the entire mesh.
55: **end for**

Probably, the most notable feature of Algorithm 2.3 is the unique way in which a numeric computation (the sum of n bits) is performed in the third step exclusively through the manipulation of buses. The last step also demonstrates that an arbitrary permutation of n objects can be performed in constant time on an $n \times n$ mesh of processors, if reconfigurable buses are supported. Overall, since each step in the algorithm can be executed in constant time, we can sort a sequence of n numbers in constant time. This shows the power and flexibility that reconfigurable buses bestow upon a standard mesh of processors, but at the high cost of $O(n^3)$ processors employed. However, more involved sorting algorithms on a mesh with reconfigurable buses have been developed [3] to alleviate the exorbitant cost induced by the large number of processors required to sort in $O(1)$ parallel time in the algorithm described in this section.

2.7 Parallel Algorithms for Unconventional Computations

When it comes to tackling a computational problem with unconventional characteristics, a parallel approach may yield even better results than those we have seen so far in this chapter. In what follows, we describe a series of particular computing environments for which parallel processing proves extremely beneficial, if not critical. These relatively new computational paradigms, although unconventional, are still realistic and many real-world applications can be identified as instances of such unconventional computations.

In some cases, the performance of the parallel algorithm, expressed in terms of the speedup or quality-up achieved, is superlinear in the number of processors used in the parallel computer, relative to what the best sequential algorithm can offer. In other cases, the inherently parallel nature of the problem renders a sequential machine (or even one with insufficient processing elements) useless, and consequently,

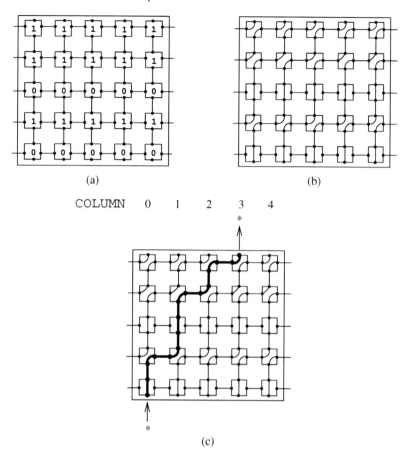

COLUMN 0 1 2 3 4

(c)

Fig. 2.10 Ranking in MESH i: **(a)** Each processor in the leftmost column broadcasts the value contained in its R register to all processors in its row; **(b)** A processor containing a 0 connects its N and S ports, whereas a processor containing a 1 connects W with N and S with E; **(c)** The processor in the *bottom left corner* sends a *marker symbol* on the bus to which its S port is connected.

a sequential solution fails altogether. When this occurs, the improvement in performance obtained through the use of a parallel model can be considered unbounded.

2.7.1 Computations that Can be Simulated Sequentially

The two examples of unconventional paradigms presented in this section belong to the class of *real-time computation*. The concept of *real time* is difficult to capture formally [4] despite the broad spectrum of real-time systems encountered in everyday life: air traffic control, process control (e.g., in a chemical plant), nuclear reactors, telecommunications, integrated vision/robotics/AI systems, etc. From the algorithmic viewpoint adopted herein, we are interested mainly on the relationship between data and time, particularly on the *deadlines* imposed on when input data should be processed after it becomes available and when the outputs (or results) are to be produced. These deadlines are considered *tight* (measured in terms of a few time units) and *firm* (missing a deadline causes the computation to fail).

A parallel approach may prove to be very useful when computing with deadlines, by helping the computation to terminate on time or by computing a better solution than it is possible sequentially, in the time allowed. Thus, a sequential computer can also reach a solution by trying to adapt (or simulate) the parallel algorithm on the single processor it is endowed with, but the consequences may be dramatic in terms of either speed or quality. Furthermore, the superlinear improvement in performance that we are about to describe is *consistent* and *provable*, in the sense that it occurs in every instance of the computational problem under consideration. In particular, this improvement is independent of any discrepancies between the sequential and parallel computers used and it is not an occasional artifact due to an inefficient sequential algorithm, a restricted memory size on the sequential computer, or expensive context switching that has to be performed on the sequential machine.

2.7.1.1 Superlinear Performance in Speed

Consider the following computation taking place in a real-time environment.

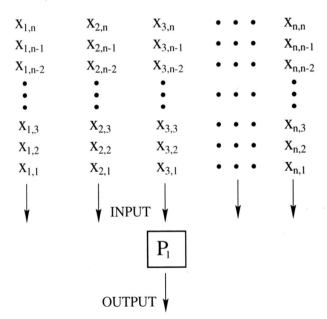

Fig. 2.11 A sequential computer can monitor only one stream.

Problem

n sources provide data to solve a certain problem. Each source provides n data, one per time unit. The problem can be solved using the data set provided by any of the n sources. However, only one source allows the problem to be solved in n time units, while data from all other sources lead to a solution in 2^n time units. Moreover, it is impossible to tell a priori which source leads to a quick solution. This can only be determined once all n data from a source become available. Finally, there is a deadline on the availability of each datum produced by each source: if the current datum generated by a source is not acquired and stored by a processor during that time unit, the datum is irretrievably lost, being overwritten at the beginning of the next time unit by the subsequent piece of data.

Sequential Solution

A sequential computer can monitor only one stream with its single processing unit, so it chooses one arbitrarily (third stream in the example depicted in Fig. 2.11) and computes the solution using the n data supplied by that stream. In the worst case, the time required to complete the computation sequentially is $T_1 = n + 2^n$ time units.

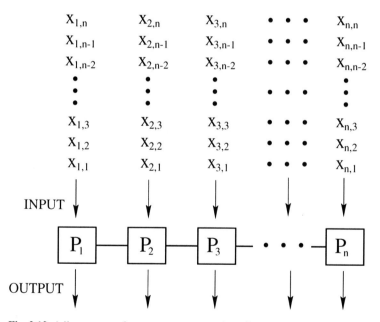

Fig. 2.12 A linear array of n processors can monitor all streams.

Parallel Solution

The parallel model we choose to deal with this problem is a linear array of processors (see Fig. 2.12), arguably the weakest model of parallel computation. Even so, if the array comprises n processors, then each of them can monitor one of the n streams and attempt to solve the computational problem using the n data provided by that stream (Fig. 2.12). One (and only one) of the processors attains a solution n time units after the arrival of the last datum in its chosen stream. Therefore, the time required in parallel to complete the computation is $T_n = n + n = 2n$ time units.

Analysis

The ratio between the sequential and parallel running times gives us the following speedup:

$$S(1,n) = \frac{T_1}{T_n} = \frac{n + 2^n}{2n} = \theta \left(\frac{2^{n-1}}{n} \right). \tag{2.12}$$

This speedup is exponential in the number of processors employed in the linear array, thus contradicting the "speedup theorem." Similarly, Brent's theorem does not apply either. If the linear array contains only q processors, where $2 \leq q < n$, then we find ourselves again in the situation where we cannot monitor all streams. The q processors can choose only at random q streams to monitor. In the worst case, none of the processors chooses a stream whose data lead to a solution in n time units and the running time $T_q = n + 2^n$ time units is no better than the sequential running time T_1. This worst case occurs with probability $1 - (q/n)$, so when n is large and q is small, it is almost certain that the computation will require exponential time to complete. By comparison with the n-processor solution, the slowdown incurred is

$$s(n,q) = \frac{T_q}{T_n} = \frac{n + 2^n}{2n} = \theta \left(\frac{2^{n-1}}{n} \right). \tag{2.13}$$

This slowdown is superlinear in $\lceil n/q \rceil$, regardless of the value of q, and the Folk Slowdown theorem does not hold.

Although the processors that equip the parallel model are always considered identical with the one employed in the sequential computer, it is interesting to note that in the paradigm described above, a superlinear speedup in n can still be achieved even if each processor of the linear array is n times slower than the processor of the sequential machine. More precisely, assume that a processor of the parallel computer requires n time units to execute the same (arithmetic and logical) operations performed by the processor of the sequential computer in one time unit. Then the parallel completion time becomes $n + n^2$ time units, while the sequential completion time remains $n + 2^n$, in the worst case. Their ratio confirms that the parallel computer continues to achieve a speedup superlinear in n, despite the unreasonable assumption made about its processors in favor of the sequential computer.

Similar real-time computational paradigms have been uncovered for which the speedup obtained exceeds any conventional bounds (e.g., data accumulation, one-way functions [5,6]).

2.7.1.2 Superlinear Performance in Quality

When computing under the pressure of a deadline, it is sometimes the case that both the sequential and parallel computer succeed in arriving at a solution before the specified deadline. However, for some classes of problems, the solution computed in parallel is far superior in quality to the best one obtained sequentially. This is especially true for numerical computations, due to their particular characteristics. Numerical methods always yield approximate results because of roundoff and truncation errors, so a qualitative measure can easily be assigned to the solution of a numerical problem, depending on the amount of error it contains.

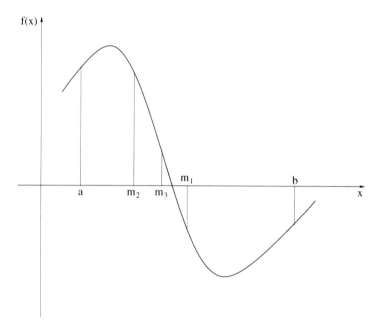

Fig. 2.13 Computing a zero of a continuous function f in the interval $[a,b]$ using the bisection method.

Suppose, for example, that we wish to compute a *zero* of a continuous function f (a value x_{exact}, such that $f(x_{exact}) = 0$) in a given interval $[a,b]$ (with $f(a) \times f(b) < 0$). The bisection method repeatedly halves the interval where the solution can be found by computing the middle point $m = (a+b)/2$ and focusing only on the half-interval for which the images through f of the two extremities have opposite signs (see Fig. 2.13). After r such iterations, the error contained in the computed solution x_{approx} is bounded by:

$$|x_{exact} - x_{approx}| \leq \frac{|b-a|}{2^r}. \qquad (2.14)$$

Now consider the following real-time computational environment for the bisection method.

Problem

At the beginning of each time unit i, a new 3-tuple $\langle f, a, b \rangle_i$ is received for which a zero is to be found. It is required that each 3-tuple be processed as soon as it is received and an appropriate solution x_{approx} be produced as output as soon as it is computed. Furthermore, one output must be produced at the end of each time unit, with possibly an initial delay before the first output is produced. It is assumed that reading a tuple $\langle f, a, b \rangle$, performing one iteration of the bisection algorithm and producing x_{approx} as output once it has been computed can all be performed within one time unit.

Sequential Solution

Here, there is a single processor whose task is to read each incoming 3-tuple, to compute x_{approx} and to produce the latter as output. Since each 3-tuple must be processed immediately upon arrival, it follows that the sequential computer has only one time unit to try to compute an approximate solution for the current tuple, before having to move on to the next input. Consequently, the sequential algorithm can perform no more than one iteration on each input $\langle f, a, b \rangle$ and produce the solution $x_{approx} = m = (a+b)/2$. This being the only option available, it is by default the best solution possible sequentially.

Fig. 2.14 n processors arranged in a linear array can compute a better approximation than a single processor. P_i applies iteration i of the bisection method for each 3-tuple received.

Parallel Solution

To tackle the problem in parallel we are using again a linear array of n processors arranged as illustrated in Fig. 2.14. Processor P_1 is designated to receive the successive input 3-tuples, while it is the responsibility of P_n to produce x_{approx} as output. The same strict deadline forcing the sequential computer to execute only one iteration of the bisection method for each received tuple applies also to P_1, which must turn its attention to a new input every time unit. Unlike the sequential solution, however, the parallel algorithm can perform additional iterations as follows.

Once P_1 has executed its single iteration on $\langle f, a_1, b_1 \rangle$, it sends $\langle f, a_2, b_2 \rangle$ to P_2 and deals with the next 3-tuple arriving as input. Now P_2 can execute an additional iteration before sending $\langle f, a_3, b_3 \rangle$ to P_3. This continues until $x_{approx} = (a_n + b_n)/2$ is produced as output by P_n. Meanwhile, $n - 1$ other 3-tuple inputs coexist in the array (one in each of $P_1, P_2, \ldots, P_{n-1}$) at various stages of processing. One time unit after P_n has produced its first x_{approx}, it produces a second, and so on, such that an output emerges from the array every time unit, after an initial delay of n time units. In this way, each output x_{approx} is the result of applying n iterations of the bisection algorithm, since there are n processors and each executes one iteration.

Analysis

For a numerical computation, the quality of a solution is most appropriately defined as its accuracy, which in turn can be taken to be the inverse of the maximum error. In the sequential case, when only one iteration can be applied ($r = 1$ in Eq. (2.14)), the maximum error is $|b - a|/2$. In contrast, each parallel solution undergoes n iterations before being output by P_n and therefore the maximum error is $|b - a|/2^n$. By defining quality-up as the ratio of the parallel accuracy to the sequential accuracy, we have

$$\text{quality-up}(1, n) = \frac{1/\text{parallel error}}{1/\text{sequential error}} = \frac{2^n}{2} = 2^{n-1}. \tag{2.15}$$

According to this result, the improvement in quality is exponential in the number of processors employed in the parallel model. Other numerical algorithms may also yield a superlinear quality-up when executed on a parallel machine. It is the case, for example, of the trapezoidal method to compute an approximation for the value of a definite integral [7]. Extending the range of applications, discrete optimization problems and cryptographic methods may lead to a superlinear improvement in quality as well, with respect to their sequential counterparts, if the computations take place in the presence of deadlines [7, 8].

2.7.2 Computations that Cannot be Simulated Sequentially

In the examples presented above (Sects. 2.7.1.1 and 2.7.1.2), the various time constraints placed upon the data (whether it is input or output) make the simulation of the parallel solution attempted by the sequential computer extremely inefficient, in terms of either speed or quality of the result. But, inefficient as it may be, a sequential solution can still be produced. In this section, we explore a different class of unconventional paradigms, whose inherent parallelism renders a sequential simulation pointless. The main attribute of these computing paradigms is their dynamic nature. Their characteristics change during the computational process itself, whether it is the data or the complexity of the algorithm that evolves with time. Evolution, or merely change, is a fundamental property of many systems that we observe and investigate, whether they are physical, biological, economic, social or of any other kind. Evolving computational processes occurring in nature are, in fact, the main motivation and inspiration behind studying inherently parallel computing paradigms.

At an abstract level, the following generic problem needs to be solved: a set of n input variables $x_0, x_1, \ldots, x_{n-1}$ have to be read and a certain function $\mathscr{F}(x_0, x_1, \ldots, x_{n-1})$ must be computed and the result reported. In some of the instantiations that follow, what evolves during the computation is the complexity of each step in the algorithm, while in others it is the input variables that determine the dynamics of the system.

2.7.2.1 Time-Varying Variables

In this paradigm (and the one that follows), time plays the main role. Each argument of function $\mathscr{F}$ is itself a function of time: $x_0(t), x_1(t), \ldots, x_{n-1}(t)$. At each time unit, the values assumed by the input variables change in such a way that the new value cannot be predicted from the former, nor the former recovered from the latter. Certainly, this makes the computation of $\mathscr{F}(x_0(t_0), \ldots, x_{n-1}(t_0))$ at the precise moment $t = t_0$ a challenging task, in case we do not have the capability of reading all n input variables, in parallel, at the right moment.

Assuming that reading the value of one input variable requires one time unit, a sequential computer can only read the value of one variable at moment t_0. By the time this is done, one time unit has elapsed and all other variables have changed values in a random, unstoppable and irreversible manner. Clearly, the sequential computer fails to perform the computation. A machine endowed with n processors, on the other hand, can instruct each processor to read the value of a distinct variable, simultaneously, at the required moment t_0. The computation of $\mathscr{F}$ can then proceed as necessary. Quantum bits affected by errors over time or genes in a living cell changing over time as the cell ages or becomes infected are possible examples of variables in a computation, whose values are affected by the relentless passage of time.

2.7.2.2 Time-Varying Computational Complexity

When analyzing the computational complexity of a given algorithm, we usually focus on how this quantity varies as a function of the problem size, without paying too much attention to how the complexity of each step in the algorithm varies throughout the computation. Though in many cases the complexity of each step is a constant, there are computations for which the cost of executing essentially similar steps is different from one step to another.

The passage of time not only can influence the values assumed by various variables in a computation, but it can also directly influence the computational complexity of a given step in an algorithm. If the cost of executing step S_j is a function of the particular moment in time when that step is executed, then what we have is a procedure with steps of time-varying computational complexity. For example, if the computational complexity of S_j is described by the function $c(t) = 2^{2^t}$, then the computational resources required to complete that step are rapidly growing with the moment in time when S_j is actually executed.

Moreover, if the function $c(t)$ describes the computational complexity of any step composing a certain algorithm, then a sequential computer may quickly run out of resources (time, memory, etc.) when trying to execute that algorithm, regardless of the order in which the steps are executed. In contrast, a parallel computer equipped with sufficient processors, such that one processor is in charge of executing only one step, can complete the computation after just two time units by having all steps executed simultaneously at time $t = 0$, each by a different processor (assuming, of course, that the algorithm allows a full parallelization).

Time-varying computational complexity is everywhere around us. As software viruses spread with time they become more difficult to deal with, a spaceship racing away from Earth becomes ever harder to track, and so on.

2.7.2.3 Rank-Varying Computational Complexity

Another factor that can dictate the complexity of a step is its *rank*, defined as the order of execution of that step. Examples of this kind are hardly new. Euclid's algorithm for computing the greatest common divisor of two numbers executes the same basic operation (a division) at each step, but the size of the operands (and implicitly the complexity of the operation) decreases continually. Algorithms for which an amortized analysis can be applied also make good examples of rank-varying computational complexity. Incrementing a binary counter [9] is a procedure in which the number of bit flips at each step is not constant, though it is neither strictly increasing nor strictly decreasing with the rank.

Suppose now that the cost of executing the *ith* step of an algorithm is $c(i) = 2^i$ elementary operations or time units. Since the computational complexity of a step grows exponentially with its rank, a sequential machine may again have difficulties keeping up, while a parallel approach can avoid the whole problem of dealing with steps of ever-increasing complexity by executing them in parallel. Thus, all steps in the parallel algorithm have rank 1.

The difference between a rank-driven and a time-driven computational complexity can probably be synthesized best in the following manner. If the cost of executing step S_j depends only on the state of the system after executing the previous $j - 1$ steps, regardless of how much time was consumed to reach that state, then we clearly have an example of rank-varying computational complexity.

2.7.2.4 Interacting Variables

In the next paradigm that we describe, it is the interactions among mutually dependent variables caused by an interfering agent (performing the computation) that is the origin of the evolution of the system under consideration. Thus, a relationship exists between $x_0, x_1, \ldots, x_{n-1}$ that connects them together. Any attempt to read the value of any one variable will inevitably and unpredictably disturb the values of the remaining variables. More precisely, the act of reading x_i, for any $i \in \{0, 1, \ldots, n-1\}$, causes the system to make a transition from state $(x_0, x_1, \ldots, x_i, \ldots, x_{n-1})$ to $(x_0', x_1', \ldots, x_i', \ldots, x_{n-1}')$. In this way, some of the values needed in the computation of $\mathscr{F}$ may be lost without possibility of recovery. This is the hallmark of the *interacting variables* paradigm and the only way to deal with it is to read the values of all variables at the same time, so that eventual disturbances after a simultaneous reading no longer affect the computation of $\mathscr{F}$. Examples of physical systems with interacting variables are:

- a quantum register made up of n entangled quantum bits. Reading (measuring) any one of them necessarily causes the others to assume a state compatible with the outcome of the measurement.
- n living organisms housed in a closed environment and depending on one another for survival. Performing any operation on one of the organisms in exclusion of the others may have the effect of disturbing the equilibrium sufficiently to provoke a serious adverse effect (or even death) on the remaining organisms.
- a chemical system under stress (viewed as a change in pressure, temperature or concentration). According to Le Châtelier's principle, if a system at equilibrium is subjected to a stress, the system will shift to a new equilibrium in an attempt to reduce the stress.

2.7.2.5 Variables Obeying a Global Condition

Finally, the relationship among the input variables may take the form of a global property $\mathscr{P}(x_0, x_1, \ldots, x_{n-1})$ that characterizes the initial state of the system and which must be maintained throughout the computation. In particular, if the effect of the computation is to change x_i to x_i' at some point, then $\mathscr{P}(x_0, x_1, \ldots, x_i', \ldots, x_{n-1})$ must be true for the new state of the system. If the property $\mathscr{P}$ is not satisfied at a given moment of the computation, the latter is considered to have failed.

There exist problems for which acting on a single variable (regardless of which particular one) will inevitably lead to a violation of the global condition and the only way to reach a final state is to act simultaneously on all variables, during each step of the computation. Geometric flips, map recoloring and rewriting systems are three examples of transformations that may be constrained by a global mathematical condition [10].

2.8 Non-Universality in Computation

Finally, we relate the inherently parallel computing paradigms presented in the previous section with the hypothetical notion of a Universal Computer. Such a machine must be able to follow (execute) the steps of any program made up of basic input, output and internal processing operations. The Universal Computer is intended to be the most general possible model of computation, encompassing all existing or imagined computational paradigms. It must also have a means of communicating with the outside world at any time during a computation, either for receiving input or for producing output (results). The machine is endowed with the ability to acquire input data through measurements on outside-world systems, performed by a set of probes (or sensors). The program, the input data (either received or acquired), the output and all intermediate results are stored in (and can be retrieved from) a memory which is generously allowed to be unlimited.

To make this Universal Computer a "realistic" model of computation, it is subjected to the *finiteness condition*: In one step, requiring one time unit, the Universal Computer can execute a finite and fixed number of basic operations (arithmetic, logic, read, write, measure, etc.). It is precisely this limitation (quite natural and reasonable) that makes the Universal Computer a utopian concept. Specifically, five classes of computable functions $\mathscr{F}$ are described in Sect. 2.7.2, which cannot be computed by any machine obeying the finiteness condition. This condition restricts the number of input variables upon which the Universal Computer can operate in parallel (read, measure, transform and so on, depending on the particular paradigm under discussion). So, if the Universal Computer is able to act on n variables in parallel, during one step, where n can be arbitrarily large, but finite, then the Universal Computer will fail to solve the same problem for $n+1$ variables. In other words, the Universal Computer cannot simulate a computation that is perfectly possible for another machine. However, it is exactly the principle of *simulation* that lies at the heart of *universality*.

Choosing a machine endowed with $n+1$ processing units as the Universal Computer is not a solution. By an adversary argument, we can construct a problem instance involving $n+2$ variables and the Universal Computer will fail once again to compute the required function $\mathscr{F}$, although it can be trivially computed by a machine with $n+2$ processors. This argument is valid for any given Universal Computer having a fixed (and finite) number of processing elements and therefore a limited degree of parallelism to tackle such inherently parallel tasks. An infinite hierarchy of parallel models is thus formed (see Fig. 2.15), in which a machine on a certain level can simulate any machine below it in the hierarchy, but none above it. And since the principle of simulation is the cornerstone for the concept of a Universal Computer, we must conclude that the existence of such a machine is impossible.

We wish to draw the reader's attention on the remarkable similarity between the underlying formal structures of this negative result regarding universality in computing, on one hand, and Gödel's incompleteness theorem in mathematics, on the other hand. The similarity becomes apparent in the light of the following analogies. An arbitrary machine U_i in the hierarchy depicted in Fig. 2.15 corresponds to a

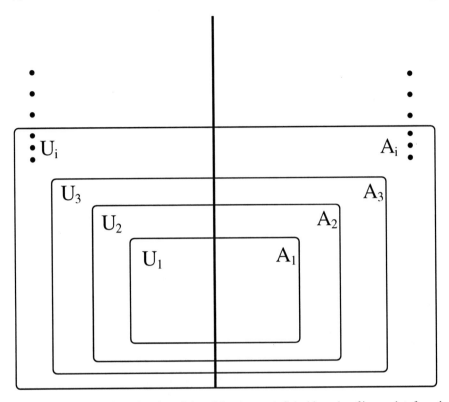

Fig. 2.15 An infinite hierarchy of parallel models mirror an infinite hierarchy of incomplete formal systems.

certain formal system A_i for which a self-referencing proposition G_i can be constructed such that the proposition, although true, cannot be proved within A_i. Since A_i allows the formulation of a true proposition that cannot be proved within A_i, it follows that A_i is incomplete, just as U_i is not universal.

To go up in the hierarchy, we add the recalcitrant proposition as a new axiom of A_i, thus obtaining a new system A_{i+1}. For the new system, G_i does not create problems anymore, but we can now construct a new proposition G_{i+1}, not provable within A_{i+1}. We can prove G_{i+1} in a new system A_{i+2}, which in turn has its own problem proposition G_{i+2} not provable within it, and so on forever. To complete the analogies, an unprovable proposition corresponds to a problem instance unsolvable by a machine on the current level. Thus, just as no complex-enough formal system can be considered complete, there is no computing machine that can be considered universal.

2.9 Conclusion

Parallel computing emerged from the need to speed up computations that, otherwise, would necessitate an impractical amount of time to complete. Even today, reducing the running time remains the main motivation behind using a parallel algorithm for many practical applications.

For decades, parallel computing has been an active field of research and it is even more so nowadays. With every new technological advancement, an opportunity presented itself to try to incorporate the new technology into more efficient parallel models of computation. An illustrative example is the way linear arrays and meshes of processors were augmented with electronic buses (at first, fixed, and later, reconfigurable) and optical buses, as the technologies developed and matured. A more recent example is the way parallel computing can harness the immense computational power offered by the Internet in order to accomplish tasks that, not long ago, were thought to be beyond the capabilities of any computer.

Lenstra, who was the first together with several colleagues to factor the ninth Fermat number $2^{2^9} + 1$ (155 decimal digits), also played an instrumental role in cracking both RSA-129 and RSA-130. More precisely, a 129-digit number used as a public key to encrypt a message by applying the RSA algorithm was factored in 1994 after a gargantuan computational effort that lasted about eight months and involved some 1600 computers distributed over the Internet. Improvements in factoring technology made possible a much quicker factorization of a 130-digit RSA key two years later. Moreover, Lenstra believes that RSA codes depending on 512-bit numbers (corresponding to 155 decimal digits) are within the reach of cypherpunks, if they could get hundreds of thousands of machines computing for them (see [11] page 167).

These results clearly showed the huge potential offered by models relying on massive parallelism and further encouraged research into how to better exploit this potential. The field of distributed computing, a very attractive research field nowadays, can trace back its roots to those early efforts. Models such as clusters and grids (see Sects. 2.3.4 and 2.3.5 at the beginning of the chapter) will certainly continue to play an increasing role in the landscape of parallel computing. This is facilitated by the continued increase in availability and affordability of relatively cheap processors and simple computers, compared with the sophistication and cost of supercomputers with similar performances as grids or clusters.

The current renewed interest in parallel computing methods we are witnessing today is also due to a certain shift in the type of applications dominating the field of computing. With ubiquitous, embedded and mobile computing devices becoming more prevalent, the design of efficient on-line and real-time algorithms becomes more and more important. In this context, the benefits of a parallel approach can be tremendous, not only in terms of speed but also for the quality of the solution computed, especially for numerical applications. Also, unconventional computing paradigms have been uncovered for which a parallel solution is the only viable option for a successful approach.

Finally, the study of parallel methods of computation is worth pursuing in its own respect for the valuable insights offered into the theory of computing, in general, as it is the case, for example, with the negative result on universality presented in Sect. 2.8.

References

1. S. G. Akl, Parallel Computation: Models and Methods, Prentice-Hall, Upper Saddle River, NJ, 1997.
2. J. JáJá, An Introduction to Parallel Algorithms, Addison-Wesley, Reading, MA, 1992.
3. Y. Ben-Asher, D. Peleg, R. Ramaswami, A. Schuster, The power of reconfiguration, Journal of Parallel and Distributed Computing 13 (1991) 139–153.
4. C. M. Krishna, K. G. Shin, Real-Time Systems, McGraw-Hill, New York, 1997.
5. S. D. Bruda, S. G. Akl, On limits on the computational power of data-accumulating algorithms, Information Processing Letters 86 (4) (2003) 221–227.
6. S. G. Akl, Superlinear performance in real-time parallel computation, The Journal of Supercomputing 29 (1) (2004) 89–111.
7. S. G. Akl, Parallel real-time computation: Sometimes quality means quantity, Computing and Informatics 21 (5) (2002) 455–487.
8. S. G. Akl, S. D. Bruda, Improving a solution's quality through parallel processing, The Journal of Supercomputing 19 (2001) 219–231.
9. T. H. Cormen, C. E. Leiserson, R. L. Rivest, C. Stein, Introduction to Algorithms, MIT Press, Cambridge, MA, 2001.
10. S. G. Akl, Evolving computational systems, in: S. Rajasekaran, J. H. Reif (Eds.), Parallel Computing: Models, Algorithms, and Applications, CRC Press, Boca Raton, FL 2007, a modified version is available as Technical Report No. 2006-526, School of Computing, Queen's University, Kingston, Ontario, Canada.
11. J. Brown, The Quest for the Quantum Computer, Touchstone Edition, Simon & Schuster, New York, 2001.

Chapter 3
Tools for Parallel and Distributed Computing

Thomas Fahringer

Abstract

Performance engineering of parallel and distributed applications is a complex task that iterates through various phases, ranging from modeling and prediction, to performance measurement, experiment management, data collection, and bottleneck analysis. There is no evidence so far that all of these phases should/can be integrated in a single monolithic tool. Moreover, the emergence of Cloud computing as well as established Grid infrastructures as a wide-area platform for high-performance computing raises the idea to provide tools as interacting Web services that share resources, support interoperability among different users and tools, and most important provide omni-present services over Grid or Cloud infrastructures.

We have developed the ASKALON tool set to support performance-oriented development of parallel and distributed applications. ASKALON comprises four tools, coherently integrated into a Web service-based distributed architecture. SCALEA is a performance instrumentation, measurement, and analysis tool of parallel and distributed applications. ZENTURIO is a general purpose experiment management tool with advanced support for multi-experiment performance analysis and parameter studies. AKSUM provides semi-automatic high-level performance bottleneck detection through a special-purpose performance property specification language. The Grid-Prophet enables the user to model and predict the performance of parallel and distributed applications at early development stages.

In this chapter we describe the overall architecture of the ASKALON tool set and outline the basic functionality of the four constituent tools. The structure of each tool is based on the composition and sharing of remote Web services, thus enabling tool interoperability. In addition, a Data Repository allows the tools to share common application performance and output data which has been derived by the individual tools. A Service Repository is used to store common portable Web

Thomas Fahringer
Institute for Computer Science, University of Innsbruck, Technikerstr. 21a,
A-6020 Innsbruck, Austria, e-mail: Thomas.Fahringer@uibk.ac.at

R. Trobec et al. (eds.), *Parallel Computing*, DOI 10.1007/978-1-84882-409-6_3,

service implementations. A general-purpose Factory service is employed to create service instances on arbitrary remote computing sites. Discovering and dynamically binding to existing remote services is achieved through a Registry service. The ASKALON visualization diagrams support both online and post-mortem visualization of performance and output data. We demonstrate the usefulness and effectiveness of ASKALON by applying the tools to a variety of real-world applications.

3.1 Introduction

Computational Grids, Cloud infrastructures [1], and parallel computers have become an important asset that enable application developers to aggregate resources scattered around the globe for large-scale scientific and engineering research. However, developing applications that can effectively utilize Grid/Cloud infrastructures and parallel computers still remains very difficult due to the lack of high-level tools to support application programmers. To this date, many individual efforts have been devoted to support performance-oriented development of parallel and distributed applications. Commonly users must write programs at low level of abstractions such as MPI [2] or Web services technology [3] which are both error-prone and time-consuming. Other languages such as OpenMP [4] allow incremental parallelization at a higher level but at the cost of controlling locality which is an important aspect on most parallel computers. Porting existing software tools on the Grid and Cloud infrastructures poses additional challenges. Portability and interoperability of software tools on the Grid and on Clouds are critical issues which have not been thoroughly addressed by the scientific community. We believe that this situation has been caused by the heterogeneous and often machine-dependent nature of tools, complex operating system and compiler dependencies, as well as differences and incompatibilities in tool functionality, interfaces, and other proprietary solutions.

Languages, software, system, and network neutrality have been successfully addressed over the past 10 years by well-known distributed object-oriented component technologies such as the Java Remote Method Invocation (RMI [5]), the Common Object Request Broker Architecture (CORBA [6]), Microsoft's Distributed Component Object Model (DCOM [7]), Enterprise Java Beans [8], Jini [9], Web services [10], or JavaSymphony [11].

Web services address heterogeneous distributed computing by defining techniques for describing software components, methods for accessing them, and discovery methods that enable the identification of relevant service providers. A key advantage of Web services over previous distributed technology approaches is their programming language, model, network, and system software neutrality.

Following the advantages offered by Web services, the Open Grid Services Architecture (OGSA) [12] builds on the Web services technology mechanisms to uniformly expose Globus Grid services semantics, to create, name, and discover transient Grid service instances, to provide location transparency and multiple protocol bindings for service instances, and to support integration with underlying native platform facilities. The Web Service Resource Framework (WSRF – [13]) is the technical specification which defines extensions and specializations to the Web services technology to standardize and ease the development of Grid services as required by OGSA.

In this chapter we describe the ASKALON tool set for parallel and distributed computing. ASKALON integrates four interoperable tools: SCALEA for instrumentation and performance analysis, ZENTURIO for automatic experiment management, AKSUM for automatic bottleneck analysis, and the Grid-Prophet for performance prediction. The tool-set has been designed as a distributed set of stateful Web services (WSRF – [13]), exporting a platform independent standard API. Platform dependent and proprietary services are preinstalled on specific appropriate sites and can be remotely accessed through a portable interface. A service repository is employed to store implementations of public portable Grid services. Each tool provides its own graphical user portal to be accessed by the user in a friendly and intuitive way. Remote services are created by a general purpose *Factory* service using the information from the Service Repository. On the other hand, the portals discover and bind to existing service instances by means of advanced lookup operations invoked on a *Registry* service [14]. Interoperability between tools is naturally achieved by allowing multiple clients to connect and share the same service instances from the initial design phase. Furthermore, a Data Repository with a standard schema definition allows tools to share performance and output data of Grid and Cloud infrastructures and applications.

This chapter is organized as follows. The next section discusses related work. Section 3.3 presents an overall Grid service-based architecture of the ASKALON tool-set. Sections 3.4, 3.5, 3.6, and 3.7 describe the basic functionality of each tool in brief. Various experiments conducted by each individual tool on several real-world applications are reported in Sect. 3.8. Concluding remarks and future work are presented in Sect. 3.9.

3.2 Related Work

Early work at the Technical University of Munich developed THE TOOL-SET [15], consisting of a mixture of performance analysis and debugging tools for parallel computing. Attempts to accommodate these tools into a single coherent environment produced the On-line Monitoring Interface Specification (OMIS) [16]. In contrast to this effort, ASKALON focuses on performance analysis for parallel, Cloud, and Grid applications, whose tools are integrated through a distributed Grid service-based design.

Significant work on performance measurement and analysis has been done by Paradyn [17], TAU [18], Pablo toolkit [19], and EXPERT [20]. SCALEA differs from these approaches by providing a more flexible mechanism to control instrumentation for code regions and performance metrics of interest. Although Paradyn enables dynamic insertion of probes into a running code, Paradyn is currently limited to instrumentation of subroutines and functions, whereas SCALEA can instrument – at compile-time only – arbitrary code regions including single statements. repositories, options, and high-level analysis. Paradyn also supports experiment management [21] through a representation of the execution space of performance experiments and techniques for quantitative comparison of several experiments. In contrast to ZENTURIO, experiments (by varying problem and machine size parameters) have to be set up manually under Paradyn.

The National Institute of Standards and Technology (NIST) developed a prototype for an automated benchmarking tool-set [22] to reduce the manual effort in running and analyzing the results of parallel benchmarks. Unlike in ZENTURIO, experiment specification is restricted to predefined parameters available through a special purpose graphical user interface.

Various groups have developed performance tools that are oriented toward automatic analysis. Paradyn [9] performs an automatic online analysis by searching for performance bottlenecks based on thresholds and a predefined (but immutable) set of hypotheses. The European working group APART [23] defined a specification language for performance properties of parallel programs based on which JavaPSL, the language for performance property specification used in AKSUM, has been designed. Performance properties defined by APART also inspired some of the predefined properties AKSUM provides. Kappa-Pi [24] and Earl/Expert [20] are postmortem tools that search for performance properties in message passing trace files in combination with source code analysis. Expert also covers OpenMP and mixed parallel programs and uses the concept of performance properties organized in a hierarchy. Performance properties are also used in the Peridot [25] project.

There have been several approaches employing machine learning methods in different fields, ranging from simple activities execution time [26], compiler performance optimization and prediction [27], scheduling [28] to networks [29]. There is also a series of related work for single activity execution time predictions using application profiles [30], analytical methods [31], soft benchmarks [32], historical data [26, 33] etc. But to the best of our knowledge, no effort has been made to predict the performance of workflows with different variations of problem sizes, Grid sites, and high-level middleware services.

Another effort closer to our focus is by Glatard et al. in [34]. The authors use probabilistic models to analyze workflow performance in the Grid, by considering execution times of individual activities, and data transfers between the activities and modeling various other execution phases as random variables. Gelenbe et al. [35] and Mussi et al. [36] also considered the execution time of a task graph as a random variable and determined its distribution from the graph parameters. These approaches assume very simple application workflows, ignoring complex control flows between activities, and loops over different (sets of) activities – which are completely driven by input problem-size, and are of core importance for overall workflow performance predictions. Moreover, variations in execution time due to input data set are not taken into account. In contradiction, we take into account the workflow structure attributes as well as the problem-size used for application workflow execution. Furthermore our work also considers the optimizations done by different schedulers used for mapping workflow activities on different Grid sites.

Authors in [33,37,38] have used different attributes to define similarity of execution of single activities. Lee et al. [38] have also described attributes to define similarity of resource states and policies. In contrast to these approaches which focus on single activities execution time predictions, our work emphasizes execution time prediction of full workflows. We consider (almost) all major workflow attributes describing its execution at different Grid infrastructural levels (such as Grid site, network etc.), in particular the workflow structure attributes to consider workflow structure similarities defined by Wombacher et al. [39]. Another major difference in our approach from the existing approaches (for activity execution time prediction) is the inclusion of problem-size attribute to describe a workflow execution.

Gibbons et al. [37] and Lee et al. [26] use a fixed set of templates, and Smith et al. [33] employ greedy search and genetic search to find effective templates. Compared to these efforts, we employ dynamic methods of supervised exhaustive search and evolution programming to find suitable template sets. Supervised exhaustive search is better than the greedy search in a way that it uses a notion of external supervision, by assigning probabilities of selection to different attributes and thus guiding the selection process. Similarly, our method of evolution programming is more efficient than genetic search used in [33] in implementation as well as in computation. In contrast to these approaches, we also introduce a notion of attribute relationships to decide their inclusion while selecting attributes for suitable templates.

3.3 ASKALON Architecture

The ASKALON tool set consists of four performance tools that we identified as important for performance analysis of parallel, Grid, or Cloud applications, integrated into a single coherent environment. Each tool consists of a single user portal and a set of distributed Grid services (see Fig. 3.1). The functionality of each tool is implemented by shared use of own services, together with the ones provided by the other integrated tools. In this section we show how such a distributed service design helps the integration of tools and enables their interoperability. The design solutions adopted are, however, fully generic for a Grid environment and are not constrained to our performance analysis goal. The concrete scenarios by means of which each performance tool interoperates with others through shared use of common services is described in the corresponding sections of this section (see Sects. 3.4, 3.5, 3.6, and 3.7). Furthermore, such a generic design will be beneficial for the integration of new Grid tools that we envision for the future.

The services are based on the WSRF-technology [13] and expose a platform independent standard API, expressed in the standard Web Services Description Language (WSDL) [40]. Platform dependent and proprietary services are preinstalled on specific appropriate sites from where they can be remotely accessed in a portable way, via the Simple Object Access Protocol (SOAP) [41] over HTTP. By isolating

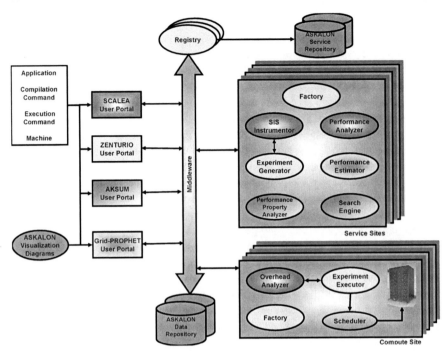

Fig. 3.1 The ASKALON tool set architecture.

platform dependencies on critical resources, extra flexibility for the installation and management of the tools is achieved. Each tool provides its own graphical user portal to be accessed in a friendly and intuitive way. The user portals are light-weight clients, easy to be installed and managed by the end-users. User portals reside on the user's local machine (e.g., a notebook) and provide gateways to performance tools by dynamically creating and connecting to remote services. ASKALON services can be persistent (e.g., Factory) or transient, as specified by WSRF. All services can be accessed concurrently by multiple clients, which is an essential feature in a Grid or Cloud environment and enables tool interoperability. The Grid Security Infrastructure (GSI) [42] based on single sign-on, credential delegation, and Web services security [43] through XML digital signature and XML encryption is employed for authentication across ASKALON user portals and Grid services.

Remote service instances are created by a general-purpose *Factory* service using the information from the Service Repository. Additionally, the *Data Repository* (see Sect. 3.3.1) with a common standard schema definition stores and shares common performance and output data of the applications under evaluation. It thus provides an additional mode of integration and interoperability among tools. To increase reliability of the system by avoiding single point of failures, multiple Service and Data Repository instances are replicated on multiple sites and run independently.

An WSRF-based asynchronous event framework enables Grid services to notify clients about interesting system and application events. ASKALON services support both push and pull event models, as specified by the Grid Monitoring Architecture (GMA) [44]. Push events are important for capturing dynamic information about running applications and the overall Grid or Cloud infrastructure on-the-fly and avoid expensive continuous polling. Pull events are crucial for logging important information, for instance in cases when tools like ZENTURIO run in off-line mode, with disconnected off-line users.

ASKALON classifies the Grid/Cloud sites on which the services can run into two categories (see Fig. 3.1):

(1) *Compute sites* are Grid/Cloud locations where end applications run and which host services intimately related to the application execution. Such services include the Experiment Executor of ZENTURIO, in charge of submitting and controlling jobs on the local sites and the Overhead Analyzer of SCALEA, which transforms raw performance data collected from the running applications into higher-level more meaningful performance overheads.

(2) *Service sites* are arbitrary Grid/Cloud locations on which ASKALON services are preinstalled or dynamically created by using the Factory service.

3.3.1 Data Repository

All ASKALON tools share a common Data Repository for storing information about the parallel and distributed applications under evaluation. The repository implementation is based on the PostgreSQL [45] open-source relational database system. The database schema definition reflects a layered design and has been jointly implemented by all tool developers.

Any tool can optionally store relevant experimental data including application, source code, machine information, and performance and output results into the repository. An interface with search and filter capabilities for accessing repository and leveraging the performance data sharing and tool integration [46] is provided. Tools exchange data via the Data Repository and also provide direct interfaces to subscribe for specific performance metrics, or parameter study results. Data can also be exported into XML format so that it can easily be transferred to and processed by other tools.

SCALEA stores mostly performance overheads, profiles, and metrics in the Data Repository. ZENTURIO through the Experiment Executor adds information about experiment parameters (ZEN variables) as well as output data required by parameter studies. AKSUM adds through its Property Analyzer the ZENTURIO schema definition information about high-level performance properties (inefficiency, scalability) and their severity. The Grid-Prophet can access information provided by any ASKALON tool to guide its prediction effort. Moreover, predicted performance data can be inserted into the Data Repository as well, which can be accessed by ZENTURIO and AKSUM instead of invoking SCALEA for a real program run.

3.3.2 ASKALON Visualization Diagrams

In addition to the distributed Web service-based design and the common Data Repository, ASKALON provides a Java-based package that comprises a set of generic and customizable visualization diagrams [47]. Available diagrams include linechart, barchart, piechart, surface, as well as a more sophisticated hierarchical diagram for the simultaneous visualization of a maximum of seven dimensions, which is used to graphically display performance studies of distributed and parallel programs.

Besides visualizing static post-mortem information, all diagrams accept online data streams as input for dynamic on-line visualization of parallel and distributed program behavior. The diagrams are generic and fully customizable, which enable both user and Web services to map application parameters, output results, or performance metrics onto arbitrary visualization axes. All ASKALON tools employ the ASKALON visualization diagrams.

3.4 SCALEA

SCALEA [48] is a performance instrumentation, measurement, and analysis tool for parallel programs that supports post-mortem performance analysis.

3.4.1 Instrumentation

The *Instrumentation Service* provides support to instrument services and applications. We support three approaches: command line options, directives, and dynamic instrumentation. In the first approach, the SCALEA Instrumentation System (SIS) provides automatic instrumentation of Fortran MPI, OpenMP, HPF, and mixed OpenMP/MPI programs. The user can select (by directives or command-line options) code regions (loops, procedures, arbitrary code regions, I/O statements, HPF INDEPENDENT loops, OpenMP PARALLEL loops, OpenMP SECTIONS, OpenMP CRITICAL, MPI barrier statements, etc.) and performance metrics (wallclock, cpu time, communication overhead, cache misses, synchronization overhead, etc.) of interest for which SCALEA automatically generates instrumentation code and determines the desired performance values during or after program execution. Furthermore, SIS provides a mechanism in order to control tracing/profiling by turn on/off the measurement process. Moreover, SIS offers an interface for other tools to traverse and annotate an abstract syntax tree to specify code regions for which performance metrics should be obtained. Based on preselected code regions and/or performance metrics, SIS automatically analyzes source codes and inserts probes (instrumentation code) in the code which will collect all relevant performance information during execution of the program on a target architecture. The source code level approach, however, requires all the source files to be available. In addition, instrumentation and measurement metrics can not be configured at runtime. To overcome these problems, we are currently exploiting the dynamic instrumentation mechanism based on Dyninst [49]. In order to enable dynamic instrumentation, we implement a *mutator service* which contains Dyninst API calls, the code that implements the runtime compiler and the utility routines to manipulate the application process. A mutator is responsible for controlling the instrumentation of an application process on the machine where the process is running. We developed an XML-based instrumentation request language (IRL) to allow users and services to specify code regions for which performance metrics should be determined and to control the instrumentation process.

With the command-line options, performance metrics and code regions for instrumentation are specified through the command-line parameters when invoking the instrumentation system. Command-line options can be used along with directives.

3.4.2 Overhead Analyzer

SCALEA provides a novel classification of performance overheads for parallel programs that include data movement, synchronization, control of parallelism, additional computation, loss of parallelism, and unidentified overheads [48]. The Overhead Analyzer Service is used to investigate performance overheads of a parallel program based on the overhead classification.

This analysis allows the user to examine sources of overheads simultaneously with code regions causing the overhead in the parallel application. Based on that, the user can further launch other functions to analyze the code regions in detail or can deploy suitable methods to tune the code in order to reduce the overhead.

The types of overheads and portion of identified overhead within total overhead determined are dependent on the instrumentation and measurement. In an overhead analysis phase, we conduct two tasks: (1) to determine the total overhead T_o and (2) to determine detailed types of overheads for each code region.

Given a code region r, let $T_s(r)$ and $T_p(r)$ be the execution time of a sequential version and a parallel version with p processors, respectively. The total overhead $T_o(r)$ of code region r when executed with p processors can be computed as follows:

$$T_o(r) = T_p(r) - \frac{T_s(r)}{p}$$

The total overhead can be determined only when (1) both sequential and parallel version of r exist or (2) r is an addition parallel programming-dependent code region. In the latter case, r is necessarily required for the parallelization of programs, e.g., a code region used to send data (e.g., MPI_SEND); r is not introduced in the sequential version.

The total overhead provides a value of how much overhead occurs in a code region; with total overhead we can determine whether a performance problem exists or not. However, in order to examine more detailed information about the sources that contribute on the total overhead, we need to determine sub overhead categories of the total overhead. These categories can provide more insightful information, which can be used to reveal the causes of performance problems.

To this end, we may need to divide a code region into subregions, to measure subregions and to determine types of overheads of individual subregions besides measuring the code region.

Performance overheads of code region instances of a given experiment are computed, displayed, and stored into the Data Repository.

The SIS measurement library supports profiling of parallel applications, collecting timing, counter information, as well as hardware parameters via the PAPI library [50]. The Overhead Analyzer computes performance overheads and stores them into the Data Repository.

3.4.3 Performance Analyzer

The Performance Analyzer Service evaluates the raw performance data collected during program execution and stores them into the Data Repository. All requested performance metrics are computed. Several analyses (e.g., Load Imbalance Analysis, Inclusive/Exclusive Analysis, Metric Ratio Analysis, Overhead Analysis, Summary Analysis) are provided.

While most performance tools investigate the performance for individual experiments one at a time, SCALEA goes beyond this limitation by supporting also performance analysis for multiple experiments (e.g., Speedup/Improvement Analysis, Scalability Analysis, Multi-Region Analysis, Multi-Set Experiment Analysis). The user can select several experiments, code regions, and performance metrics of interest whose associated data are stored in the Data Repository. The outcome of every selected metric is then analyzed and visualized for all experiments. SCALEA supports the following multi-experiment analyses:

- *performance comparison for different sets of experiments*: The overall execution of the application across different sets of experiments can be analyzed; experiments in a set are grouped based on their characteristics (e.g., problem sizes, communication libraries, platforms).
- *overhead analysis for multi-experiments*: Various sources of performance overheads across experiments can be examined.
- *parallel speedup and efficiency at both program and code region level*: Commonly, these metrics are applied only at the level of the entire program. SCALEA, however, supports examination of scalability at both program and code region level ranging from a single statement to the entire program.

3.5 ZENTURIO

ZENTURIO [51] is a tool to automatically generate and conduct a large number of experiments in the context of large-scale performance and parameter studies on cluster, Grid, and Cloud architectures. ZENTURIO uses the ZEN language to specify a large set of performance and parameter study experiments in a compact and user friendly manner. Thereafter, it automatically generates, conducts, and analyzes the performance and output data through a distributed service-oriented Grid architecture shielded from the end-user by means of a graphical user portal. ZENTURIO systematically organizes the performance and output data produced by all experiments into a Data Repository for post-mortem analysis.

3.5.1 ZEN Experiment Specification Language

Existing parameter study tools provide support to specify value ranges for application parameters of interest, e.g., by means of external scripting languages [52], or through graphical annotation of input files [53]. All of these approaches, however, force the user to export the application parameters to global input files or program arguments, which often require undesired source code adaptation for using the tool. Additionally, there are no tools that combine the experiment specification and management with cross-experiment performance analysis.

In contrast, ZENTURIO defines a directive-based language called ZEN [54] to annotate arbitrary application files. ZEN directives are used to assign value sets to so called *ZEN variables*. A ZEN variable can represent any problem, system, or machine parameter, including program variables, file names, compiler options, target machines, machine sizes, scheduling strategies, data distributions, etc. The value set represents the list of interesting values for the corresponding parameter. The advantage of the directive-based approach over an external script [52] is the ability to specify more detailed experiments (e.g., associate local scopes to directives, restrict parametrization to specific local variables, evaluate different scheduling alternatives for individual loops, etc.).

ZEN defines four kinds of ZEN directives as follows:

- *Substitute directives* assign a set of values to an application parameter. Each value from the set represents an experimental value for the parameter that shall be used by the application scientist in a separate experiment. The parameter instantiation is performed through plain string substitution that replaces all occurrences of the parameter name with its experimental value (in the scope of the directive);
- *Assignment directives* have analogous specification semantics as the substitute directive with the difference that the parameter instantiation is performed by inserting an assignment statement in place of the directive, which assigns the experimental value to the parameter name (as program variable);

- *Constraint directives* define a boolean condition over multiple parameters which restricts the set of possible experiments to a meaningful subset;
- *Performance directives* are used to request a wide variety of performance metrics for specific code regions of the program. The scope of the ZEN language is therefore not restricted to parameter studies.

A file/application annotated with ZEN directives is called ZEN file/application. A ZEN transformation system generates all ZEN file instances for a ZEN file, based on the ZEN directives inserted. The SCALEA instrumentation engine, which is based on a complete Fortran90 OpenMP, MPI, and HPF front-end and unparser, is used to instrument the application for performance metrics. The ZEN performance behavior directives are translated to SCALEA SIS directives and compiler command-line options.

3.5.2 Experiment Generator

The Experiment Generator is in charge of generating the experiments defined by an input ZEN application. Each ZEN file of the ZEN application is first parsed using the scanner and parser modules of the ZEN Transformation System which produces an abstract syntax tree. The abstract syntax trees of all ZEN files are then given as input to the experiment generation algorithm which generates a set of ZEN application instances which corresponds to an experiment.

We use SCALEA (see Sect. 3.4) as an instrumentation engine for Fortran 90, OpenMP, MPI, and HPF to instrument the application for performance metrics based on ZEN performance directives.

Additionally, we provide an interface to logically insert ZEN directives into the abstract syntax tree of each parsed ZEN file for situations when it is not practical to insert the directives manually.

We provide as part of the Experiment Generator service four methods for generating the experiments of a ZEN application:

1. *synchronous* by means of a single method invocation. This approach is rather primitive since the synchronous invocation can be very expensive and produces blocking (i.e., non-responsive) clients;
2. *iterative*, compliant with the pull event model, each experiment being returned by an iterator upon synchronous request;
3. *asynchronous*, compliant with the push event model, each experiment being sent to the client using an asynchronous callback as soon as it is generated;
4. *random*, by instantiating each ZEN variable (or a subset of them) with a random ZEN element. This method is used for implementing randomized optimization algorithms as in genetic algorithms.

In the case of using ZENTURIO in online Grid mode, the Experiment Generator automatically transfers the experiments to the target Grid execution site using the GridFTP protocol. In the case of using DUROC as job manager, the experiments are copied to multiple destination Grid sites which we retrieve from the RSL description of the application.

3.5.3 *Experiment Executor*

The *Experiment Executor* is a generic service with a high-level interface for executing and managing experiments on target Grid execution sites. We designed the Experiment Executor as a stand-alone Grid service independent of ZENTURIO that can be deployed for experiment management purposes in other infrastructures too. The Experiment Executor assumes a properly installed application on the target execution site(s). The Experiment Executor interacts at the back-end with a batch job scheduler, which in the current implementation can be Condor [55], LoadLeveler, LSF, PBS, and Sun Grid Engine for cluster, and GRAM [56] or DUROC for Grid computing.

After each experiment has completed, the application output results and performance data are stored into the ASKALON Data Repository (see Sect. 3.3.1). High-level performance overheads are computed by the Overhead Analyzer service of SCALEA. An *Application Data Visualizer* portlet of the user portal, developed on top of the ASKALON visualization diagrams (see Sect. 3.3.2), automatically generates visualization diagrams that display the variation of performance and output data across multiple sets of experiments.

3.6 AKSUM

AKSUM [57] has been designed to be a multi-experiment analysis tool, to a high degree independent of hardware and programming paradigms; it provides the user with a uniform and highly customizable interface to instrument an application, access and analyze performance data relative to several experiments, define how experiments are generated and executed, control the end of the search process, and define the search output. Once this information has been provided (or the default values have been accepted), AKSUM automatically conducts performance analysis without any user interference.

Figure 3.2 depicts the architecture of AKSUM. Through its user portal, the user inputs hypotheses that should be tested, machine and problem sizes for which performance analysis should be done (application input parameters), files that compound the application, and possibly conditions to stop the analysis process. The user portal displays, while the search process is going on, which hypotheses were evaluated to be true for the machine and problem sizes tested.

The experiment engine (implemented by ZENTURIO) launches the experiments considering the platform where the application will run. The instrumentation and monitoring engine is responsible for monitoring and instrumenting the application

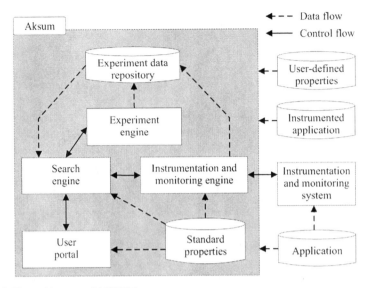

Fig. 3.2 The architecture of AKSUM.

independently of the language or paradigm utilized; it relies on an instrumentation and monitoring system to instrument the user's application and generate raw performance data, which is processed and stored in the experiment Data Repository, where the experiment engine also stores data. Currently, we use SCALEA (see Sect. 3.4) and Twilight [58] as instrumentation and monitoring systems. SCALEA is responsible for instrumenting Fortran programs, while Twilight instruments Java programs. AKSUM also uses the abstract syntax tree generated by the front end of VFC [59], which allows AKSUM to traverse the structure of Fortran programs and to inform SCALEA which code regions must be instrumented.

The search engine coordinates the entire search process and, using the data in the experiment Data Repository, tries to detect performance problems (called performance properties) in the application. The user-provided data, which influence the search process, flow from the user portal to the search engine, while the output of the search process flows from the search engine to the user portal.

3.6.1 Search Engine

The search engine coordinates the search process; it tries to detect performance properties in the application using the data generated by the other engines and stored in the experiment Data Repository. Properties are hierarchically organized into tree structures called property hierarchies, which are used to tune and prune the search for performance properties. For example, one may assume that, if an application is efficient, there is no need to compute its load imbalance. This assumption can be encoded in a specific property hierarchy by placing the property LoadImbalance under the property inefficiency. Another example would be the definition of a property hierarchy without any communication properties when it is known that the application is encoded as an OpenMP code and runs on a shared memory machine.

Each node in the property hierarchy represents a performance property and is described by two elements:

- Performance property name: the name of the performance property associated with this node; the property definition is stored in a property repository (defined by the user or provided by AKSUM).
- Threshold: a value that is compared against the severity value of each instance of the property represented by this node; if the severity value is greater than or equal to this value, then the property instance is critical and will be included in the list of critical properties.

Figure 3.3 shows a property hierarchy with six properties, and how the property LoadImbalance is customized. There are four standard property hierarchies provided by AKSUM, covering message passing, shared memory, mixed parallel programs, and distributed Java programs, but the user can define and store new property hierarchies from scratch or based on these predefined hierarchies. The reference code region for every property node in the predefined property hierarchies is per default set to the main program.

Users can use the set of properties provided by AKSUM and also specify in Java their own properties and add them to AKSUM in order to extend it. Any new property must be a class defining the following three methods:

- *boolean holds*( ): returns true if the property (class) instance holds (that means, the "negative performance behavior" is present).
- *float getSeverity*( ): returns a value between 0 and 1 indicating how severe a property instance is (the closer to 1, the more severe the property instance is).
- *float getConfidence*( ): returns a value between 0 and 1 that indicates the degree of confidence in the correctness of the value returned by *holds*.

AKSUM comes with a library, called JavaPSL [60], to help with the specification of performance properties, as it allows easy access to the performance data (timing information, overheads, and hardware counters) that SCALEA provides.

The process of searching for performance properties usually finishes when all application instances have been executed. In addition, AKSUM supports the definition of checkpoints to stop the search for properties before the end of the last experiment. A checkpoint is a Boolean function defined as follows:

- *op(severity(property, code region, number of experiments)) relop value* where *op* $\in$ *maximum, minimum, average, standard deviation*
- *relop* $\in$ { $>$, $\geq$, $<$, $\leq$, =, $\neq$ }. Any property and any code region are also valid values for property and code region.

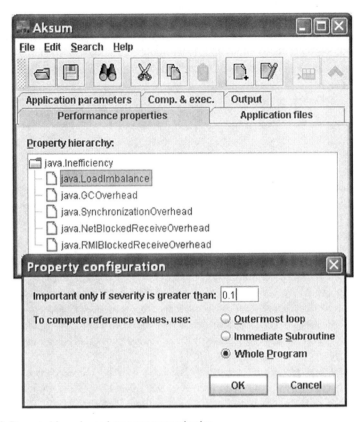

Fig. 3.3 Property hierarchy and property customization.

The following checkpoint, for instance, means that the search must stop if the severity of the any property in any code region is greater than 0.6.

$$maximum(severity(any\ property,\ any\ code\ region,\ 1)) > 0.6$$

The user-supplied input data is provided to the *search engine*, which is in the center of AKSUM and controls the entire search process. By issuing requests to the *instrumentation engine*, the search engine determines the performance information to be collected for application code regions and problem and machine sizes. The instrumentation engine of AKSUM invokes the SCALEA Instrumentation service for the actual code instrumentation, that is, it is a layer that enables the search engine to access and traverse application files in a machine independent way, to instrument them, and to transparently modify makefiles, scripts, and the compilation command line in order to link the instrumented application with the instrumentation library provided by SCALEA.

The instrumented code is submitted to ZENTURIO's Experiment Generator service, which changes the execution parameters according to the input parameters provided by the user and transfers the files to the appropriate Grid sites where ZEN-TURIO's Experiment Executor service will compile and execute the experiments, as well as transfer performance data to the Data Repository after each experiment has been executed.

The search engine evaluates the performance data in the Data Repository by invoking a *Performance Property Analyzer* service, which determines all critical performance properties (i.e., property instance whose value returned by the method *getSeverity* is greater than a certain threshold). A cycle consisting of consecutive phases of application execution and property evaluation is continued until all experiments are done or some system or user-defined condition stops the search process. Under the user portal, every performance property that has been determined to be critical is dynamically displayed (together with the source code) to the user during the search process and stored in a Data Repository.

3.6.2 Reinforcement Learning for Performance Analysis

While the analysis techniques used in AKSUM are effective to find performance problems, it still lacks a more formal approach that can be used to explain the decisions taken during the analysis and to justify their correctness. A well-established theory to model the performance analysis problem was needed which, when implemented, performed as good as or better than the original implementation of AKSUM. For this purpose reinforcement learning was used as part of AKSUM to model the performance analysis problem for two reasons: The trial-and-error nature of reinforcement learning resembles closely the empirical character of performance analysis, and, differently of other forms of learning, no expert teacher is required to tell the agent the correct actions to take.

Performance analysis can be seen as a reinforcement problem where the goal is to find in a short time many performance problems and with as little as possible interference in the application's behavior. Because reinforcement learning is based on trial and error, it would take too much time to learn the right actions for a given state if the performance analysis were post-mortem and the instrumentation static, since the reward would come only after the application finished executing. For this reason, for AKSUM only the dynamic performance analysis has been modeled as a reinforcement learning problem.

As usual, the first challenge when modeling real-world problems is deciding which elements are significant when solving the problem and which are not. An excessive number of variables added to the problem definition may slow down the resolution: reinforcement learning may have more signals to perceive and process, and it may take some time until the agent finally realizes that a variable has little or no significance for the problem. On the other hand, a model represents a type of biased knowledge, where a learning agent is told which signals can be safely ignored according to someone's point of view (points of view, however, are not always right). Another challenge is converting the result of actions to a scalar value that the agent can use as reinforcement, which may also contain a biased view of the problem and therefore will be transferred to the agent. Details of how reinforcement learning has been applied as part of AKSUM to support the search for performance analysis can be found in [61].

3.7 Grid-Prophet

The Grid-Prophet is a performance estimator for Grid workflow applications. An application workflow is a sequence of activities (small parts of an application) connected through control flows and/or data flows. The performance prediction of a Grid workflow is defined as the time it takes to execute all workflow activities. The workflow activities may be mapped on different Grid sites in the Grid, with/without the support of high-level Grid-services (such as resource broker, advance reservations etc.). The performance of a workflow in the Grid mainly depends upon the selected set of Grid sites, the problem size of the application, and states of the Grid resources, besides many other factors.

Performance prediction of a workflow is required to support dynamic execution of a Grid workflow. In addition, it helps in

- providing a decisive base for selection of Grid sites regarding time and cost.
- analyzing the overall performance and diagnosing performance bottlenecks, (e.g., a guide for scheduler and enactment engine).
- comparisons of different meta-scheduling policies/algorithms used to guide the workflow execution in the Grid, etc.

The prediction of the workflow execution time can be speculated in two dimensions: *reciprocal workflow performance prediction* that predicts workflow execution time for one iteration of activities in the critical path of the workflow (including one iteration of each loop in the workflow structure), and *absolute workflow performance prediction* that includes several repeating iterations of workflow until its termination. Here, we focus on absolute workflow execution time prediction.

For different perspectives of decision makings in the Grid, there may be different requirements of workflow execution time predictions in terms of *optimistic predictions* as the most likely performance or the last (nearest neighbor) similar execution performance; *pessimistic predictions* in terms of upper and lower bounds of execution time, and *accuracy of predictions* in terms of confidence intervals for the predictions. We address the most likely performance and the accuracy of the predictions here.

On the other hand, performance of a workflow is very profound to predict due to several phases in workflow execution and involvement of resources from almost all levels of the Grid infrastructure. Performance of the whole workflow is comprehended from the performance of individual activities along the critical path of the workflow, different Grid services involved in automatic workflow execution such as meta scheduler, resource broker etc., and the network transfers. The critical path of the workflow may include some control flow and parallel or sequential loops over some/all activities. Moreover, performance of individual activities is shaped from architecture specific execution of individual activities, local resource manager (LRM) policies, queue wait times, and external load (memory and CPU) at the time of their execution. The different phases of a workflow execution in the Grid are depicted in Fig. 3.4.

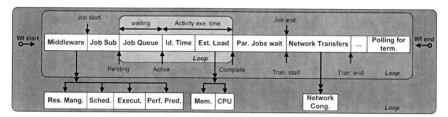

Fig. 3.4 Different execution phases of a Grid workflow.

Different Grid infrastructural components affecting workflow performance include Grid middleware, wide area network, Grid site, and local network [62] – these are shown in Fig. 3.5.

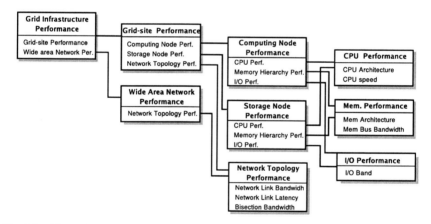

Fig. 3.5 Different Grid infrastructural components affecting workflow performance in the Grid.

3.7.1 Prediction Techniques

It is very difficult to formulate the performance of a workflow execution in the Grid from characterized performance of the individual Grid components involved in the execution of the workflow, due to involvement of the several components (such as

CPU, memory, network etc.) as shown in Fig. 3.5. Moreover, finding and including their combinatorial relative effects on workflow performance is even more complex. In order to take into account the effects of all the major components affecting performance of application workflow and include their combinatorial effects on workflow performance, we employ two methods from the domain of machine learning to predict performance of application workflows. First, the *Similarity Templates* and second, the *Local Learning Framework*. These methods are addressed in Sects. 3.7.1.2 and 3.7.1.3 respectively.

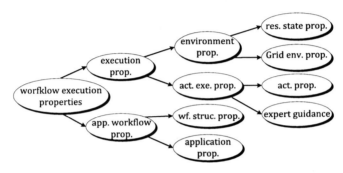

Fig. 3.6 Workflow properties composition.

3.7.1.1 Characterizing Workflow Performance

To exploit maximum strength of machine learning methods and get a higher prediction accuracy, we characterize the application workflow execution at the levels of different phases (see Fig. 3.4). At a coarse-grain level we define workflow execution in terms of application workflow properties and execution properties. Application workflow properties can be considered in terms of activity properties (like problem size, executables, versions, etc.) and workflow structure properties (like position of the activities, their dependencies etc.). The workflow structural properties need to be defined and considered in workflow similarity definitions [39] to incorporate the effects of structural differences (like dependencies between activities) in the executions of the workflows.

The execution properties are defined in terms of activity execution properties and execution environment properties (like Grid middleware, background load on the Grid sites etc.). Furthermore, activity execution properties are defined in terms of activity properties (activity description [63], parameters for execution etc.) and expert execution guidance like scheduling policy/algorithm used by a meta scheduler. Likewise, the execution environment properties include the Grid sites selected for execution, resource state properties (like jobs in the queue, jobs running, free memory etc.), and Grid environment properties (like Grid-middleware, high-level services used like meta-scheduling). For application workflows, the set of activities in the workflow and the Grid sites selected to map these activities are considered. This composition of properties is shown in the Fig. 3.6. The detailed workflow execution attributes at these levels are summarized in Table 3.1.

Table 3.1 Workflow execution attributes in the Grid.

	Attributes	Type
Workflow-level	Workflow name	Nominal
	Set of activities	Nominal vector
	Activity position	Numeric vector
	Workflow start time	Numeric
Application-level	Application name, desc	Nominal
	Problem size	Nominal vector
	Executables	Nominal vector
	Versions	Nominal vector
	File sizes	Numeric vector
Execution-level	Set of Grid-sites	Nominal vector
	Activities mapping	Nominal vector
	Time	Numeric vector
	Grid middleware	Nominal
	Scheduling strategy	Nominal
Resource-level	Jobs already in the queue	Nominal vector
	CPUs req. in the queue	Numeric
	Jobs already running	Nominal vector
	Already occupied CPUs	Numeric
	Jobs running in parallel	Nominal vector
	CPUs occupied in parallel	Numeric
Policy-level	User-name	Nominal
	Group (VO)	Nominal
	Queues	Nominal vector
Network-level	Bandwidth	Numeric
	Latency	Numeric

Type *nominal* represents non-numeric values; vector represents more than one value.

3.7.1.2 Similarity Templates

It is well known that similar applications' executions with similar setups are more likely to have similar execution time than the executions which have nothing or little in common [37]. We employ similarity templates (sets of attributes to define similarity) using a more sophisticated definition of similarity of application workflow execution than in [26, 33], and argue that carefully selected properties of similarity can lead to significant improvements in dynamic prediction accuracy, particularly employing workflow structure attributes. This is why we achieve higher prediction accuracy in our results (see Sect. 3.8.4) than related work [26, 33].

One of the core issues for the effectiveness of the predictions generated from the templates is to define a suitable set of templates and evaluate them quantitatively by using historical traces. On the one hand, putting fewer or unrelated attributes in the templates will lead to the generation of classes which will classify unrelated jobs (having very few attributes in common) together. On the other hand, putting too many attributes in a template will result in too many classes with fewer jobs classified in each. None of these classes will truly be a candidate for predictions for all the classes and thus predictions with poor accuracy will be the consequence.

We introduce here two search techniques to construct a suitable and meaningful set of templates τ: the *Supervised Exhaustive Search* algorithm and *Evolution Programming* [64]. The τ contains a set of workflow attributes α_j (described in Table 3.1), and may be any subset of $\mathbb{P}\{(\alpha_1^1), (\alpha_1^2, \alpha_2^2), ..., (\alpha_1^j, \alpha_2^j, ..., \alpha_j^j)\}$, where $\mathbb{P}$ represents the power set, and α_i^j corresponds to the ith attribute in the jth template.

The template generation phase is followed by generating classes from these templates by assigning different respective possible values to attributes in the templates. These classes are then used to generate predictions. We select the class with the minimum standard deviation of execution times of its data instances (historical data categorized in that class), for generating predictions. To generate predictions from the selected class, we use three induction models. The first is simple by taking the mean of all the instances in the class. The second technique exploits an additional information, the total number of CPUs (from different Grid sites) associated with historical run times, by performing regressions to compute the coefficients a and b for linear regression ($r = an + b$), logarithmic regression ($\log r = a \log n + b$) and inverse regression ($r = \frac{n}{a} + \frac{1}{b}$) between number of CPUs n and job run time r. The third technique we use is the *least mean square error* of the predictions generated based on previous induction models.

3.7.1.3 Local Learning Framework

We employ the *Instance Based Learning* (IBL) in the framework of local learning. The main advantage of IBL is that it is sensitive to even small variations in the attributes of workflow execution and thus can yield higher accuracy. Typically, it makes predictions for the query of given specifications of data attributes by finding its *nearby* data instances and then applying some induction model. The four major components of IBL framework are (1) the distance function, (2) the number of nearest neighbors, (3) the weighting function for the neighbors, and (4) the induction model to fit the neighbors.

The IBL algorithm assumes all the data instances correspond to points (representing data) in the n-dimensional space R_n (real numbers), where n is the number of data attributes (here workflow execution attributes). To measure the *nearness* of the data (described by the attributes described in Table 3.1), we employ the Heterogeneous Euclidean-Overlap Metric (HEOM) [65] as a distance function. This distance function can easily be used on nominal and numeric scalar attributes and we also extended it for numeric vector attributes.

We employ three induction models to generate predictions. First, the mean of the *k-nearest neighbors*, which considers k data points closest to the data point queried for prediction is computed. Second, the *weighted average (k-WA)* that considers k nearest neighbors and takes their weighted average based on their distance from the queried specifications (weight nearer neighbors higher and vice versa) is determined. Third, the linear *locally weighted regression* is computed, where closest points are weighted by proximity to the prediction data point using a kernel and then computing linear regression using the weighted points.

The application of IBL requires tuning for some of the parameters for its effective functioning; for example, different attributes in the distance function must be assigned some weights, to differentiate between their relativeness to data items, in order to measure the "better" *nearness*. Furthermore, selection of a good induction model, history size, and neighbor size must be tuned too. We exploit evolution programming [64] to optimize these parameters by minimizing average prediction error on the historical data set. We compare the optimization of different attributes through Bias-Variance Analysis [66] and adaptively select a set of optimized parameter values.

3.8 Experiments

In this section, we present numerous experiments to demonstrate the usefulness and effectiveness of the ASKALON tool set for a variety of real-world applications.

3.8.1 Performance Analysis with SCALEA

We illustrate SCALEA by applying it to a mixed OpenMP/MPI Fortran program that solves the two-dimensional Stommel Model of Ocean Circulation using a five-point stencil and Jacobi iteration. This code has been automatically instrumented, executed, measured, and analyzed for several problem and machine sizes based on user-provided SIS directives inserted in the source code. Our experiments have been conducted on Gescher but nodes are run with Linux 2.4.17-SMP patched with `perfctr` for hardware counters measurement. Gescher is a cluster multiprocessor system that consists of 16 nodes; each comprises four Intel Pentium III Xeon 700 MHz CPUs with 1MB full-speed L2 cache, 2Gbyte ECC RAM, Intel Pro/100+Fast Ethernet, Ultra160 36GB hard disk is run with Linux. We use MPICH 1.2.3 and `pgf90` compiler version 3.3 from the Portland Group Inc. The problem size is set to 200×200 points.

SCALEA supports the programmer in the effort to examine detailed performance overheads for an experiment of a given program. Two modes are provided for this analysis. First, *Region-to-Overhead* mode (see Fig. 3.7) allows the programmer to select any code region instance for which all detected performance overheads are displayed. Second, the *Overhead-to-Region* mode (see Fig. 3.8) enables the programmer to select the performance overhead of interest, based on which SCALEA displays the corresponding code region(s) in which the selected overhead occurs. This selection can be limited to a specific code region instance, thread or process. For both modes the source code of a region is shown only if the code region instance is selected by a mouse click.

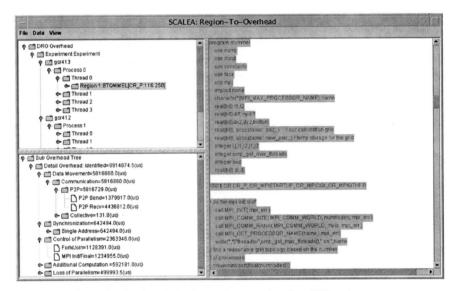

Fig. 3.7 Region to Overhead mode for Stommel executed on four SMP nodes.

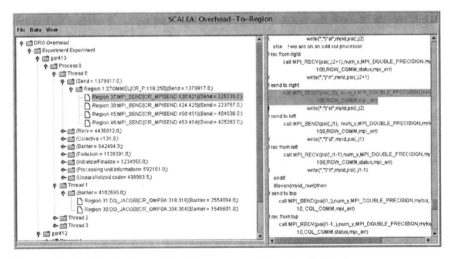

Fig. 3.8 Overhead to Region mode for Stommel executed on four SMP nodes.

Figure 3.9 presents the execution time of Stommel in six experiments. Overall, Stommel does not scale well. The reason is mostly due to the high overhead (especially communication) as presented in Fig. 3.10.

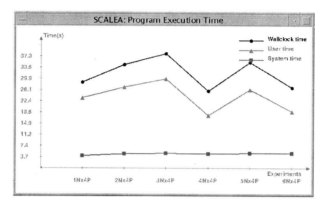

Fig. 3.9 Execution time of Stommel in six experiments. $1N \times 4P$ means 1 SMP node with four processors.

Experiments	1Nx4P	2Nx4P	3Nx4P	4Nx4P	5Nx4P	6Nx4P
Data movement	3.341	5.584	8.056	5.817	5.293	7.173
Synchronization	1.04	0.625	0.642	0.642	0.636	0.631
Control of parallelism	1.139	2.394	2.381	2.363	3.86	2.372
Loss of Parallelism	0.501	0.506	0.507	0.5	0.5	0.502
Additional Overhead	0.269	0.698	0.707	0.592	0.708	0.627
Total identified overhead	6.29	9.808	12.292	9.915	10.996	11.305
Total execution time(s)	28.071	33.873	37.396	25.22	34.587	26.346

SCALEA: Overhead Table

Fig. 3.10 Performance overheads of Stommel in six experiments.

3.8.2 Performance and Parameter Studies of a Three-Dimensional Particle-In-Cell Application with ZENTURIO

The *three-Dimensional Particle-In-Cell (3DPIC)* [67] is a Fortran 90 MPI application that simulates the interaction of high intensity ultrashort laser pulses with plasma in three-dimensional geometry. In this section we present a 3DPIC performance study based on the following parameter annotations:

1. The *machine size* is restricted by the peculiarities of this application to 1, 4, 9, 12, 16, 25, and 36 parallel processes which we have expressed through the count

argument of the GRAM RSL script shown in Example 3.1. Based on the number
of processes of one experiment, GRAM allocates the correct number of dedicated
SMP nodes using PBS as back-end local job manager. We set the job type to
single which gave us flexibility in selecting the local interconnection network.
We started the application using the shell script illustrated in Example 3.2 which
assigns to the MPIRUN ZEN variable the path to the mpirun script;

2. The *interconnection network* is studied by annotating an application Makefile.
Similarly, a constraint directive associates the implementation specific mpirun
command with the correct MPI library;

3. The *performance metrics* of interest are the execution time and the communication overhead.

Example 3.1 (Globus RSL script).

```
(*ZEN$ SUBSTITUTE count\=4 = { count={1,1,3,3,4,7,9} }*)
& (count=4)
   (jobtype=single)
   (directory="/home/radu/APPS/LAPW0/znse_6")
   (executable="script.sh") )
```

Example 3.2 (Shell script – script.sh).

```
#!/bin/sh
cd $PBS_O_WORKDIR
n = `wc -l < $PBS_NODEFILE`
#ZEN$ ASSIGN MPIRUN ={ /opt/local/mpich/bin/mpirun,
                       /opt/local/mpich_gm/bin/mpirun.ch_gm }
$(MPIRUN) -np $n -machinefile $PBS_NODEFILE lapw0
```

We inserted five ZEN directives into four files to generate a total of 14 experiments. Figure 3.11 indicates a good scalability behavior of the 3DPIC application.
The use of the Myrinet network yields approximately 50% performance improvement compared to the Fast Ethernet, which is explained by the reduced communication time (see Fig. 3.11) over the faster Myrinet network with lower latency and
higher bandwidth. Fig. 3.11 shows a relatively low ratio between the application
execution time (i.e., one full pie) and the MPI overheads measured, which explains
the good application scalability.

3.8.3 Performance Analysis for a Backward Pricing Application with AKSUM

The backward pricing application [68] implements the backward induction algorithm to compute the price of an interest rate-dependent product. The backward induction algorithm has been implemented as an HPF code based on which the VFC
compiler [59] generates a mixed OpenMP/MPI code. Based on the user provided

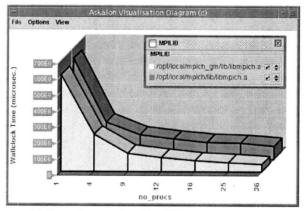

(a) Network comparison (Fast Ethernet versus Myrinet).

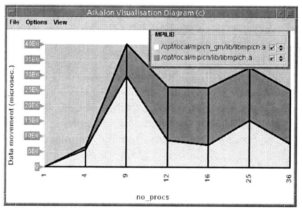

(b) Communication overhead comparison (Fast Ethernet versus Myrinet).

(c) Contribution of the Myrinet communication overheads to the wall-clock time.

Fig. 3.11 3DPIC performance results for various machine sizes.

input data, the search engine of AKSUM automatically determines that seven performance properties in the property hierarchy are critical for this code (see Fig. 3.12), where the properties are presented in ascending order of severity. As usual, the user portal displays initially only the property names for those instances whose severity is above the user-defined threshold (we set it to 0.01). The property instances can be shown by expanding each property name. For every instance the corresponding program unit and severity value is indicated. In the backward pricing application, the most serious performance property is ExecutionTimeLoadImbalance, which has an instance that holds for the main (entire) program with severity value 0.80 (see the entry BW_HALO_3 0.80). The same property holds for the subregion of the main program indicated by the entry BW_HALO_2 0.80. The severity of the ExecutionTimeLoadImbalance property instances for the entire application increases with the number of execution threads (not shown in Fig. 3.12), from 0.01 for 2 CPUs to 0.80 for 64 CPUs. This behavior also explains the increasing severity values for the inefficiency property (varying from 0.05 for 2 CPUs to 0.79 for 64 CPUs) All other properties in the property hierarchy have lower severity values (SynchronizationOverhead: 0.01, MessagePassingOverhead: 0.17 with 64 CPUs, for the other machine sizes 0.0).

The main program calls the subroutine BW, which calls subroutine COMPUTE_SLICE. As the properties' inefficiency and nonscalability are not critical for COMPUTE_SLICE, and since the critical instances of these properties have always approximately the same value for both the main program and the subroutine BW, we conclude that performance tuning should mainly be concentrated on subroutine BW.

3.8.4 Workflow Performance Prediction with the Grid-Prophet

We present results from our experiments to demonstrate the effectiveness of our approach for application workflow performance prediction for a real-world application workflow MeteoAG [69] in the Austrian Grid [70]. The structure of MeteoAG workflow is shown in Fig. 3.13. Here we present our prediction accuracy results as

$$average\ normalized\ absolute\ error = \frac{|ActExeTime - PredExeTime|}{ActExeTime}$$

where $ActExeTime$ represents actual execution time and $PredExeTime$ represents predicted execution time. Accuracy of the predictions is evaluated across different problem sizes of the workflow and also along different number of Grid sites used to execute the workflow.

Normalized absolute error for three problem sizes of MeteoAG and for different combinations of Grid sites from one to nine is shown in Fig. 3.14. The average normalized absolute errors in our predictions for the three problem sizes over different combinations of Grid sites were respectively 18, 30, and 23%. The same across different number of Grid sites was 24%. Maximum error remained at different number of Grid sites for the three problem sizes. The overall maximum error (61%)

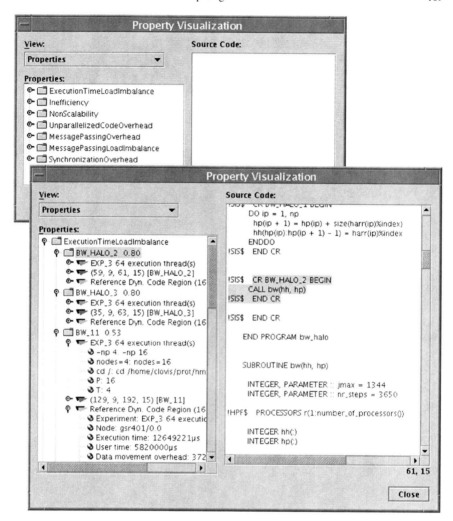

Fig. 3.12 AKSUM property visualization for the backward pricing application.

was found against problem size of three when executed on different combinations of seven Grid sites, and the minimum error (10%) was found for problem size of one when executed on different combinations of eight Grid sites. We find our accuracy results much better than related works using similar techniques for single activity execution time predictions [26, 33], both quantitatively and considering that our predictions are for a set of activities potentially with complex dependencies (workflow).

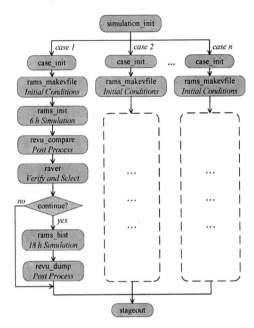

Fig. 3.13 MeteoAG workflow structure, activities, control flows, and data flows.

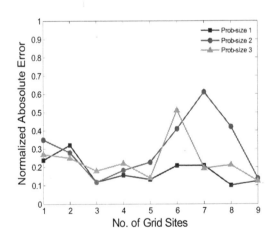

Fig. 3.14 Normalized absolute error in performance predictions for MeteoAG workflow for different combinations of problem sizes and Grid sites.

3.9 Conclusions

The development of the ASKALON tool set has been driven by the need of scientists and engineers to perform performance analysis, experiment management, parameter studies, modeling, and prediction of parallel and distributed applications for cluster and Grid infrastructures. ASKALON supports these functionalities through the provision of four sophisticated tools: SCALEA for instrumentation and performance analysis; ZENTURIO for experiment management, performance, and parameter studies; AKSUM for automatic bottleneck detection and performance interpretation; and the Grid-Prophet for performance modeling and prediction. Each tool can be accessed and manipulated via advanced user portals. ASKALON has been designed as a distributed Grid service-based architecture and implemented on top of the WSRF technology and Globus toolkit. Designing each tool as a composition of remote Grid service provides a series of advantages: (1) isolates platform dependencies on specific critical sites under a well-defined portable API; (2) enables light-weight clients, easy to be installed and managed by users on local sites (e.g., on notebooks); (3) allows the interaction of multiple tools by accessing resources concurrently through common shared services. The ASKALON tools exchange information through a common Data Repository or interoperate through the underlying Grid services. A generic visualization package that supports a wide variety of portable diagrams in both post-mortem and on-line modes is employed by the user portals of all tools.

Currently, we are working on a more elaborate integration and interoperability of all tools to reflect the continuously evolving Globus, WSRF, and Web service-based Grid specifications. Moreover, the ASKALON is also adjusted to be applied to applications on Cloud computing infrastructures.

Acknowledgments I am very thankful to Radu Prodan, Farrukh Nadeem, and the entire DPS (Distributed and Parallel Systems) team from the Institute of Computer Science at the University of Innsbruck as well as to Hong-Linh Truong from the Vienna University of Technology and Clovis Seragiotto, formerly with the University of Vienna. This work has been accomplished as part of the AURORA project funded by the Austrian Science Fund, the Austrian Grid project funded by the Austrian Federal Ministry for Education, Science and Culture, and the K-WF Grid project funded by the EU 6th FP under contract IST-2002-511385.

References

1. R. Ramakrishnan, Cloud computing – was thomas watson right after all?, in: ICDE, IEEE, (2008), p. 8, http://dx.doi.org/10.1109/ICDE.2008.4497407.
2. Message Passing Interface Forum, MPI: A Message-Passing Interface Standard, University of Tennessee, Knoxville, TN, (1995).
3. E. R. Harold, XML: EXtensible Markup Language, IDG Books, San Mateo, CA, USA, (1998), http://www.amazon.com/exec/obidos/ISBN=0764531999/cafeaulaitA/.

4. L. Dagum, R. Menon, OpenMP: An industry-standard API for shared-memory programming, IEEE Computational Science and Engineering 5(1) (1998) 46–55, http://www.computer.org/cse/cs1998/c1046abs, http://dlib.computer.org/cs/books/cs1998/pdf/c1046.pdf.

5. W. Grosso, Java RMI, O'Reilly, Associates, Inc., 981 Chestnut Street, Newton, MA 02164, USA, 2002, designing and building distributed applications.

6. D. S. Linthicum, CORBA 2.0?, Open Comput. 12 (2) (1995) 68–70.

7. N. Brown, C. Kindel, Distributed Component Object Model Protocol: DCOM/1.0, Microsoft Corporation and Redmond, WA, 1998.

8. B. Roth, An introduction to Enterprise Java Beans technology, Java Report: The Source for Java Development 3.

9. W. K. Edwards, Core Jini, IEEE Micro 19 (5) (1999) 10–10.

10. W3C, Web Services Activity, http://www.w3.org/2002/ws/.

11. T. Fahringer, A. Jugravu, JavaSymphony: New Directives to Control and Synchronize Locality, Parallelism, and Load Balancing for Cluster and GRID-Computing, in: ACM Java Grande – ISCOPE 2002 Conference, ACM, Seattle, 2002, pp. 8–17.

12. I. Foster, C. Kesselman, J. Nick, S. Tuecke, The Physiology of the Grid: An Open Grid Services Architecture for Distributed Systems Integration, The Globus Project and The Global Grid Forum, http://www.globus.org/research/papers/OGSA.pdf (November 2002).

13. G. A. WSRF, Web services resource framework, http://www.globus.org/wsrf.

14. M. Siddiqui, T. Fahringer, Gridarm: Askalon's grid resource management system, in: EGC, (2005), pp. 122–131.

15. R. Wismüller, T. Ludwig, THE TOOL-SET – An Integrated Tool Environment for PVM, in: H. Lidell, A. Colbrook, B. Hertzberger, P. Sloot (Eds.), Proc. High-Performance Computing and Networking, Vol. 1067 of Lecture Notes in Computer Science, Springer-Verlag, Brussels, Belgium, (1996), pp. 1029–1030.

16. R. Wismüller, J. Trinitis, T. Ludwig, OCM – A Monitoring System for Interoperable Tools, in: Proc. 2nd SIGMETRICS Symposium on Parallel and Distributed Tools SPDT'98, ACM Press, Welches, OR, USA, (1998), pp. 1–9, http://www.in.tum.de/\~{}wismuell/pub/spdt98.ps.gz.

17. B. Miller, M. Callaghan, J. Cargille, J. Hollingsworth, R. Irvin, K. Karavanic, K. Kunchithapadam, T. Newhall, The Paradyn Parallel Performance Measurement Tool. IEEE Computer 28 (1995) 37–46.

18. A. Malony, S. Shende, Performance technology for complex parallel and distributed systems, in: In G. Kotsis and P. Kacsuk (Eds.), Third International Austrian/Hungarian Workshop on Distributed and Parallel Systems (DAPSYS 2000), Kluwer Academic Publishers, Dortdrect, (2000), pp. 37–46.

19. D. A. Reed, R. A. Aydt, R. J. Noe, P. C. Roth, K. A. Shields, B. W. Schwartz, L. F. Tavera, Scalable Performance Analysis: The Pablo Performance Analysis Environment, in: Proc. Scalable Parallel Libraries Conf., IEEE Computer Society, (1993), pp. 104–113.

20. F. Wolf, B. Mohr, Automatic Performance Analysis of Hybrid MPI/OpenMP Applications, in: Proceedings of the Eleventh Euromicro Conference on Parallel, Distributed and Network-based Processing (PDP-11), IEEE Computer Society Press, (2003), pp. 13–22.

21. K. L. Karavanic, B. P. Miller, Experiment management support for performance tuning, in: ACM (Ed.), Proceedings of the SC'97 Conference, ACM Press and IEEE Computer Society Press, San Jose, California, USA, (1997).

22. M. Courson, A. Mink, G. Marcais, B. Traverse, An automated benchmarking toolset, in: HPCN Europe, (2000), pp. 497–506, citeseer.nj.nec.com/424567.html.

23. APART – IST Working Group on Automatic Performance Analysis: Real Tools, Aug 2001 until July 2004, http://www.kfa-juelich.de/apart.

24. A. Espinosa, T. Margalef, E. Luque, Integrating Automatic Techniques in a Performance Analysis Session, in: Proceedings of the 6th International Euro-Par Conference 2000, Lecture Notes in Computer Science 1900, Springer, (2000), pp. 173–177.

25. M. Gerndt, A. Schmidt, M. Schulz, R. Wismueller, Performance Analysis for Teraflop Computers – A Distributed Automatic Approach, in: Proceedings of 10th Euromicro Workshop on Parallel, Distributed, and Network-based Processing (EUROMICRO-PDP 2002), Canary Islands, SPAIN, (2002), pp. 23–30.
26. H. Li, D. Groep, J. Templon, L. Wolters, Predicting job start times on clusters, in: CCGRID '04: Proceedings of the 2004 IEEE International Symposium on Cluster Computing and the Grid, IEEE Computer Society, Washington, DC, USA, (2004), pp. 301–308.
27. J. Cavazos, C. Dubach, F. V. Agakov, E. V. Bonilla, M. F. P. O'Boyle, G. Fursin, O. Temam, Automatic performance model construction for the fast software exploration of new hardware designs, in: CASES, (2006), pp. 24–34.
28. P. Priore, D. D. L. Fuente, A. Gomez, J. Puente, A review of machine learning in dynamic scheduling of flexible manufacturing systems, Artificial Intelligence for Engineering Design 15 (3) (2001) 251–263.
29. A. Eswaradass, X.-H. Sun, M. Wu, Network bandwidth predictor (nbp): A system for online network performance forecasting, in: CCGRID '06: Proceedings of the Sixth IEEE International Symposium on Cluster Computing and the Grid, (2006), pp. 265–268.
30. V. Taylor, X. Wu, J. Geisler, R. Stevens, Using kernel couplings to predict parallel application performance, in: HPDC '02: Proceedings of the 11th IEEE International Symposium on High Performance Distributed Computing, (2002), p. 125.
31. D. A. Bacigalupo, S. A. Jarvis, L. He, D. P. Spooner, D. N. Dillenberger, G. R. Nudd, An investigation into the application of different performance prediction methods to distributed enterprise applications, Journal of Supercomput. 34 (2) (2005) 93–111.
32. F. Nadeem, M. M. Yousaf, R. Prodan, T. Fahringer, Soft benchmarks-based application performance prediction using a minimum training set, in: E-SCIENCE '06: Proceedings of the Second IEEE International Conference on e-Science and Grid Computing, IEEE Computer Society, Washington, DC, USA, (2006), p. 71, http://dx.doi.org/10.1109/E-SCIENCE.2006.131.
33. W. Smith, I. Foster, V. Taylor, Predicting application run times with historical information, Journal of Parallel and Distrib. Comput.
34. T. Glatard, J. Montagnat, X. Pennec, A probabilistic model to analyse workflow performance on production grids, in: CCGRID, (2008), pp. 510–517.
35. E. Gelenbe, E. Montagne, R. Suros, C. M. Woodside, A performance model of block structured parallel programs, in: Proceedings of the international workshop on Parallel algorithms & architectures, (1986), pp. 127–138.
36. P. Mussi, P. Nain, Evaluation of parallel execution of program tree structures, in: ACM SIGMETRICS Performance Evaluation Review, 12(3) (1984), 78–87.
37. R. Gibbons, A historical application profiler for use by parallel schedulers, in: Job Scheduling Strategies for Parallel Processing, Springer Verlag, London, (1997), pp. 58–65.
38. H. Li, J. Chen, Y. Tao, D. Gro, L. Wolters, Improving a local learning technique for queuewait time predictions, in: CCGRID '06: Proceedings of the Sixth IEEE International Symposium on Cluster Computing and the Grid (CCGRID'06), (2006), pp. 335–342.
39. A. Wombacher, M. Rozie, Piloting an empirical study on measures forworkflow similarity., in: IEEE SCC, (2006), pp. 94–102.
40. E. Christensen, F. Curbera, G. Meredith, S. Weerawarana, Web Services Description Language (WSDL), http://www.w3.org/TR/wsdl (March 2001).
41. A. Ryman, Simple Object Access Protocol (SOAP) and Web Services, in: Proceedings of the 23rd International Conference on Software Engeneering (ICSE-01), IEEE Computer Society, Los Alamitos, California, (2001), pp. 689–689.
42. I. Foster, C. Kesselman, G. Tsudik, S. Tuecke, A security architecture for computational grids, in: Proceedings of the 5th ACM Conference on Computer and Communications Security (CCS-98), ACM Press, New York, (1998), pp. 83–92.
43. B. Atkinson, G. Della-Libera, S. Hada, M. Hondo, P. Hallam-Baker, J. Klein, B. LaMacchia, P. Leach, J. Manferdelli, H. Maruyama, A. Nadalin, N. Nagaratnam, H. Prafullchandra, J. Shewchuk, D. Simon, Web Services Security (WS-Security), Specification,

Microsoft Corporation, http://www-106.ibm.com/developerworks/library/ ws-secure/ (Apr. 2002).

44. B. Tierney, R. Aydt, D. Gunter, W. Smith, V. Taylor, R. Wolski, M. Swany, A Grid Monitoring Architecture, The Global Grid Forum, http://www-didc.lbl.gov/GGF-PERF/ GMA-WG/papers/GWD-GP-16-2.pdf+ (January 2002).

45. R. Herzog, PostgreSQL – the Linux of databases, Linux Journal 46, ftp://ftp.ssc. com/pub/lj/listings/issue46/2245.tgz.

46. H.-L. Truong, T. Fahringer, On Utilizing Experiment Data Repository for Performance Analysis of Parallel Applications, in: 9th International Europar Conference(EuroPar 2003), Lecture Notes in Computer Science, Springer-Verlag, Klagenfurt, Austria, (2003), pp. 27–37.

47. T. Fahringer, ASKALON Visualization Diagrams, http://www.dps.uibk.ac.at/ projects/askalon/visualization.

48. H.-L. Truong, T. Fahringer, SCALEA: A Performance Analysis Tool for Parallel Programs, Concurrency and Computation: Practice and Experience 15 (11–12) (2003) 1001–1025.

49. B. Buck, J. K. Hollingsworth, An API for Runtime Code Patching, The International Journal of High Performance Computing Applications 14 (4) (2000) 317–329.

50. S. Browne, J. Dongarra, N. Garner, K. London, P. Mucci, A Scalable Cross-Platform Infrastructure for Application Performance Tuning Using Hardware Counters, in: Proceedings SC'2000, (2000).

51. R. Prodan, T. Fahringer, ZENTURIO: A Grid Middleware-based tool for experiment management of parallel and distributed applications, Journal of Parallel and Distributed Computinghttp://www.cs.iit.edu/~sun/jpdc/.

52. D. Abramson, R. Sosic, R. Giddy, B. Hall, Nimrod: A tool for performing parameterised simulations using distributed workstations high performance parametric modeling with nimrod/G: Killer application for the global grid?, in: Proceedings of the 4th IEEE Symposium on High Performance Distributed Computing (HPDC-95), IEEE Computer Society Press, Virginia, (1995), pp. 520–528.

53. M. Yarrow, K. M. McCann, R. Biswas, R. F. V. der Wijngaart, Ilab: An advanced user interface approach for complex parameter study process specification on the information power grid, in: Proceedings of Grid 2000: International Workshop on Grid Computing, ACM Press and IEEE Computer Society Press, Bangalore, India, (2000).

54. R. Prodan, T. Fahringer, ZEN: A Directive-based Language for Automatic Experiment Management of Parallel and Distributed Programs, in: Proceedings of the 31st International Conference on Parallel Processing (ICPP-02), IEEE Computer Society Press, Vancouver, Canada, (2002).

55. M. J. Litzkow, M. Livny, M. W. Mutka, Condor : A hunter of idle workstations, in: 8th International Conference on Distributed Computing Systems, IEEE Computer Society Press, Washington, D.C., USA, (1988), pp. 104–111.

56. K. Czajkowski, I. Foster, N. Karonis, S. Martin, W. Smith, S. Tuecke, A resource management architecture for metacomputing Systems, in: D. G. Feitelson, L. Rudolph (Eds.), Job Scheduling Strategies for Parallel Processing, Springer Verlag, London, (1998), pp. 62–82, lect. Notes Comput. Sci. vol. 1459.

57. T. Fahringer, C. Seragiotto, Automatic search for performance problems in parallel and distributed programs by using multi-experiment analysis, in: International Conference On High Performance Computing (HiPC 2002), Springer Verlag, Bangalore, India, (2002).

58. C. Seragiotto Jr., T. Fahringer, Performance analysis for distributed and parallel java programs with aksum, in: CCGRID, (2005), pp. 1024–1031.

59. S. Benkner, VFC: The Vienna Fortran Compiler, Scientific Programming, IOS Press, The Netherlands 7 (1) (1999) 67–81.

60. T. Fahringer, C. Seragiotto, Modeling and detecting performance problems for distributed and parallel programs with JavaPSL, in: Proceeding SC'2001, Denver, USA, (2001).

61. C. S. Junior, Systematic Performance Analysis and Interpretation for Parallel and Distributed Programs with Aksum, Ph.D. thesis, Vienna University of Technology (2005).

62. F. Nadeem, R. Prodan, T. Fahringer, A. Iosup, Benchmarking grid applications, in: CoreGRID Workshop on Middleware, Springer Verlag, Dresden, Germany, (2007).

63. M. Siddiqui, A. Villazon, J. Hofer, T. Fahringer, Glare: A grid activity registration, deployment and provisioning framework, in: SC '05: Proceedings of the 2005 ACM/IEEE conference on Supercomputing, IEEE Computer Society, Washington, DC, USA, (2005), p. 52, http://dx.doi.org/10.1109/SC.2005.30.
64. Z. Michalewicz, Genetic Algorithms + Data Structures = Evolution Programs (3rd ed.), Springer-Verlag, London, (1996).
65. D.R. Wilson, T. M. Martinoz Improved heterogenous distance functions., J. Artificial Intelligence Research 6 (1997) 1–34.
66. G. Valentini, T. Dictterich Bias-variance analysis of support vector machines for the development of SVM-based ensemble methods, Journal of Machine Learning Research 5 (2004) 725–775.
67. M. Geissler, Interaction of High Intensity Ultrashort Laser Pulses with Plasmas, Ph.D. thesis, Vienna University of Technology (2001).
68. E. Dockner, H. Moritsch, Pricing Constant Maturity Floaters with Embeeded Options Using Monte Carlo Simulation, Technical Report AuR_99-04, AURORA Technical Reports, University of Vienna (January 1999).
69. S. Felix., J. Qin, F. Nadeem, Performance, Scalability and Quality of the Meteorological Grid Workflow MeteoAG, in: 2nd Austrian Grid Symposium, Innsbruck, Austria, (2006).
70. T A G Consortium, http://www.austriangrid.at.

Chapter 4
Grid Computing

Uroš Čibej, Anthony Sulistio and Rajkumar Buyya

Abstract

The vision of Grid computing is to develop a platform which gathers geographically distributed resources (such as computational power, data, and equipment) into one very powerful and easy to use system. In this chapter, we present the main motivations behind this technology. Furthermore, we outline the challenges that researchers need to face when constructing such a complex distributed system. To demonstrate the practical impact, we describe various tools and applications which are already been extensively used to solve real problems. Finally, we give some pointers to the future directions in which Grid computing will evolve.

4.1 Introduction

Because of lack of adequate and, more importantly, standardized solutions for distributed computing on a larger (geographically distributed) scale, a new paradigm was coined in the late 1990s which is now known as Grid computing. The goal of Grid computing is to gather various resources into one, simple-to-use, secure, and quality assuring environment. As it is best described in [1]:

Uroš Čibej
Faculty of Computer and Information Science, University of Ljubljana, Tržaška 25, 1000 Ljubljana, Slovenia,
e-mail: uros.cibej@fri.uni-lj.si

Anthony Sulistio
Grid Computing and Distributed Systems Laboratory, The University of Melbourne, Australia,
e-mail: anthony@csse.unimelb.edu.au

Rajkumar Buyya
Grid Computing and Distributed Systems Laboratory, The University of Melbourne, Australia,
e-mail: raj@csse.unimelb.edu.au

R. Trobec et al. (eds.), *Parallel Computing*, DOI 10.1007/978-1-84882-409-6_4,
© Springer-Verlag London Limited 2009

> Grid is a type of parallel and distributed system that enables the sharing, selection, and aggregation of geographically distributed "autonomous" resources dynamically at runtime depending on their availability, capability, performance, cost, and users' quality-of-service requirements.

The resources can be anything from computational power, storage, network capability, scientific equipment, or even people. The development of this technology was driven by the current state of technology, and by the increasing need of users for new types of applications. Therefore, let us first elaborate the emerging requirements of the users:

- Collaboration. The functioning of organizations has changed drastically in the last decade. People are becoming much more mobile, due to their organizations participating in international projects. Thus, there is a need for sophisticated remote collaboration tools to share data, storage, and project specific services and equipment.
- Computing power. Modern users require more and more computing power. However, they might not need it all the time, only on specific occasions (e.g., when a project deadline is approaching). A purchase of a special high-performance system, which could provide the required computing power for such occasions, is simply not economically feasible.
- Enormous amounts of data. Scientists are generating enormous amounts of data from their experiments, and various organizations are gathering more and more data for future analysis. Such generated or gathered data need to be made accessible to many people and applications in the organization (or to even broader audience).

On the other hand, a lot of resources nowadays remain under-utilized:

- Ubiquitous connectivity. Most of the computers and computing systems have a network connection and are therefore easily accessible from anywhere in the world.
- Under-utilized computing power. Most of the modern computing systems are used for very simple tasks and usually these systems remain mainly unused. Fig. 4.1 shows an example of the CPU usage during a working day of a desktop computer. The CPU is completely unused during night time, but also during working hours the load is very low.
- Increase in connection speeds. It is commonly known that the number of elements in an integrated circuit doubles every 18 months (Gordon Moore's law [2]). Despite the age of this prediction, it is still very accurate. It is less commonly known, however, that the speed of computer network components doubles every 6 months [3]. In practice, this means that communication between computer processes over the network is not such a big bottleneck anymore. This is a great opportunity for new types of data-intensive applications, which were not possible until now.

As shown in Fig. 4.2, Grid computing can be a solution to the increasing needs of users by gathering and exploiting the resources that are already available.

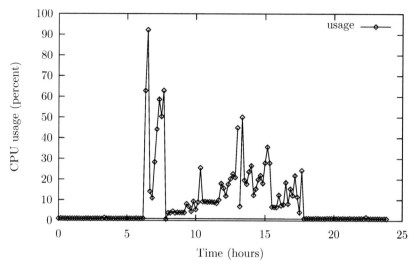

Fig. 4.1 A example graph of a CPU utilization for a desktop PC during a working day. The word processor, mail, browser, and other typical application hardly load the CPU. The peak utilization shown in the figure is the result of virus scanner checking the system.

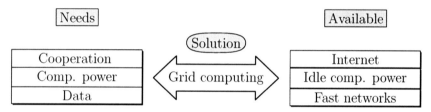

Fig. 4.2 Grid computing as a solution for the increasing requirements of users.

Besides these user requirements, Grid computing makes it possible to explore various new aspects and applications of distributed computing, and to bring this innovative approach to a much broader group of users. In what follows, we present various categories of Grid computing depending on the main solution they provide.

4.1.1 Grid Categorization

Grid computing represents a significant achievement toward the aggregation of clusters and/or other networked resources for solving large-scale data-intensive or compute-intensive applications [4]. Depending on the target application domain and purpose, Grids can be classified into several categories [5], as shown in Fig. 4.3.

1. Computational Grids. These provide distributed computing facilities for executing compute-intensive applications, such as Monte Carlo simulations [6],

Fig. 4.3 Types of Grids and their focus. The Grids on higher level use the functionality of the Grids on the lower levels. The focus also shifts from dealing with infrastructural problems to providing advanced services to the users.

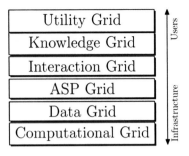

and Bag-of-Tasks (BoT) applications [7], where each consists of a collection of independent tasks or jobs. Some projects such as Nimrod-G [8], SETI@home [9], and MyGrid [10] utilize Grids to schedule these applications on available resources.

2. `Data Grids`. These provide the infrastructure to access, transfer, and manage large datasets stored in distributed repositories [11, 12]. In addition, Data Grids focus on satisfying requirements of scientific collaborations, where there is a need for analyzing large collections of data and sharing the results. Such applications are commonly found in the area of astronomy [13], climate simulation [14], and high energy physics [12]. There are several projects involved in Data Grids, namely LHCGrid [15], Biogrid [16], Virtual Observatory [17], and Avaki EII [18].

3. `Application Service Provisioning (ASP) Grids`. These concentrate on providing access to remote applications, modules, and libraries hosted on data centers or Computational Grids, e.g., NetSolve [19].

4. `Interaction Grids`. These provide services and platforms for users to interact with each other in a real-time environment, e.g., AccessGrid [20]. Thus, this type of Grid is suitable for multimedia applications, such as video conferencing, and those that require fast networks.

5. `Knowledge Grids`. These work on knowledge acquisition, data processing, and data management. Moreover, they provide business analytics services driven by integrated data mining services. Some projects in this field are Knowledge Grid [21] and the EU Data Mining Grid [22].

6. `Utility Grids`. These focus on providing one or more of the above Grid services to end-users as information technology (IT) utilities on a pay-to-access basis. In addition, they set up a framework for the negotiation and establishment of contracts, and allocation of resources based on user demands. Existing projects in this area are Utility Data Center [23], at the enterprise level and Gridbus [24] at the global level.

These types of Grids can be logically realized as a layer of services with one building on top of the other. A Grid on a higher layer utilizes the services of Grids that operate at lower layers in the design. For example, a Data Grid utilizes the services of Computational Grid for data processing, and hence builds on it. Moreover, lower-layer Grids focus heavily on infrastructural aspects, whereas higher-layer

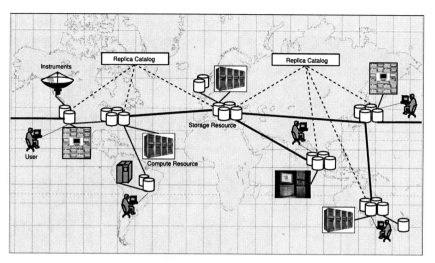

Fig. 4.4 A typical grid usage scenario.

ones focus on users and Quality of Service (QoS) delivery, such as deadline and cost.

A typical usage scenario of Grid activities, in this case for a Data Grid, is shown in Fig. 4.4. Scientific instruments, e.g., a satellite dish, generate large datasets which are stored in a Storage Resource. The Storage Resource then notifies a Replica Catalogue (RC) about a list of available datasets. The RC acts as an indexing server for handling registrations, notifications, and queries from resources and users. Next, this RC will synchronize its information with other RCs in the Grid. When a user submits his/her jobs, a Compute Resource communicates to the nearest RC to find out the location of the required datasets (if not stored locally). Then, the Compute Resource requests replicas or copies of these datasets from the Storage Resource. The RCs may be arranged in different topologies depending on the requirements of the application domain, the size of the collaboration around the application, and its geographical distribution [25]. Moreover, various replication techniques [26–28] may be applied to minimize the transfer time and bandwidth costs.

Based on this usage scenario, from the user's perspective, Grid computing can be considered as creating a *virtual* computer aggregating large hardware and storage infrastructures that are managed by different organizations across the world [4]. This scenario also identifies several key functionalities or components that need to be addressed by Grid resource providers:

- user interface, where users can submit and track jobs by using a command-line interface or a remote login, a graphical user interface (QMON for Sun Grid Engine [29]) or a web-based portal, such as the P-GRADE Portal [30] and the BioGrid Portal [31].
- security and access management, where users need to be authenticated and authorized before submitting jobs and using the resources respectively.

- administration and monitoring, where resource administrators can control and monitor the current state of resources, and users can track or see the progress of their jobs through an interface.
- resource discovery, where resources register their status and availability to a central server or a Replica Catalogue, as shown in Fig. 4.4. Thus, users can query about these resources.
- data management, where resources manage queries, replication, and deletion of datasets. In addition, various replication techniques are applied.
- resource management, where resources are allocated, assigned, and accessed according to QoS criteria, such as advance reservation, deadline, and cost.
- job scheduling, where a local resource scheduler, such as Maui [32], executes waiting jobs in a queue based on the QoS criteria, as mentioned above.

4.1.2 Comparison Between Clusters and Grids

Grid computing has a lot of common points with similar technologies, such as clusters. For understanding Grid computing better, it is necessary to know the similarities and, even more importantly, the differences between these systems. In this section, we will point out the essential differences which also determine different approaches to design and implementation of these technologies.

The first difference we can point out is the expected size of the system. A typical size of cluster can be measured at most in hundreds of nodes, whereas Grid systems are expected to have thousands of nodes. This is the basic motivation for a more scalable design of services which are mainly centralized in clusters.

Another crucial difference is the ownership and administration. In clusters, the nodes are owned and administered by a single organization. However, in Grids, every node can have a different owner and therefore different policies of resource usages. This raises many security issues which need to be handled very differently from security in cluster systems. The connection between nodes is another distinction which makes Grids different from clusters. Clusters are connected by dedicated high-speed networks, whereas Grids mostly use public internet. Because of this fact, in Grids we cannot make strong reliability assumptions, and the large latency of connections makes them unsuitable for the applications that prosper in cluster (communication-intensive applications). These and some other points are summarized in Table 4.1.

4.1.3 Putting It All Together

In this introduction part, we have presented the motivation behind Grid Computing. In addition, we have outlined the most relevant features these systems have to offer and the differences between grids and clusters.

Table 4.1 Key characteristics of clusters and grids.

Characteristics	Clusters/data centers	Grids
Population	Commodity computers	High-end computers
Size/scalability	100s	1000s
Ownership	Single	Multiple
Discovery	Membership service	Centralized indexing and De-centralized Info Services
Service negotiation	Yes	Yes, SLA based
User management	Centralized	Decentralized and also VO (virtual organization)- based
Resource management	Centralized	Distributed
Allocation/scheduling	Centralized	Decentralized
Standards/interoperability	VIA based	Web services-based and Open Grid forum efforts
Single system image	Yes	No
Capacity	Stable and guaranteed	Varies, but high
Throughput	Medium	High
Interconnection network	Dedicated, high-end	Mostly public internet, Some used high-end networks
Speed (Latency, Bandwidth)	Low, high	High, low
Application drivers	Science, business, enterprise computing, web applications, data centers	e-Science, e-business, multi-party conferencing, integration of scientific instruments

When building Grids, lessons learned from other types of distributed systems are relevant. However, to be able to obtain a common infrastructure with all the reliability, security, and QoS that the Grid computing vision demands, other challenges need to be faced. In what follows, we will describe some of difficult problems researchers and developers of grid systems have to deal with, and outline the solutions which are already available.

4.2 Challenges in Grid Computing

Managing various resources and applications scheduling in highly dynamic Grid environments is a complex and challenging process. Resource and data management are not only about scheduling large and compute- and data-intensive applications, but also the manner in which resources are allocated, assigned, and accessed. In this Section, we list several challenges that are currently being addressed in the research community.

4.2.1 Resource Sharing

As mentioned earlier, there is an increasing need of international collaboration among users and institutions. In Grid computing, a virtual organization (VO) allows users and institutions to gain access to their accumulated pool of resources to run applications from a specific field [33], such as high-energy physics or aerospace design. Common resources that can be shared are computers, networks, software licenses, storages, or scientific instruments.

The operational phase of a VO is a complex task because institutions may involve in one or more VOs. Therefore, resource sharing in VOs is conditional and rules-driven by means of authentication and authorization. For authentication, well-known protocols are usually used, such as Community Authorization Service (CAS) [34], Kerberos [35], and X.509 [36].

With regards to the authorization in a VO, Alfieri et al. [37] propose a centralized authorization service, named Virtual Organization Membership Service (VOMS). VOMS stores a list of groups, and each user's roles and accesses in a database. Thus, VOMS is used to bind authorization information to a user identity [37]. On the other hand, Sinnott et al. [38] suggest a federated model for scalability, where sites manage their own security information and authorization policies.

VO management is another challenging task, where institutions and users can join/leave a VO at any time. Thus, VOs have a short lifespan. Nasser et al. [39] propose a methodology to dynamically build a VO, based on an Organization Based Access Control (OrBAC) model. The OrBAC [40] models a multi-administered environment, where users, actions, and objects can be represented as role, activity, and view, respectively. This abstraction allows the OrBAC policy rules to be independent of the physical underlying infrastructure [40]. As a result, runtime binding of available resources can be done according to pre-defined criteria.

In a recent work, the XtreemOS project [41] is developing a Linux-based operating system that supports a VO model and management. Therefore, XtreemOS provides interoperability with other VO frameworks and security models, customizable access control, a scalable dynamic VO management, and dynamic mapping between VO and operating system entities [42].

4.2.2 Guaranteed Quality of Service

To ensure resources are available for applications when required, several researchers have proposed the need for *advance reservation* (AR) [43–46]. Common resources that can be reserved or requested are compute nodes (CNs), storage elements (SEs), network bandwidth, or a combination of any of these.

In general, reservation of the aforementioned resources can be categorized into two: *immediate* and *advance*. However, the main difference between these two reservations is the starting time. Immediate reservation acquires the resources to be utilized straight away, whereas advance reservation defers their usage later in the future.

Advance reservation can be useful for several applications, as described below:

- parallel applications, where each task requires multiple compute nodes simultaneously for execution.
- workflow applications, where each job may depend on the execution of other jobs in the application. Hence, it needs to wait for all of the dependencies to be satisfied before it can be executed.

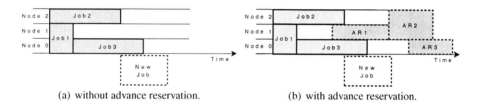

(a) without advance reservation.   (b) with advance reservation.

Fig. 4.5 Comparison of scheduling without and with advance reservation.

However, there are challenges in adopting advance reservation into Grids. Some of these are:

1. Significantly more complex operations and algorithms are needed for scheduling jobs, as shown in Fig. 4.5. A reservation-based system needs to handle incoming bookings and queries with respect to available spaces in the current and future time, as depicted in Fig. 4.5(b). Note that without AR, the future time is not considered, as illustrated in Fig. 4.5(a).
2. Possibly longer waiting time for other jobs in the queue, and lower resource utilization due to fragmentations or idle time gaps, as illustrated in Fig. 4.5. For example, in Fig. 4.5(a), in a system without AR, a new job that requires two compute nodes can be scheduled after *Job2*. However, in a system that uses AR, this new job can be executed only after *AR2*, as depicted in Fig. 4.5(b).
3. Potentially more negotiations between the resource and users due to their requests being rejected. Hence, the system needs to manage the overheads of many requests for reserving future availability.
4. Regulating resource supplies and reservation demands during busy and non-busy periods, as this has an impact on utilization, income revenue, number of rejections, and waiting time for local jobs in the system queue.
5. Possible loss of income due to cancellations and no-shows of existing reservations, since unused AR slots can not be sold to other jobs.

To address some of the challenges, Sulistio [43] presents a system model for scheduling task graphs with advance reservation and interweaving to increase resource utilization, and proposes a new data structure, named Grid advance reservation Queue (GarQ), for administering reservations in the Grid system efficiently. In addition, Sulistio [43] provides a case for an elastic reservation model, where users can self-select or choose the best option in reserving their jobs, according to their QoS needs, such as deadline and budget. With the adaptation of an on-line strip packing (OSP) algorithm into the elastic model, it manages to reduce the number of rejections and *fragmentations* (idle time gaps) caused by having reservations in the Grid system.

Similarly, the fuzzy model introduced by Roeblitz et al. [44] provides a set of parameters when requesting a reservation, and applies speedup models for finding the alternative solutions. Moreover, their model requires additional input conditions, such as the gap between two consecutive time slots and maximum number of time slots. However, no optimization on the resource utilization is considered.

The model proposed by Siddiqui et al. [45] uses a three-layered negotiation protocol, where the allocation layer deals with flexible reservations on a particular Grid resource. In this layer, the authors also used the strip packing method. However, the resources are dynamically partitioned into different shelves based on demands or needs, where each shelf is associated with a fixed time length, number of CNs, and cost. Thus, the reservation request is placed or offered into an adjacent shelf that is more suitable.

Venugopal et al. [47] present an alternate offers protocol to handle negotiations of previously rejected reservation requests. The protocol allows each party to submit a counter proposal until an agreement has been reached. Thus, it enables the user to fulfill its QoS requirements. Note that the last two challenges of AR are discussed next through the use of economy model.

4.2.3 Resource Regulation

Buyya et al. [48] introduced the Grid economy concept that provides a mechanism for regulating supply and demand, and calculates pricing policies based on these criteria. Thus, Grid economy offers an incentive for resource owners to join the Grid, and encourages users to utilize resources optimally and effectively, especially to meet the needs of critical applications.

Regulating supply and demand of resources is an important issue, as a study by Smith et al. [49] showed that providing AR capabilities increases waiting times of applications in the queue by up to 37% with backfilling. This study was conducted, without using any economy models, by selecting 20% of applications using reservations across different workload models. This finding implies that without economy models or any set of AR policies, a resource accepts reservations based on a first-come-first-serve basis and is subject to availability. Moreover, it also means that these reservations are treated similarly to high priority jobs in a local queue.

In order to address the above problem, Sulistio [43] proposes the use of Revenue Management (RM) techniques. The main objective of RM is to maximize profits by providing the right price for every product to different customers, and periodically update the prices in response to market demands [50]. Therefore, a resource provider can apply RM techniques to *shift demands* requested by budget conscious users to off-peak periods as an example. Hence, more resources are available for users with tight deadlines in peak periods who are willing to pay more for the privilege. As a result, the resource provider gains more revenue in this scenario.

Apart from RM, numerous economic models for resource management have been proposed in the literature. These include commodity market models (e.g., Nimrod-G [8] and Gridbus Broker [51]), tendering or contract-net models (e.g., JaWS [52] and Mariposa [53]), auction models (e.g., POPCORN [54], Spawn [55], and OCEAN [56]), bid-based proportional resource sharing models (e.g., Tycoon [57]), and cooperative bartering models [58].

In Nimrod-G, a user specifies QoS parameters, such as deadline and budget to a broker. Then, the broker schedules user tasks to resources with different allocation systems. In Tycoon, a user specifies his/her preference of each resource by giving a weight to it manually. Then, the Tycoon's agent selects which resource to bid on based on the user weight and total bid of other users for each resource. In contrast, Sulistio and Buyya [59] propose a time optimization algorithm that schedules a user application in auction-based proportional share systems across multiple VOs. Thus, the algorithm adjusts a user bid periodically on these systems in order to finish the application on time. Broberg et al. [60] provide detailed and comprehensive descriptions on other economy models.

4.2.4 Data Management

Another important aspect in Grid Computing is data management. As we pointed out in the Introduction, Data Grids are a type of grids which focus on the efficient management and retrieval of distributed datasets. Data Grids enable applications to produce and manipulate large amounts of data. In modern production Grids, the amount of data is expected to be measured in Terabytes and even Petabytes [61]. Therefore, the two most important components of a Data Grid are

- a reliable high-performance data transfer mechanism, and
- a scalable replica management and discovery system.

Data transfer involves not only moving data between Grid resources, but also other aspects of data access such as security, access control, and management of data transfers. Security can be divided into three categories: authentication, authorization, and encryption of data transfers.

For authentication, well-known protocols are usually used, such as Kerberos [35] and X.509 [36]. Authorization mechanisms used in Data Grids are typically similar to mechanism used in UNIX systems for file access permissions, but more fine-grained mechanisms have been developed and used, such as Role-Based Access Control [62] and Task-Based Authorization Control [63]. For data encryption, the predominant method in production Grids is data encryption through Secure Sockets Layer (SSL).

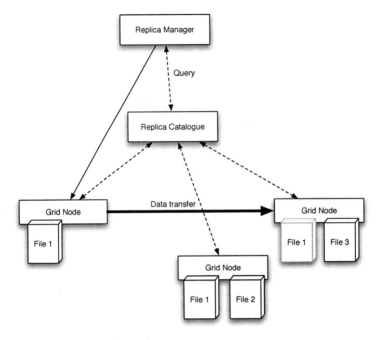

Fig. 4.6 Replica manager architecture.

Because of the large amount of data, data replication is crucial to provide fault-tolerance, scalability, and minimization of the use of bandwidth. A replica management system ensures access to the required data, while managing the underlying storage. Typical architecture of a replication system is shown in Fig. 4.6. The two components that orchestrate the replications are Replica Manager and Replica Catalogue.

Replica Catalogue is used as an information system which maps logical names of files or data objects to their physical names, i.e., the locations of the replicas of a certain data. They also store metadata about replicas, so users can query Replica Catalogues looking for specific features of the data. Grid nodes update the Replica Catalogue when any change of the data or metadata occurs. Replica Catalogue can be completely centralized or organized in a certain topology. The catalogue can be organized hierarchically, such as Globus Replica Catalog [64], or it can use data hashes to catalogue entries similarly to P2P networks [65].

The second component is Replica Manager, which can differ in terms of its distribution and type of replica strategy. The Replica Manager can be a centralized entity of the grid system, or it can be distributed – usually every storage element has its own Replica Manager. Replica Managers use different strategies to decide how many replicas of data to make and where to put them. These strategies can be divided into static and dynamic. Static replication strategies do not adapt to changes in the system, since the location of the data stays unchanged during runtime of the system. For a highly dynamic grid system, static replication might sound unreasonable; however, this approach offers some advantages [66]. Algorithms for static placement can be much more rigorously analyzed and can find a robust placement which will provide satisfactory service for most cases.

However, most of the developed replication strategies are dynamic. A lot of work in dynamic replication strategies has been done in the project EU Data Grid. This project was focusing on developing an infrastructure for analysis of results of the Large Hadron Collider. Bell et al. [67] presented two simple methods which are derived from caching in operating systems, namely, Least Recently Used (LRU), and Least Frequently Used (LFU). The methods work as follows: scheduler submits jobs to a certain grid node. The Replica Manager then transfers the required files to the local storage and, if there is enough space, creates a replica of the file. However, if there is not enough space, the manager deletes some data based on the LRU or LFU criterion.

Carman et al. [68] presented an economic strategy of replication. In this model, the files are handled as economic goods, and nodes make replicas of data based on the estimated profit (i.e., how many times the replica will be actually used). The estimated profit is calculated based on the previous usage pattern of the file. Recently Nicholson [69] compared the LRU, LFU, and economic replication on larger Data Grids and demonstrated that LRU and LFU strategies perform better on larger grids than the sophisticated economic model.

Independently of EU Data Grid project, other authors presented innovative replication strategies. Ranganathan et al. [70] compared different combinations of replication and scheduling algorithms and demonstrated that these two problems are highly interconnected.

We have to point out that all the mentioned strategies do not address a very important issue, and that is data synchronization. For current applications this was not critical, but for applications that also update the available data, replication strategies need to take into account the time required for updating the replicas. This is an important open question for future research in this area.

4.3 Tools and Applications

In this Section, we demonstrate the versatility of Grid computing by describing various very different fields in which they have already been successfully applied. We also describe some of the most important toolkits that make it possible to easily deploy large-scale Grid system.

4.3.1 Middleware

The elementary building blocks of a Grid system are gathered in what is called middleware. Grid middleware is typically a "bag of tools," which can be combined together in many different ways in order to implement the vision described in the introduction of this chapter. Many different toolkits exist nowadays. We give a short overview of three most influential and most widely used.

4.3.1.1 Globus Toolkit

Globus toolkit [71] is one of the first middleware toolkits and many production grids are based on it. This toolkit is being developed as a part of Globus alliance, an international collaboration for development of future Grid technologies. The "bag of tools" which Globus provides can be divided into five main groups:

1. Common Runtime. The common runtime includes libraries for different languages (currently Python, C, and Java) which enable building new services and applications.
2. Security. The tools for authorization, authentication, delegation of jobs, and community authorization are the core of the security of Globus Toolkit. The security is based on Grid Security Infrastructure (GSI).
3. Execution Management. These components are responsible for management of jobs in a grid. They consist of schedulers, monitoring, and coordination tools.
4. Data Management. Data management in Globus consists of Reliable File Transfer, Data Replication tool, and OGSA-DAI which enables various types of databases to be accessed via the grid.
5. Information Services. Information service include tools for monitoring the resources and jobs, and tools for discovery of needed resources. Currently, there are three components which enable this functionality: WebMDS, Index, and Trigger.

All the described services are implemented as Web (Grid) Services, but Globus includes also the components which are not based on Web Services, since older releases included them. However, it is expected that in the future releases only Web Service-based components will remain.

4.3.1.2 gLite

The gLite middleware is produced by the EGEE (Enabling Grids for E-sciencE) project. EGEE is one of the most important grid projects in the EU, which builds on the legacy of preceding projects such as EU DataGrid [72]. The goal of this project is to construct a production-quality infrastructure for e-Science, to attract new fields (from science as well as from industry) to this exciting technology, and to develop and maintain the gLite middleware.

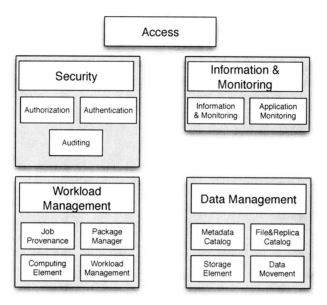

Fig. 4.7 Components of gLite.

The target application of this project is primarily high-energy physics, more specifically the Large Hadron Collider and the analysis of data this experiment will produce. But the applications are definitely not limited to this field and EGEE will disseminate the results to other scientific disciplines as well.

Similarly to Globus, gLite tools can be divided into five groups (also shown in Fig. 4.7):

1. Access. Different set of command line tools and APIs for accessing other components of gLite and building new services and applications.
2. Security. Authentication, Authorization, and Auditing components provide the required security in gLite.
3. Information and Monitoring. This functionality is split into two tools, one is basic Generic Information and Monitoring system and the other is Application monitoring.
4. Data Management is done by four components: the Metadata Catalog, File and Replica Catalog, an abstraction of a Storage Resource which is called Storage Element, and a tool for data transfers called File Transfer Service.
5. Workload management. Also for workload management, there are four core components: an abstraction of a Computing Element, Job Provenance tool, a Package Manager and a general Workload Management component.

All the main components of gLite have been built to be interoperable with Globus toolkit.

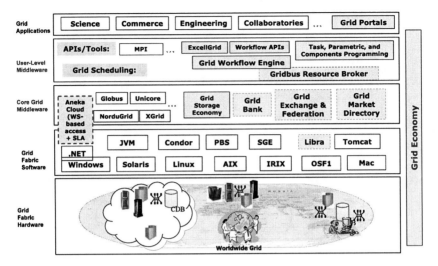

Fig. 4.8 Gridbus components in a broader context.

4.3.1.3 Gridbus

The Gridbus Project is engaged in the design and development of grid middleware technologies to support eScience and eBusiness applications. The main guidance for the development of this toolkit is the Grid Economy Model. The components of Gridbus middleware can be divided into three groups:

- Grid Fabric Software,
- Core Grid Middleware,
- and User-Level Middleware.

For the first group, Gridbus offers Libra, an economy-based scheduler for clusters. The second group of components consists of basic grid management components, such as Grid Storage Economy, Grid Exchange and Federation, Grid Bank, and Grid Market Directory. A completely novel tool which is based on .NET is Aneka. Aneka is a complete computing platform and is currently moving toward a commercial release. The third group of Gridbus components consists of scheduling components, such as Grid Workflow Engine and Gridbus Resource Broker, and programming tools such as ExelGrid, Workflow APIs and other Task, Parametric and Components programming kits. Some of the Gridbus technologies have been developed by making use of Web Services technologies and services provided by low-level Grid middleware, particularly Globus Toolkit. Gridbus components are depicted in a broader context in Fig. 4.8.

4.3.2 Tools for Computationally Intensive Applications

The first goal of Grid computing was to gather large amounts of distributed computing resources to be able to solve even larger computationally intensive tasks. Many communities that are dealing with this type of applications are already used to certain tools and applications. Bringing the grid to these communities means extending well-known tools and enabling them to use Grids without changing the user experience. This brings only the benefits of the new technology and no overhead of learning new tools and new skills.

We present a few such tools, which already have a large community of users, and were successfully extended to use geographically distributed resources with the use of Grid computing.

4.3.2.1 MPICH-G2

MPICH-G2 [73] is an implementation of the Message Passing Interface (MPI) framework, which is especially popular in clusters. More specifically, MPICH-G2 implements the MPI v1.1 specification. This framework enables the developers to write applications which consist of tightly-coupled processes. The processes coordinate their execution by passing messages to each other. MPI provides an abstraction for starting and stopping the processes and transferring the data among processes.

MPICH-G2 uses Grid services for manipulating processes on distributed resources, and for security. MPICH-G2 also adjusts the communication method, depending on whether the processes communicate locally or between geographically distributed resources.

4.3.2.2 Cactus

Cactus [74] is an open-source environment for developing parallel, high-performance simulations and visualizations. The goal users are scientists and engineers from a variety of fields; currently it is most widely used in research collaborations of physicists and computer scientists. The structure of Cactus is highly modular; its name is basically derived from its design. The base code infrastructure is provided by the central core (or flesh), all the other functionality is extended from the flesh with application modules (which are called thorns).

Thorns implement the functionality from different fields of science and engineering, such as general relativity equation solver, elliptic equation solver, or simulations of a 3D scalar field. But thorns also provide other functionality for orchestrating the computation, such as parallel I/O, data distribution, and access to grid resources. Since Cactus is designed to run on many different architectures, from desktop computers, to supercomputers, the heterogeneity of grids presents no problems. Cactus provides thorns which make it possible to execute Cactus computations on

Globus-enabled resources. Furthermore, various Grid portals have been developed, which make it possible to easily start and monitor Cactus applications.

4.3.2.3 Condor-G

Condor-G [75] is an extension of the well-known tool Condor, which is used for High Throughput Computing on large collections of distributive-owned computing resources. Condor provides facilities, such as job management, scheduling policies, and monitoring.

The architecture of Condor includes a collector which is responsible for gathering the information about worker nodes in the system. The users submit the requirements for their jobs in the form of ClassAds. Based on the ClassAd, the collector matches the job with the suitable nodes. Furthermore, Condor provides the infrastructure for monitoring the jobs, fault-tolerance, credential management, and job-interdependencies (which is most suitable for workflow execution). Condor-G has been developed to be fully compatible with Globus Toolkit and has already been extensively used in several grid projects.

4.3.2.4 GridSolve

GridSolve [76] is an extension of NetSolve, an environment that enables aggregating distributed computational resources. It is based on RPC and it is oriented to computationally intensive scientific applications. NetSolve autonomously chooses suitable computational resources on the network, based on the task requirements. It also enables some fault-tolerance, since the failed jobs are restarted.

The architecture of this system includes three entities: a lightweight client, an agent, and a server. The agent is an additional layer between the client and the server which is responsible for resource discovery, scheduling, load balancing, and fault-tolerance. GridSolve includes interfaces to many widely used languages for computational tasks like Fortran, MATLAB, Octave, C, Mathematica, and others.

4.3.3 Tools for Workflow Composition and Execution

Workflows are a natural paradigm to describe applications that arise in many disciplines from astronomy, bioinformatics, chemistry to economy, and business analysis. A lot of effort has been made to introduce workflow execution to grids, and overview can be found in [77].

A workflow execution can be divided into three stages [78]: (1) Workflow Generation, (2) Mapping, and (3) Execution system. Workflow Generation consists of tools for constructing the workflow, e.g., visual development environments for drawing workflow templates. It also consists of tools which populate a template workflow with data and tools for gathering and analysis of workflow results. In recent years a lot of user-friendly tools have been developed for these tasks, some of the most widely used are Triana [79], Karajan [80], and Taverna [81]. For popularization of solving problems with workflows, a social network like myExperiment [82] has been created for scientists to exchange workflows they use in their research.

The second component is the workflow management system, which is responsible for mapping and orchestrating the workflow. In this state an abstract workflow is mapped the specific sites on which it is going to be executed. The Workflow management system therefore includes a scheduler for the workflows, but also other components, which enable a reliable, scalable, and optimized execution. Examples of such systems include Pegasus [78] and Gridbus workflow engine [83].

4.3.4 Tools That Support Advance Reservation

In this Section, we present a brief description on some advance reservation projects or systems for job and resource management in Grids.

4.3.4.1 PBS Pro

Portable Batch System, Professional Edition (PBS Pro) [84, 85], is a local resource manager that supports scheduling of batch jobs. It is the commercial version of PBS with added features such as advance reservation, security (e.g., authentication and authorization), cycle harvesting of idle workstations, information management (e.g., up-to-date status of a resource and its queue length), and automatic input/output file staging. PBS Pro can be installed on Unix/Linux and Microsoft Windows operating systems.

PBS Pro consists of two major component types: user-level commands and system daemons or services (i.e., Job Server, Job Executor and Job Scheduler) [85]. Commands, such as submit, monitor and delete jobs, can be first submitted through a command-line interface or a graphical user interface. These commands are then processed by the Job Server service. These jobs are eventually executed by the Job Executor service or MOM. In addition, PBS Pro enables these jobs to be submitted to Globus [86] via the Globus MOM service. Finally, the Job Scheduler service enforces site policies for each job, such as job prioritization, fairshare, job distribution or load balancing, and preemption. By default, the Job Scheduler uses the First In First Out (FIFO) approach to prioritize jobs, however, it can also use a Round Robin or fairshare approach, where jobs are ordered based on the group's usage history and resource partitions.

Reservations are treated as jobs with the highest priority by the Job Scheduler service. Hence, reservation requests need to be checked for possible conflicts with currently running jobs and existing confirmed reservations, before they are being accepted. Requests that fail this check are denied by the Job Scheduler service.

4.3.4.2 Sun Grid Engine (SGE)

Sun Grid Engine (SGE) is an advanced resource management tool for distributed computing environments [29]. It is deployed in a cluster and/or campus Grid testbed, where resources can have multiple owners, but they can also belong to a single site and organization. SGE enables the submission, monitoring, and control of user jobs through a command line interface or a graphical user interface via QMON. In addition, SGE supports checkpointing, resource reservation, and Accounting and Reporting Console (ARCo) through a web browser.

In SGE, resources need to be registered or classified into four types of hosts. The master host controls the overall resource management activities (e.g., job queues and user access list), and runs the job scheduler. The execution host executes jobs, while the submit host is used for submitting and controlling batch jobs. Finally, the administration host is given to other hosts, apart from the master host, to perform administrative duties. By default, the master host also acts as an administration host and a submit host.

To manage resource reservations, each job is associated with a usage policy or priority, the user group, waiting time, and resource sharing entitlements [29]. Thus, the earliest available nodes will be reserved for pending jobs with higher priority by the SGE scheduler automatically. This reservation scenario is mainly needed to avoid the job starvation problem for large (parallel) jobs. On the other hand, SGE can leverage an external scheduler, such as Maui Scheduler [32] to provide more comprehensive reservation functionalities.

4.3.4.3 Highly-Available Resource Co-Allocator (HARC)

Highly-Available Resource Co-Allocator (HARC) [87], developed by the Center of Computation & Technology (CCT) at Louisiana State University (USA), is an open-source system for managing multiple reservations of various resources. This can be done by users sending reservation requests to HARC via its Java API or a command-line interface. Then, the requests are managed by HARC Acceptors. These Acceptors are responsible for interacting with an individual Resource Manager of a specific type, similar to GARA's LRAM. Next, the Resource Manager communicates with a local scheduler to determine the resource availability in the future for a particular request. Finally, the Resource Manager sends a message to users via Acceptors, whether it accepts or rejects the given reservation request. If the request is accepted, then it needs to be committed afterwards [87].

From the above description, HARC employs a two-phase commit protocol. To ensure the reliability of Acceptors and to prevent any missing messages, HARC uses Paxos Commit [88], a transaction commit protocol, where it uses multiple Acceptors for the same user to communicate with Resource Managers. With this approach, each Resource Manager will send the same message to multiple Acceptors. If the head or lead Acceptor fails, then other Acceptors will take its place automatically.

In HARC, new types of resource can be integrated easily by creating new Resource Managers. To reserve compute nodes, the HARC Compute Resource Manager works with a local batch scheduler that supports advance reservation, such as Maui Scheduler [32] or Moab Workload Manager [89]. To reserve network bandwidth, the HARC Network Resource Manager acts as a centralized scheduler that oversees the overall management of network traffic for the entire testbed [90].

4.3.5 G-Lambda Grid Scheduling System

The Grid scheduling system, developed as part of the G-lambda project, is a Web Service system that is able to allocate resources (compute nodes and network) in advance [91]. The aim of the G-lambda project is to build a standard web service interface among resource management systems in Grid and network computing [92]. The Grid scheduling system consists of two main components: the Grid Resource Scheduler (GRS) and the Network Resource Management System (NRM).

The GRS is developed using Globus Toolkit 4 [93], a Java implementation of Web Services Resource Framework (WSRF). It handles reservation requests from applications or Grid portals. To reserve compute nodes, the GRS interacts with Computing Resource Manager (CRM) on each site. To reserve network bandwidth, the GRS communicates with Network Resource Management System (NRM). The NRM provides optical paths on a GMPLS-controlled network infrastructure. GMPLS is a generalization of the MPLS architecture, where it supports multiple types of switching other than label switching, such as lambda and fibre (port) [94].

To satisfy the user's QoS requirements, the scheduling module inside the GRS interacts with the CRM and/or NRM to locate available reservation slots using a depth-first search scheme [91]. However, new scheduling techniques can be easily incorporated into the module without affecting the rest of the system.

4.3.6 Application Fields

Until now we described some tools for writing general user applications. In what follows, we outline some specific disciplines where Grid computing has already proved very useful, and we give examples of successful projects which helped applying Grids into practice.

4.3.6.1 Medicine and Healthcare

Medicine offers plenty of possibilities for Grid computing applications. The first area is medical research, where analysis and visualization of large amount of data can lead to new discoveries. Examples of this type of application are: Neuro-Grid [95], MediGRID [96], and KidneyGrid [97].

The second area where medicine can benefit from Grid computing is the introduction of this technology directly into the healthcare system. The examples of usage in healthcare include:

- easy collaboration between physicians,
- easier diagnostics, e.g., MammoGrid [98],
- virtual surgery,
- and simulation of treatments.

Healthcare applications need a lot of data and computational power, but due to high sensitivity of data in the healthcare system, security is the most important aspect. Grids offer a flexible but strong model of security and could therefore be trustworthy enough to be widely implemented.

4.3.6.2 Astronomy

Astronomers are producing large amounts of data from sources like regular telescopes and radio telescopes. The data is gathered on only a few places on Earth (or in space if we are talking about the Hubble telescope), but astronomers are spread all over the planet and require the access to this data. Furthermore, the analysis of data requires a lot of computational power. Thus, Grid computing is a very well-suited platform for this type of application.

Some of the very successful projects have already tested the benefits of Grid, such as GRIST [99], AstroGrid-D [100], and Australian Astronomy Grid [101].

4.3.6.3 Finance

The financial industry is another field of possibilities for Grid computing. Businesses acquire unprecedented amounts of data which need to be analyzed as fast as possible. Speed and accuracy bring the competitive edge to the companies and is therefore of vital importance. Examples of such applications include value-at-risk calculations, real-time trading, and risk modeling. A lot of these applications require Monte Carlo simulations and a scalable computing and data platform would bring the needed speed and accuracy to the companies. Projects such as GriFin [102] are striving to build such a platform which will satisfy very demanding financial companies.

4.3.6.4 Digital Libraries

The main goals of traditional libraries are the following:

- production of new knowledge,

- preservation of the produced knowledge,
- and access to the knowledge for many generations.

The modern libraries are having more and more problems to achieve these goals due to an incredible explosion of information that needs to be archived and also made easily accessible. The libraries are therefore forced to shift to digital applications to be able to cope with this problem. The effort of implementing digital libraries is an ongoing process which is sometimes denoted as Digital Library Technologies. The solutions which are available now are mostly static archives, which provide only simple services. The next step is building highly dynamic digital libraries, which would provide more fault-tolerance, optimization, and also very flexible services which can adapt to specific needs of libraries and users. And this is where Grids can bring the needed flexibility and dynamics. Libraries will become global and the goals mentioned above will be served even much better than before. The most influential project that is trying to bring the advantages of Grid computing to digital libraries is DILIGENT [103].

4.4 Conclusions and Future Trends

In this chapter, we have presented the technology that enables revolutionary applications, gathers unused computing power, facilitates complex collaborations and access to distributed data and equipment.

It has been more than a decade of intensive work of making the Grid computing vision a reality. Grids have progressed from highly experimental environments, which were mainly used as a playground for new ideas, to trustworthy environments for critical applications.

We have presented some of the challenges that researchers had/have to deal with when building quality Grid systems. We also gave an outline of middleware, tools, and applications which are already being successfully used in practice.

As this technology becomes more mature, the industry is starting to embrace the possibilities it offers. The big players in IT industry (e.g., IBM, Sun, and HP) developed their own Grid computing solutions. The tools to set up world-wide grid systems are much closer to "out of the box" solutions, which is crucial for the wide acceptance of grids.

Grid technologies are also becoming an integral part of operating systems with projects like XtreemOS [41] developing a Linux-based operating system with the most advanced grid functionality already included. We are facing a (r)evolution of how computers are being used. Business models are shifting from "selling software" to "lending services" which will hopefully also bring a less monopolized software/service market.

New paradigms have emerged and the lessons learned from Grid computing and other breakthrough ideas from distributed computing are being used to develop even more flexible, extendible, and scalable platforms for future applications. Some of the paradigms which have been developed in this decade include P2P computing,

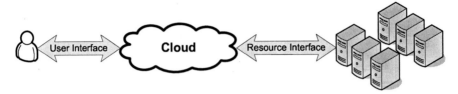

Fig. 4.9 The vision of cloud computing.

service computing, market-oriented computing, utility computing, and the most recent and most perspective paradigm is cloud computing [104].

Cloud computing has been nominated as the "next big thing" in computing. This paradigm shares most of the vision of Grid Computing, i.e., to collect distributed resources and present them to the users as a uniform service as shown in Fig. 4.9. However, there are also fundamental differences.

Some of the most well-known cloud computing solutions include: Amazon Elastic Compute Cloud, Google App Engine, Microsoft Azure, and Sun Network.com. As it can be seen, cloud computing is coming mostly from industry and is moving slowly to the research community, whereas Grid computing on the other hand was born in the academic community and is progressing to the industry. At the moment, the academic effort in cloud computing can be mostly seen in the Aneka framework [105], which was basically started as a Grid Computing solution and is now progressing to include cloud computing ideas.

The industrial origin of cloud computing brings also a more commercial view to the technology. Therefore one of the main goals is to develop simple and efficient economic models for selling cloud services (mostly computing and data resources). Another crucial difference between Grid Computing and cloud computing is technological, i.e., the extensive usage of virtualization technologies in clouds. Virtualization facilitates the development of services on heterogeneous platforms, which was one of the key problems in grids, since most of the existing middlewares run on a limited variety of platforms. With virtualization technologies, the security also improves, since the applications can run isolated in virtual machines. Therefore, these applications do not have a lot of influence on the host system.

The emergence of so many different paradigms besides Grid computing is promising a very interesting future for this technology, and the impact will definitely be substantial on all areas of society.

References

1. Grid computing info center, http://www.gridcomputing.com (2008).
2. G. E. Moore, Cramming more components onto integrated circuits, Electronics (1965), pp. 114–117.
3. L. Roberts, Beyond moore's law: Internet growth trends, Computer 33 (1) (2000) 117–119.
4. I. Foster, C. Kesselman (Eds.), The Grid: Blueprint for a Future Computing Infrastructure, Morgan Kaufmann Publishers, San Francisco, USA, 1999.

5. C. S. Yeo, R. Buyya, M. D. de Assuncao, J. Yu, A. Sulistio, S. Venugopal, M. Placek, Utility computing on global grids, in: H. Bidgoli (Ed.), The Handbook of Computer Networks, Vol. III Part 1, John Wiley & Sons, New York, USA, 2007.

6. D. Abramson, J. Giddy, L. Kotler, High performance parametric modeling with nimrod/G: killer application for the global grid?, in: Proceedings of the 14th International Parallel and Distributed Processing Symposium(IPDPS'00), Cancun, Mexico, (2000), pp. 520–528.

7. W. Cirne, F. Brasileiro, J. Sauve, N. Andrade, D. Paranhos, E. Santos-Neto, R. Medeiros, Grid computing for bag of tasks applications, in: Proceedings of the 3rd IFIP Conference on E-Commerce, E-Business and E-Government, Sao Paolo, Brazil, (2003), pp. 591–609.

8. R. Buyya, D. Abramson, J. Giddy, Nimrod-G: An architecture for a resource management and scheduling system in a global computational grid, in: Proceedings of the 4th International Conference & Exhibition on High Performance Computing in Asia-Pacific Region (HPC Asia'00), Beijing, China, (2000), pp. 283–289.

9. D. P. Anderson, J. Cobb, E. Korpela, M. Lebofsky, D. Werthimer, SETI@home: An experiment in public-resource computing, Communications of the ACM 45 (11) (2002) 56–61.

10. L. B. Costa, L. Feitosa, E. Araujo, G. Mendes, R. Coelho, W. Cirne, D. Fireman, MyGrid: A complete solution for running bag-of-tasks applications, in: Proceedings of the Simposio Brasileiro de Redes de. Computadores (SBRC'04), Gramado, Brazil, (2004).

11. A. Chervenak, I. Foster, C. Kesselman, C. Salisbury, S. Tuecke, The data grid: Towards an architecture for the distributed management and analysis of large scientific datasets, Network and Computer Applications 23 (2001) 187–200.

12. W. Hoschek, F. J. Jaén-Martínez, A. Samar, H. Stockinger, K. Stockinger, Data management in an international data grid project, in: Proceedings of the 1st International Workshop on Grid Computing (Grid'00), Bangalore, India, (2000), pp. 77 – 90.

13. J. C. Jacob, D. S. Katz, T. Prince, G. B. Berriman, J. C. Good, A. C. Laity, E. Deelman, G. Singh, M.-H. Su, The montage architecture for grid-enabled science processing of large, distributed datasets, in: Proceedings of the Earth Science Technology Conference (ESTC'04), (2004).

14. M. J. Mineter, C. H. Jarvis, S. Dowers, From stand-alone programs towards grid-aware services and components: A case study in agricultural modelling with interpolated climate data, Environmental Modelling and Software 18 (4) (2003) 379–391.

15. LCG Computing Fabric Area, http://lcg-computing-fabric.web.cern.ch (2008).

16. Biogrid Project, http://www.biogrid.jp/e/project/index.html (2008).

17. International Virtual Observatory Alliance, http://www.ivoa.net (2008).

18. Avaki EII - Enterprise Data Integration Sofware, http://www.sybase.com/products/allproductsa-z/avakieii (2008).

19. K. Seymour, A. YarKhan, S. Agrawal, J. Dongarra, NetSolve: Grid enabling scientific computing environments, in: L. Grandinetti (Ed.), Grid Computing and New Frontiers of High Performance Processing, Vol. 14 of Advances in Parallel Computing, Elsevier, Netherlands, (2005), pp. 33–51.

20. L. Childers, T. Disz, R. Olson, M. E. Papka, R. Stevens, T. Udeshi, Access grid: Immersive group-to-group collaborative visualization, in: Proceedings of the 4th International Immersive Projection Technology Workshop, Ames, USA, (2000).

21. M. Cannataro, D. Talia, The knowledge grid, Communications of the ACM 46 (1) (2003) 89–93.

22. EU Data Mining Grid, http://www.datamininggrid.org (2008).

23. S. Graupner, J. Pruyne, S. Singhal, Making the utility data center a power station for the enterprise grid, Tech. Rep. HPL–2003–53, HP Labs, Palo Alto, USA (2003).

24. R. Buyya, S. Venugopal, The gridbus toolkit for service oriented grid and utility computing: An overview and status report, in: Proceedings of the 1st International Workshop on Grid Economics and Business Models (GECON'04), Seoul, Korea, (2004), pp. 19– 66.

25. S. Venugopal, R. Buyya, K. Ramamohanarao, A taxonomy of data grids for distributed data sharing, management and processing, ACM Computing Surveys 38 (1) (2006) 1–53.

26. H. Stockinger, Database replication in world-wide distributed data grids, Ph.D. thesis, Fakultät für Wirtschaftswissenschaften und Informatik, Universität Wien (2001).
27. M. Tang, B.-S. Lee, C.-K. Yeo, X. Tang, Dynamic replication algorithms for the multi-tier data grid, Future Generation Computer Systems 21 (5) (2005) 775–790.
28. V. Agarwal, G. Dasgupta, K. Dasgupta, A. Purohit, B. Viswanathan, DECO: Data Replication and Execution CO-Scheduling for Utility Grids, in: Proceedings of the 4th International Conference on Service Oriented Computing, Chicago, USA, (2006), pp. 52–65.
29. Sun Grid Engine, http://gridengine.sunsource.net (2008).
30. G. Sipos, P. Kacsuk, Multi-grid, multi-user workflows in the P-GRADE portal, Journal of Grid Computing 3 (3–4) (2005) 221–238.
31. H. Gibbins, K. Nadiminti, B. Beeson, R. Chhabra, B. Smith, R. Buyya, The Australian BioGrid Portal: Empowering the molecular docking research community, in: Proceedings of the 3rd APAC Conference and Exhibition on Advanced Computing, Grid Applications and eResearch (APAC'05), Gold Coast, Australia, (2005), pp. 26–30.
32. Maui Cluster Scheduler, http://www.clusterresources.com/pages/products/maui-cluster-scheduler.php (2008).
33. I. Foster, C. Kesselman, S. Tuecke, The anatomy of the grid: enabling scalable virtual organizations, High Performance Computing Applications 15 (3) (2001) 200–222.
34. L. Pearlman, V. Welch, I. Foster, C. Kesselman, S. Tuecke, A community authorization service for group collaboration, in: Proceedings of IEEE 3rd International Workshop on Policies for Distributed Systems and Networks, Monterey, USA, (2002).
35. B. Neuman, T. Ts'o, Kerberos: An authentication service for computer networks, IEEE Communications Magazine 32 (9) (1994) 33–38.
36. R. Housley, W. Polk, W. Ford, D. Solo, Internet X. 509 Public Key Infrastructure Certificate and Certificate Revocation List (CRL) Profile (2002).
37. R. Alfieri, R. Cecchini, V. Ciashini, L. dell'Agnello, A. Frohner, K. Lorentey, F. Spataro, VOMS, an authorization system for virtual organizations, in: Proceedings of the 1st European Across Grids Conference, Santiago de Compostela, Spain, (2003).
38. R. O. Sinnott, D. W. Chadwick, J. Koetsier, O. Otenko, J. Watt, T. A. Nguyen, Supporting decentralized, security focused dynamic virtual organizations across the grid, in: Proceedings of the 2nd IEEE International Conference on e-Science and Grid Computing, Amsterdam, Netherlands, (2006).
39. B. Nasser, R. Laborde, A. Benzekri, F. Barrere, M. Kamel, Dynamic creation of inter-organizational grid virtual organizations, in: Proceedings of the 1st IEEE International Conference on e-Science and Grid Computing, Melbourne, Australia, (2005).
40. A. A. E. Kalam, R. E. Baida, P. Balbiani, S. Benferhat, F. Cuppens, Y. Deswartes, A. Miege, C. Saurel, G. Trouessin, Organization based access control, in: Proceedings of the 4th International Workshop on Policies for Distributed Systems and Networks, Lake Como, Italy, (2003), pp. 120–131.
41. C. Morin, XtreemOS: A grid operating system making your computer ready for participating in virtual organizations, in: 10th IEEE International Symposium on Object/Component/Service-Oriented Real-Time Distributed Computing (ISORC 2007). IEEE, (2007), pp. 393 – 402.
42. M. Coppola, Y. Jégou, B. Matthews, C. Morin, L. P. Prieto, Ó. David Sánchez, E. Y. Yang, H. Yu, Virtual organization support within a grid-wide operating system, IEEE Internet Computing 12 (2) (2008) 20–28.
43. A. Sulistio, Advance reservation and revenue-based resource management for grid systems, Ph.D. thesis, The University of Melbourne, Australia (2008).
44. T. Roeblitz, F. Schintke, A. Reinefeld, Resource reservations with fuzzy requests, Concurrency and Computation: Practice & Experience (CCPE) 18 (13) (2006) 1681–1703.
45. M. Siddiqui, A. Villazon, T. Fahringer, Grid capacity planning with negotiation-based advance reservation for optimized QoS, in: Proceedings of the 2006 ACM/IEEE conference on Supercomputing (SC'06), Florida, USA, (2006), p. 21.
46. S. Naiksatam, S. Figueira, Elastic reservations for efficient bandwidth utilization in lambda-grids, Future Generation Computer Systems 23 (1) (2007) 1–22.

47. S. Venugopal, X. Chu, R. Buyya, A negotiation mechanism for advance resource reservation using the alternate offers protocol, in: Proceedings of the 16th International Workshop on Quality of Service (IWQoS'08, Twente, The Netherlands, (2008), pp. 40–49.

48. R. Buyya, D. Abramson, S. Venugopal, The grid economy, Proceedings of the IEEE 93 (3) (2005) 698–714.

49. W. Smith, I. Foster, V. Taylor, Scheduling with advanced reservations, in: Proceedings of the International Parallel and Distributed Processing Symposium (IPDPS'00), Cancun, Mexico, (2000), pp. 127–132.

50. R. L. Phillips, Pricing and Revenue Optimization, Stanford University Press, Pala Alto, CA, (2005).

51. S. Venugopal, R. Buyya, L. Winton, A grid service broker for scheduling e-science applications on global data grids: Research articles, Concurrency and Computation: Practice and Experience (CCPE) 18 (6) (2006) 685–699.

52. S. Lalis, A. Karipidis, JaWS: An open market-based framework for distributed computing over the internet, in: Proceedings of the 1st IEEE/ACM International Workshop on Grid Computing (Grid'00), Bangalore, India, (2000), pp. 87–106.

53. M. Stonebraker, R. Devine, M. Kornacker, W. Litwin, A. Pfeffer, A. Sah, C. Staelin, An economic paradigm for query processing and data migration in Mariposa, in: Proceedings of the 3rd International Conference on Parallel and Distributed Information Systems (PDIS'94), Austin, USA, (1994), pp. 58 – 68.

54. O. Regev, N. Nisan, The POPCORN Market – An online market for computational resources, in: Proceedings of the 1st International Conference on Information and Computation Economies (ICE'98), Charleston, USA, (1998), pp. 148 – 157.

55. C. A. Waldspurger, T. Hogg, B. A. Huberman, J. O. Kephart, W. S. Stornetta, Spawn: A distributed computational economy, Software Engineering 18 (2) (1992) 103–117.

56. P. Padala, C. Harrison, N. Pelfort, E. Jansen, M. Frank, C. Chokkareddy, OCEAN: The open computation exchange and arbitration network, a market approach to meta computing, in: Proceedings of the 2nd International Symposium on Parallel and Distributed Computing (IS-PDS'03), Ljubljana, Slovenia, (2003), pp. 185–192.

57. K. Lai, B. A. Huberman, L. Fine, Tycoon: A distributed market-based resource allocation system, Tech. Rep. arXiv:cs.DC/0404013, HP Labs, Palo Alto, USA (April 2004).

58. B. F. Cooper, H. Garcia-Molina, Bidding for storage space in a peer-to-peer data preservation system, in: Proceedings of the 22nd International Conference on Distributed Computing Systems (ICDCS'02), Vienna, Austria, (2002), pp. 372–381.

59. A. Sulistio, R. Buyya, A time optimization algorithm for scheduling bag-of-task applications in auction-based proportional share systems, in: Proceedings of the 17th International Symposium on Computer Architecture and High Performance Computing, Rio de Janeiro, Brazil, (2005), pp. 235–242.

60. J. Broberg, S. Venugopal, R. Buyya, Market-oriented grids and utility computing: The state-of-the-art and future directions, Journal of Grid Computing 6 (3) (2008) 255–276.

61. R. Moore, C. Baru, R. Marciano, A. Rajasekar, M. Wan, Data-intensive computing, the Grid: Blueprint for a new computing infrastructure, Morgan Kaufmann (1999) 105–129.

62. R. Sandhu, E. Coyne, H. Feinstein, C. Youman, Role-based access control models, computer (1996) 29(2) 38–47.

63. R. Thomas, R. Sandhu, Task-based authorization controls (TBAC): a family of models for active and enterprise-oriented authorization management, Database Security 11 (1998) 166–181.

64. B. Allcock, J. Bester, J. Bresnahan, A. Chervenak, I. Foster, C. Kesselman, S. Meder, V. Nefedova, D. Quesnel, S. Tuecke, Data management and transfer in high-performance computational grid environments, Parallel Computing 28 (5) (2002) 749–771.

65. A. Chervenak, M. Cai, Applying peer-to-peer techniques to grid replica location services, Journal of Grid Computing 4 (1) (2006) 49–69.

66. U. Čibej, B. Slivnik, B. Robič, The complexity of static data replication in data grids, Parallel Comput. 31 (8+9) (2005) 900–912.

67. W. H. Bell, D. G. Cameron, L. Capozza, A. P. Millar, K. Stockinger, F. Zini, Simulation of dynamic grid replication strategies in optorsim, in: Proc. IEEE Workshop on Grid Computing (Grid'2002), Springer Verlag, Lecture Notes in Computer Science, (2002), pp. 46–57.

68. W. H. Bell, D. G. Cameron, R. Carvajal-Schiaffino, A. P. Millar, K. Stockinger, F. Zini, Evaluation of an economy-based file replication strategy for a data grid, in: Proc. International Workshop on Agent based Cluster and Grid Computing, IEEE Computer Society Press, (2003), p. 661.

69. C. Nicholson, D. G. Cameron, A. T. Doyle, A. P. Millar, K. Stockinger, Dynamic data replication in lcg 2008, in: Proc. UK e-Science All Hands Meeting, (2006), pp. 1259–1271.

70. K. Ranganathan, I. Foster, Decoupling computation and data scheduling in distributed data-intensive applications, in: Proc. International Symposium on High Performance Distributed Computing, (2002), pp. 352–358.

71. I. Foster, Globus toolkit version 4: Software for service-oriented systems, Journal of Computer Science and Technology 21 (4) (2006) 513–520.

72. The European DataGrid Project, http://eu-datagrid.web.cern.ch/eu-datagrid (2008).

73. N. Karonis, B. Toonen, I. Foster, MPICH-G2: a Grid-enabled implementation of the message passing interface, Journal of Parallel and Distributed Computing 63 (5) (2003) 551–563.

74. M. Ripeanu, A. Iamnitchi, I. Foster, Cactus application: Performance predictions in grid environments, Lecture Notes in Computer Science (2001) 807–816.

75. J. Frey, T. Tannenbaum, M. Livny, I. Foster, S. Tuecke, Condor-G: A computation management agent for multi-institutional grids, Cluster Computing 5 (3) (2002) 237–246.

76. A. YarKhan, J. Dongarra, K. Seymour, GridSolve: The evolution of a network enabled solver, International Federation for Information Processing-Publications-IFIP 239 (2007) 215.

77. J. Yu, R. Buyya, A taxonomy of workflow management systems for grid computing, Journal of Grid Computing 3 (3) (2005) 171–200.

78. E. Deelman, J. Blythe, Y. Gil, C. Kesselman, G. Mehta, S. Patil, M. H. Su, K. Vahi, M. Livny, Pegasus: Mapping scientific workflow onto the grid, in: Across Grids Conference 2004, Nicosia, Cyprus, (2004), pp. 11–20.

79. I. Taylor, M. Shields, I. Wang, A. Harrison, Visual grid workflow in triana, Journal of Grid Computing 3 (3) (2005) 153–169.

80. G. von Laszewski, Java CoG kit workflow concepts for scientific experiments, Technical Report, Argonne National Laboratory, Argonne, IL, USA, (2005).

81. T. Oinn, M. Greenwood, M. Addis, M. Alpdemir, J. Ferris, K. Glover, C. Goble, A. Goderis, D. Hull, D. Marvin, et al., Taverna: lessons in creating a workflow environment for the life sciences, Concurrency and Computation 18 (10) (2006) 1067.

82. C. Goble, D. De Roure, myExperiment: social networking for workflow-using e-scientists, in: Proceedings of the 2nd workshop on Workflows in support of large-scale science, ACM Press New York, NY, USA, (2007), pp. 1–2.

83. Gridbus workflow homepage, http://www.gridbus.org/workflow/ (2008).

84. PBS Pro, http://www.pbsgridworks.com/ (2008).

85. B. Nitzberg, J. M. Schopf, J. P. Jones, PBS Pro: Grid computing and scheduling attributes, in: Grid Resource Management: State of the Art and Future Trends, Kluwer Academic Publishers, Norwell, MA, USA, (2004), pp. 183–190.

86. I. Foster, C. Kesselman, Globus: A metacomputing infrastructure toolkit, Supercomputer Applications 11 (2) (1997) 115–128.

87. J. MacLaren, HARC: The highly-available resource co-allocator, in: Proceedings of the International Conference on Grid Computing, High-PerformAnce and Distributed Applications (GADA'07), Vilamoura, Algarve, Portugal, (2007), pp. 1385–1402.

88. J. Gray, L. Lamport, Consensus on transaction commit, ACM Transactions on Database Systems (TODS) 31 (1) (2006) 133–160.

89. Moab workload manager, http://www.clusterresources.com/pages/products/moab-cluster-suite/workload-manager.php (2008).

90. J. MacLaren, Co-allocation of compute and network resources using HARC, in: Proceedings of Lighting the Blue Touchpaper for UK e-Science: Closing Conference of the ESLEA Project, Edinburgh, UK, 2007, p. 16.
91. A. Takefusa, M. Hayashi, N. Nagatsu, H. Nakada, T. Kudoh, T. Miyamoto, T. Otani, H. Tanaka, M. Suzuki, Y. Sameshima, W. Imajuku, M. Jinno, Y. Takigawa, S. Okamoto, Y. Tanaka, S. Sekiguchi, G-lambda: Coordination of a grid scheduler and lambda path service over GMPLS, Future Generation Computer Systems 22 (8) (2006) 868–875.
92. G-lambda, http://www.g-lambda.net (2008).
93. I. Foster, Globus toolkit version 4: Software for service-oriented systems, in: IFIP International Conference on Network and Parallel Computing (NPC'06)), Tokyo, Japan, (2006), pp. 2–13.
94. E. Mannie, RFC 3945: Generalized Multi-Protocol Label Switching (MPLS) Architecture, http://www.ietf.org/rfc/rfc3945.txt (Oct. 2004).
95. J. Geddes, S. Lloyd, A. Simpson, M. Rossor, N. Fox, D. Hill, J. Hajnal, S. Lawrie, A. McIntosh, E. Johnstone, et al., NeuroGrid: Using grid technology to advance neuroscience, in: Computer-Based Medical Systems, 2005. Proceedings. 18th IEEE Symposium on, (2005), pp. 570–572.
96. Medigrid project homepage, http://www.medigrid.de/ (2008).
97. X. Chu, A. Lonie, P. Harris, S. R. Thomas, R. Buyya, A service-oriented grid environment for integration of distributed kidney models and resources, Concurrency and Computation: Practice and Experience 20 (9) (2008) 1095–1111.
98. S. Amendolia, M. Brady, R. McClatchey, M. Mulet-Parada, M. Odeh, T. Solomonides, MammoGrid: Large-scale distributed mammogram analysis, The New Navigators: From Professionals to Patients (2003).
99. J. Jacob, R. Williams, J. Babu, S. Djorgovski, M. Graham, D. Katz, A. Mahabal, C. Miller, R. Nichol, D. Berk, et al., Grist: Grid data mining for astronomy, Astronomical Data Analysis Software and Systems (ADASS) XIV (2004).
100. Astrogrid-d project homepage, http://www.gac-grid.de/ (2008).
101. Australian virtual laboratory, http://aus-vo.org/ (2008).
102. Project grifin homepage, http://www.grifin.eu/ (2008).
103. D. Castelli, DILIGENT project homepage http://www.diligentproject.org/ (2008).
104. R. Buyya, C. Yeo, S. Venugopal, Market-oriented cloud computing: vision, hype, and reality for delivering IT services as computing utilities, in: Proceedings of 10th IEEE International Conference on High Performance Computing and Communications, (2008).
105. X. Chu, K. Nadiminti, C. Jin, S. Venugopal, R. Buyya, Aneka: Next-generation enterprise grid platform for e-Science and e-Business applications, in: e-Science and Grid Computing, IEEE International Conference on, (2007), pp. 151–159.

Chapter 5
Parallel Structured Adaptive Mesh Refinement

Jarmo Rantakokko and Michael Thuné

Abstract

Parallel structured adaptive mesh refinement is a technique for efficient utilization of computational resources. It reduces the computational effort and memory requirements needed for numerical simulation of complex phenomena, described by partial differential equations. Structured adaptive mesh refinement (SAMR) is applied in simulations where the domain is divided into logically rectangular patches, where each patch is discretized with a structured mesh. The purpose of adaptive mesh refinement is to automatically adapt the mesh to the resolution required to represent important features of the simulated phenomenon in different subdomains. In a parallel computing context, an important consequence of the adaptation is that the dynamically changing resolution leads to a dynamically changing work load, data volume, and communication pattern at run-time. This calls for dynamic load balancing and has implications for data placement as well as parallelization granularity.

This chapter gives an overview of structured adaptive mesh refinement approaches. After a brief introductory survey of SAMR techniques and software packages, the main part of the chapter addresses various issues related to implementation of SAMR on parallel computers. In particular programming models, data placement and load balancing are discussed, for shared memory as well as distributed memory platforms. Various approaches and algorithms are presented. The appropriate choice of dynamic load balancing algorithm, data placement strategy, programming model, etc., depends on both the application state and the computer platform. There

Jarmo Rantakokko
Department of Information Technology, Uppsala University, Sweden,
e-mail: Jarmo.Rantakokko@it.uu.se

Michael Thuné
Department of Information Technology, Uppsala University, Sweden,
e-mail: Michael.Thune@it.uu.se

R. Trobec et al. (eds.), *Parallel Computing*, DOI 10.1007/978-1-84882-409-6_5,
© Springer-Verlag London Limited 2009

is no single best alternative under all circumstances. Consequently, the chapter ends with an account of ongoing research where the objective is to equip SAMR-based simulation software with additional adaptivity, e.g., automatic selection of load balancing algorithms and automatic decision about level of parallelization granularity using a hybrid MPI/OpenMP programming model.

5.1 Introduction

Adaptive mesh refinement (AMR) is one of the core techniques in high-performance scientific computing. AMR is indispensable for simulation of phenomena of interest in science and engineering that exhibit large variation in scales [1]. Simulations where high-performance computers are applied to compute numerical solutions to mathematical models of various phenomena are used in numerous important contexts, such as climate modeling, vehicle design, combustion modeling, to mention a few.

AMR makes numerical simulation software self-configuring by automatically adjusting the accuracy of the simulation to a specified level. The "mesh" is the set of data points used in the simulation. With AMR the set of data points is repeatedly adapted at run-time. Points are added or removed to heuristically minimize the resource utilization for attaining a required numerical accuracy. This means that the mesh evolves dynamically at run-time.

Figure 5.1 shows an example from computational systems biology. This is a simulation of a biochemical "clock" known as the Circadian rhythm. Such clocks regulate the cell biochemistry in many organisms. The axes in Fig. 5.1 correspond to two molecular species, X and Y, respectively. The point (x, y) represents a state where there are x molecules of species X and y molecules of species Y. Figure 5.1 shows contour plots and underlying mesh points. The different contour plots correspond to different points in time. For each state (x, y), the graph for time t shows the probability that the biological system is in that state at that point in time. The series of graphs exhibit a periodicity pattern that constitutes the Circadian rhythm. This example is taken from [2] where the interested reader can find more details.

Presently, we focus on the adaptive mesh depicted in Fig. 5.1. Two things are of particular interest to note. First, the mesh is finer where the probability is dense. In this region there are steep gradients in the solution and consequently more mesh points are required for accurate resolution, compared to the regions with low probability density. Second, the refinements move with time, in response to the time-dependent behavior of the probability density. The adaptive mesh refinement algorithm presented in [2] achieves this automatically, without any need for human intervention. This is typical for adaptive mesh refinement techniques.

Realistic PDE models of phenomena related to nature, technical artifacts, economic and social systems, etc., are too complicated to be solved analytically. Software for simulation of such phenomena uses numerical algorithms to com-

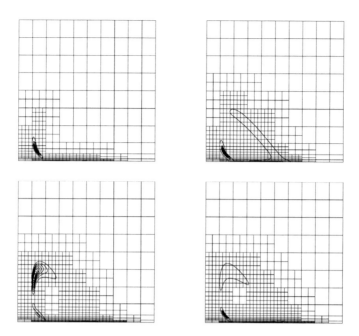

Fig. 5.1 Snap-shots from a simulation of the Circadian rhythm, a biological "clock." The contours show probability densities. The underlying rectilinear grid depicts the adaptive mesh used in the simulation. The mesh points are at the intersection of the grid lines. To make the picture clearer, we only show every eighth grid line in each dimension. For details, see [2].

pute approximate, numerical solutions to the PDEs. Approximate solution values are computed at each mesh point.

In the biological clock example described above and in numerous other, important applications, *structured* meshes are used. Here, "structured" means that mesh points can be indexed in such a way that neighbor relations between points can be inferred from the indices. By making use of this property, it is possible to design particularly efficient adaptive mesh refinement techniques for structured meshes. This is known as Structured Adaptive Mesh Refinement (SAMR). Such algorithms are the topic of this chapter. In particular we will discuss computer science issues related to the implementation and execution of SAMR-based simulations on parallel computers. Parallelization issues have always been central in the context of SAMR, since computationally demanding simulations of complex phenomena require execution on parallel high-performance computing platforms. The focus on parallel SAMR has been further accentuated in recent years, with the emergence of multicore processors.

5.2 An Introduction to SAMR

We begin by a broad overview of the issues involved. This will serve as a background for the more detailed presentation in later sections.

5.2.1 Approaches to Structured Adaptive Mesh Refinement

The purpose of adaptive mesh refinement is twofold: to reduce the amount of computational resources needed by numerical PDE solvers for the simulation of complex phenomena and to control the numerical errors in the simulation. A "numerical PDE solver" is a computer program that solves a set of PDEs numerically. The program's resource utilization in terms of computational effort and memory requirements depends on the number of mesh points. By using fewer mesh points in the simulation we get shorter execution time and need less memory. However, with fewer mesh points we may also get lower accuracy in the computed, approximate solution.

The challenge addressed by AMR is to use as few mesh points as possible to attain a given, desired accuracy. A numerical PDE solver computes numerical solutions to the PDEs for mesh points in a certain domain. By adding more mesh points in the domain we increase the resolution of the simulation. The key to AMR is the observation that the resolution required to attain the desired accuracy differs between different regions of the computational domain. Higher resolution is required in regions where the solution exhibits steep gradients, than in areas with relatively constant solution. Also, in the simulation of a time-dependent phenomenon the required resolution in a certain region can vary with time. An AMR algorithm will *automatically* add mesh points in regions where higher resolution is required and remove mesh points where less resolution is needed.

Figure 5.2 provides an illustration of how an AMR algorithm works. This example shows a snap-shot of a computation based on a so called block-wise SAMR algorithm that will be described in more detail later in this chapter. In this case the computational domain is covered by a regular pattern of rectangular blocks of mesh points. The simulated phenomenon is a pulse that is moving through the domain. As can be seen in Fig. 5.2, the blocks in the vicinity of the pulse have a higher resolution than the blocks where the solution is almost constant. The AMR algorithm has automatically set an appropriate resolution in each block, to meet the accuracy requirements defined by the user. Moreover, in this time-dependent simulation, the AMR algorithm will automatically increase the block resolution as the pulse moves into a new region and decrease the block resolution in the region that the pulse has left.

So what is particular about *structured* adaptive mesh refinement? Normally, a structured mesh takes the form of a logically rectangular grid.[1] A numerical PDE solver based on such grids can be implemented using array data structures to rep-

---

[1] In the following, "mesh" and "grid" will be used as synonymous concepts.

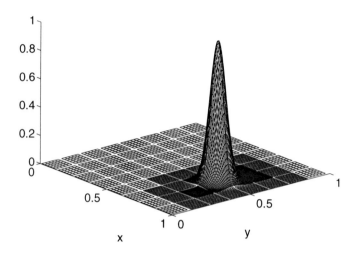

Fig. 5.2 A snap-shot of a SAMR-based simulation of a pulse moving through a domain. Note that the resolution is higher in the region containing the pulse than in the surrounding region with almost constant solution.

resent the mesh points and their associated solution values. The key to efficiency in PDE solvers based on structured grids is that the neighbor relations between mesh points can be inferred from the array indices in the data structure used to store the mesh. Due to this property the operation to retrieve the solution values at neighboring mesh points can be particularly efficiently implemented for structured meshes. This is essential in algorithms for solving PDEs numerically, since the solution at one mesh point depends on solution values at neighboring mesh points. From this point of view structured meshes are advantageous in simulations where the execution time is a critical bottle neck.

When adaptive mesh refinement is applied to numerical PDE solvers on structured meshes, it is important that the mesh refinement procedure preserves the structure of the mesh. This will preserve the efficiency of the execution of the numerical PDE solver. Also, the AMR procedure itself can be more efficiently implemented if it takes advantage of the mesh structure.

The general idea of AMR is to begin computing on a coarse grid. When a preliminary solution has been computed, the point-wise errors in the computed solution are estimated and mesh points where the accuracy is too low relative to a user-defined threshold are flagged. Subsequently, the grid is refined in areas around the flagged points. Then, a new, preliminary solution is computed on the refined grid and the refinement procedure is repeated recursively until a sufficiently accurate solution has been obtained.

In a time-dependent simulation, the AMR procedure is applied after each simulated time step. In this way, the grid continues to adapt to the evolving solution.

The most common approach to *structured* AMR is to adapt the grid by adding refined rectangular grid patches in areas where higher resolution is required and to remove such patches from areas where higher resolution is no longer needed. This approach was developed by Berger and Oliger [3] and was later modified by Berger and Colella [4].

Berger's approach consists of four steps:

1. Point-wise estimation of the errors in the computed solution.
2. Flagging of points where the accuracy is insufficient.
3. Clustering of flagged points.
4. Grid-fitting, i.e., insertion of higher resolution grid patches around such clusters.

The refinements are inserted as new, logically rectangular grid patches on top of the underlying coarser grid. When the refinement procedure is iteratively repeated, even finer patches may be superimposed on the first level of refinement patches, etc.

In the original procedure by Berger and Oliger [3], the superimposed grid patches were allowed to be arbitrarily oriented with respect to the underlying patches. This allows for patches being aligned to spatially directed phenomena of interest, such as shocks. However, this freedom of patch alignment introduces additional overhead, since the transfer of information between underlying and superimposed patches becomes more complicated. Consequently, Berger and Colella introduced the restriction that the boundaries of the various patches had to be parallel to each other, the so called Berger–Colella approach [4].

The result of the SAMR procedure described above is a hierarchical, composite, structured grid [5]. A numerical PDE solver based on SAMR needs to be able to compute the solution on such a grid. Most commonly, the simulations concern time-dependent phenomena. The solution is then computed at different, discrete time levels.

The steps in the SAMR procedure described above contribute to the overall execution time of the simulation. There is a trade-off between the gain in execution time by reducing the number of grid points and the loss in terms of overhead costs for the mesh adaptation.

In order to reduce the overhead, some authors have suggested a *block-wise* approach to SAMR [6–11], with the purpose to avoid the clustering and grid-fitting steps in the Berger–Colella approach. The initial, coarse grid is divided into a number of blocks. Subsequent refinements are carried out with respect to *entire* blocks. That is, if some points in a block are flagged for refinement, then the whole block is refined.

Finally, the Trompert–Verwer approach can be mentioned as an example of a structured adaptive mesh refinement algorithm that does *not* operate on logically rectangular patches [12]. Their basic idea is to avoid the clustering and grid fitting steps by refining only at flagged points. This leads to refinement patches in the form of *stair-shaped* grids. Consequently, a numerical PDE solver to be used with this kind of refinement has to be able to operate on such grids, which requires some additional book-keeping compared with calculations on rectangular patches. In our understanding, this approach has not been widely adopted, so it will not be further discussed here.

5.2.2 SAMR in a Computer Science Perspective

From a computer science point of view the challenges in SAMR are related to the data structures. Both the Berger–Colella approach and the block-wise approach result in hierarchical grids. Since the block-wise approach is a simplified version of the Berger–Colella approach we will base our discussion on the latter. Differences between the two approaches will be pointed out where this is of importance.

The key data structure in the Berger–Colella version of SAMR is the grid hierarchy. The hierarchy consists of logically rectangular, structured grid patches. Each patch is said to be on a certain level of refinement. Level 0 has the coarsest resolution. Moving from level l to level $l + 1$ means that the resolution increases by some factor.

Let us assume for simplicity that the computational domain can be covered by one structured grid. [2] Under this assumption there is only one level 0 patch. It covers the entire computational domain and constitutes the "base grid." In regions where finer resolution is needed, level 1 patches are superimposed on the base grid. The procedure is recursive, so that level $l + 1$ patches are superimposed on level l patches where further resolution is required.

The resulting grid hierarchy can be regarded as a tree structure. Its root is the base grid, level 0, and each patch on level $l + 1$ is the child of a level l patch.

Given the hierarchic grid data structure, we need to understand how to use it for solving PDEs numerically. Assume that we have computed a sufficiently accurate approximate solution at discrete time level t and that the solution is stored in a hierarchical grid resulting from the SAMR procedure indicated above. In order to advance the solution to the next discrete time level, $t + k$, we have to compute the solution on each of the refinement levels in the grid hierarchy. The computations begin on the base grid and then move to the finer levels in order of refinement.

Before we begin to update the solution on level $l + 1$, data from level l are interpolated to provide boundary data for level $l + 1$. When we have computed a more accurate solution on a level $l + 1$ patch, that solution will be projected onto the corresponding region of the underlying level l patch, to replace the less accurate solution there.

This procedure introduces data dependencies. *Within* a level of refinement, data at neighboring mesh points depend on each other. In addition, there are data dependencies *between* levels of refinement. In regions where different level patches intersect, the same PDE solution is represented on each of the different patches in

---

[2] More general cases can be handled in a similar way by the introduction of multi-block grids.

the intersection, with different resolution. This data dependency is manifested via the interpolation and projection operations discussed above.

Both the intra-level and inter-level data dependencies have consequences for serial data placement and parallel data distribution. This will be discussed in more detail in Sect. 5.4.1.

Existing parallel implementations of SAMR use a single program, multiple data approach. That is, all processors or threads execute the same program but on different partitions of the data. This means that the central step in parallelizing the SAMR computations is to partition the data and then to distribute the data over the available processors.

The base grid will initially be partitioned and distributed over the processors. The partitions will be generated by a grid partitioning algorithm with the objectives to balance the load and minimize the amount of synchronization and communication between the processors.

As the simulation proceeds, the adaptive mesh refinement will result in the addition of grid points in some regions of the grid and possibly in the removal of grid points from other regions. This will disturb the load balance significantly. For an example, see the case study in [13, Sect. 3].

To maintain parallel efficiency of the SAMR-based simulation, dynamic rebalancing of the load is required. Typically, this has to be done frequently during a simulation. As a consequence, parallel SAMR requires special dynamic load balancing algorithms that are considerably faster than algorithms for static load balancing. See Sect. 5.4.2 for a survey of algorithms for dynamic load balancing of SAMR applications.

The parallel implementation of SAMR applications can be realized in a distributed memory environment using explicit message passing (e.g., MPI) or in a global shared memory using a thread model (OpenMP or Pthreads). In either case, the grid patches or parts of the grid patches are assigned to different processes or threads using the SPMD approach, as described above. The parallelization models can also be combined into a hybrid approach using both the MPI and thread model. Then different grid patches are assigned to different MPI processes and within a MPI process the computations over a patch are further parallelized using threads.

5.2.3 Software Frameworks for SAMR

A large number of frameworks for solving partial differential equations using the technique of structured adaptive mesh refinement have been developed. Many of these are also freely available for downloading on the Internet. We will here consider some of these that are frequently mentioned in the literature.

The vast part of these are implementing the Berger–Colella algorithm with a hierarchy of refinement levels on top of each other. The parallelization model is a distributed memory model using MPI for message passing. In this group we have, e.g., CHOMBO [14], AMROC [15, 16], ENZO [13], GrACE [17], AMRLib [18], and SAMRAI [19].

In using the global shared memory parallelization model we have AMRCLAW [20] that has been parallelized in two levels with OpenMP [21]. Also the Racoon framework has been parallelized in two levels but using a hybrid of MPI and POSIX-Pthreads [6].

While AMRCLAW is implementing the Berger–Colella approach Racoon is using the block-wise approach, i.e., a flat grid divided into a number of blocks. At refinement of a block the block is split into smaller blocks which then are using a higher resolution. PARAMESH [10] is another framework with this block-wise approach. PARAMESH was originally developed on the Cray T3E using the SHMEM communication library but is now converted into using MPI as well.

In the examples above the applications and the adaptivity algorithms are integrated in the software framework. In the AGRIF framework [22] the application and the model independent parts are more loosely coupled. Here, the user specifies the application model in a specification file and then the model-dependent functions are created by the framework. The model independent parts, i.e., the adaptivity algorithms are included in AGRIF and coupled to the created model-dependent functions when building the AMR-solver.

5.3 Details of SAMR

We will now add some detail to the description given above. In particular, we elaborate on aspects of the SAMR algorithm that are of importance from a computer science point of view.

5.3.1 Advancing the Solution on a Structured Adaptive Grid Hierarchy

In the case of structured grids, the numerical PDE solver will typically be based on finite difference or finite volume approximations. Assume that the simulation has been going on for some time, and that the mesh adaptation algorithm has generated a grid hierarchy consisting of $L + 1$ refinement levels, where level 0 is the base grid, and level L contains the most refined patches. For simplicity, we first consider the example $L = 2$ and assume that each refinement increases the resolution by a factor of 2 in each space dimension.

In this case, the following procedure advances the solution from time t to $t + k$. First, one time step of size k is taken on the base grid. Then, the base grid solution is interpolated to the first level of refinement, to provide boundary data for the level 1 patches. Subsequently, one time step of size $k/2$ is taken on the level 1 patches.

Then, the level 1 solution is interpolated to the boundaries of the level 2 patches where two time steps of size $k/4$ are taken.

After one such pass through the grid hierarchy, the solution has been advanced to time $t+k$ on refinement level 0, and to time $t+k/2$ on level 1 and level 2. Since the solution on level 2 is expected to be more accurate than the one on level 1, it is projected to the level 1 patches, before the computation continues.

Next, another time step of size $k/2$ is taken on level 1, which brings the level 1 solution to time $t+k$. Interpolating from level 1 to provide boundary data for level 2, and then taking another two steps of size $k/4$ on level 2 brings also the finest level solution to time $t+k$. Subsequently, there is a projection of the solution from level 2 to level 1 and then from level 1 to level 0.

More generally, L is a non-negative integer and the resolution is refined by an integer constant $r(l)$ when moving from level l to level $l+1$. Then, the procedure advance(l, k) for advancing the solution on grid level l from time t to time $t+k$ can be recursively expressed as shown in Algorithm 5.1.

Algorithm 5.1 Basic steps of Structured Adaptive Mesh Refinement (SAMR)

```
advance(l, k):
  take one step of size k on level l
  if l=L then
    return
  else
    interpolate from level l to l+1
    for i=1 to r(l):
      advance(l+1,k/r(l))
    endfor
    project from level l+1 to level l
  endif
```

The procedure call advance(0, k) advances the solution from time t to time $t+k$ in the entire grid hierarchy.

Note that we have described the procedure in principle. Actual implementations may use variants of the algorithm. In particular, in the block-wise approach to SAMR, there is no grid hierarchy as in the general Berger–Colella approach. Each block has a certain level of refinement and covers a particular subdomain. The block contains the solution computed at its level of refinement on that subdomain. In addition the same block contains the corresponding solution computed at the next coarser level of refinement. This allows for a considerable simplification of the algorithm for advancing the solution in time.

In a parallel computing context, the interpolation and projection steps constitute synchronization points that can lead to serialization bottlenecks, degrading the parallel performance. This will be further discussed in Sect. 5.4.1.

5.3.2 The Algorithmic Key Components of SAMR

From a numerical analysis point of view, there are three key components in SAMR: *error estimation*, *mesh refinement*, and *conditions at internal boundaries*. We will now briefly indicate how each of these can be handled.

The point-wise error estimate is typically computed via Richardson extrapolation or similar techniques. Algorithmically, this means that to obtain error estimates on a grid patch $\mathscr{G}$, we compute two approximate solutions on $\mathscr{G}$, one using all mesh points and the other one using only every second mesh point. The information contained in the two solutions is combined to derive the error estimate.

In general, the error for a fixed mesh size will be largest where the solution changes most rapidly. Consequently, to attain a given accuracy, higher resolution is required in areas with large gradients than in regions where the solution changes more slowly.

When error estimates are available, the next step in the SAMR procedure is to flag all points where the point-wise error exceeds a given threshold. The threshold value is typically set by the user as an input parameter.

In the Berger–Colella approach to SAMR, the procedure continues with clustering of the flagged points and grid-fitting around clusters. Normally, the rectangular bounding box around a cluster will contain both flagged and non-flagged points. If the fraction of non-flagged points is large, then it is inefficient to cover the entire area inside the bounding box with a refined grid patch. Consequently, when initial clusters have been created, they will be subdivided into smaller ones. The standard way of dividing a cluster into two smaller clusters is to use the Berger–Rigoutsos algorithm [23]. The clusters are subdivided repeatedly until the ratio between non-flagged and flagged points is small enough. Finally, the area inside each bounding box is covered with a new, superimposed, refined grid patch.

In the block-wise approach to SAMR, the base grid is initially divided into equally sized blocks. Whenever one single point is flagged in a box, the entire box is refined. This simplification of the procedure avoids the overhead costs for clustering and grid-fitting.

An additional advantage of the block-wise approach is to allow for simpler data structure and data dependencies, compared to the tree structure in the Berger–Colella case.

In a variant of the block-wise approach, blocks are split after refinement to keep the number of mesh points equal in all blocks. This leads to a slightly more complicated data structure, but simplifies dynamic load balancing, as will be discussed in more detail below.

Finally, the insertion of patches or division of the base grid into blocks introduces artificial internal boundaries. It is necessary that the interpolation and projection procedures mentioned above preserve both the stability and accuracy of the numerical PDE solver and important properties of the solution across the internal boundaries. These are central research issues concerning SAMR methods from a numerical analysis point of view. From a computer science perspective, the interpolation across internal boundaries will introduce data dependencies between blocks.

These dependencies must be taken into account in data placement, load balancing, etc., in order to avoid performance bottlenecks.

5.4 Computer Science Aspects of SAMR

In Sect. 5.2 we indicated some computer science aspects of SAMR, related to data dependencies, dynamic load balancing, and parallelization models. After the introduction provided in the previous sections, we now have the background required for a more detailed discussion about these aspects.

5.4.1 Data Dependencies

For efficient execution, data should be placed in memory in such a way that locality properties due to data dependencies are respected. This is important for both serial and parallel execution.

In the serial case, data placement that allows for exploitation of spatial and temporal locality properties will lead to efficient utilization of the cache memory. This is also a prerequisite for high performance on each of the processors in a parallel computing environment.

What are the consequences of these observations in the case of SAMR? As described above, the structured adaptive mesh refinement results in a hierarchy of grid patches. This hierarchy is a tree structure. The root of the tree is the coarse base grid, level 0. Each patch on level l is the child node of a level $l-1$ patch. In other words, each level of mesh refinement constitutes a level of nodes in the tree structure.

The data dependencies between the nodes are as follows. Data need to be interpolated from parent nodes to child nodes in the tree, and data need to be projected from child nodes to parent nodes. Nodes on the same tree level need to exchange data if the corresponding patches share a boundary segment. In all these cases, nodes with data dependencies between them will be close to each other in the tree.

Consequently, for efficient execution, data should be placed in memory in such a way that the locality properties of the tree structure are preserved. Since storage in computer memory is linear, the tree structure has to be mapped to a linear representation that exhibits approximately the same locality pattern. One way to achieve this is to use *inverse space filling curves* for the mapping [24].

In a parallel computing context there are additional issues related to data dependencies. Most importantly, if the data dependencies described above are not handled properly they can easily lead to *serialization* of the parallel execution. This is due to the control flow in Algorithm 5.1, where one level at a time is updated. Before level $l+1$ can be updated, it has to receive data via interpolation from the recently updated level l, etc. As a consequence, if all level l patches are mapped to the same processor, P_k, then all other processors will be idle while P_k is busy updating the level l patches. Likewise, when the other levels are being updated, P_k will be idle.

Obviously, it is not a good idea to distribute the data as in this example. It is necessary that grid patches are mapped to processors in such a way that serialization bottlenecks are avoided. This can be done in different ways, with different pros and cons, as will become clear when we now turn to the issue of load balancing.

5.4.2 Dynamic Load Balancing

Due to the need for frequent re-balancing of the load in parallel SAMR, the challenge is to design dynamic load balancing algorithms that are very fast but still yield a data distribution of acceptable quality. There are several issues involved. The arithmetic work load should be evenly distributed over the processors. Synchronization and communication overhead should be low. Moreover, the overhead costs for the load balancing procedure itself, including the redistribution of data between processors, should be small.

For fast re-balancing of the load in parallel SAMR, it is advantageous to exploit the fact that the grids are structured. It is straightforward to subdivide a structured, rectangular grid patch across gridlines so that the result will be a set of smaller, structured, rectangular blocks of grid points. With this approach, partitioning of structured grids will result in partitions, where each partition consists of a single structured grid block or is a union of several such blocks. All the load balancing algorithms discussed below have this property.

Structured grids are rectangular. If the grid is Cartesian, it is strictly rectangular. However, the structured grid may also be curvilinear. Then, there will be a corresponding Cartesian grid on which the actual computations are carried out and a mapping from the strictly rectangular computational grid to the logically rectangular physical grid.

A single structured grid can only cover one, logically rectangular domain. Normally, real-life applications involve more complex domains. They are treated by decomposing the domain into a number of logically rectangular, possibly overlapping subdomains, each of which can be covered with a structured grid. It is then said that the complex domain is covered by a "composite, structured grid." The multiblock grids that are used, e.g., in aircraft simulations constitute an important special case.

Assume that we are going to carry out a simulation based on structured adaptive mesh refinement, using a composite, structured base grid. Then, there are various alternatives for the initial load balancing of the base grid (for an overview, see [25, 26]). All of these alternatives are relatively straightforward generalizations of load balancing techniques for single, structured grids. Consequently, without significant loss of generality, the remainder of this subsection will focus on the situation where there is a *single*, structured base grid.

Moreover, we consider structured adaptive mesh refinement of the Berger–Colella type (see Sect. 5.2.1). This means that the data structure to be distributed – so as to attain load balance – is a grid hierarchy consisting of an underlying base grid with one or several levels of superimposed refinement patches.

There are two basic strategies for addressing the load balancing problem in this context. One is *patch based*, which means that each patch is distributed over a group of processors, in a way that does not explicitly preserve the hierarchical relationship between patches (see, e.g., [10, 13, 27–29]). The second alternative is *domain based*, in the sense that the underlying *domain*, represented by the base grid, is partitioned and distributed over the processors (see, e.g., [30, 31]). Superimposed patches are distributed accordingly, so that each part of a patch is placed in the processor where it "geographically" belongs, according to the domain distribution.

Using a patch-based approach, the load balancing due to the introduction of new patches can be handled straightforwardly, by distributing each new patch over all processors, or over an appropriate subset of processors.

This alternative is not available in the domain-based case. There, a repartitioning of the entire grid hierarchy will typically be required in order to rebalance the load. Such domain-based repartitioning will amount to "cutting" through all levels of the grid hierarchy, as if there were only one single grid. In practice, only the base grid is partitioned. The impact of the superimposed patches is taken into account by theoretically mapping their work loads down to the base grid. Consequently, domain-based repartitioning of a structured grid hierarchy is equivalent to the partitioning of a single structured grid with inhomogeneous work load.

The two basic strategies are complementary to each other in terms of strengths and weaknesses. With the patch-based strategy it is relatively easy to balance the arithmetic work load. However, this approach will typically yield a significant communication overhead, due to the interaction between different refinement levels in the grid hierarchy. The domain-based approach, on the other hand, will avoid this inter-level communication, by preserving the hierarchical relationship between the grid patches. This comes at the price of getting a more difficult load balancing problem for the arithmetic work load.

As a consequence of the complementary properties of patch-based and domain-based approaches, *hybrid* techniques, combining elements of both approaches, are also of interest to explore (see, e.g., [26, 31, 32]).

We begin by considering the patch-based approach. There, the grid hierarchy can be regarded as a collection of n patches. A straightforward way of distributing such a collection over p processors is to divide each patch into p rectangular boxes, distributing one box to each processor. This means that the "partition" for each processor will consist of n boxes, one from each patch. It has been shown theoretically that this simple strategy works well in cases where the number of processors and patches is modest, and the individual patches are close to square in shape [33].

Another kind of patch-based strategy is to distribute the n patches according to some bin-packing algorithm or similar strategies such as greedy and round-robin algorithms. The simplest alternative is to assign complete patches to processors. However, this is likely to yield a bad arithmetic work load distribution and also scales badly to large numbers of processors. More sophisticated alternatives include splitting patches into smaller rectangular blocks, as described above, in order to balance the arithmetic work load properly. Finally, there are also ways to take the communication work load into account in a patch-based context. See [34] and [13] for examples of elaborate algorithms along these lines.

The domain-based alternative can be rephrased as the partitioning of a single, structured grid with inhomogeneous work load, as explained above. A classic way of treating such cases is the binary dissection algorithm proposed by Berger and Bokhari [35]. Here, the grid is divided into precisely p partitions, one for each processor, in such a way that the arithmetic load is well balanced. Later developments of the algorithm introduce a parameterization in order to take into account the trade-off between the arithmetic and the communication work loads [36]. In the original algorithm, the number of processors was assumed to be 2^v for some integer v. However, the generalization to arbitrary numbers of processors is straightforward [37, Sect. 4.1].

A more general way of handling the inhomogeneous partitioning problem was proposed by Rantakokko [38]. He introduced a framework within which a wide variety of algorithms can be conceived as special cases. The framework consists of three phases. The first phase is a structured division of the grid into a number of boxes. Here, the Berger–Bokhari algorithm can be used as an alternative. In general, any blocking scheme can be used. Moreover, it is not necessary to generate exactly p boxes. It is often advantageous to create $m > p$ boxes in order to get a better balanced arithmetic work load. In the case of recursive blocking, the recursion can be carried to different depth in different parts of the grid, depending on the variation of work load within the grid. The result of the blocking phase is a number of boxes, with neighborhood relations between them. This can be represented as a graph, where nodes correspond to boxes, and edges correspond to neighborhood relations. This graph will typically be *unstructured*. The second phase of Rantakokko's framework is to partition this box-graph into p partitions, using any technique for general graph partitioning. The result will be a number of partitions, where each partition consists of a collection of boxes. The number of boxes may vary between partitions. The partitions are mapped onto processors. Finally, as a third phase, a post processing step is carried out, where boxes residing on the same processor are being merged if possible, in order to avoid unnecessary overhead due to the box management.

This framework can be used in many different ways to yield specific algorithms. For example, Rantakokko used his framework for investigating a variety of algorithms for the partitioning of *static*, structured grids, both single grids [38] and multi-block grids [25].

Parashar and Brown [17, 30] and Steensland et al. [31, and references therein] used Rantakokko's framework for investigating domain-based algorithms for dynamic load balancing of grid hierarchies emanating from structured adaptive mesh refinement. The crucial difference between the static and the dynamic case is that the partitioning of the box-graph needs to be done very rapidly in the dynamic case. Techniques based on inverse space-filling curves constitute a promising alternative in this context [24,30,39,40]. The conclusion of Steensland's work on strictly domain-based algorithms is that they work well for "shallow" grid hierarchies, with up to three levels of refinement [41]. For deeper hierarchies, it becomes increasingly difficult to balance the arithmetic work load with a domain-based approach [31].

With a patch-based approach, the load balance can be kept under control even if there are many levels of refinement. On the other hand, with such an approach each new level of refinement will add to the cost for synchronization due to inter-level data dependencies.

Since patch-based and domain-based algorithms have complementary strengths and weaknesses, and both kinds of algorithms are less suitable for deep grid hierarchies, Steensland has suggested a hybrid approach [31]. There, the initial domain is coarsely decomposed into "natural regions." Subsequently, each natural region is partitioned with a partitioning algorithm that has been proven to be successful for that kind of region. For example, for a coarse grid region a regular blocking scheme may be used. For a region with few refinement levels, a strictly domain-based algorithm may be preferable, and for a region with many refinement levels, some blend of domain-based and patch-based approach can be applied. Steensland implemented a parameterized framework for this hybrid approach and reported promising results [31, Chapter 9]. The further exploration of these ideas is subject to continuing research.

Block-wise SAMR algorithms (see Sect. 5.2.1) can be considered as a simplified version of the Berger–Colella approach. The simplification can be exploited for dynamic load balancing. There are two variants. In one variant of block-wise SAMR [7–9], the number of blocks is kept fixed during the entire execution of the parallel SAMR code, but the work load associated with a block will vary as an effect of mesh refinement. In this case, the SAMR grid can be described as a very coarse and very regular graph of blocks, with varying node weights representing the work load associated with the blocks. Due to the small size of the graph it can be partitioned very quickly with inverse space-filling curve-based methods. It is even feasible to use more advanced graph partitioning algorithms that would be prohibitively expensive for dynamic load balancing in more general kinds of SAMR. In a comparison of different partitioning algorithms for this variant of block-wise SAMR, Steensland et al. found that an alternative based on Spectral Recursive Bisection actually gave the shortest overall execution time in some cases [42].

In the other variant of block-wise SAMR a block is subdivided whenever it is refined [6,10,11]. The effect is that the number of blocks grows but the number of grid points per block is constant. The latter simplifies load balancing. To partition grid level l, the level l blocks can be ordered in a sequence according to an inverse space filling curve index. Subsequently, a load balanced partitioning can be conveniently obtained by dividing the sequence of blocks into equally sized subsequences. Since all blocks have the same number of mesh points, all level l blocks will have the same workload, and equally sized partitions of the sequence of blocks will give a balanced workload among the partitions. For an example of this approach, see [6].

In summary, a large number of dynamic load balancing algorithms, patch based as well as domain based and hybrids, have been tried by various researchers. An important conclusion to be drawn is that *no single algorithm is the best alternative in all cases*. For this reason, Steensland et al. [31] proposed the development of a *meta-partitioner* for the dynamic load balancing of dynamic grid hierarchies. The meta-partitioner will be equipped with a variety of dynamic load balancing algorithms. During a simulation, whenever dynamic load balancing is called for, the meta-partitioner will assess the current state of the application and computer system. Based on that assessment, an appropriate load balancing strategy will be selected. All of the algorithms mentioned above are potential candidates for inclusion in such a tool. The actual selection of algorithms could either be based on ideas from the area of recommender systems (see, e.g., [43, 44]) or use simpler tabular mappings based on coarse classifications of load balancing algorithms with respect to various application and system state characteristics (see, e.g., [42,45]). The meta-partitioner idea will be further discussed in Sect. 5.5.5.

5.4.3 Parallelization Models

All the dynamic load balancing algorithms discussed above partition the grid hierarchy in such a way that each partition consists of one or several structured, rectangular blocks of grid points. The most common parallelization model for SAMR is in local name space using MPI. Then, the parallelization is coarse grained over entire blocks, i.e., each processor is responsible for one or more blocks. The computations are performed locally within each processor and block. The data dependencies between the partitions are handled with communication calls in MPI. To make the communication efficient and to avoid unnecessary communication/synchronization overheads the number of messages can be minimized by collecting data from several blocks that are going to the same processor into one large message. The communication can then be performed asynchronously with MPI_ISEND and MPI_PROBE in a *first-come-first-serve* order.

An alternative parallelization model is to use global name space and OpenMP directives. A straightforward implementation is to use a fine grained loop level parallelism within each block, i.e., each block is parallelized over all threads. Unfortunately, the parallel overheads very quickly become large and the performance degrades significantly [46]. A better alternative is to have a coarse grained parallelization over the blocks, i.e., processing locally one or more blocks in each thread. OpenMP gives some support for load balancing through the *schedule* directive in loops (e.g., schedule dynamic) but as we also need to keep the data locality this approach is not sufficient for our goals [47]. A remedy is to mimic the MPI parallelization, i.e., to only use the *parallel* directive and check if the particular block belongs to this thread's partition before executing the code on the thread. The partitions can be computed with an explicit load balancing algorithm as in the MPI case.

Combining the MPI approach and the OpenMP approach exploits both the coarse-grain and the fine-grain parallelism. We can parallelize with MPI over the blocks and use OpenMP within the blocks for loop-level parallelism. A similar approach has been implemented in Racoon [6] but by using a combined MPI-Pthreads parallelization. The two-level parallelization approach has further been developed for dynamic allocation of resources giving further improvements of performance, see Sect. 5.5.2. Finally, in [21] they have a two-level parallelization in OpenMP using nesting of the threads. The first level is a parallelization over the blocks and the second over the loops within a block, similarly to the combined MPI-OpenMP approach described above.

The different parallelization models are preferable in different settings depending on the application state (the grid hierarchy, refinement levels, number of patches, etc.) and the computer state (number of available processes and threads, memory characteristics, etc.). Generally, the coarse-grain parallelization model with MPI is preferable for cases with many small patches while the fine-grain loop-level parallelism is preferable for cases with few large patches. Note that the application state changes during run-time in SAMR applications, giving preference to different parallelization models in different phases within one run. For further discussion on this topic see Sects. 5.5.2 and 5.6.

5.5 Some Results

So far, we have given a general overview of parallel SAMR, citing results by various research groups, including some of our own work. Next, we will briefly mention some additional research results specifically from our own research group at Uppsala University.

5.5.1 An Integrated Decomposition and Partitioning Approach for Irregular Block-Structured Applications

Irregular block decompositions are commonly used in scientific applications where partial differential equations are solved numerically. For example, in structured multi-block methods the computational grid is decomposed into blocks and the blocks are fitted around or within an object. The blocks may then be of different sizes and connected to each other in an irregular fashion. In structured adaptive mesh refinement techniques we have irregular regions with high error. The flagged points, i.e., the high error points, are clustered together and a new refined level of grids with an irregular block decomposition is created. Similar techniques are also used in Ocean modeling. Here, we have an irregular geometry of water points but we still use a rectangular structured grid covering both land and water. The land points are then masked out in the computations. Still, the inactive points will consume both processor power and memory. It is then necessary to have an irregular block decomposition to cover the active points as efficiently as possible, minimizing the overheads associated with the inactive points.

Rantakokko developed an integrated block decomposition and partitioning method for irregularly structured problems arising in Ocean modeling [48]. The approach is not limited to Ocean modeling but is also suitable for other similar applications, e.g., structured adaptive mesh refinement applications. The algorithm consists of three steps or phases. The idea is to first cluster the water points in "dense" blocks, striving to get a block efficiency – i.e., the fraction of active points – above a given threshold. The next step is to distribute the blocks onto the processors with a small load imbalance ratio and a low number of inter-processor dependencies. The final step is to try to merge blocks on the same processor into larger rectangular blocks. The objective of the last step is to reduce the total number of blocks, since there is a small cost associated with each block, e.g., starting up loops, calling functions, and updating the block boundaries.

The domain decomposition method has been compared with the Berger–Rigoutsos grid clustering algorithm, [23], and the new method gives better results for the applications here. The new approach is simpler and faster but still gives a higher block efficiency, i.e., a higher fraction of active points in the blocks. A result is that it is not necessary to introduce the complexity of the Berger–Rigoutsos algorithm to get comparable or even better results. The distribution method has also been compared with other algorithms found in the literature, namely, the k-way partitioning algorithm in Metis, an inverse space filling curve-based method, and a bin-packing algorithm. A second result is that the new method outperforms the other algorithms. It gives a better load balance with fewer blocks and then less serial overhead in the solver, resulting in a shorter parallel solver execution time. The irregular block decomposition and partitioning algorithm has successfully been used by the Swedish Meteorological and Hydrological Institute to partition the Baltic Sea [49,50] in their operational Ocean model.

5.5.2 A Hybrid Dynamic MPI-OpenMP Model

All the previously mentioned parallelization approaches (see Sect. 5.4.3) have some drawbacks making them non-optimal in different settings. A parallelization on the block level has limited parallelism dictated by the number of blocks. A parallelization on the grid point level, i.e., within the blocks, has many synchronization points. Moreover, some blocks may be too small to parallelize giving excessive parallel overheads. A static mixed coarse-fine level parallelization increases the parallelism compared to block-level parallelization and decreases the parallel overheads compared to parallelization on the grid point level. Still, some blocks may be too small to parallelize with a fixed number of threads while some other blocks may be large enough to benefit from using more threads than were given from start. These conditions also change during run-time, e.g., at some state none of the blocks is suitable for loop-level parallelism while at another state most of the blocks would benefit from additional threads. This makes it hard for static parallelization/programming models to be efficient for structured adaptive mesh refinement methods.

Rantakokko suggested a dynamic MPI-OpenMP approach, [51], that sets the number of threads for each block individually depending on the size of the block. With this approach, the blocks are assigned to MPI-processes and then an appropriate number of threads are spawned for each block at run-time (using the function OMP_SET_NUM_THREADS before each parallel region). The most obvious strategy is to set the number of threads proportional to the number of grid points in the blocks. This approach gives the same work load per thread and each block can be processed in unit time, approximately. But, the parallel overhead grows significantly with the number of threads and the benefit of exploiting the low-level parallelism is lost if too many threads are used. Thus, a better strategy is to use proportionally fewer threads on the largest blocks and to give the blocks appropriate weights before load balancing between processes. What the optimal number of threads per block is depends on the absolute block sizes (not relative) and the computer system (synchronization overhead, cache performance, thread allocation time, etc.). While static hybrid approaches, with fixed number of processes and threads throughout the whole simulation, can fail to give good performance for some settings and stages of the grid hierarchy, the dynamic strategy adapts the number of threads continuously to the grid hierarchy and gives a high parallel efficiency throughout the whole simulation. This is shown for the test cases in [51].

5.5.3 Geographical Locality

The global name space model assumes that the grid hierarchy is placed in a shared memory to which all threads have equally fast access. Under this assumption, all threads with the same computational work load will execute with the same speed. In reality, the assumption is only fulfilled in shared memory systems with *uniform* memory access time, so called UMA systems.

Larger shared memory computers do not have the UMA property. In order to scale to large systems, the logically shared memory has to be physically distributed. Such a computer is typically built as a network or cluster of nodes, where each node can be regarded as a UMA system in itself. Since the computer has shared memory, all threads will be able to directly access all memory units, but the memory access time will be non-uniform. Accessing a memory item in the thread's local memory is faster than making an access to a remote memory unit. In such a non-uniform memory access (NUMA) system, threads with the same computational work load can execute with different speed, because of the differences in memory access time.

As a consequence, *geographical* locality becomes important for efficient execution of parallel SAMR codes on NUMA systems. Geographical locality means that a thread's partition of data is placed in the local memory of the node on which the thread is executing.

Markus Nordén et al. made an experimental study under carefully controlled circumstances, to measure the effect of geographical locality on the performance of a parallel, block-wise SAMR code [47]. For the experiment, four nodes of a

Sun Fire 15K system were used in dedicated mode. Each node was a four processor UMA system. In one part of the experiment, the parallel SAMR code was executed using four processors on a single node. This constituted a small UMA system. In another part of the experiment, the same code was executed on four different nodes, using one processor on each node. This was a NUMA system. Finally, a third part of the experiments used the same NUMA system, but added data migration directives to achieve geographical locality. Below this case will be denoted by NUMA-MIG.

In all parts of the experiment, care was taken to bind threads to specific CPUs, not allowing for automatic thread migration. The parallel SAMR code was written in Fortran 90 with OpenMP. A first-touch strategy was used to ensure that the initial placement of data preserved geographical locality. Each time the grid was adapted, dynamic load balancing was invoked to repartition and remap data to threads. In the NUMA case, the repartitioning and remapping operations were not followed by migration of data. Consequently, geographical locality was destroyed. In the NUMA-MIG case, on the other hand, a migration directive was inserted after each instance of dynamic load balancing, so that data were relocated in accordance with the new partitioning and mapping, to preserve geographical locality.

In short, the results of the experiments were as follows: while the execution time on the UMA system was ca. 4 hours, the same parallel SAMR execution on the NUMA system took more than 6.6 hours, whereas the NUMA-MIG case executed in ca. 4 hours. The difference in execution time between the UMA and the NUMA case depends on the NUMA ratio, i.e., the quotient between remote and local memory access time. On the computer system used for these experiments the NUMA ratio was ca. 2.

It was expected that the UMA system would give the shortest execution time. Nordén et al. were primarily interested in two issues: How much longer would the execution time be on the NUMA system, due to the lack of geographical locality? Would the active migration of data in the NUMA-MIG case pay off, or would the overhead for migration be prohibitively large?

The experimental results summarized above show that there is a significant penalty in not preserving geographical locality. Most importantly, the results demonstrate the viability of active data migration to preserve the locality. The NUMA-MIG experiments gave the same execution time as those carried out in UMA mode, implying that the overhead for data migration was negligible.

The conclusion of these experiments is that for OpenMP-based parallel SAMR implementations on large shared memory machines it would be highly desirable to be able to give directives about data migration. In the absence of such directives, multi-threading via POSIX is the alternative for those who wish to have full control over data placement to preserve geographical locality [6].

It is not necessary to have *explicit* data placement directives. In the NUMA-MIG experiments we used Sun Microsystem's `migrate-on-next-touch` directive that was available in the Sun Fire 15K system. The directive was inserted after the dynamic load balancing operation, with the effect that after each instance of load balancing, data items were migrated according to the first-touch principle. Only data that had actually been assigned to a new thread were moved. This is a convenient way of preserving geographical locality. With such a directive added to OpenMP, the competitiveness of OpenMP-based parallel SAMR implementations would increase significantly.

5.5.4 A Hybrid Patch/Domain-Based Partitioner Framework

Both patch-based and domain-based partitioners have inherent shortcomings for partitioning different grid hierarchies in SAMR, as described above. In particular, deep grid hierarchies including many refinement levels are difficult to partition. In [31] a hybrid partitioner *Nature+Fable* for SAMR is presented. The hybrid approach combines both patch-based and domain-based algorithms to take advantage of their complementary strengths. Nature+Fable is a partitioning framework providing a number of parameters that can be tuned to represent different partitioning algorithms.

First a pre-partitioning step is used to generate coarse partitions that are mapped to a group of processors. The coarse partitions are then further partitioned within the processor groups. Here, the grid is separated into un-refined and refined regions. The un-refined regions are partitioned with a patch-based approach while the refined regions are further separated into bi-levels, i.e., two overlaying refinement levels that are grouped together. The bi-levels are then partitioned with a domain-based approach. A set of parameters governs the partitioning process yielding different partitioning algorithms for different settings of the parameters.

Partitioning outcomes from a suite of test cases show that the hybrid approach gives results that are comparable with the domain-based approach in terms of communication volumes and comparable to the patch-based approach in terms of load balance [31]. These results are very encouraging as the hybrid approach can be further tuned for each setting of the application state [52].

5.5.5 A Meta-Partitioner for Structured Grid Hierarchies

The meta-partitioner was briefly mentioned in Sect. 5.4.2. It is based on the observation that no single partitioning algorithm will be the best alternative in all circumstances. On the contrary, the state of the grid hierarchy may change so much due to the mesh adaptation that different partitioning algorithms will be preferable at different instances of dynamic load balancing during a single run of a parallel SAMR code.

The idea, then, is that when dynamic load balancing is required, the parallel SAMR code will call the meta-partitioner. Based on information about the current state of the application and computer system the meta-partitioner will automatically select an appropriate partitioning algorithm that it will subsequently invoke to carry out the re-partitioning of the SAMR grid hierarchy. At the next instance of dynamic load balancing during the same run of the parallel SAMR code, the meta-partitioner will be called again, possibly invoking another partitioning algorithm, etc.

Henrik Johansson has recently implemented the meta-partitioner as a collection of CCA components [53]. [3] One of the meta-partitioner components is a data base with stored application states collected from a series of benchmarking experiments. For each of the stored states, performance data for a variety of partitioning algorithms are available in the data base. Another meta-partitioner component matches the current application state to the stored application states in the data base. The meta-partitioner selects the partitioning algorithm that gave the best performance for those stored application states that were closest to match the current application state [55].

Johansson's meta-partitioner is currently being tested using partitioning algorithms generated within the Nature+Fable framework. The preliminary results are promising.

5.6 Conclusions and Future Work

In this chapter, we have given a brief overview of structured adaptive mesh refinement techniques, with particular emphasis on issues related to their implementation on parallel computers. Two definite conclusions can be drawn from the existing literature on these topics.

First, methods based on adaptive mesh refinement significantly enhance the possibilities to simulate phenomena that exhibit large variation in scales. Climate modeling, vehicle design, and combustion in 3D are three examples from the very long list of cases with this property. Due to their usefulness in simulations of such phenomena, AMR-based methods are among the core techniques in computational science and engineering.

The second conclusion from the existing research on parallel SAMR is that the appropriate choice of dynamic load balancing algorithm, data placement strategy, programming model, etc., depends on both the application state and the computer platform. This is because the characteristics of the application change dynamically during the course of the simulation. In particular, in a parallel SAMR-based simulation there may be significant changes in data volume and communication-to-computation ratio at run-time. As a consequence, the overall execution time would decrease if different load balancing algorithms were used at different stages of the simulation.

The dynamically changing characteristics of parallel SAMR-based simulations are the rationale for the meta-partitioner project discussed above. The meta-partitioner adds an element of *self-configuration* to the parallel SAMR code in selecting the load balancing algorithm automatically at run-time. In addition, the meta-partitioner implies *self-optimization*, since different load balancing algorithms

---

[3] CCA is the Common Component Architecture, see [54].

will be selected at different instances of load re-balancing during a single run of the parallel SAMR software, with the purpose to reduce the execution time.

Self-configuration and self-optimization are two of the key aspects of *autonomic computing* [56]. In our continued work we aim to equip parallel SAMR software with additional elements of autonomicity. Primarily, we will consider two further kinds of self-optimization. The first is to make the software able to automatically select an appropriate amount of parallel resources. The second is to make the parallel SAMR software able to automatically select details of the parallelization model. The software would ideally be able to make adjustments in parallel resource allocation and parallelization model at run-time, in response to the dynamically changing characteristics of the simulation.

Making parallel SAMR codes increasingly autonomic will contribute to making them even more attractive. First, self-configuration and self-optimization will be instrumental in ensuring efficient execution for a variety of simulation cases and parallel computing platforms without special hand-tuning of the code for each case and/or platform. Second, a certain degree of autonomicity that ensures good performance without human intervention is essential to make parallel SAMR accessible to a broad community of scientists and engineers who want to simulate phenomena of interest in their areas of research but are not experts in parallel computing.

In conclusion, parallel SAMR is by now a core technique in computational science and engineering. Appropriately implemented it significantly widens the scope for simulations of important phenomena, for example in combustion and climate modeling. Current and future research activities aiming to add elements of autonomicity to the software will further increase the potential of parallel SAMR techniques.

Acknowledgments Our research on parallel SAMR has been conducted in cooperation with Ph.D. students and colleagues at the Department of Information Technology, Uppsala University. In particular, we want to thank former and present Ph.D. students Henrik Johansson, Henrik Löf, Markus Nordén, Stefan Söderberg, and Johan Steensland for their contributions. We also thank our colleagues Lars Ferm, Erik Hagersten, Sverker Holmgren, and Per Lötstedt.

References

1. E. Steinthorsson, D. Modiano, Advanced methodology for simulation of complex flows using structured grid systems, Tech. Rep. 95-28, ICOMP, NASA Lewis Research Center, Cleveland, OH (1995).
2. L. Ferm, P. Lötstedt, P. Sjöberg, Conservative solution of the Fokker–Planck equation for stochastic chemical reactions, BIT 46 (2006) 561–583.
3. M. J. Berger, J. Oliger, Adaptive mesh refinement for hyperbolic partial differential equations, Journal of Computational Physics 53 (1984) 484–512.
4. M. J. Berger, P. Colella, Local adaptive mesh refinement for shock hydrodynamics, Journal of Computational Physics 82 (1989) 64–84.
5. M. J. Berger, Data structures for adaptive grid generation, SIAM Journal on Scientific and Statistical Computing 7 (1986) 904–916.

6. J. Dreher, R. Grauer, Racoon: A parallel mesh-adaptive framework for hyperbolic conservation laws, Parallel Computing 31 (2005) 913–932.
7. L. Ferm, P. Lötstedt, Blockwise adaptive grids with multigrid acceleration for compressible flow, AIAA J. 37 (1999) 121–123.
8. P. Lötstedt, S. Söderberg, Parallel solution of hyperbolic pdes with space-time adaptivity, in: D. H. R. Vilsmeier, F. Benkhaldour (Ed.), Finite Volumes for Complex Applications II, Hermes Science, Paris, 1999, pp. 769–776.
9. P. Lötstedt, S. Söderberg, A. Ramage, L. Hemmingsson-Fländén, Implicit solution of hyperbolic equations with space-time adaptivity, BIT 42 (2002) 128–153.
10. P. MacNeice et al, PARAMESH: A parallel adaptive mesh refinement community toolkit, Computer Physics Communications 126 (2000) 330–354.
11. K. G. Powell et al., A solution-adaptive upwind scheme for ideal magnetohydrodynamics, Journal of Computational Physics 154 (1999) 284–309.
12. R. A. Trompert, Local uniform grid refinement for time-dependent partial differential equations, Ph.D. thesis, University of Amsterdam (1994).
13. Z. Lan, V. E. Taylor, G. Bryan, A novel dynamic load balancing scheme for parallel systems, Journal of Parallel and Distributed Computing 62 (2002) 1763–1781.
14. P. Colella, D. T. Graves, N. D. Keen, T. J. Ligocki, D. F. Martin, P. W. McCorquodale, D. Modiano, P. O. Schwartz, T. D. Sternberg, B. V. Straalen, Chombo software package for AMR applications: Design document, Available at the Chombo website: http://seesar.lbl.gov/ANAG/chombo/ (September 2008).
15. R. Deiterding, Parallel adaptive simulation of multi-dimensional detonation structures, Ph.D. thesis, Brandenburgische Technische Universität Cottbus (2003).
16. R. Deiterding, Detonation structure simulation with AMROC, in: L. Y. et. al. (Ed.), High Performance Computing and Communications, No. 3726 in Lecture Notes in Computer Science, Springer, Berlin Heidelberg, (2005), pp. 916–927.
17. M. Parashar, J. Browne, System engineering for high performance computing software: The HDDA/DAGH infrastructure for implementation of parallel structured adaptive mesh refinement, in: Structured Adaptive Mesh Refinement Grid Methods, Volume 117 of IMA Volumes in Mathematics and its Applications, Springer-Verlag, Berlin (2000), pp. 1–18.
18. C. Rendleman, V. Beckner, M. Lijewski, W. Crutchfield, J. Bell, Parallelization of structured, hierarchical adaptive mesh refinement algorithms, Computing and Visualization in Science 3 (2000) 147–157.
19. A. Wissink, R. Hornung, S. Kohn, S. Smith, N. Elliott, Large scale parallel structured AMR calculations using the SAMRAI framework, in: Proceedings of Supercomputing 2001, Denver, USA, (2001).
20. M. Berger, R. LeVeque, Adaptive mesh refinement using wave-propagation algorithms for hyperbolic systems, SIAM Journal of Numerical Analysis 35 (1998) 2298–2316.
21. R. Blikberg, T. Sørevik, Load balancing and OpenMP implementation of nested parallelism, Parallel Computing 31 (2005) 984–998.
22. L. Bebreu, C. Vouland, E. Blayo, AGRIF: Adaptive grid refinement in Fortran, Computers and Geosciences 34 (2008) 8–13.
23. M. J. Berger, I. Rigoutsos, An algorithm for point clustering and grid generation, IEEE Transactions on Systems, Man and Cybernetics 21 (1991) 1278–1286.
24. J. Pilkington, S. Baden, Dynamic partitioning of non-uniform structured workloads with spacefilling curves, IEEE Transactions on Parallel and Distributed Systems 7 (3) (1996) 288–300.
25. J. Rantakokko, Partitioning strategies for structured multiblock grids, Parallel Computing 26 (12) (2000) 1161–1680.
26. M. Thuné, Partitioning strategies for composite grids, Parallel Algorithms and Applications 11 (1997) 325–348.
27. D. Balsara, C. Norton, Highly parallel structured adaptive mesh refinement using language-based approaches, Journal of parallel computing 27 (2001) 37–70.
28. R. D. Hornung, S. Kohn, The SAMRAI homepage, structured adaptive mesh refinement applications infrastructure, http://www.llnl.gov/CASC/SAMRAI/.

29. J. J. Quirk, A parallel adaptive grid algorithm for computational shock hydrodynamics, Applied Numerical Mathematics 20 (1996) 427–453.

30. M. Parashar, J. C. Browne, On partitioning dynamic adaptive grid hierarchies, presented at HICSS-29 (1996).

31. J. Steensland, Efficient partitioning of dynamic structured grid hierarchies, Ph.D. thesis, Uppsala University (2002).

32. Z. Lan, V. Taylor, G. Bryan, Dynamic load balancing of SAMR applications on distributed systems, in: Proceedings of Supercomputing 2001, (2001).

33. M. Thuné, Straightforward partitioning of composite grids for explicit difference methods, Parallel Computing 17 (1991) 665–672.

34. H. Johansson, A. Vakili, A patch-based partitioner for parallel SAMR applications, accepted for publication in the proceedings of the IASTED International Conference on Parallel and Distributed Computing and Systems, November 2008.

35. M. J. Berger, S. Bokhari, A partitioning strategy for non-uniform problems on multiprocessors, IEEE Transactions on Computers 85 (1987) 570–580.

36. S. H. Bokhari, T. W. Crockett, D. M. Nicol, Binary dissection: Variants & applications, Tech. Rep. ICASE Report No. 97-29, NASA Langley Research Center, Hampton, VA (1997).

37. J. Rantakokko, Strategies for parallel variational data assimilation, Parallel Computing 23 (1997) 2017–2039.

38. J. Rantakokko, A framework for partitioning structured grids with inhomogeneous workload, Parallel Algorithms and Applications 13 (1998) 135–152.

39. C.-W. Ou, S. Ranka, Parallel remapping algorithms for adaptive problems, Journal of Parallel and Distributed Computing 42 (1997) 109–121.

40. J. Steensland, Dynamic structured grid hierarchy partitioners using inverse space-filling curves, Tech. Rep. 2001-002, Uppsala University, Department of Information Technology, Uppsala, Sweden (2001).

41. J. Steensland, M. Thuné, S. Chandra, M. Parashar, Towards an adaptive meta-partitioner for parallel SAMR applications, in: Proceedings of the IASTED International Conference on Parallel and Distributed Computing Systems, Las Vegas, (2000), pp. 425–430.

42. J. Steensland, S. Söderberg, M. Thuné, Comparison of dynamic load balancing techniques for a parallel SAMR algorithm, in: T. Sørevik, F. Manne, R. Moe, A. H. Gebremedhin (Eds.), Applied Parallel Computing—New Paradigms for HPC in Industry and Academia, Springer-Verlag, Heidelberg, (2001), pp. 160–169, (Lecture Notes in Computer Science, Vol. 1947).

43. E. N. Houstis et al., PYTHIA-II: A knowledge/database system for managing performance data and recommending scientific software, ACM TOMS 26 (2000) 227–253.

44. N. Ramakrishnan, C. J. Ribbens, Mining and visualizing recommendation spaces for elliptic PDEs with continuous attributes, ACM TOMS 26 (2000) 254–273.

45. S. Chandra, Armada: A framework for adaptive application-sensitive runtime management of dynamic applications, Master's Thesis, Graduate School, Rutgers University, NJ (2002).

46. J. Rantakokko, Comparison of parallelization models for structured adaptive mesh refinement, in: M. Danelutto, D. Laforcena, M. Vanneschi (Eds.), Lecture Notes in Computer Science 3149, Springer-Verlag, Heidelberg (2004), pp. 615–623.

47. M. Nordén, H. Löf, J. Rantakokko, S. Holmgren, Geographical locality and dynamic data migration for OpenMP implementations of adaptive PDE solvers, in: Lecture Notes in Computer Science 4315, (2008), pp. 382–393.

48. J. Rantakokko, An integrated decomposition and partitioning approach for irregular block-structured applications, in: J. Romlin et al. (Ed.), Proceedings of the IEEE International Parallel and Distributed Processing Symposium, IPDPS 2000, Springer–Verlag, Berlin, (2000), pp. 485–496, Lecture Notes in Computer Science, Vol. 1800.

49. T. Wilhelmsson et al., Increasing resolution and forecast length with a parallel ocean model, in: Proceedings of the Second EuroGOOS International Conference, (1999).

50. T. Wilhelmsson, J. Schüle, Running an operational baltic sea model on the T3E, in: Proceedings of the Fifth European SGI/Cray MPP Workshop, Cineca, Bologna, (1999).

51. J. Rantakokko, A dynamic MPI-OpenMP model for structured adaptive mesh refinement, Parallel Processing Letters 15 (2005) 37–47.

52. H. Johansson, Performance characterization and evaluation of parallel PDE solvers, Licentiate Thesis 2006-010, Department of Information Technology, Uppsala University (2006).
53. H. Johansson, Design and implementation of a dynamic and adaptive meta-partitioner for parallel SAMR grid hierarchies;, Technical Report 2008-017, Department of Information Technology, Uppsala University (2008).
54. The Common Component Architechture, http://www.cca-forum.org/.
55. L. Li, B. Norris, H. Johansson, L. C. McInnes, J. Ray, Component infrastructure for managing performance data and runtime adaptation of parallel applications, accepted for publication in the Proceedings of PARA2008, Trondheim, Norway, 2008.
56. J. O. Kephart, D. M. Chess, The vision of autonomic computing, IEEE Computer 36 (1) (2003) 41–50.

Chapter 6
Applications and Parallel Implementation of QMC Integration

Peter Jez, Andreas Uhl and Peter Zinterhof

Abstract

In this chapter we deal with numerical computation of integrals over the domain $\mathbb{R}^s$ ($s > 1$) with respect to a positive weight function. For one-dimensional integrals Gauss Hermite formulas compute integrals with respect to a Gaussian weight with quite high accuracy but for high-dimensional integrals the effort increases exponentially. For integrals over the s-dimensional unit cube probabilistic methods like Monte Carlo (MC) are not affected by this so-called "curse of dimensions," but the convergence rate is rather poor. If the integration nodes are not pure random points but special deterministic point sequences (the method is called Quasi Monte Carlo (QMC) due to this fact) this rate can be significantly improved. These low-discrepancy sequences appear also in the computation of integrals over $\mathbb{R}^s$.

In the first part of the chapter we review some theoretic results about QMC integration over $[0, 1)^s$. Then the integration over $\mathbb{R}^s$ is considered for the case of a Gaussian type weight function. In the second part we discuss the execution of these computations in a parallel environment. Due to the convergence rate of QMC integration (which of course depends on the smoothness of the integrand also) the integrand must be evaluated on a huge amount of integration nodes. To speed up this calculation it is split into (more or less) independent tasks which run on different processing elements (PES). After reviewing the state of the art in parallel and distributed QMC, we execute experiments for three different concepts of parallelization: Blocking, leaping and parametrization.

Peter Jez
Department of Computer Sciences, University of Salzburg, J.-Haringer-Strasse 2,
5020 Salzburg, Austria, e-mail: peter.jez@sbg.ac.at

Andreas Uhl
Department of Computer Sciences, University of Salzburg, J.-Haringer-Strasse 2,
5020 Salzburg, Austria, e-mail: uhl@cosy.sbg.ac.at

Peter Zinterhof
Department of Computer Sciences, University of Salzburg, J.-Haringer-Strasse 2,
5020 Salzburg, Austria, e-mail: peter.zinterhof@sbg.ac.at

R. Trobec et al. (eds.), *Parallel Computing*, DOI 10.1007/978-1-84882-409-6_6,

In the last part an error estimation for the computation in a parallel environment based on the diaphony is presented and applied to one of the test functions used in the experiments.

6.1 Introduction

For the numerical computation of integrals over an interval or the real line several methods are available. The most simple one is the computation of an approximative Riemann sum (the so-called rectangular rule). The most accurate methods are the Gauss quadrature formulas based on the zeroes of orthogonal polynomials. Whereas the usability of these methods for one-dimensional integration is without doubt, they fail for higher dimensions. The reason of this fact is the exponentially growing effort for increasing the dimensionality. To minimize this so-called "curse of dimensions" other integration methods were developed: The so-called Monte Carlo (MC) and Quasi Monte Carlo (QMC) methods [1]. Whereas in the MC case the integration nodes are produced by a random number generator (RNG), low-discrepancy point sets and sequences (e.g., (t,m,s)-nets or (t,s)-sequences [2]) are employed in QMC algorithms. QMC techniques improve the probabilistic error bounds of MC techniques especially in higher dimensions. Nevertheless, these techniques are related [3] since a full-period random number sequence may be seen as a low-discrepancy point set (e.g., a rank-1 lattice rule in the case of a linear congruential generator) as well.

In applications, integration problems over the s-dimensional unit cube often arise. So we review the basics about these methods over this important domain and recall ways to extend the results to integration problems over the real line. A famous error estimation for QMC integration is the Hlawka–Koksma inequality. Unfortunately, the application of this estimator often fails due to the nonexistent total variation in the sense of Hardy and Krause or its difficult computation. For integrands of special function classes, the so-called reproducing kernel Hilbert spaces (RKHS), it is possible to get analogous error estimations, which are easy to compute. They also deliver a measure of the quality of a point sequence, the so-called diaphony. We give an example of an RKHS the basis of which is related to Hermite polynomials. This space delivers also a generalized diaphony as an error estimation for integrands from this RKHS with respect to a weight function of Gaussian type.

High-dimensional numerical integration problems may require a significant amount of computation power. Therefore, substantial effort has been invested in finding techniques for performing these computations on all kinds of parallel architectures (see [4–7] for an exhaustive overview). In order to minimize the communication within a parallel system, each processing element (PE) requires its own source of integration nodes. Therefore, the aim is to investigate QMC techniques for using separately initialized and disjoint sets of integration nodes on a single PE.

In this chapter, we propose and evaluate techniques to compute QMC integrals over the real line (in case of a Gaussian-type weight function) on parallel systems. Sect. 6.2 reviews QMC integration methods over $[0, 1]^s$; these results are extended to the case of integration over the real line in Sect. 6.3 and examples from Physics and Financial Engineering are given. In Sect. 6.4 we provide an overview of using QMC techniques on parallel systems. Experimental results are provided in Sect. 6.5 employing Good Lattice Points (GLP) and Zinterhof sequences as QMC point sets for which different distribution strategies are evaluated. Finally, Sect. 6.6 shows an application of the concept of diaphony to provide analytic error estimations for heterogeneous parallel systems and Sect. 6.7 concludes this chapter.

6.2 Monte Carlo and Quasi Monte Carlo Methods in Numerical Integration Over $[0, 1)^s$

The problem is as follows: Suppose $f(x) : [0, 1)^s \to \mathbb{R}$ is given. The task is the computation of

$$I = \int_{[0,1)^s} f(x)dx. \tag{6.1}$$

The MC quadrature formula is quite simple: Choose N random points $\{x_k\}_{k=1}^N$ (the sequence of integration node points) uniformly distributed in the s-dimensional unit cube. Then, an approximation of the integral is given by

$$I \approx I_{approx} = \frac{1}{N} \sum_{k=1}^N f(x_k). \tag{6.2}$$

The dimension of the domain does not change the formula. The problem is shown by the so-called "main theorem of MC integration": The error $|I - I_{approx}|$ has the magnitude

$$|I - I_{approx}| = O\left(\frac{1}{\sqrt{N}}\right),$$

which can be shown by the central limit theorem of probability. Now the following question arises: Can this poor convergence rate be improved, at least for special classes of integrands? The answer to this question is given by the QMC integration methods. The difference between MC and QMC is that not arbitrary sequences of random points are used but sequences of points with special properties. A measure of the "quality" of a point sequence is the so-called discrepancy of a sequence [8].

Definition 6.1. Let $\{x_k\}_{k=1}^N = \left\{\left(x_k^{(1)}, \ldots, x_k^{(s)}\right)\right\}_{k=1}^N$ be a finite sequence of points in $\mathbb{R}^s$. Then the number

$$D_N\left(\{x_k\}_{k=1}^N\right) := \sup_{I \subseteq [0,1)^s} \left| \frac{\#(x_k : x_k \in I)}{N} - \int_I dx \right| \tag{6.3}$$

is the discrepancy of the sequence $\{x_k\}_{k=1}^N$. I runs through all subintervals of the form $[\alpha_1, \beta_1) \times \ldots \times [\alpha_s, \beta_s)$ of the unit cube.

The term of discrepancy delivers a characterization of the distribution of the sequence [8]:

Theorem 6.1. *A sequence $\{x_k\}_{k=1}^N$ is uniform distributed modulo 1 if and only if*

$$\lim_{N \to \infty} D_N\left(\{x_k\}_{k=1}^N\right) = 0$$

If we permit only intervals of the form

$$I = [0, a_1) \times \ldots \times [0, a_s)$$

in (6.3), the resulting number is called the *-discrepancy $D_N^*\left(\{x_k\}_{k=1}^N\right)$ of the sequence $\{x_k\}_{k=1}^N$. Another type of discrepancy is given by the so-called L_2 discrepancy $D_N^{(2)}\left(\{x_k\}_{k=1}^N\right)$:

$$D_N^{(2)}\left(\{x_k\}_{k=1}^N\right) :=$$

$$= \left(\int_{[0,1)^s} \left| \frac{\#(x_k : x_k \in [0,y_1) \times \ldots \times [0,y_s))}{N} - y_1 y_2 \ldots y_s \right|^2 dy_1 \ldots dy_s \right)^{\frac{1}{2}}$$

The square of the L_2 discrepancy can be computed explicitly by Warnock's formula (see [9]):

$$\left(D_N^{(2)}\left(\{x_k\}_{k=1}^N\right)\right)^2 = \frac{1}{3^s} - \frac{1}{N 2^{s-1}} \sum_{k=1}^N \prod_{j=1}^s \left(1 - \left(x_k^{(j)}\right)^2\right) +$$

$$+ \frac{1}{N^2} \sum_{k,l=1}^N \prod_{j=1}^s \left(1 - \max\left(x_k^{(j)}, x_l^{(j)}\right)\right)$$

An efficient algorithm to compute the L_2 discrepancy was provided by Heinrich [10]. For detailed reference regarding discrepancies see [8]. The importance of discrepancies in numerical integration is given by the famous Hlawka–Koksma inequality. This inequality estimates the error of the numerical computation of the integral (6.1) by a quadrature formula (6.2) by a product where one factor depends only on the integrand (to be more precise, it is the total variation in the sense of Hardy and Krause of the integrand) and the second factor is the *-discrepancy of the used sequence:

$$\left| \sum_{k=1}^N f(x_k) - \int_{[0,1)^s} f(x)dx \right| \leq V(f) D_N^*\left(\{x_k\}_{k=1}^N\right)$$

The disadvantage of this estimation is the existence and the computation of the total variation. To get an idea of the quantity of the discrepancy we present some results here. A lower bound is provided by Roth [11]: Every sequence of N points in $\mathbb{R}^s$ with $s \geq 2$ satisfies

$$D_N^* > \frac{1}{2^{4s}} \sqrt{\frac{1}{((s-1)log2)^{s-1}} \frac{(\log N)^{\frac{s-1}{2}}}{N}},$$

and every infinite sequence satisfies the inequality

$$D_N^* > C_s \frac{(\log N)^{\frac{s}{2}}}{N},$$

where C_s is a constant only depending on the dimension s. An upper bound is given by the famous Erdös–Turan inequality: For $h \in \mathbb{Z}$ denote $\bar{h} = \max(|h|, 1)$. Then an upper bound of the discrepancy is given by the following inequality:

$$D_N \leq 2s^2 3^{s+1} \left(\frac{1}{m} + \sum_{h_1,\ldots,h_s=-m;h_i \neq 0}^{m} \frac{1}{\bar{h}_1 \ldots \bar{h}_s} \left| \frac{1}{N} \sum_{n=1}^{N} e^{2\pi i(h_1 x_k^{(1)} + \ldots + h_s x_k^{(s)})} \right| \right)$$

In one dimension the *-discrepancy can be computed by a formula provided by Niederreiter:

$$D_N^* = \frac{1}{2N} + \max_{i=1,\ldots,N} \left| x_i - \frac{2i-1}{2N} \right|$$

For the proof and further details to discrepancies we refer to [12]. Sequences which satisfy

$$D_N^* = O\left(\frac{\log^s N}{N} \right)$$

are called "low-discrepancy sequences." They deliver the best results in numerical integration. Famous examples of these sequences are

- Halton sequence [13]: The *-discrepancy of this sequence satisfies

$$D_N^* \leq C_s \frac{\log^s N}{N} \tag{6.4}$$

with a constant C_s depending on the dimension. If the dimension increases the constant grows super exponentially. Niederreiter [2] showed

$$\lim_{s \to \infty} \frac{\log C_s}{s \log s} = 1$$

- Faure sequence [14]: The *-discrepancy satisfies an inequality analog to (6.4), but the constant decreases very fast when $s \to \infty$.
- Sobol sequence [15]
- (t,m,s)-nets of Niederreiter [2]: These sequences are generalized Sobol sequences
- "Good Lattice Points" by Korobow [16] and Hlawka [17]

- The monothetic or Weyl sequences [18]: Let $\theta = (\theta_1, \ldots, \theta_s) \in \mathbb{R}^s$ with θ_i algebraic numbers independent over $\mathbb{Q}$. Then Niederreiter [19] showed that the sequence $\{x_n\}_{n=1}^N = (n\theta_1, \ldots, n\theta_s)$ has a discrepancy $D_N = O\left(\frac{1}{N^{1-\varepsilon}}\right)$. For more information about these sequences see also [20].

A special case of the Weyl sequences are the Zinterhof sequences [21], which we use in our experiments: Let $(r_1, r_2, \ldots, r_s) \in \mathbb{Q}^s$ with $r_i \neq r_j \neq 0$ for all $i, j = 1, 2, \ldots, s$ and $\theta = (e^{r_1}, e^{r_2}, \ldots, e^{r_s})$. For computational aspects of these low-discrepancy sequences applied to financial engineering, that is, implementation examples and computational results see [22].

We use also the Good Lattice Points by Korobow and Hlawka and that's why we give a more detailed description of these point sequences here. A different name for them is "optimal coefficients" which was introduced by Korobow (see [16]), the name "Good Lattice Points" was introduced by Hlawka (see [17]). Let $\{x\}$ be the fractional part of x and recall the definition of this sequence (as above we use $\overline{m} = \max(|m|, 1))$ [23]:

Definition 6.2. Let $p > 1$ be an integer and let $a_v(p), v = 1, 2, \ldots, s$ be integers with $gcd(a_v(p), p) = 1$ for all $v = 1, \ldots, s$. Let $\delta_p(m)$ be defined by

$$\delta_p(m) = \begin{cases} 1 & m \equiv 0 \bmod p \\ 0 & else \end{cases}$$

If there are constants $\beta = \beta(s)$ and $C = C(s)$ with

$$\sum_{k_1, k_2, \ldots, k_s = -(p-1)}^{p-1} \frac{\delta_p(k_1 a_1 + \ldots + k_s a_s)}{\overline{k_1} \ldots \overline{k_s}} \leq C \frac{\log^\beta p}{p}$$

for infinitely many values p then we call $a_1, \ldots, a_s$ optimal coefficients with index β. For prime N the sequence $(x_k)_{k=1}^N \in [0,1)^s$ of the GLP is now given by

$$x_k = \left(\left\{\frac{a_1 k}{N}\right\}, \ldots, \left\{\frac{a_s k}{N}\right\}\right)$$

Korobow showed that for all prime numbers p there are integers $a_1, \ldots, a_s$ with

$$\sum_{k_1, k_2, \ldots, k_s = -(p-1)}^{p-1} \frac{\delta_p(k_1 a_1 + \ldots + k_s a_s)}{\overline{k_1} \ldots \overline{k_s}} \leq \frac{2(3 + 2\log p)^s}{p}$$

and provided a method for the computation of optimal coefficients for $N = p_1 p_2$ with p_1, p_2 prime numbers.

6.2.1 Application of Reproducing Kernel Hilbert Spaces

A very important aspect of numerical integration is the error estimation. The preferred estimator has the shape of the Hlawka–Koksma estimator: One term depends only on the integrand, the second one rates the quality of the point sequence. The disadvantage of the Hlawka–Koksma estimator is the difficult computation of the total variation in the sense of Hardy and Krause. For special function classes it is very easy to get these desired error estimations. These function classes are the RKHS. We recall their definition here [24].

Definition 6.3. Let H be a Hilbert space of complex-valued functions defined on a set F with inner product $< .,. >$. If there is a function $K(x,y) : F \times F \rightarrow \mathbb{C}$ with the properties

1. the functions $g_y(x) := K(x,y)$ are elements of H for all $y \in F$
2. $K(x,y) = \overline{K(y,x)}$
3. for all $f(x) \in H$ and all $y \in F$ we have $f(y) = < f(x), K(x,y) >$ (reproducing property)

Then H is called an RKHS with kernel $K(x,y)$.

In the following we assume that there is an element $g \in H$ which fulfills the equation

$$\int_{[0,1)^s} f(x)dx = < f,g > \tag{6.5}$$

for all $f \in H$. The following computation leads to an error estimation for numerical integration of functions from an RKHS H. Let $f \in H$ and consider

$$\left| \frac{1}{N} \sum_{k=1}^{N} f(x_k) - \int_{[0,1)^s} f(x)dx \right|$$

The reproducing property and (6.5) leads to

$$\left| \frac{1}{N} \sum_{k=1}^{N} f(x_k) - \int_{[0,1)^s} f(x)dx \right| = \left| < f(y), \frac{1}{N} \sum_{k=1}^{N} K(y,x_k) - g(y) > \right| \leq$$

$$\leq \|f\| \cdot \left\| \frac{1}{N} \sum_{k=1}^{N} K(y,x_k) - g(y) \right\| =$$

$$= \|f\| \cdot \left(\frac{1}{N^2} \sum_{k,l=1}^{N} K(x_k,x_l) - \frac{2}{N} \sum_{k=1}^{N} \Re g(x_k) + \|g\|^2 \right)^{\frac{1}{2}} = \|f\| \cdot r_N$$

The factor r_N is called the g-diaphony of the sequence $\{x_k\}_{k=1}^{N}$ [25].

We mention one famous example: For $v = (n_1, \ldots, n_s) \in \mathbb{Z}^s$ denote $\overline{v} = \overline{n_1} \ldots \overline{n_s}$, let a be a positive integer and

$$H := \left\{ f(x) : [0,1)^s \to \mathbb{C}, f(x) = \sum_{v \in \mathbb{Z}^s} a_v \frac{e^{2\pi i v x}}{\overline{v}^a}, \sum_{v \in \mathbb{Z}^s} |a_v|^2 < \infty \right\}$$

We define a scalar product on H by

$$< \frac{e^{2\pi i v x}}{\overline{v}^a}, \frac{e^{2\pi i \mu x}}{\overline{\mu}^a} >= \delta_{v\mu}$$

Then a reproducing kernel is given by

$$K_a(x,y) = \sum_{v \in \mathbb{Z}^s} \frac{e^{2\pi i v(x-y)}}{\overline{v}^{2a}}$$

By Hurwitz' representation of Bernoulli polynomials $B_k(x)$ (for more information about Bernoulli polynomials see [26] and the references there) the kernel can be written in closed form. With $\{x\} = x \bmod 1$ we have

$$K_a(x,y) = \prod_{j=1}^{s} \left(1 + (-1)^a \frac{(2\pi)^{2a}}{(2a)!} B_{2a}(\{x_j - y_j\}) \right)$$

In this particular case the element g in (6.5) is equal to 1 and for $a = 1$ the resulting term for r_N is equal to the weighted spectral test of the sequence $\{x_k\}_{k=1}^N$ [27].

6.3 QMC Methods for Integrals over $\mathbb{R}^s$ with a Weight Function

QMC integration methods are used most often over the domain $[0,1)^s$. They can also be applied to integrals over $\mathbb{R}^s$ with a weight function. This problem was studied in [28]. In our experiments we consider integrals with the weight function

$$w(x) : \mathbb{R}^s \to \mathbb{R}, w(x) = \prod_{j=1}^{s} e^{-x_j^2}, \tag{6.6}$$

where $w(x)$ is the density of a Gaussian measure up to a constant and it is separable. Integrals of this type arise quite often in statistics, mathematical finance and physics. The method can also be applied to other weight functions which are separable and positive. For QMC integration we need a sequence of points in $\mathbb{R}^s$ with suitable distribution properties. A well-known method to get such a sequence for separable weight functions is the inversion method [29]. We demonstrate this method on our weight function (6.6): Our task is the computation of

$$I = \int_{\mathbb{R}^s} f(x) e^{-|x|^2} dx \tag{6.7}$$

The strategy is to transform (6.7) to an integral over the unit cube. Set

$$y_i = F(x_i) = \frac{1}{\sqrt{\pi}} \int_{-\infty}^{x_i} e^{-t^2} dt \qquad i = 1, \ldots, s \qquad (6.8)$$

Then we get

$$I = \int_{\mathbb{R}^s} f(x) e^{-|x|^2} dx = \sqrt{\pi}^s \int_{(0,1)^s} f\left(F^{-1}(y_1), \ldots, F^{-1}(y_s)\right) dy$$

The domain of integral on the right-hand side is the unit cube and therefore we can apply the classical QMC methods: Let $\{y_k\}_{k=1}^N = \left\{\left(y_k^{(1)}, \ldots, y_k^{(s)}\right)\right\}_{k=1}^N$ be a sequence of points in the unit cube. Then an approximation of (6.7) is given by

$$I \approx \frac{\sqrt{\pi}^s}{N} \sum_{k=1}^N f\left(F^{-1}\left(y_k^{(1)}\right), \ldots, F^{-1}\left(y_k^{(s)}\right)\right) \qquad (6.9)$$

In the following we give an example of an RKHS where we can deduce an analog to the classical diaphony. The kernel is delivered by the well-known Mehler formula. We use the abbreviation: For $\lambda = (\lambda_1, \ldots, \lambda_s) \in \mathbb{R}^s$ and $v = (v_1, \ldots, v_s) \in \mathbb{N}_0^s$ we use

$$\lambda^v := \lambda_1^{v_1} \ldots \lambda_s^{v_s}$$

Let $H_n(x)$ be the the Hermite polynomial of degree n defined by

$$H_n(x) := (-1)^n e^{x^2} \frac{d^n}{dx^n} e^{-x^2}$$

These polynomials form an orthogonal system with respect to the weight function $w(x) = e^{-x^2}$. From these polynomials we get an orthonormal basis of the $L^2(\mathbb{R}^s)$: Let $v = (n_1, \ldots, n_s) \in \mathbb{N}_0^s$ and

$$\Phi_v(x) = \Phi_v(x_1, \ldots, x_s) = \prod_{j=1}^s \frac{H_{n_j}(x_j)}{\sqrt{2^{n_j} n_j! \sqrt{\pi}}}$$

Then by the orthogonality of the Hermite polynomials and

$$\int_{-\infty}^{+\infty} H_n^2(x) e^{-x^2} dx = 2^n n! \sqrt{\pi},$$

we have orthonormality

$$\int_{\mathbb{R}^s} \Phi_v(x) \Phi_\mu(x) e^{-|x|^2} dx = \delta_{n_1 m_1} \ldots \delta_{n_s m_s}.$$

We consider the following function space H_λ depending on an s-dimensional parameter $\lambda = (\lambda_1, \ldots, \lambda_s)$ with $|\lambda_i| < 1$ given by

$$H_\lambda = \left\{ f(x) : \mathbb{R}^s \to \mathbb{C} : f(x) = \sum_{v \in \mathbb{N}_0^s} a_v \Phi_v(x) \lambda^v, \sum_{v \in \mathbb{N}_0^s} |a_v|^2 < \infty \right\} \qquad (6.10)$$

with the inner product defined by

$$< \Phi_v(x)\lambda^v, \Phi_\mu(x)\lambda^\mu > = \delta_{n_1 m_1} \dots \delta_{n_s m_s}$$

The reproducing kernel $K(x,y)$ is now delivered by the s-dimensional variant of Mehler's formula [30]:

$$K(x,y) := \sum_{v \in \mathbb{N}_0^s} \Phi_v(x)\Phi_v(y)\lambda^{2v} = \frac{1}{\sqrt{\pi^s}} \prod_{j=1}^{s} \frac{1}{\sqrt{1-\lambda_i^4}} e^{\frac{2x_j y_j \lambda_j^2 - \lambda_j^4(x_j^2 + y_j^2)}{1-\lambda_j^4}}$$

We will apply now the reproducing property to the error estimation for the approximate computation of the integral (6.7) for a function $f(x) \in H_\lambda$ by the formula (6.9):

$$\left| \frac{\sqrt{\pi}^s}{N} \sum_{k=1}^{N} f(x_k) - I \right| =$$

$$= \left| < f(y), \frac{\sqrt{\pi}^s}{N} \sum_{k=1}^{N} K(y,x_k) > - a_{(0,\dots,0)} \sqrt{\sqrt{\pi}}^s \right| =$$

$$= \left| < f(y), \frac{\sqrt{\pi}^s}{N} \sum_{k=1}^{N} K(y,x_k) - \sqrt{\sqrt{\pi}}^s \, \Phi_{(0,\dots,0)}(y) > \right| \le$$

$$\le \|f\| \cdot \left\| \frac{\sqrt{\pi}^s}{N} \sum_{k=1}^{N} K(y,x_k) - \sqrt{\sqrt{\pi}}^s \, \Phi_{(0,\dots,0)}(y) \right\|$$

The last inequality is a consequence of the Cauchy–Schwarz inequality. $\|.\|$ denotes the norm induced from the inner product $< .,. >$. The last factor can be computed to

$$\left\| \frac{\sqrt{\pi}^s}{N} \sum_{k=1}^{N} K(y,x_k) - \sqrt{\sqrt{\pi}}^s \, \Phi_{(0,\dots,0)}(y) \right\| =$$

$$= \left[\frac{\pi^s}{N^2} \sum_{k,l=1}^{N} K(x_k,x_l) - \sqrt{\pi}^s \right]^{\frac{1}{2}} =: R_N\left(\{x_k\}_{k=1}^{N} \right)$$

This is an analogon to the classical diaphony defined on the s-dimensional unit cube.

Before we present the numerical experiments we show two applications of integrals of this type.

6.3.1 Feynman's Path Integrals

We start with the following problem (see [31]): Let $\kappa > 0$ and s be a given initial time. Consider the initial value problem for the function $T(x,t) : \mathbb{R}^3 \times \mathbb{R} \to \mathbb{R}$ with

$$\frac{\partial T}{\partial t} = -\kappa \Delta T - U(x)T \qquad t > s \tag{6.11}$$
$$T(x,s) = T_0(x)$$

The solution can be written in the form

$$T(x,t) = \int_{\mathbb{R}^3} K(x,t;y,s)T_0(y)dy$$

The kernel $K(x,t;y,s)$ is expressed via the discrete action $S(x_0,x_1,\ldots,x_{n+1}) : \mathbb{R}^3 \times \ldots \times \mathbb{R}^3 \to \mathbb{R}$ defined by

$$S(x_0,\ldots,x_{n+1}) = \sum_{j=0}^{n} \left(\frac{1}{4\kappa} \left(\frac{x_{j+1}-x_j}{\Delta t} \right)^2 + U(x_j) \right) \Delta t \tag{6.12}$$

So we have Feynman's formula for the kernel $K(x,t;y,s)$:

$$K(x,t;y,s) = \int_{\mathbb{R}^{3n}} e^{-S(x_0,x_2,\ldots,x_{n+1})} \frac{dx_1 dx_2 \ldots dx_n}{(4\pi\kappa\Delta t)^{\frac{3n}{2}}} \tag{6.13}$$

with $x_0 = x$ and $x_{n+1} = y$. For the time interval $[s,t]$ we use the equidistant composition $s = t_0 < t_1 \ldots < t_{n+1} = t$ with $t_j = s + j\Delta t$. We observe the following: For small Δt the dimensionality of the integrand in (6.13) is quite high. With a slight modification we get the initial value problem for Schrödingers equation from (6.11). Therefore we must replace the time t by the imaginary time $\frac{it}{\hbar}$ and the constant $\kappa = \frac{\hbar^2}{2m}$ with $\hbar = \frac{h}{2\pi}$. h denotes Planck's constant of action and has the value

$$h = 6.626 \times 10^{-34} Js$$

To transform the integral (6.13) we introduce new variables $a_1, a_2, \ldots, a_n$ by

$$a_{i+1} = x_{i+1} - x_i \qquad i = 1,\ldots,n-1$$
$$a_1 = x_1$$

or equivalently

$$x_i = \sum_{k=1}^{i} a_k \qquad i = 2,\ldots,n$$
$$x_1 = a_1$$

Remark The new variables a_i are of course elements of $\mathbb{R}^3$: $a_i = (a_i^{(1)}, a_i^{(2)}, a_i^{(3)})$. For easier writing we use $a_i^2 = |a_i|^2$ and $a_i + a_j = (a_i^{(1)} + a_j^{(1)}, a_i^{(2)} + a_j^{(2)}, a_i^{(3)} + a_j^{(3)})$. The discrete action (6.12) is written in these new variables in the following form:

$$S(x, a_1, \ldots a_n, y) = \left(\frac{1}{4\kappa} \left(\frac{a_1}{\Delta t} \right)^2 + U(x) \right) \Delta t +$$

$$+ \sum_{j=2}^{n} \left(\frac{1}{4\kappa} \left(\frac{a_j}{\Delta t} \right)^2 + U \left(\sum_{k=1}^{j} a_k \right) \right) \Delta t +$$

$$+ \left(\frac{1}{4\kappa} \left(\frac{y - \sum_{k=1}^{n} a_k}{\Delta t} \right)^2 + U \left(\sum_{k=1}^{n} a_k \right) \right) \Delta t$$

Now we can write the integral (6.13) in the form

$$K(x, t; y, s) = \int_{\mathbb{R}^{3n}} g(a_1, a_2, \ldots, a_n) w(a_1, a_2, \ldots, a_s) da_1 \ldots da_n,$$

with weight function

$$w(a_1, \ldots, a_n) = \frac{1}{(4\pi\kappa\Delta t)^{\frac{3n}{2}}} e^{-\sum_{j=1}^{n} \frac{a_j^2}{4\kappa\Delta t}}$$

and with integrand

$$g(a_1, a_2, \ldots, a_n) = e^{-\Delta t U(x) - \Delta t \sum_{j=2}^{n} U \left(\sum_{k=1}^{j} a_k \right) + \left(\frac{1}{4\kappa} \left(\frac{y - \sum_{k=1}^{n} a_k}{\Delta t} \right)^2 + U \left(\sum_{k=1}^{n} a_k \right) \right) \Delta t}$$

Remark The weight function has the following property:

$$\int_{\mathbb{R}^{3n}} w(a_1, \ldots, a_n) da_1 \ldots da_n = 1$$

It can be interpreted as the measure of the set of all piecewise linear paths between x and y. For $\Delta t \to 0$ this measure converges to the well-known Wiener measure. A very detailed and comprehensive exposition of the usage of path integrals in the various fields of their applications is available in [32].

6.3.2 Application in Financial Engineering

At the beginning of our discussion we repeat some terms from mathematical finance. We start with the definition of a Brownian motion [22]:

Definition 6.4. A stochastic process

$$W(t) = (W_1(t), W_2(t), \ldots, W_d(t))$$

with $0 \leq t \leq T$ is called a standard linear Brownian motion on $\mathbb{R}^d$ if it has $W(0) = 0$, continuous sample paths, independent increments and $W(t) - W(s) \sim N(0, (t-s)I)$.

A generalization of the standard linear Brownian motion is given by the following (we formulate it for one dimension, the definition for multiple dimension is analogous) [22]:

Definition 6.5. Let $\mu \in \mathbb{R}$ and $\sigma > 0$. A process $B(t)$ is called a Brownian motion with drift μ and covariance σ (written in the form $B(t) \sim BM(\mu, \sigma)$) if $B(t)$ has continuous paths and independent increments with

$$B(t) - B(s) \sim N\left((t-s)\mu, (t-s)\sigma\right)$$

Remark The process $B(t)$ from the previous definition satisfies the following stochastic differential equation:

$$dB(t) = \mu dt + \sqrt{\sigma} dW(t)$$

where $W(t)$ is a standard linear Brownian motion.

A disadvantage of the Brownian motion is that the values of a Brownian motion can attain negative values, an undesirable fact in modeling of prices. Therefore Paul Samuelson [33] introduced the geometric Brownian motion (GBM) as a model in finance: Suppose we have a Brownian motion $B(t) \sim BM(\mu, \sigma^2)$ on $\mathbb{R}$. The stochastic process $S(t) := S(0)exp(B(t))$ satisfies the stochastic differential equation

$$dS(t) = S(t)\left(\mu + \frac{1}{2}\sigma^2\right)dt + S(t)\sigma dW(t)$$

Such a process is called geometric Brownian motion. An alternative definition is the following [22]:

Definition 6.6. Let $\mu \in \mathbb{R}$, $\sigma > 0$ and $W(t)$ be a standard Brownian motion. A process $S(t)$ satisfying

$$\frac{dS(t)}{S(t)} = \mu dt + \sigma dW(t)$$

is called a geometric Brownian motion with drift μ and volatility parameter σ (written as $S(t) \sim GBM(\mu, \sigma^2)$).

A process $S(t) \sim GBM(\mu, \sigma^2)$ with initial value $S(0)$ is given by

$$S(t) = S(0)\exp\left(\left(\mu - \frac{1}{2}\sigma^2\right)t + \sigma W(t)\right)$$

or for arbitrary $u < t$

$$S(t) = S(u)\exp\left(\left(\mu - \frac{1}{2}\sigma^2\right)(t-u) + \sigma(W(t) - W(u))\right)$$

This formula provides a recursive procedure for simulating values of $S(t)$ at discrete values $0 = t_0 < t_1 < \ldots < t_n = T$:

$$S(t_{i+1}) = S(t_i) \exp \left(\left(\mu - \frac{1}{2}\sigma^2 \right) (t_{i+1} - t_i) + \sigma \sqrt{t_{i+1} - t_i} Z_{i+1} \right)$$

with independent $Z_i \sim N(0,1)$.

Consider now a call option: The buyer of the call has the right to buy the underlying security at time T for a price K (= "strike price"). Of course the buyer of the option earns profit only if the actual price of the security at time T is higher than the strike price. In financial terms this is called "in the money." We consider especially the case of an Asian call option. The payoff of this path-dependent option is now given by a function $G(Z_1, \ldots, Z_n)$ of the form

$$G(Z_1, \ldots, Z_n) = e^{-\mu T} \left(\frac{1}{n} \sum_{k=1}^{n} S(t_k) - K \right) H \left(\frac{1}{n} \sum_{k=1}^{n} S(t_k) - K \right)$$

where $H(x)$ denotes the Heaviside function: $H(x) = 0$ for $x \leq 0$ and $H(x) = 1$ for $x > 0$. Pricing the option now means to evaluate $E(G(Z_1, \ldots, Z_n))$ with respect to the standard normal distribution of $Z_1, \ldots, Z_s$. In other words we must evaluate the integral

$$E(G(Z_1, \ldots, Z_n))$$
$$= \frac{1}{\sqrt{2\pi}^n} e^{-\mu T} \int_{\mathbb{R}^n} \left(\frac{1}{n} \sum_{k=1}^{n} S(t_k) - K \right) H \left(\frac{1}{n} \sum_{k=1}^{n} S(t_k) - K \right) e^{-\frac{|z|^2}{2}} dz$$

This shows a significant increase of the dimensionality of the problem if the time discretization gets finer.

For more details on the usage of uniform distribution in finance and special sequences see [34, 35].

6.4 QMC Integration on Parallel Systems

Different types of parallel or distributed systems require specific attention towards their specific properties. For example, in systems with heterogeneous PE computing capacities, this variety in computing speed requires dynamic load balancing capability. Grid environments are the worst possible application environment since we additionally face heterogeneous network capacity, failure of hardware resources, additional hardware resources becoming available during the computation and many more specific properties which require highest possible flexibility of the parallel QMC techniques employed (see [21, 36, 37] for some examples).

In addition to that error bounds and computation results should preferably carry over from sequential execution. If the QMC point sets differ between sequential and parallel execution, the quality of the results needs to be investigated thoroughly. Reproducibility is as well an important issue to be considered.

So far, two entirely different strategies have been discussed in literature to employ QMC sequences in parallel and distributed environments.

1. Splitting a given QMC sequence into separately initialized and disjoint parts which are then used independently on the PEs. This strategy comes in two flavors:

   - **Blocking**: p disjoint contiguous blocks of maximal length l of the original sequence are used on the PEs. This is achieved by simply using a different starting point on each PE (e.g., PE_i, $i = 0, \ldots, p - 1$, generates the vectors $x_{il}, x_{il+1}, x_{il+2}, \ldots, x_{il+l-1}$). In case a large number of smaller blocks is used index j is assigned dynamically to PE_i which generates the vectors $x_j, x_{j+1}, \ldots, x_{j+l-1}$ (where j is incremented in steps of size l to avoid overlap).
   - **Leaping**: interleaved streams of the original sequence are used on the PEs. Each PE skips those points consumed by other PEs (*leapfrogging*) (e.g., employing p PEs, PE_i, $i = 0, \ldots, p - 1$, generates the vectors $x_i, x_{i+p}, x_{i+2p}, \ldots$).

2. Using inherently independent sequences on the different PEs (denoted as "parametrization" which can be realized, e.g., by randomizations of a given QMC sequence).

Blocking has been suggested in many application-focused papers. Mascagni and Karaivanova [38] propose to use disjoint contiguous blocks from Halton, Faure, and Sobol' sequences in the context of solving sparse systems of linear algebraic equations. Numerical experiments are carried out on a homogeneous cluster using static load distribution. In a second paper [39] the same authors use the suggested techniques for computing extremal eigenvalues, again a QMC sequence is "neatly broken into same-sized subsequences" by blocking. The authors point out that this simple strategy cannot be employed in general for all types of simulation settings. Alexandrov et al. [40] use scrambled Sobol' and Halton sequences to solve certain linear algebra systems. They discuss static and dynamic load balancing and point out the importance of efficient dynamic load balancing in GRID environments. Load balancing is done by dynamically distributing chunks (i.e., blocks) of relatively small size to avoid unevenly sized chunks. Techniques for efficiently generating nonadjacent chunks an a single PE are discussed in this paper. Tests are carried out on homogeneous and heterogeneous systems; in the latter case MPICH over Globus-2 GRID software is used. Li and Mascagni [41] propose to extend techniques used in GRID-based MC methods, for example, the *N-out-of-M* scheduling strategy, to QMC sequences by using scrambled quasi random sequences. Furthermore, known statistical properties of MC carry over to scrambled quasi random sequence, thus allowing partial result validation and intermediate value checking. Wan et al. [42] present a parallel strategy for pricing multidimensional American options. In the first stage, the QMC sequence is generated by independently computing equally sized blocks on the PEs using static load distribution. For the second stage two strategies, one being the stochastic mesh method which involves a backward recursion, for data distribution are compared, both of which correspond to distributing the original sequence in blocks of different size in different manner across the PEs. Tests are conducted on an SGI Onyx machine. Schürer [43] employs equally sized blocks of (t,m,s)-nets on the PEs when comparing QMC integration techniques to adaptive cubature rules. An SGI Power Challenge is used as a test platform. In previous work [44] we have conducted experiments with blocking Niederreiter (t,s)-sequences where large disjoint blocks are used on the PEs. Good reliability of the results has been observed in homogeneous and (simulations of) heterogeneous environments (tests conducted on an SGI Power Challenge). We have also provided theoretical evidence for this good behavior by showing that discrepancy estimates of arbitrary blocks do not degrade as compared to estimates of entire (t,s)-sequences [45].

Leaping has been discussed much more controversially in literature than blocking. Bromley [46] describes a leapfrog parallelization technique to break up the Sobol' sequence into interleaved substreams in an efficient manner. We have generalized this idea to all types of binary digital (t,s)-sequences [45] in earlier work. Based on these techniques, Li and Mullen [47] use a leapfrog scheme for (t,m,s)-nets to solve financial derivative problems. However, severe problems occur with leapfrog parallelization especially in case of processor speed heterogeneity which results in QMC point sets that do not correspond to sequential computation. Initial results showed that single (t,s)-sequence substreams with leaps of the form 2^n lead to extremely poor numerical integration results whereas this is not the case for leaps of the form 2^{n+1} [44]. Using leaped substreams parallelization in a heterogeneous processor speed environment may therefore lead to severely degraded results as compared to sequential execution when this form of leaping is employed. Different PEs consume a different number of integration nodes and so the poor results of using single substreams are propagated to the parallel results if no synchronization among PEs is performed [44,45,48]. We have also provided theoretical evidence for the observed effects by showing the discrepancy estimated of leaped substreams to be significantly larger as compared to the original sequences [45]. It has also turned out that not only 2^n type substreams are affected by poor quality but these effects occur for many forms of leaps and are highly unpredictable [45,48].

Parametrization has been proposed as a QMC parallelization strategy by two groups independently. DeDoncker et al. [49–51] propose randomized (Korobov) lattice and Richtmyer rules (which are a special type of Weyl sequences) and discuss load distribution strategies for homogeneous and heterogeneous architectures [52]. Results are provided for both, homogeneous and heterogeneous environments, and in both cases result accuracy and execution efficiency was reported to be very well. Ökten and Srinivasan [53] propose to use Halton and scrambled Halton sequences with leaped base sequences on different PEs. Excellent theoretical error estimations are provided and also experimental results for homogeneous as well as for heterogeneous environments exhibit high quality. Parametrization is also compared to blocking and leaping in this work and advantages and disadvantages of the three schemes are analyzed for different application scenarios. Srinivasan [54] confirms the findings of the latter paper and refines the comparison of the three parallelization strategies based on simulation results for pricing financial derivatives.

6.5 Numerical Experiments

6.5.1 Sequential Computations

In this section we investigate the behavior of the computation of integrals given in (6.7). In the following experiments the dimension s is chosen to be 10. We use the notation

$$erf(x) = \frac{2}{\sqrt{\pi}} \int_0^x e^{-t^2} dt$$

for the error function, and by $H_n(x)$ we denote the Hermite polynomial of degree n. Let $n = (n_1, \ldots, n_s) \in \mathbb{N}_0^s$, $\beta = (\beta_1, \ldots, \beta_s) \in \mathbb{R}^s$, $\gamma = (\gamma_1, \gamma_2, \ldots, \gamma_s) \in (\mathbb{R}_+)^s$ and $\delta = (\delta_1, \delta_2, \ldots, \delta_s) \in (\mathbb{R}_+)^s$ be arbitrary parameters. As test functions we use

$$f_1^{(n)}(x_1, x_2, \ldots, x_{10}) = \prod_{j=1}^{10} \frac{H_{n_j}(x_j)}{\sqrt{2^{n_j} n_j! \sqrt{\pi}}}$$

$$f_2^{(\beta)}(x_1, x_2, \ldots, x_{10}) = \prod_{j=1}^{10} erf(\beta_j x_j)$$

$$f_3^{(\gamma)}(x_1, x_2, \ldots, x_{10}) = \prod_{j=1}^{10} e^{-\gamma_j |x_j|} - \prod_{j=1}^{10} e^{\frac{\gamma_j^2}{4}} \left(1 - erf\left(\frac{\gamma_j}{2}\right)\right)$$

$$f_4^{(\delta)}(x_1, x_2, \ldots, x_{10}) = \prod_{j=1}^{10} e^{-\delta_j x_j^2} - \prod_{j=1}^{10} \sqrt{\frac{1}{1 + \delta_j}}$$

$f_1^{(n)}(x)$ is an unbounded polynomially growing function, $f_2^{(\beta)}(x)$ is bounded $\left(\left|f_2^{(\beta)}(x)\right| < 1\right)$, $f_3^{(\gamma)}(x)$ and $f_4^{(\delta)}(x)$ are also bounded. In the applications functions of these type occur often. We have

$$\int_{\mathbb{R}^s} f_i(x) e^{-|x|^2} dx = 0$$

for $i = 1, 2, 3, 4$. The functions $f_3^{(\gamma)}(x_1, \ldots, x_{10})$ and $f_4^{(\delta)}(x_1, \ldots, x_{10})$ are not separable in contrast to $f_1^{(n)}(x_1, \ldots, x_{10})$ and $f_2^{(\beta)}(x_1, \ldots, x_{10})$. In our experiments we use the following parameter values:

$$n = (8, 5, 4, 4, 10, 10, 8, 7, 8, 9)$$
$$\beta = (2.45, 9.67, 0.67, 1.78, 6.45, 1.67, 0.67, 20.78, 1.45, 9.67)$$
$$\gamma = \delta = (2.0, 0.78, 3.78, 4.7, 1.3, 7.0, 1.4, 0.4, 2.0, 0.78)$$

These values can be chosen arbitrarily.

Remark The inversion of the Gauss distribution function (6.8) cannot be done in closed form. We must use numerical approximations to do this. In finance often the following method is used: The standard normal distribution

$$\Phi(x) = \frac{1}{\sqrt{2\pi}} \int_{-\infty}^{x} e^{-\frac{t^2}{2}} dt$$

can be inverted by solving the equation

$$\Phi(x) - y = 0$$

using Newton's method. This gives the following approximation sequence $x_k^{(approx)}$ for a given y:

$$x_{k+1}^{(approx)} = x_k^{(approx)} + \left(y - \Phi\left(x_k^{(approx)}\right)\right) \exp\left(0.5\left(x_k^{(approx)}\right)^2 + \log\sqrt{2\pi}\right)$$

Marsaglia et al. (see [55]) suggest now the following starting point:

$$x_0^{(approx)} = \pm\sqrt{\left|-1.6\log\left(1.0004 - (1-2y)^2\right)\right|}$$

where the sign depends on $y > \frac{1}{2}$ or not.

In our computations we prefer a direct method. The aim is to express the integral in terms of the error function $erf(x)$ defined by

$$erf(x) := \frac{2}{\sqrt{\pi}} \int_0^x e^{-t^2} dt$$

and the complementary error function $erfc(x) := 1 - erf(x)$. To get a sequence following the Gaussian distribution we must solve the equation

$$\frac{1}{\sqrt{\pi}} \int_{-\infty}^{x} e^{-t^2} dt = y \tag{6.14}$$

where y is a coordinate from a point of a (low discrepancy) sequence in the unit cube. At first we investigate the case $y > \frac{1}{2}$: From the normal distribution we know now that $x > 0$. In this case we can write Eq. (6.14) in the form

$$\frac{1}{\sqrt{\pi}} \int_{-\infty}^{0} e^{-t^2} dt + \frac{1}{\sqrt{\pi}} \int_0^x e^{-t^2} dt = \frac{1}{2}(1 + erf(x)) = y$$

or

$$erfc(x) = 2(1 - y)$$

Now we can use an approximation for $erfc(x)$:

$$erfc(x) \approx \frac{2}{2+x} \exp\left(-x^2 + P\left(\frac{2}{2+x}\right)\right)$$

$(x > 0)$. $P(x)$ is a Tschebyscheff polynomial. The inversion is then done by two iterations of a modified Newton scheme suggested by the astronomer Edmund Halley (a friend of Isaac Newton, the famous comet carries his name): If $f(x)$ is twice differentiable and an equation of the form

$$f(x) = 0$$

must be solved, then Halley suggests the iteration

$$x_{i+1} = x_i - \frac{f(x_i)}{f'(x_i)\left(1 - \frac{f(x_i)f''(x_i)}{2(f'(x_i))^2}\right)}$$

The case $y < \frac{1}{2}$ is reduced to the previous case in the following way. We solve the equation

$$\frac{1}{\sqrt{\pi}} \int_{-\infty}^{1-x} e^{-t^2} dt = 1 - y$$

For $y = 0$ the solution of (6.14) is given by $x = \frac{1}{2}$. For more information about this algorithm see [56]. In Figs. 6.1 and 6.2 the error of the computation of the integral (6.7) is plotted for the Zinterhof sequence and GLP respectively. The range of the integration nodes is $[10^6 : 10^7]$ for the Zinterhof sequence. For the sequence of the GLP the range is $[10^5 : 10^7]$. Due to the high differences of the integration results the ordinate shows the logarithm of the error. We observe that the GLP give much

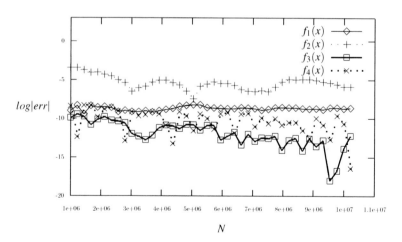

Fig. 6.1 Integration error of the test functions in case of Zinterhof sequence.

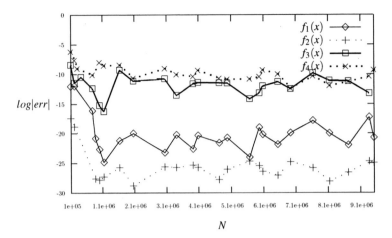

Fig. 6.2 Integration error of the test functions in case of the Good Lattice Points.

better results for the test functions $f_1^{(n)}(x)$ and $f_2^{(\beta)}(x)$. For $f_3^{(\gamma)}(x)$ and $f_4^{(\delta)}(x)$ the error for both computations is of the same magnitude.

6.5.2 Parallel Case

We consider the three different QMC parallelization strategies discussed in Sect. 6.4: Leaping, blocking and parametrization.

If the environment consists of almost identical computers with identical load the result is not different from the computation on a single system considering the whole sequence. Problems can arise if the environment is inhomogeneous (due to load of some nodes of the environment, different hardware of the nodes, etc.). To achieve a defined inhomogeneity we consider two different cases: One PE uses much less points (factor 10^{-3}) than the other PEs (the one slow case) and that one PE is much faster than the other PEs (factor 10^3), the one fast case.

We execute the parallel computation on a Linux cluster. Each system has two AMD Athlon MP2800+ processors and the systems are connected via 6x6 SCI torus. We use a subset of the cluster consisting of 10 systems. On each system we start 2 processes, so we have 20 PEs.

6.5.3 Experimental Results

6.5.3.1 The Leaped Zinterhof Sequence

We start with the s-dimensional Hermite polynomial $f_1^{(n)}(x)$. In Fig. 6.3 the integration error for $f_1^{(n)}(x)$ is plotted for the one slow case and for the one fast case. In comparison also the error of the sequential computation is shown. As an interesting

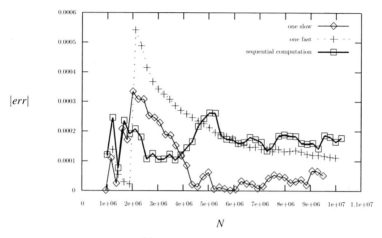

N

Fig. 6.3 Integration error for $f_1^{(n)}(x)$ for the Zinterhof sequence with leaping.

fact for this polynomial we observe better results for the inhomogeneous cases than for the sequential computation.

In Fig. 6.4 we consider the computation of the integral over the bounded function $f_2^{(\beta)}(x)$. The main difference to $f_1^{(n)}(x)$ is that $f_2^{(\beta)}$ is bounded by 1. For this test function we observe that the one slow case shows quite the same behavior as the sequential computation (oscillating behavior). The one fast case behaves even better. This function shows a stable behavior under the considered inhomogeneities.

In the next experiment we investigate the behavior of computation of the integral over the function $f_3^{(\gamma)}(x)$. We expect a high stability of the computation and rather good results. These are plotted in Fig. 6.5. In the considered range of integration nodes we observe a fast decrease of the error for the sequential computation. The one slow case behaves quite similar. As an interesting fact the one fast computation shows an increasing error.

In the next experiment we investigate the behavior of $f_4^{(\delta)}(x)$. As seen in Sect. 6.3.1 integrals over functions of this type arise in mathematical physics. We expect an even higher stability of the computation as for $f_3^{(\gamma)}(x)$. Indeed this is true as seen in Fig. 6.6. For an increasing number of integration nodes all error curves

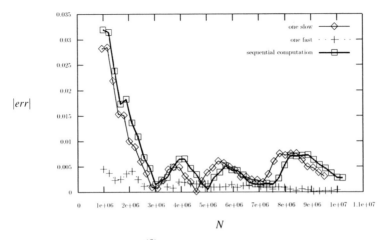

Fig. 6.4 Integration error for $f_2^{(\beta)}(x)$ for the Zinterhof sequence with leaping.

show qualitatively the same behavior which indicates a high stability under the considered inhomogeneities.

The most stable results are delivered by $f_2^{(\beta)}(x)$ and $f_4^{(\delta)}(x)$. The error of the computation in the inhomogeneous environment delivers comparable errors or even better results as the sequential case. $f_3^{(\gamma)}(x)$ is affected by the inhomogeneities: For increasing number of integration nodes the error does not decrease any more.

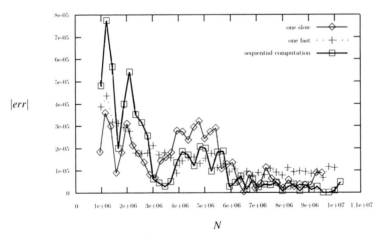

Fig. 6.5 Integration error for $f_3^{(\gamma)}(x)$ for the Zinterhof sequence with leaping.

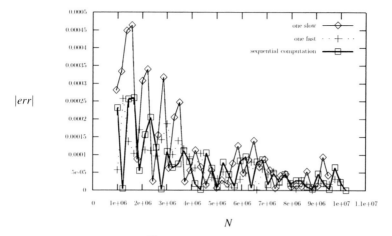

$$N$$

Fig. 6.6 Integration error for $f_4^{(\delta)}(x)$ for the Zinterhof sequence with leaping.

6.5.3.2 The Leaped Sequence of Good Lattice Points

In the next series of experiments we investigate the computation of the integral (6.7) for the GLP as integration nodes. As parallelization strategy we choose the leaping strategy.

The transformed sequence of GLP shows a quite high sensitivity with respect to inhomogeneities of the computation environment. The reason of this fact is the dependency of the optimal coefficients on the number of integration node points. The observed range of integration nodes is from 10^5 to 10^7. Due to the differences between the sequential computation and the parallel computation the ordinate shows the logarithm of the error.

Figure 6.7 shows the plots of the errors for the computation of the integral over $f_1^{(n)}(x)$. As expected the sequential computation results are better than the other cases. In case of the Zinterhof sequence we observed the reverse.

Now we investigate the computation of the integral over the bounded function $f_2^{(\beta)}(x)$. In Fig. 6.8 we observe a similar behavior: The sequential computation delivers quite accurate results. The error rate of the sequential computation is smaller than the one in the previous computation, but the parallel computation error is of the same magnitude. So the impact of inhomogeneity is even higher than for the polynomial considered before.

In the following experiments we investigate the functions $f_3^{(\gamma)}(x)$ and $f_4^{(\delta)}(x)$, see Figs. 6.9 and 6.10 respectively. As an interesting fact the errors of the parallel computations are comparable with the sequential one or even better: The one slow case gives more accurate results than the sequential and the one fast case. But all error curves show an oscillating behavior with quite high amplitudes. Glasserman

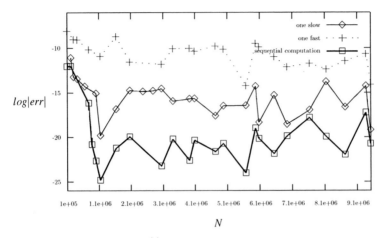

Fig. 6.7 Integration error of $f_1^{(n)}(x)$ for the Good Lattice Points with leaping.

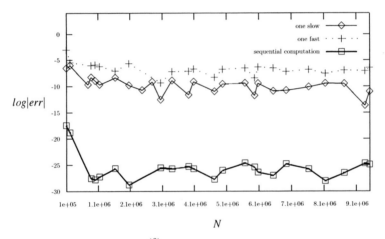

Fig. 6.8 Integration error of $f_2^{(\beta)}(x)$ for the Good Lattice Points with leaping.

[22] observed in his sequential and especially financial tests an erratic behavior of these sequences. We observe that the error in the case of $f_3^{(\gamma)}(x)$ and $f_4^{(\delta)}(x)$ is larger than for $f_1^{(n)}(x)$ and $f_2^{(\beta)}(x)$.

6.5.3.3 The Blocked Zinterhof Sequence

We start with the computation result of $f_1^{(n)}(x)$. In Fig. 6.11 the errors of the parallel computations are plotted vs. the sequential computation. We observe that the one slow case delivers a very similar result to the sequential computation. The one fast

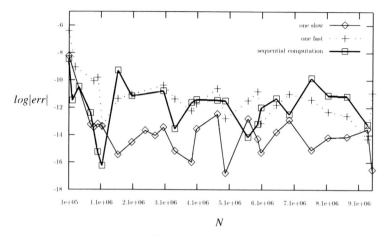

Fig. 6.9 Integration error of $f_3^{(\gamma)}(x)$ for the Good Lattice Points with leaping.

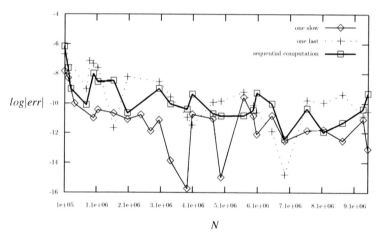

Fig. 6.10 Integration error of $f_4^{(\delta)}(x)$ for the Good Lattice Points with leaping.

case delivers much better results over the considered range of integration nodes. This is different to the leaping case, where we saw the best results for the one slow case (see Fig. 6.3).

Figure 6.12 shows the computation results for $f_2^{(\beta)}(x)$. The magnitude of the error is in the range of the sequential computation and the leaped computation. One difference is visible for the one fast case: Whereas in the leaping strategy the error decreases for growing N (see Fig. 6.4) it does not show this behavior for the blocking strategy.

In the next experiments (Fig. 6.13 and 6.14) we investigate the behavior of the computation of the integral over $f_3^{(\gamma)}(x)$ and $f_4^{(\delta)}(x)$. We observe again a very high

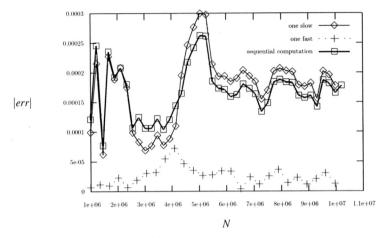

Fig. 6.11 Integration error for $f_1^{(n)}(x)$ for the Zinterhof sequence with blocking.

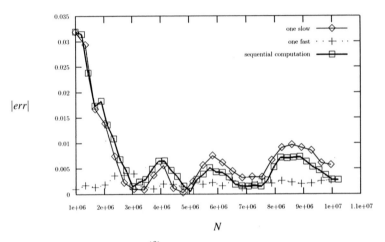

Fig. 6.12 Integration error for $f_2^{(\beta)}(x)$ for the Zinterhof sequence with blocking.

stability of the computations. The sequential computation and the one slow case show a very similar behavior. But for these functions also the one fast case shows a decrease of the error. For these functions the inhomogeneity does not harm the computation.

6.5.3.4 The Blocked Sequence of Good Lattice Points

In this paragraph we consider experiments regarding the blocked sequence of GLP. In the case of the leaping method we discovered a significant loss of accuracy of

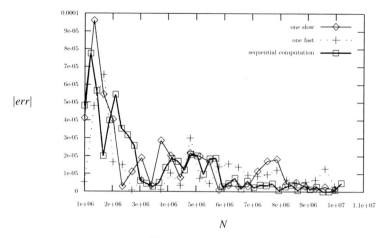

Fig. 6.13 Integration error for $f_3^{(\gamma)}(x)$ for the Zinterhof sequence with blocking.

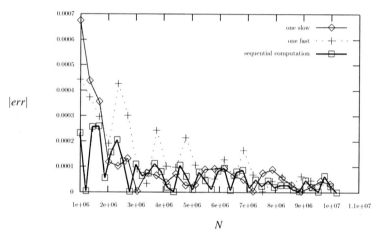

Fig. 6.14 Integration error for $f_4^{(\delta)}(x)$ for the Zinterhof sequence with blocking.

the results for the test functions $f_1^{(n)}(x)$ and $f_2^{(\beta)}(x)$. All test functions showed a sensibility with respect to the considered inhomogeneities. The next experiments will show the behavior of the computation with the blocking strategy.

At first we investigate the polynomial $f_1^{(n)}(x)$. Fig. 6.15 shows the integration errors of the sequential computation versus the parallel computations with inhomogeneity. The error of the parallel computation is higher than the error of the sequential one and qualitatively the same as in Fig. 6.7.

An analogous behavior is seen in Fig. 6.16 for $f_2^{(\beta)}(x)$. Again we observe an error curve comparable with Fig. 6.8.

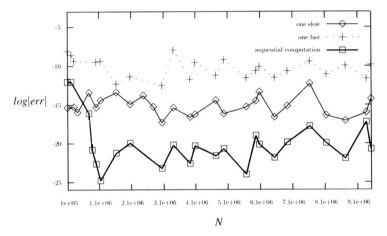

Fig. 6.15 Integration error for $f_1^{(n)}(x)$ for the Good Lattice Points with blocking.

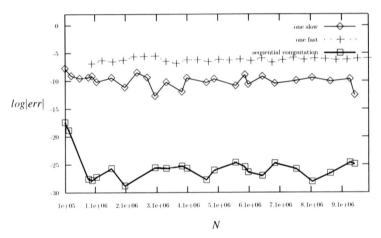

Fig. 6.16 Integration error for $f_2^{(\beta)}(x)$ for the Good Lattice Points with blocking.

The error rates of $f_3^{(\gamma)}(x)$ and $f_4^{(\delta)}(x)$ are plotted in Figs. 6.17 and 6.18 respectively. As an interesting fact the best results are achieved for these functions in the one slow case.

6.5.3.5 Parametrization for the Zinterhof Sequence

In this paragraph we discuss the usage of the parametrization method [57] on our test functions. In our case we use the Zinterhof sequence $x_n = \{n\theta\}$ with

$$\theta = \left(e^1, e^{\frac{1}{2}}, \ldots, e^{\frac{1}{s}} \right) \in \mathbb{R}^s$$

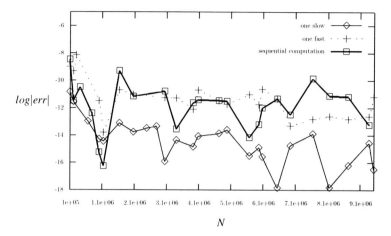

Fig. 6.17 Integration error for $f_3^{(\gamma)}(x)$ for the Good Lattice Points with blocking.

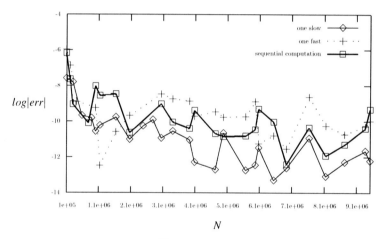

Fig. 6.18 Integration error for $f_4^{(\delta)}(x)$ for the Good Lattice Points with blocking.

To get the different sequences of integration nodes we proceed in the following way: In the computation environment consisting of M PEs the K-th PE $(0 \leq K \leq M-1)$ uses the sequence with

$$\theta_K = \left(e^{\frac{1}{sK+1}}, e^{\frac{1}{sK+2}}, \ldots, e^{\frac{1}{(K+1)s}} \right)$$

It can be shown that these sequences are asymptotically independent [58]: The correlation matrix of the M sequences of N integration nodes differs from an identity matrix by a matrix with entries of $O\left(N^{\varepsilon-1}\right)$.

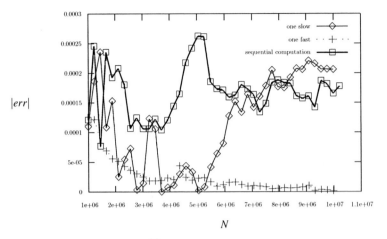

Fig. 6.19 Integration error for $f_1^{(n)}(x)$ with parametrization method.

Our first experiment deals with the polynomial $f_1^{(n)}(x)$. Fig. 6.19 shows a plot of the error curves of the parallel computation with parametrization method vs. the sequential computation. We observe the best results for the one fast case. The magnitude of the error is comparable with the leaping strategy.

The next experiment deals with the bounded function $f_2^{(\beta)}(x)$. Fig. 6.20 shows the plot of the error curves. The one slow case gives a much better result than for the leaping case, but the one fast inhomogeneity is worse (for the leaping case see Fig. 6.4).

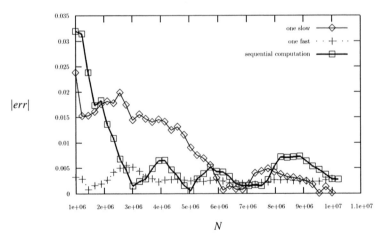

Fig. 6.20 Integration error for $f_2^{(\beta)}(x)$ with parametrization method.

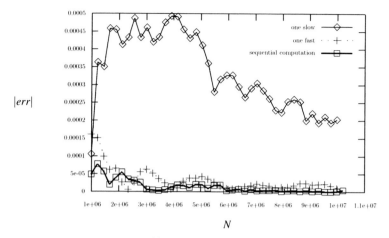

Fig. 6.21 Integration error for $f_3^{(\gamma)}(x)$ with parametrization method.

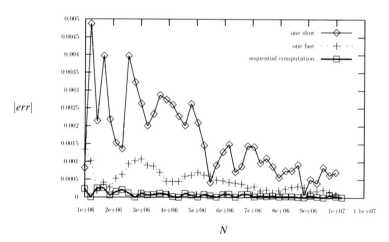

Fig. 6.22 Integration error for $f_4^{(\delta)}(x)$ with parametrization method.

The functions $f_3^{(\gamma)}(x)$ and $f_4^{(\delta)}(x)$ are quite sensitive for the one slow inhomogeneity. Whereas in the leaping strategy the one slow case does not harm the computation (see Figs. 6.5 and 6.6) it has much more impact in the parametrization case (Figs. 6.21 and 6.22).

6.5.4 *Overall Comparison*

In this section we compare the behavior of the parallel computation of the integral (6.7) in the case of our test functions. We investigate the one slow case. Due to the fact of the quite different results we have the logarithm of the error on the ordinate. For all test functions the most accurate results are delivered from the sequence of

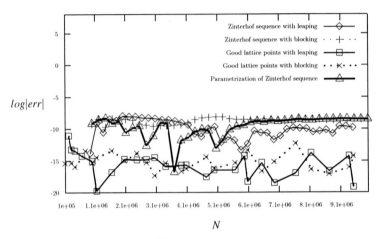

Fig. 6.23 Integration error for $f_1^{(n)}(x)$.

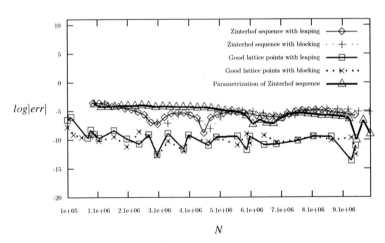

Fig. 6.24 Integration error for $f_2^{(\beta)}(x)$.

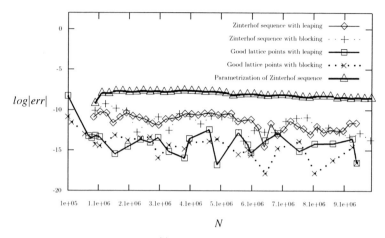

Fig. 6.25 Integration error for $f_3^{(\gamma)}(x)$.

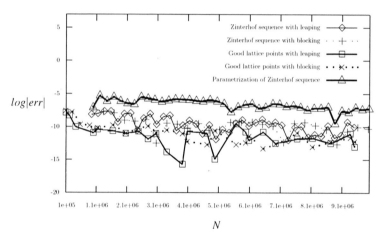

Fig. 6.26 Integration error for $f_4^{(\delta)}(x)$.

the GLP. Although their error curves show an oscillating behavior the result is always better than the results for the Zinterhof sequence with leaping, blocking or parametrization. For the test function $f_1^{(n)}(x)$ the quality of the good results of the GLP is much better than the other results. For the (bounded) test function $f_2^{(\beta)}(x)$ also the other methods show a decrease of the error. The function $f_3^{(\gamma)}(x)$ shows a quite slow decrease of the error for the parametrization method whereas the other computations show much better results. Also the function $f_4^{(\delta)}(x)$ should be computed with the GLP due to the slightly better results, although the results for the Zinterhof sequence exhibit the same order of magnitude except for the parametrization approach. The parametrization method shows a faster decrease of the error but the starting value (=error for $N = 10^6$) is much higher than for the function $f_3^{(\gamma)}(x)$. Except for the polynomially growing function $f_1^{(n)}(x)$ all methods show a decrease of the error for an increasing number of integration nodes and can be used for the computation.

The one fast case shows the same behavior: The best results are achieved for the GLP, but the differences are not so significant as in the one slow case. The error in case of the GLP is higher than in the one slow case. The reason is the dependency of the GLP on the number of integration nodes. The results for the function $f_4^{(\delta)}(x)$ have the same order except the parametrization (see Fig. 6.27). In Figs. 6.23, 6.24,

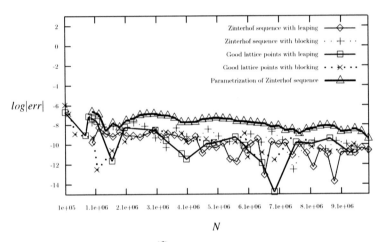

Fig. 6.27 Integration error for $f_4^{(\delta)}(x)$ for one fast case.

6.25, 6.26 and 6.27 we observe that the parametrization method of the Zinterhof sequence delivers always the worst results. The blocking and leaping strategy of the Zinterhof sequence and the GLP respectively give errors of the same magnitude. For $f_1^{(n)}(x)$ and $f_2^{(\beta)}(x)$ the best results are delivered from the GLP whereas for $f_4^{(\delta)}(x)$ the error for the Zinterhof sequence is of the same order as the error for the GLP.

6.6 Application of the Diaphony in Parallel Computation

In this section we apply the concept of the diaphony to estimations of the errors of QMC integration: QMC of functions from an RKHS caused by inhomogeneities of the parallel computation environment. Suppose we have a collection consisting of M PEs. We use a sequence of node points $(x_k)_{k=1}^N \in \mathbb{R}^s$. The PE i, $i = 1, \ldots, M$, uses a subsequence of the original one denoted by $(x_k), k \in A_i$ where A_i are disjoint subsets of $\{1, 2, \ldots, N\}$. Let $c_i = |A_i|$. Let $D\left(\{x_k\}_{k=1}^N\right)$ be the diaphony [25] of the sequence. An error estimation is given by (see [58])

$$\left| \frac{\sqrt{\pi}^s}{N} \sum_{k=1}^N f(x_k) - I \right| = \left| \frac{\sqrt{\pi}^s}{N} \sum_{i=1}^M \frac{c_i}{c_i} \sum_{k \in A_i} f(x_k) - I \right| = \left| \sum_{i=1}^M \frac{c_i}{N} \frac{\sqrt{\pi}^s}{c_i} \sum_{k \in A_i} f(x_k) - I \right|$$

Usage of the diaphony allows us to separate the error estimator into two terms:

$$\left| \sum_{i=1}^M \frac{c_i}{N} \frac{\sqrt{\pi}^s}{c_i} \sum_{k \in A_i} f(x_k) - I \right| =$$

$$= \left| \sum_{i=1}^M \frac{c_i}{N} \left(\frac{\sqrt{\pi}^s}{c_i} \sum_{k \in A_i} f(x_k) - I \right) \right| \leq \|f\| \cdot \sum_{i=1}^M \frac{c_i}{N} D\left(\{x_k\}_{k \in A_i}\right)$$

This means the error caused by the sequence is given by a weighted mean of the diaphony of the subsequences of each PE.

In our next consideration we deal with estimations of errors caused by inhomogeneities: To be more precise we want to estimate the difference of the sequential computation and the disturbed parallel computation. We consider the i-th PE and assume that not all integration nodes are consumed by the computation: Let $B_i \subset A_i$ with $|B_i| = d_i < c_i$. We want to estimate the difference

$$\left| \frac{\sqrt{\pi}^s}{c_i} \sum_{k \in A_i} f(x_k) - \frac{\sqrt{\pi}^s}{d_i} \sum_{k \in B_i} f(x_k) \right|$$

We will obtain 2 estimations: The first one is based on the consumed integration nodes, the second one is based on the unconsumed integration nodes.

The first estimation is quite simple:

$$\left| \frac{\sqrt{\pi}^s}{c_i} \sum_{k \in A_i} f(x_k) - \frac{\sqrt{\pi}^s}{d_i} \sum_{k \in B_i} f(x_k) \right| =$$

$$= \left| \frac{\sqrt{\pi}^s}{c_i} \sum_{k \in A_i} f(x_k) - I + I - \frac{\sqrt{\pi}^s}{d_i} \sum_{k \in B_i} f(x_k) \right| \leq$$

$$\leq \|f\| \cdot \left(D\left(\{x_k\}_{k \in A_i}\right) + D\left(\{x_k\}_{k \in B_i}\right) \right)$$

This means we can estimate the error caused by the parallel computation environment by the diaphony of all integration nodes which should be processed by the i-th PE and the diaphony of the really processed integration nodes.

The second estimation is achieved in the following way:

$$\left| \frac{\sqrt{\pi}^s}{c_i} \sum_{k \in A_i} f(x_k) - \frac{\sqrt{\pi}^s}{d_i} \sum_{k \in B_i} f(x_k) \right| =$$

$$= \left| \left(\frac{1}{c_i} - \frac{1}{d_i} \right) \sqrt{\pi}^s \sum_{k \in B_i} f(x_k) + \frac{\sqrt{\pi}^s}{c_i} \sum_{k \in A_i \backslash B_i} f(x_k) \right| =$$

$$= \left| \frac{d_i - c_i}{c_i} \frac{\sqrt{\pi}^s}{d_i} \sum_{k \in B_i} f(x_k) + \frac{\sqrt{\pi}^s}{c_i} \sum_{k \in A_i \backslash B_i} f(x_k) \right| =$$

$$= \frac{c_i - d_i}{c_i} \left| \frac{\sqrt{\pi}^s}{d_i} \sum_{k \in B_i} f(x_k) - I + I - \frac{\sqrt{\pi}^s}{c_i - d_i} \sum_{k \in A_i \backslash B_i} f(x_k) \right| \le$$

$$\le \|f\| . \frac{c_i - d_i}{c_i} \left(D\left(\{x_k\}_{k \in B_i} \right) + D\left(\{x_k\}_{k \in A_i \backslash B_i} \right) \right)$$

We can also get an estimation of the difference of the computation in the inhomogeneous environment and the exact value of the integral: Let $N_1 = d_1 + \ldots + d_M$. Then we get

$$\left| \sum_{i=1}^M \frac{d_i}{N_1} \frac{\sqrt{\pi}^s}{d_i} \sum_{k \in B_i} f(x_k) - I \right| \le$$

$$\le \left| \sum_{i=1}^M \frac{d_i}{N_1} \frac{\sqrt{\pi}^s}{d_i} \sum_{k \in B_i} f(x_k) - \sum_{i=1}^M \frac{c_i}{N} \frac{\sqrt{\pi}^s}{c_i} \sum_{k \in A_i} f(x_k) \right| +$$

$$+ \left| \sum_{i=1}^M \frac{c_i}{N} \frac{\sqrt{\pi}^s}{c_i} \sum_{k \in A_i} f(x_k) - I \right| \le$$

$$\le \left| \sum_{i=1}^M \frac{d_i}{N_1} \frac{\sqrt{\pi}^s}{d_i} \sum_{k \in B_i} f(x_k) - I \right| + 2 \left| \sum_{i=1}^M \frac{c_i}{N} \frac{\sqrt{\pi}^s}{c_i} \sum_{k \in A_i} f(x_k) - I \right| \le$$

$$\le \|f\| . \sum_{i=1}^M \left[\frac{d_i}{N_1} D\left(\{x_k\}_{k \in B_i} \right) + 2 \frac{c_i}{N} D\left(\{x_k\}_{k \in A_i} \right) \right] \tag{6.15}$$

In a similar way an estimation based on the unconsumed points is given by

$$\left| \sum_{i=1}^M \frac{d_i}{N_1} \frac{\sqrt{\pi}^s}{d_i} \sum_{k \in B_i} f(x_k) - I \right| \le$$

$$\le \|f\| \sum_{i=1}^M \frac{c_i}{N} D(\{x_k : k \in A_i\}) +$$

$$+ \frac{N - N_1}{N} \|f\| \sum_{i=1}^M \left(\frac{d_i}{N_1} D(\{x_k : k \in B_i\}) + \frac{c_i - d_i}{N - N_1} D(\{x_k : k \in A_i \backslash B_i\}) \right)$$

We will apply now the estimation (6.15) to the one slow blocking case for the Zin-terhof sequence. By the well-known expansion (see [59])

$$e^{-a^2x^2} = \sum_{n=0}^{\infty} \frac{(-1)^n a^{2n}}{2^{2n} n! (1+a^2)^{n+\frac{1}{2}}} H_{2n}(x)$$

for $x \in \mathbb{R}$ and $\Re a^2 > -1$ we can show that the test function

$$f_4^{(\delta)}(x_1,\ldots,x_{10}) = \prod_{j=1}^{10} e^{-\delta_j x_j^2} - \prod_{j=1}^{10} \sqrt{\frac{1}{1+\delta_j}}$$

is an element of an RKHS of the form (6.10) if

$$\lambda_i^4 > \frac{\delta_i^2}{(1+\delta_i)^2}$$

for $1 \leq i \leq 10$. For our parameter δ this is guaranteed if $\lambda_i \equiv \lambda = 0.95$. The norm (induced from the scalar product in H_λ) of $f_4^{(\delta)}(x)$ is given by

$$\left\| f_4^{(\delta)} \right\| = \left[\sum_{(n_1,\ldots,n_{10})\in\mathbb{N}_0^{10}}' \prod_{j=1}^{10} \binom{2n_j}{n_j} \left(\frac{\delta_j^2}{4\lambda_j(1+\delta_j)^2} \right)^{n_j} \right]^{\frac{1}{2}}$$

The prime indicates that the term with index $(0,\ldots 0)$ is omitted. A numerical com-putation gives the value

$$\left\| f_4^{(\delta)} \right\| \approx 0.8189266$$

In Fig. 6.28 the estimated error vs. the error of the one slow case is plotted. Due to the high difference the logarithm of the error is plotted. The reason for this overes-timation is that the diaphony is the worst case error for all functions of the RKHS for the considered point sequence. Our estimation uses the diaphony of the really considered integration nodes which is of course greater than the diaphony of the whole point sequence.

6.7 Conclusion

We have shown that parallel QMC techniques are a sound way to compute inte-grals over the real line in case of Gaussian type weight functions. Even in case of significant inhomogeneities of the PEs computing capacities we observe rather stable results in most cases. While for Zinterhof sequences we even observe im-provements of the sequential integration results for some settings and test functions, the results obtained with GLP degrade with respect to sequential accuracy in the same scenarios. Overall, the latter point sets deliver better results as compared to

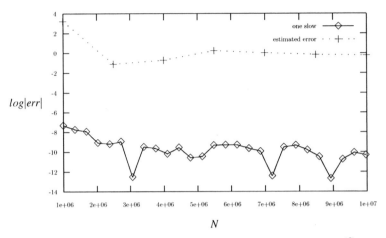

Fig. 6.28 Estimated error vs. the computed error for the one slow case of $f_4^{(\delta)}(x)$.

Zinterhof sequences; however, for most scenarios the integration error is within the same order of magnitude (except for parametrization, which gives the worst results of all techniques considered). Taking the ease of construction of Zinterhof sequences into account, their use employing blocking and leaping in parallel environments can specifically be recommended.

Acknowledgments This work has been partially supported by the Austrian Grid Project 2. Austrian Grid is a project funded by the *bm:bwk (Federal Ministry for Education, Science and Culture)* after recommendation by the Austrian Council for Research and Technology Development.

References

1. G. Evans, Practical Numerical Integration, Wiley, Chichester, (1993).
2. H. Niederreiter, Random Number Generation and Quasi-Monte Carlo Methods, CBMS-NSF Regional Conference Series in Applied Mathematics 62, Society for Industrial and Applied Mathematics (SIAM), 1992.
3. K. Entacher, P. Hellekalek, and P. L'Ecuyer, Quasi-Monte Carlo node sets from linear congruential generators, Monte Carlo and Quasi-Monte Carlo Methods 1998, Springer (2000) 188–198.
4. E. deDoncker, R. Zanny, and K. Kaugars, Distributed numerical integration algorithms and applications, in: Proceedings of the 4th World Multiconference on Systemics, Cybernetics, and Informatics (SCI'00) (2000) 244–249.
5. A.R. Krommer and C.W. Überhuber, Numerical Integration on Advanced Computer Systems, Lecture Notes in Computer Science 848, Springer, Berlin, (1994).
6. A.R. Krommer and C.W. Überhuber, Computational Integration, SIAM, Philadelphia, (1998).
7. R. Schürer and A. Uhl, An evaluation of adaptive numerical integration algorithms on parallel systems, Parallel Algorithms and Applications 18 (1–2) (2003) 13–26.
8. M. Drmota and R. Tichy, Sequences, Discrepancies and Applications, Lecture Notes in Computer Science 1651, Springer, Berlin, (1997).
9. T. T. Warnock, Computational investigations of low discrepancy point sets, Applications of Number Theory to Numerical Analysis, Academic Press, New York (1972) 319–343.
10. S. Heinrich, Efficient algorithms for computing L_2-discrepancy, Mathematics of Computation 65 (216) (1996) 1621–1633.
11. K. F. Roth, On irregularities of distribution, Mathematika 1 (1953) 73–79.
12. H. Niederreiter, Uniform Distribution of Sequences, Interscience, New York, 1974.
13. J. H. Halton, On the efficiency of certain quasi-random sequences of points in evaluating multi-dimension integrals, Numerical Mathematics 2 (1960) 84–90; Berichtigung ibid., 196
14. H. Faure, Discrépance de suites associées à un système de numération (en dimension s), Acta Arithmetica 41 (1982) 337–351.
15. I. M. Sobol, On the distribution of points in a cube and the approximate evaluation of integrals, U.S.S.R. Computational Mathematics and Mathematical Physics 7 (4) (1967) 86–112.
16. H. M. Korobow, Approximate calculation of multiple integrals with the aid of methods in the theory of numbers, Dokl. Akad. Nauk SSSR 115 (1957) 1062–1065.
17. E. Hlawka, Zur angenäherten Berechnung mehrfacher Integrale, Monatsh. Mathematik 66 (1962) 140–151.
18. H. Weyl, Über die Gibbssche Erscheinung und verwandte Konvergenzphänomene, Rend. Circ. Mat. Palermo 30 (1910) 377–407.
19. H. Niederreiter, Methods for estimating discrepancy, Applications of Number Theory to Numerical Analysis, Academic Press, New York (1972) 203–236.
20. P. Zinterhof, Einige zahlentheoretische Methoden zur numerischen Quadratur und Interpolation, Sitzungsberichte der Österreichischen Akademie der Wissenschaften, math.-nat.wiss. Klasse Abt. II 177 (1969) 51–77.
21. H. Hofbauer, A. Uhl, and P. Zinterhof, Zinterhof sequences in GRID-based numerical integration, Monte Carlo and Quasi-Monte Carlo Methods 2006, Springer, Berlin (2008) 495–510.
22. P. Glasserman, Monte Carlo Methods in Financial Engineering, Springer, Berlin, 2004.
23. H. M. Korobov, Anwendung zahlentheoretischer Methoden auf Probleme der Numerischen Mathematik (in Russian), Fismatgis, Moscow, (1963).
24. N. Aronszajn, Theory of reproducing kernels, Transactions of the American Mathematical Society 68 (1950) 337–404.
25. P. Zinterhof and C. Amstler, Uniform distribution, Discrepancy and reproducing kernel Hilbert spaces, Journal of Complexity 17 (2001) 497–515.
26. F. Costabile, F. Dell'Accio, and M. I. Gualtieri, A new approach to Bernoulli Polynomials, Rendiconti di Matematica 26 (2006) 1–12.

27. P. Hellekalek and G. Larcher, Random and Quasi-random point sets, Lecture Notes in Statistics 138, Springer, Berlin, (1998).

28. F. J. Hickernell, I. H. Sloan, and G. W. Wasilkowski, On tractability of weighted integration over bounded and unbounded regions in $\mathbb{R}^s$, Mathematics of Computation 73 (2004) 1885–1901.

29. J. S. Liu, Monte Carlo Strategies in Scientific Computing, Springer, Berlin, (2004).

30. S. Thangavelu, Hermite and Laguerre Expansions, Mathematical Notes 42, Princeton University Press, Princeton, (1993).

31. E. Zeidler, Quantum Field Theory I: Basics in Mathematics and Physics, Springer, Berlin, (2006).

32. H. Kleinert, Path Integrals in Quantum Mechanics, Statistics, Polymer Physics and Financial Markets, World Scientific Pub CO, Singapore (2006).

33. P. A. Samuelson, Proof that properly anticipated prices fluctuate randomly, Industrial Management Review 6 (1965) 41–50.

34. G. Larcher and G. Leobacher, Quasi-Monte Carlo and Monte Carlo methods and their applications in finance, Surveys on Mathematics for Industry 11 (2005) 95–130.

35. G. Larcher and F. Pillichshammer, A note on optimal point distributions in $[0, 1)^s$, Journal of Computational and Applied Mathematics 206 (2007) 977–985.

36. S. Li, K. Kaugars, and E. deDoncker, Grid-based numerical integration and visualization, Sixth International Conference on Computational Intelligence and Multimedia Applications (ICCIMA'05), IEEE Computer Society Press (2005) 260–265.

37. H. Hofbauer, A. Uhl, and P. Zinterhof, Quasi Monte Carlo Integration in GRID Environments: Further Leaping Effects, Parallel Processing Letters 16 (3) (2006) 285–311.

38. M. Mascagni and A. Karaivanova, A parallel Quasi-Monte Carlo method for solving systems of linear equations, in: P. Sloot et al. (Eds.), The 2002 International Conference on Computational Science – ICCS 2002, Springer, Berlin (2002) 598–608.

39. M. Mascagni and A. Karaivanova, A parallel Quasi-Monte Carlo method for computing extremal eigenvalues, in: K. T. Fang, F. J. Hickernell, and H. Niederreiter (Eds.), Monte Carlo and Quasi-Monte Carlo Methods 2000, Springer-Verlag, Berlin (2002) 369–380.

40. V. Alexandrov, E. Atanassov, and I. Dimov, Parallel Quasi Monte Carlo methods for linear algebra problems, Monte Carlo Methods and Applications 10 (3–4) (2004) 213–219.

41. Y. Li and M. Mascagni, Grid-based Quasi-Monte Carlo applications, Monte Carlo Methods and Applied 11 (1) (2005) 39–55.

42. J. W. L. Wan, K. Lai, A. W. Kolkiewicz, and K. S. Tan, A parallel quasi Monte Carlo approach to pricing multidimensional American options, International Journal of High Performance Computing and Networking 4 (5/6) (2006) 321–330.

43. R. Schürer, Parallel high-dimensional integration: Quasi-Monte Carlo versus adaptive cubature rules, in: V. N. Alexandrov, J. J. Dongarra, B. A. Juliano, R. S. Renner, and C. J. K. Tan (Eds.), The 2001 International Conference on Computational Science – ICCS 2001, San Francisco, CA, USA, May 2001, Lecture Notes in Computer Science 2073, Springer, Berlin, (2001).

44. W. Ch. Schmid and A. Uhl, Parallel Quasi-Monte Carlo integration using (t,s)-sequences, in: P. Zinterhof, M. Vajtersic, and A. Uhl (Eds.), Parallel Computation. Proceedings of ACPC'99, Lecture Notes on Computer Science 1557, Springer (1999) 96–106.

45. W. Ch. Schmid and A. Uhl, Techniques for parallel Quasi-Monte Carlo integration with digital sequences and associated problems, Mathematics and Computers in Simulation 55 (2001) 249–257.

46. B.C. Bromley, Quasirandom number generators for parallel Monte Carlo algorithms, Journal of Parallel and Distributed Computing 38 (1996) 101–104.

47. J. X. Li and G. L. Mullen, Parallel computing of a Quasi-Monte Carlo algorithm for valuing derivatives, Parallel Computing 26 (5) (2000) 641–653.

48. K. Entacher, T. Schell, W. Ch. Schmid, and A. Uhl, Defects in parallel Monte Carlo and Quasi-Monte Carlo integration using the leap-frog technique, Parallel Algorithms and Applications 18 (1–2) (2003) 27–47.

49. E. deDoncker, A. Genz, and M. Ciobanu, Parallel computation of multivariate normal probabilities, Computing Science and Statistics 31 (1999) 89–93.
50. E. deDoncker, R. Zanny, M. Ciobanu, and Y. Guan, Distributed Quasi-Monte Carlo methods in a heterogeneous environment, in: Proceedings of the Heterogeneous Computing Workshop 2000 (HCW'2000), IEEE Computer Society Press (2000) 200–206.
51. E. deDoncker, R. Zanny, M. Ciobanu, and Y. Guan, Asynchronous Quasi-Monte Carlo methods, in: Proceedings of the High Performance Computing Symposium 2000 (HPC'00) (2000) 130–135.
52. L. Cucos and E. deDoncker, Distributed QMC algorithms: New strategies for and performance evaluation, in: Proceedings of the High Performance Computing Symposium 2002 (HPC'02)/Advanced Simulation Techniques Conference (2002) 155–159.
53. G. Ökten and A. Srivivasan, Parallel Quasi-Monte Carlo methods on a heterogeneous cluster, in: K. T. Fang, F. J. Hickernell, and H. Niederreiter (Eds.), Monte Carlo and Quasi-Monte Carlo Methods 2000, Springer (2002) 406–421.
54. A. Srinivasan, Parallel and distributed computing issues in pricing financial derivatives through Quasi-Monte Carlo, in: Proceedings of the International Parallel & Distributed Processing Symposium 2002 (IPDPS'02), Fort Lauderdale, FL, USA, April 2002, IEEE Computer Society Press (2002) 14–19.
55. G. Marsaglia, A. Zaman, and J. C. W. Marsaglia, Rapid evaluation of the inverse of the normal distribution function, Statistics and Probability Letters 19 (1994) 259–266.
56. W. H. Press, S. A. Teukolsky, W. T. Vetterling and B. P. Flannery, Numerical Recipes, Cambridge University Press, Cambridge (2007).
57. H. Hofbauer, A. Uhl, and P. Zinterhof, Quasi Monte Carlo Integration on GRIDS: Using blocked substreams, in: D. Kranzlmüller J. Volkert, T. Fahringer and W. Schreiner (Eds.), Proceedings of the 1st Austrian Grid Symposium, Austrian Computer society (2006).
58. H. Hofbauer, A. Uhl, and P. Zinterhof, Parametrization of Zinterhof sequences for GRID-based QMC integration, in: D. Kranzlmüller J. Volkert, T. Fahringer and W. Schreiner (Eds.), Proceedings of the 2nd Austrian Grid Symposium, Austrian Computer society (2007).
59. N. N. Lebedev, Special functions and their applications (in Russian), GIFML, Moscow-Leningrad, (1963).

Chapter 7
Parallel Evolutionary Computation Framework for Single- and Multiobjective Optimization

Bogdan Filipič and Matjaž Depolli

Abstract

Evolutionary computation is an area of computer science utilizing the mechanisms of biological evolution in computer problem solving. It is concerned with theoretical studies, design and application of stochastic optimization procedures, known as Evolutionary Algorithms (EAs). EAs have proven effective and robust in solving demanding optimization problems that are often difficult if not intractable to traditional numerical methods. They are nowadays widely applied in science, engineering, management, and other domains. However, a drawback of EAs is their computational complexity which originates from iterative population-based search of the solution space. On the other hand, processing a population of candidate solutions makes EAs amenable to parallel implementation that may result in significant calculation speedup.

This chapter presents a parallel evolutionary computation framework developed for solving numerical optimization problems with one or more objectives, and evaluates its performance on a high-dimensional optimization task from industrial practice. The chapter starts with an introduction to optimization problems. It distinguishes between single- and multiobjective optimization and reviews the concepts needed to deal with multiobjective optimization problems, such as the dominance relation and Pareto optimality. Next, EAs as a general-purpose optimization method are described, with a focus on Differential Evolution (DE) which is a particular kind of EA used in our framework. Then, parallelization of EAs is discussed in view of known parallelization types and speedup calculation. The chapter continues with an

Bogdan Filipič
Department of Intelligent Systems, Jožef Stefan Institute, Jamova cesta 39, 1000 Ljubljana, Slovenia, e-mail: `bogdan.filipic@ijs.si`

Matjaž Depolli
Department of Communication Systems, Jožef Stefan Institute, Jamova cesta 39, 1000 Ljubljana, Slovenia, e-mail: `matjaz.depolli@ijs.si`

introduction to the optimization problem in industrial continuous casting, used as a test problem in this work. Afterwards, the proposed parallel evolutionary computation framework is presented. The framework is based on DE and implemented on a cluster of personal computers. It is evaluated on single- and multiobjective variants of the casting optimization problem and the results are analyzed from the perspective of the problem domain and, in particular, the achieved speedup.

7.1 Introduction

In the last decades, a number of computational techniques have been proposed that take inspiration from natural phenomena. Among them is evolutionary computation [1, 2] with the underlying idea of employing the mechanisms of biological evolution in computer problem solving. Search and optimization algorithms designed according to these principles, known as Evolutionary Algorithms (EAs), simulate the evolution of candidate solutions to a given problem, usually starting from a randomly created initial set, and iteratively improving its members until their convergence. Despite its simplicity, this approach has proved efficient and widely applicable. EAs can nowadays be found in a variety of application domains, ranging from science [3] to engineering [4] to management [5].

EAs are in many respects superior to traditional algorithms. Candidate solutions in an EA can be represented and varied in a number of ways which makes these algorithms suitable for solving radically different types of optimization problems. Their operation relies on the quality of solutions being processed and requires no additional information about the search space. As a result, noncontinuous, multimodal and time-dependent problems, hard to solve with traditional algorithms, can be successfully approached with EAs. On the other hand, the population-based search performed with EAs, as opposed to the single-point search in most other algorithms, has both advantages and disadvantages. On the positive side, it results in more than one solution produced in a single algorithm run, which provides a user with alternatives that are sometimes highly desirable. As a disadvantage comes the computational burden of processing a population of candidate solutions. What helps here is the inherent parallelism of EAs: the solutions can be evaluated independently and thus run in parallel for the entire population. This property makes EAs amenable to parallel implementation that may significantly speedup the calculation. This is particularly useful when solution evaluation is computationally expensive, which is often the case with real-world problems.

This chapter describes a parallel evolutionary computation framework developed for solving numerical optimization problems. It starts with a formal introduction to optimization problems and distinguishes between single- and multiobjective optimization. It presents the basic concepts needed to deal with multiobjective optimization problems, such as the dominance relation and Pareto optimality. It continues with a presentation of EAs in general and then focuses on Differential Evolution (DE), an EA specialized in numerical optimization. Both the original

single-objective DE and its multiobjective extension are outlined. Next, parallelization of EAs is discussed regarding the types of parallelization and the calculation
of speedups. The chapter then introduces the task of process parameter tuning in
industrial continuous casting of steel, where the goal is to satisfy the empirical metallurgical criteria formulated to increase the quality of cast steel [6]. This problem
will later be used to evaluate the proposed evolutionary computation framework.
The framework itself is explained in detail. It makes use of any number of processors available and increases the performance of the optimization procedure by
distributing the evaluation of candidate solutions among the processors. Installed on
a cluster [7] of Opteron computers running under Linux, it is empirically evaluated
on the casting optimization problem. Both single- and multiobjective variants of
the problem are exercised and the results analyzed in view of the problem domain
and, in greater detail, the achieved calculation speedup. The optimization results
are comparable to the results obtained previously on the same problem instances,
while, in accordance with predictions, high speedups are achieved. These findings
also suggest further work to enhance the performance of the parallel framework on
hardware architectures different from the one used in this work.

7.2 Optimization Problems

Numerous tasks in science, engineering and business require finding the best solution from a set of candidate solutions that can be evaluated according to a quality
measure and that have to satisfy various constraints. These tasks are called optimization problems, and the procedure of solving an optimization problem is optimization.

We focus on numerical optimization problems where candidate solutions are vectors of real decision variables (sometimes called problem parameters)

$$\mathbf{x} = [x_1, x_2, ..., x_n]^{\mathrm{T}},$$

and the quality measure is a real function $f(\mathbf{x})$ defined over $\mathbb{R}^n$. Formally, a numerical optimization problem is to find a vector,

$$\mathbf{x}^* = [x_1^*, x_2^*, ..., x_n^*]^{\mathrm{T}},$$

that fulfills boundary constraints,

$$x_i^{\mathrm{low}} \leq x_i \leq x_i^{\mathrm{up}}, \ i = 1, 2, ..., n,$$

inequality constraints,

$$g_j(\mathbf{x}) \geq 0, \ j = 1, 2, ..., J,$$

and equality constraints,

$$h_k(\mathbf{x}) = 0, \ k = 1, 2, ..., K,$$

and optimizes $f(\mathbf{x})$.

The boundary constraints restrict each decision variable x_i to take values within its lower bound x_i^{low} and upper bound x_i^{up}, and determine a *decision variable space* (or decision space, for short) of a numerical optimization problem. Solutions satisfying all boundary constraints, inequality constraints and equality constraints are called *feasible solutions*. On the other hand, solutions not satisfying all the constraints are *infeasible*. Furthermore, $f(\mathbf{x})$ is known as the *objective function* or cost function. Optimizing $f(\mathbf{x})$ means either minimizing or maximizing it.

Note that the objective function is not always given explicitly. Particularly in practical optimization problems it may be very demanding, if not impossible, to formulate it. Alternatively, candidate solutions can be evaluated empirically through experiments, measurements, computer simulation, etc.

The traditional definition of a numerical optimization problem given above assumes there is only one objective, and solving such a problem is therefore referred to as single-objective optimization. However, most real-world optimization problems involve multiple objectives, and these are often in conflict with each other, in the sense that improvement of a solution with respect to a selected objective deteriorates it with respect to other objectives. In such cases we deal with multiobjective optimization problems. These can be formally stated analogously to the single-objective ones with the exception that the task is now to optimize a vector function

$$\mathbf{f}(\mathbf{x}) = [f_1(\mathbf{x}), f_2(\mathbf{x}), ..., f_M(\mathbf{x})]^T.$$

As a result, there are two spaces associated with a multiobjective optimization problem: in addition to an N-dimensional decision variable space, there is an M-dimensional *objective space* where the objective vectors can be partially ordered using the *dominance relation*. Objective vector $\mathbf{x}$ is said to dominate objective vector $\mathbf{y}$, formally $\mathbf{x} \prec \mathbf{y}$, iff $\mathbf{x}$ is not worse than $\mathbf{y}$ in all objectives and is better than $\mathbf{y}$ in at least one objective.

Let us illustrate the dominance relation with an example. Consider a multiobjective optimization problem with two objectives, f_1 and f_2, that both need to be minimized. Fig. 7.1 shows five solutions to this problem in the objective space. Comparing solution **a** with other solutions, we can observe that **a** dominates **b** since it is better than **b** in both objectives, that is, $f_1(\mathbf{a}) < f_1(\mathbf{b})$ and $f_2(\mathbf{a}) < f_2(\mathbf{b})$. It also dominates **c** as it is better than **c** in objective f_2 and not worse in objective f_1. On the other hand, **d** outperforms **a** in both objectives, therefore **d** dominates **a** or, in other words, **a** is dominated by **d**. However, regarding **a** and **e**, no such conclusion can be made because $f_1(\mathbf{a}) < f_1(\mathbf{e})$ and $f_2(\mathbf{a}) > f_2(\mathbf{e})$. We say that **a** and **e** are incomparable.

In general, in a set of solutions to a multiobjective optimization problem, there is a subset of solutions that are not dominated by any other solution (**d** and **e** in the example from Fig. 7.1). Referring to the decision variable space, we call this subset a *nondominated set of solutions*, and in the objective space the corresponding vectors are called a *nondominated front of solutions*. The concept is illustrated in Fig. 7.2 where both objectives need to be minimized again. The nondominated set of the entire feasible search space is known as the *Pareto optimal set*, and the non-

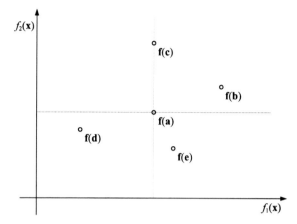

Fig. 7.1 Comparison of solutions to a multiobjective optimization problem in the objective space.

dominated front of the entire feasible search space the *Pareto optimal front* (named after Vilfredo Pareto (1848–1923), an Italian economist, sociologist and a pioneer in the field of multiobjective optimization).

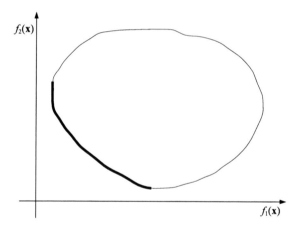

Fig. 7.2 Nondominated front of solutions in the objective space (both objectives need to be minimized).

Objective vectors from the Pareto optimal front represent different trade-offs between the objectives, and without additional information no vector can be preferred to another. With a multiobjective optimizer we search for an *approximation set* that approximates the Pareto optimal front as closely as possible. In practical multiobjective optimization, it is often important to provide a diverse choice of trade-offs. Therefore, besides including vectors close to the Pareto optimal front, the approximation set should also contain near-optimal vectors that are as diverse as possible.

7.3 Evolutionary Algorithms

Evolutionary Algorithms is a common name for a family of search and optimization procedures created and studied in the field of evolutionary computation [1, 2]. The underlying idea is to solve a given problem through computer-simulated evolution of candidate solutions. The set of candidate solutions processed by an EA is called a population, and the population members are referred to as individuals. They are represented in the form suitable for solving a particular problem. Often-used representations include bit strings, real-valued vectors, permutations, tree structures and even more complex data structures. In addition, a fitness function needs to be defined that assigns a numerical measure of quality to the individuals; it roughly corresponds to the cost function in optimization problems.

An EA, shown in pseudocode as Algorithm 7.1, starts with a population of randomly created population members, and iteratively improves them by employing evolutionary mechanisms, such as survival of the fittest individuals and exchange of genetic information between the individuals. The iterative steps are called generations, and in each generation the population members undergo selection and variation.

Algorithm 7.1 Evolutionary Algorithm (EA)

1: create the initial population $\mathbb{P}$ of random solutions;
2: evaluate the solutions in $\mathbb{P}$;
3: **while** stopping criterion not met **do**
4:     create an empty population $\mathbb{P}_{new}$;
5:     **repeat**
6:         select two parents from $\mathbb{P}$;
7:         create two offspring by crossing the parents;
8:         mutate the offspring;
9:         evaluate the offspring;
10:        add the offspring into $\mathbb{P}_{new}$;
11:    **until** $\mathbb{P}_{new}$ is full;
12:    copy $\mathbb{P}_{new}$ into $\mathbb{P}$;
13: **end while**

The selection phase of the algorithm is an artificial realization of the Darwinian principle of survival of the fittest among individuals. The higher the fitness of an individual (i.e., the quality of a solution), the higher the probability of participating in the next generation. In the variation phase, the individuals are modified in order to generate new candidate solutions to the considered problem. For this purpose, the EA applies operators, such as crossover and mutation, to the individuals. The crossover operator exchanges randomly selected components between pairs of individuals (parents), while mutation alters values at randomly selected positions in the individuals.

The algorithm runs until a stopping criterion is fulfilled. The stopping criterion can be defined in terms of the number of generations, required solution quality or as a combination of both. The best solution found during the algorithm run is returned as a result.

EAs exhibit a number of advantages over traditional specialized methods and other stochastic algorithms. Besides the evaluation of candidate solutions, they require no additional information about the search space properties. They are a widely applicable optimization method, straightforward for implementation and suitable for hybridization with other search algorithms. Moreover, it is not difficult to incorporate problem-specific knowledge into an EA in the form of specialized operators when such knowledge is available. Finally, by processing populations of candidate solutions, they are capable of providing alternative solutions to a problem in a single algorithm run. This is extremely valuable when solving multimodal, time-dependent and multiobjective optimization problems.

A somewhat more specialized EA is DE [8, 9]. It was designed for solving numerical optimization and has proved very efficient in this problem domain. In DE, candidate solutions are encoded as n-dimensional real-valued vectors. As outlined in Algorithm 7.2, new candidates are constructed through operations such as vector addition and scalar multiplication (in line 7, F denotes a predefined scalar value). After creation, each candidate is evaluated and compared with its parent and the best of them is added to the new population.

Algorithm 7.2 Differential Evolution (DE)

1: create the initial population $\mathbb{P}$ of random solutions;
2: evaluate the solutions in $\mathbb{P}$;
3: **while** stopping criterion not met **do**
4:     create an empty population $\mathbb{P}_{new}$;
5:     **for** each solution P_i, $i = 1..pop\_size$ from $\mathbb{P}$ **do**
6:         randomly select three different solutions I_1, I_2, I_3 from $\mathbb{P}$;
7:         create a candidate solution $C := I_1 + F \cdot (I_2 - I_3)$;
8:         alter C by crossover with P_i;
9:         evaluate C;
10:         **if** C is better than P_i **then**
11:             add C into $\mathbb{P}_{new}$
12:         **else**
13:             add P_i into $\mathbb{P}_{new}$;
14:         **end if**
15:     **end for**
16:     copy $\mathbb{P}_{new}$ into $\mathbb{P}$;
17: **end while**

7.3.1 Multiobjective Evolutionary Algorithms

In multiobjective optimization, finding an approximation of the Pareto optimal front in a single run requires a population-based method. Therefore, EAs are a reasonable choice for this task. However, since the objective space in multiobjective optimization problems is multidimensional, any EA originally designed for single-objective optimization needs to be extended to deal with multiple objectives. This has been done with several EAs that are now used as multiobjective optimizers and referred to as Multiobjective Evolutionary Algorithms (MOEAs) [10–12].

Based on the single-objective DE is Differential Evolution for Multiobjective Optimization (DEMO) [13]. It extends DE with a particular mechanism for deciding which solutions to keep in the population (see Algorithm 7.3). For each parent in the population, DEMO constructs a candidate solution in the same way as DE. If the candidate dominates the parent, the candidate is added to the new population. If the parent dominates the candidate, the parent is added to the new population. Otherwise, if the candidate and its parent are incomparable, they are both added to the new population. During the construction of candidates for all parents in the population, the new population possibly increases. In this case, it is truncated to the original population size using nondominated sorting and the crowding distance metric in the same manner as in the NSGA-II multiobjective algorithm [14]. These steps are repeated until a stopping criterion is met.

The serial versions of DE and DEMO described here will be used as a foundation for our parallel evolutionary computation framework to efficiently deal with single- and multiobjective optimization problems, respectively.

7.4 Parallel Single- and Multiobjective Evolutionary Algorithms

EAs are an example of inherently parallel algorithms. Fitness evaluation can be independently calculated for each individual and therefore run in parallel for the entire population at a time. This mainly results in a faster algorithm execution, that is, speedup [15], although it could in some cases also loosen hardware bottlenecks, such as memory shortage. This chapter focuses on the speedup, but also provides notes on efficiency (speedup normalized with the number of processors) and hardware bottlenecks where applicable.

7.4.1 Parallelization Types

There are four types of parallel EAs [16, 17], of which three are basic: *master–slave* (also called *global parallelization*), *island*, *diffusion* (also known as *cellular*) and *hybrid* that encompasses combinations of the basic types.

Algorithm 7.3 Differential Evolution for Multiobjective Optimization (DEMO)

```
 1: create the initial population ℙ of random solutions;
 2: evaluate the solutions in ℙ;
 3: while stopping criterion not met do
 4:     create an empty population ℙ_new;
 5:     for each solution P_i, i = 1..pop_size from ℙ do
 6:         randomly select three different solutions I_1, I_2, I_3 from ℙ;
 7:         create a candidate solution C := I_1 + F·(I_2 − I_3);
 8:         alter C by crossover with P_i;
 9:         evaluate C;
10:         if C dominates P_i then
11:             add C into ℙ_new
12:         else
13:             if P_i dominates C then
14:                 add P_i into ℙ_new;
15:             else
16:                 add both P_i and C into ℙ_new;
17:             end if
18:         end if
19:     end for
20:     if ℙ_new contains more than pop_size solutions then
21:         truncate ℙ_new;
22:     end if
23:     copy ℙ_new into ℙ;
24: end while
```

Master–slave EAs are the most straightforward type of parallel EAs and the only one that makes use of the EAs' inherent parallelism. As a consequence, they traverse the search space identically to their serial counterparts. A master–slave EA can be visualized as a master node running a serial EA with a modification in fitness evaluation. Instead of evaluating fitness serially, one individual at a time, until the entire population is evaluated, individuals are evaluated on the master and slave nodes in parallel. The highest efficiency of this parallelization type can be achieved on computers with homogeneous processors and in problem domains where the fitness evaluation time is constant and independent of the individual. When these criteria are fulfilled and the fitness evaluation time is long compared to the time required for other parts of the algorithm, near-linear speedup is possible.

Island EAs, in contrast, are multiple-population algorithms, consisting of several largely independent subpopulations that occasionally exchange a few individuals. In island EAs, each processing node represents an island, running a serial EA on a subpopulation. A new operator is introduced – migration, that handles the exchange of individuals between the islands. Migration occurs either in predefined intervals, e.g., every several generations, or after special events, e.g., when subpopulations start to converge. Communication overhead is therefore smaller compared to the master–slave parallelization type. In general, speedup increases with the number of islands, but the overall efficiency depends on how well the problem is suited for solving with multiple-population EAs compared to single-population EAs.

Diffusion EAs split population into multiple small subpopulations and divide them among the processing nodes. Every subpopulation is allowed to communicate (individuals may interact) with a predefined neighborhood of other subpopulations. These algorithms can also be considered single population with structurally constrained interactions between individuals. Parallelization of this type has large communication overhead and may be worth considering only on large computer clusters with dedicated interconnections between the neighboring processing nodes. Speedup and efficiency depend greatly on the properties of interconnections and the suitability of the problem to the structural constraints imposed by the algorithm.

Hybrid parallel EAs are an attempt to minimize the weaknesses of the basic type algorithms through their hierarchic composition. For example, the island type may be implemented on top of the master–slave type, providing possibility to use all available processing nodes, while keeping the number of islands variable. Hybrid EAs are very adaptable to the underlying hardware architecture, but their design and implementation are more complex.

7.4.2 Calculation of Speedups

Traditionally, speedup is defined as the ratio between the execution times of the best serial algorithm and the best parallel algorithm:

$$S = \frac{T_s}{T_p} \, . \tag{7.1}$$

As this definition depends on the execution times, we call it the measured speedup, to contrast it with the estimated speedup. In case of the master–slave EAs, selection of the best algorithms is trivial, since the parallel algorithm traverses the search space identically to its serial counterpart. Therefore, for a valid speedup measurement, both algorithms should be run with the same algorithm parameter setting, for the same number of generations.

More care should be taken when dealing with other types of parallel EAs. Modifications needed for the island and diffusion EAs may have a positive influence on some EAs and in some problem domains. These modifications can always be translated back into a serial algorithm, since every parallel algorithm can be trivially serialized. This way, a new, best-known serial algorithm for calculation of speedup can be obtained. Therefore, the best serial counterpart to a particular multipopulation parallel EA may either be its serial implementation or the original, single-population EA.

The only limiting factor for serialization could be hardware (e.g., multiple-population EAs require more memory than single-population EAs). In such a case, parallelization serves as a means of alleviating hardware constraints as well. The obtained speedup in such cases would be due to parallel execution and due to algorithm improvements, with either contribution unobtainable from the measurements alone.

Additionally, the island and diffusion EAs make use of additional parameters – the number of subpopulations and the size and shape of the neighborhood. In parallel implementations these parameters are to a large extent fixed to the number of processors and the computer architecture, but are free in serial implementations. Therefore, the best algorithm parameter setting may differ between serial and parallel implementations.

While measuring the parallelization speedup of the master–slave EAs is straightforward, it requires a lot of additional work for the multipopulation parallel EAs. Since the knowledge of speedup is usually not a priority to the algorithm developers, the parallel multipopulation EAs are often compared only to the original serial EAs. This technique frequently yields super-linear speedups, which are a good indication of the use of suboptimal serial algorithms.

We explore the master–slave EAs in more detail, to estimate their limitations in speedup. We start with the theoretical limit on speedup according to the Amdahl's law:

$$S_{max} = \frac{1}{(1-P) + \frac{P}{N}} \, , \qquad (7.2)$$

where P is the parallel portion of the algorithm and N is the number of processors. The actual speedup of an algorithm will depend on how well the parallel portion can be spread up among N processors. Considering the simplest master–slave parallelization type, where only fitness evaluations are parallelized, P is the portion of the serial algorithm execution time spent on fitness evaluation. It should be noted that through the process of parallelization, the interprocessor communication is added to the algorithm, which effectively decreases its parallel portion. As demonstrated later on, when the interprocessor communication is taken into consideration, P can still reach very high values if fitness evaluation is complex and time consuming. On the other hand, N is limited by the population size N_p. Only the population of a single generation can be evaluated at a time, even when more processors are available. Speedup upper bound therefore equals the population size:

$$\lim_{P \to 1} S_{max} = \lim_{P \to 1} \frac{1}{(1-P) + \frac{P}{N_p}} = N_p \, . \qquad (7.3)$$

Another important observation is that not only should $N_p \leq N$, but also $N_p \mid N$ (N_p divides N), for the algorithm to fully utilize all processors. The algorithm needs $\lceil \frac{N_p}{N} \rceil$ iterations to fully evaluate the population and therefore has $\lceil \frac{N_p}{N} \rceil \times N$ processor time slots to fill with N_p tasks (fitness evaluations). It is free to choose the best way to allocate the tasks to processor time slots over the iterations but there will always remain N_p mod N unallocated slots per generation, for which the processors will be left idle. From this we can derive the effective number of processors used by the algorithm $N_{eff} = N_p / \lceil \frac{N_p}{N} \rceil$. Finally, substituting N with N_{eff} in Eq. (7.2) we can rewrite the speedup equation as

$$S_{\max} = \frac{1}{(1-P) + \frac{P \times \lceil \frac{N_p}{N} \rceil}{N_p}} . \tag{7.4}$$

An example of $S_{\max}(N)$ for population size $N_p = 32$ and parallel fraction $P = 1$ is shown in Fig. 7.3.

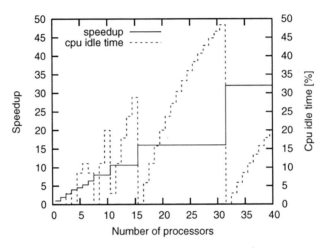

Fig. 7.3 Maximum speedup and processor idle time vs. the number of available processors for a master–slave parallel EA with $P \rightarrow 1$ and $N = 32$.

The dependence of speedup on the number of processors is alleviated by the insensitivity of EAs to the population size. Because of the stochastic nature of EAs, an approximate interval can be determined rather than an exact number for the best population size on a given problem. If the interval is larger than the number of processors, then fixing the population size to a multiple of the number of processors while keeping it inside the interval is possible. In cases when optimal selection of the population size within the interval is not possible, speedup calculation should be amended. Suppose an EA with the optimal population size in comparison to an EA with the selected population size has speedup S_{opt}. The actual maximum speedup of a parallel master–slave EA will then be

$$S_{\max}^* = \frac{S_{\max}}{S_{\text{opt}}} . \tag{7.5}$$

7.5 Casting Process Optimization Task

Continuous casting of steel is widely used at modern steel plants to produce various steel semi-manufactures. The proces is schematically shown in Fig. 7.4. In this process, liquid steel is poured into a bottomless mold which is cooled with internal water flow. The cooling in the mold extracts heat from the molten steel and initiates the formation of a solid shell. The shell formation is crucial for the support of the slab behind the mold exit. The slab then enters the secondary cooling area where additional cooling is performed by water sprays. Led by support rolls, the slab gradually solidifies and finally exits the casting device. At this stage, it is cut into pieces of predefined length.

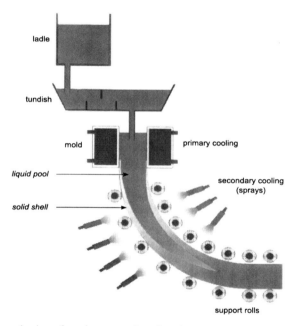

Fig. 7.4 A schematic view of continuous casting of steel.

The secondary cooling area of the casting device is divided into cooling zones and the cooling water flows in the zones can be set individually. In each zone, cooling water is dispersed to the slab at the center and corner positions. Target temperatures are specified for the slab center and corner in every zone, and the optimization task is to tune the cooling water flows in such a way that the resulting slab surface temperatures match the target temperatures as closely as possible. From metallurgical practice this is known to reduce cracks and inhomogeneities in the structure of the cast steel. Formally, an objective f_1 is introduced to measure deviations of actual temperatures from the target ones:

$$f_1 = \sum_{i=1}^{N_Z} |T_i^{\text{center}} - T_i^{\text{center}*}| + \sum_{i=1}^{N_Z} |T_i^{\text{corner}} - T_i^{\text{corner}*}|, \qquad (7.6)$$

where N_Z denotes the number of zones, T_i^{center} and T_i^{corner} the slab center and corner temperatures in zone i, and $T_i^{\text{center}*}$ and $T_i^{\text{corner}*}$ the respective target temperatures in zone i. This objective encompasses the key requirement for the process to result in high-quality cast steel. Technically, this is a single-objective version of the casting optimization task.

In addition, there is a requirement for core length, l^{core}, which is the distance between the mold exit and the point of complete solidification of the slab. The target value for the core length, $l^{\text{core}*}$, is prespecified, and the actual core length should be as close to it as possible. Shorter core length may result in unwanted deformations of the slab as it solidifies too early, while longer core length may threaten the process safety. This requirement can be treated as an additional objective, f_2:

$$f_2 = |l^{\text{core}} - l^{\text{core}*}|, \qquad (7.7)$$

and the more demanding version of the optimization task is then to minimize both f_1 and f_2 over possible cooling patterns (water flow settings). The two objectives are conflicting, hence it is reasonable to handle this optimization problem in the multiobjective manner.

In the optimization procedure, water flows cannot be set arbitrarily, but according to the technological constraints. For each zone, lower and upper bounds are prescribed for the center and corner water flows. Moreover, to avoid unacceptable deviations of the core length from the target value, a hard constraint is imposed: $f_2 \leq \Delta l_{\text{max}}^{\text{core}}$. Solutions violating the water flow constraints or the core length constraint are considered infeasible.

A prerequisite for optimization of this process is an accurate mathematical model of the casting process, capable of calculating the temperature field in the slab as a function of coolant flows and evaluating it with respect to the objectives given by Eqs. (7.6) and (7.7). For this purpose we use a numerical simulator of the process with the Finite Element Method (FEM) discretization of the temperature field and the related nonlinear heat transfer equations solved with relaxation iterative methods [6].

7.6 Parallel Evolutionary Computation Framework

We present a parallel framework for numerical single- and multiobjective optimization on homogeneous parallel computer architectures. It is based on single-objective DE and is extended to DEMO when multiobjective optimization is required.

The framework is able to utilize any number of processors by implementing the master–slave parallelization scheme for both optimization algorithms. Although designed for use on homogeneous parallel computer architectures, it can use heterogeneous architectures as well, but with lower utilization of faster processors. When a single processor is used, master–slave algorithms degenerate into their nonparallel versions, thus avoiding potential overhead of the parallelization scheme.

In the framework, the optimization procedure is performed in three stages: initialization, generational computation and finalization. The initialization consists of reading the input files and settings and the setup of initial population. Generational computation iterates over generations, where in each iteration fitness values are calculated for individuals of the current population and the EA operators are applied to them, spawning the next generation. In finalization, the results are formatted and returned to the user.

While the initialization and finalization are run by the master process, the generational computation can be run in parallel by all processes. Each iteration starts with the master process holding a vector of individuals of unknown fitness. These are then evaluated by the master and slave processes in parallel, which requires interprocess communication. For this purpose, the Message Passing Interface (MPI) [18] is used. It implements the interprocess communication in a two-part, coupled fashion. The first part distributes the data on the individuals among the slave processes, and the second part returns the fitness values to the master process. For the sake of simplicity, only the data on one individual is transferred to each slave process per communication couple. This forces the communication couple to happen more than once per generation if the population size is larger than the number of processors. The master process receives the results from the slave processes inside a blocking operation, that is, it waits for all the results before it continues execution, effectively synchronizing the processors. This, in combination with multiple communication couples per generation, causes some unnecessary synchronizations. After the fitness values for all individuals are known, the master process applies the EA operators and spawns the next generation. The slave processes are idle at this time, waiting to receive the data on individuals of the next generation.

The parallelization approach employed by the proposed framework is, in the context of multiobjective optimization, known as the Parallel Function Evaluation (PFE) variant of the single-walk parallelization [19]. It is aimed at speeding up the computations, while the basic behavior of the underlying algorithms remains unchanged.

7.6.1 Speedup Estimation

What is the expected speedup of the framework running on several processors in comparison to the framework running on a single processor, solving an optimization problem? One should be able to answer this question before starting the optimization, to use the most appropriate number of processors. To answer this question, we start with the speedup as defined in Eq. (7.1). We simplify it by only using the time for generational computation instead of the total execution time for both, the serial and parallel implementations. This is reasonable because the initialization and finalization are faster than even a single application of the EA operators, and are negligible in cases when parallelization is considered, that is, when the total execution time is expected to be long. Furthermore, because the generational computation is a series of identical single generation computations, we simplify the definition of speedup to only consider a single generation. Thus we get the initial form of the speedup equation:

$$S = \frac{T_s + T_e * N_p}{T_s + T_e \times \lceil \frac{N_p}{N} \rceil} ,$$
(7.8)

where T_e is the time required for a single fitness evaluation, T_s is the time required for the execution of a single generation, excluding the time required for fitness evaluations, N_p is the population size, and N is the number of processors. This is a good estimation if two criteria are met. The first criterion is constant time of fitness evaluation. This means that all fitness evaluations take exactly the same amount of time to complete, not depending on the input, the processor, nor any random factor. The second criterion is that parallelization produces negligible calculation overhead. In the master–slave parallelization scheme, the overhead consists of the time required for interprocess communication, including the time the master process is waiting for the results from the slave processes.

The time required for communication, T_c, can be simply added to the denominator in Eq. (7.8). It is irrelevant when it is orders of magnitude shorter than the fitness evaluation time, but when it is not, it has to be estimated, because it depends on the problem domain as well as the communication protocols and hardware. For instance, first the number of bytes used to represent the fitness function input parameters sets the base size of messages sent from the master to the slaves, and the number of bytes used to represent the evaluation results sets the base size of messages sent from the slaves to the master. Then the protocols over which the messages are sent, e.g., TCP/IP, and the library which implements message passing, e.g., MPI, increase the message sizes with their overhead. Last, the hardware determines how fast the messages of certain sizes can be sent between the processors. The speedup then equals to

$$S = \frac{T_s + T_e * N_p}{T_c + T_s + T_e \times \lceil \frac{N_p}{N} \rceil}$$
(7.9)

Eliminating the constant fitness evaluation time criterion from the equation is more complex. The master process cannot apply the EA operators until all the individuals of the population have their fitness values evaluated. The process executing the longest fitness evaluation thus forces all other processes to wait until it finishes. We define the time required for execution of n fitness evaluations in parallel, $T_{ep}(n)$, in Eq. (7.10) as the expected value of a maximum of n independent fitness evaluation times. One way of calculating the expected value is numerically, from the cumulative distribution function (CDF) of maximum time of n fitness evaluations, which equals the CDF of fitness evaluation time, raised to the power of n.

$$T_{ep}(n) = E(\max_{i=1}^{n}\{t_{e,i}\}) \tag{7.10}$$

The framework executes a series of parallel evaluations during a single generation if the population size is larger than the number of processors. Individuals are split into $\lceil \frac{N_p}{N} \rceil$ groups, with first $\lfloor \frac{N_p}{N} \rfloor$ groups of the size equal to the number of processors, and the last group (if $\lceil \frac{N_p}{N} \rceil \neq \lfloor \frac{N_p}{N} \rfloor$) of size N_p mod N. Each group is separately evaluated in parallel, adding to the total evaluation time of a population, which can now be calculated as $T_{ep}(N) \times \lfloor \frac{N_p}{N} \rfloor + T_{ep}(N_p$ mod $N)$. The final form of the estimated speedup equation can now be written as

$$S = \frac{T_s + T_e * N_p}{T_c + T_s + T_{ep}(N) \times \lfloor \frac{N_p}{N} \rfloor + T_{ep}(N_p \bmod N)} \tag{7.11}$$

7.7 Empirical Evaluation

An empirical evaluation of the proposed framework was performed on the computer cluster comprised of 17 dual processor computers. Optimization of continuous casting served as a test domain for both the single- and multiobjective optimization.

7.7.1 Experimental Setup

For the evaluation of the framework, a cluster of 17 dual-processor nodes (each node being a personal computer) was used. The nodes are all interconnected through an Ethernet switch, and, in addition, there are several direct interconnections between the nodes (see Fig. 7.5). Nodes 1 through 16 are connected by a toroidal 4-mesh, and nodes 1 through 4 are directly connected to the additional node. This node serves as a host node, through which users access the cluster. Static routing is used to direct the communication between the pairs of nodes, which are not physically interconnected, through the switch. This makes the use of any desired topology possible. In our tests, star topologies of various sizes were used.

The cluster is composed of identical personal computers, each containing two AMD Opteron 244 processors, 1024 MB of RAM, a hard disk drive and six 1000 MB/s Full Duplex Ethernet ports. On each computer, there is an independent installation of the Fedora Core 2 operating system and the MPICH v1.2.6 library that supports communication between the computers and is an implementation of the MPI.

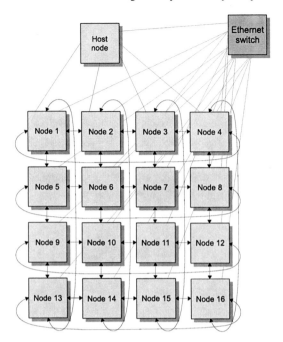

Fig. 7.5 Architecture of the cluster used in tests.

During the experiments, all nodes are required to be running only the background system processes which leaves nearly all capabilities to be used by the framework.

The parallel optimization algorithm was written in C++ and compiled with *gcc v3.3.3* for target 64-bit *Linux*, while the continuous casting simulator was compiled for 32-bit *Microsoft Windows* and was executed through an early version of *Wine* (an application providing the compatibility layer for the *Microsoft Windows* programs). There was also a layer of scripts, translating the communication between the optimization algorithm and the simulator, that is, filtering and converting input/output files of the simulator.

7.7.2 Experiments and Results

Numerical experiments in optimizing the continuous casting process were performed to analyze both the effectiveness and efficiency of the developed parallel framework. The former relates to the quality of results, while the latter refers to the speedup achieved with the parallel optimization approach.

Optimization calculations were performed for a selected steel grade and slab cross section of 1.70×0.21 m and for various casting speeds: the usually practised speed of 1.8 m/min and two lower speeds of 1.6 and 1.4 m/min that are exercised when the process needs to be slowed down to ensure the continuity of casting; for example, when a new batch of molten steel is delayed. Candidate solutions in parallel DE and DEMO were encoded as 18-dimensional real-valued vectors, representing coolant flow values at the center and the corner positions in the nine zones of the secondary cooling area. Search intervals for coolant flows at the center and the corner positions in zones 1–3 were between 0 and 50 m³/h, and in the zones 4–9 between 0 and 10 m³/h. The target core length, l^{core*}, was 27 m and the maximum allowed deviation from the target, Δl^{core}_{max}, was 7 m. Reasonable population size found in initial experiments was 30.

It turned out that for the single-objective and the two-objective versions of the task, the parallel optimization procedure was able to discover the solutions known from previous applications of serial optimization algorithms [6,20]. To illustrate the results for the more challenging two-objective version, Fig. 7.6 shows the resulting nondominated fronts of solutions (approximating Pareto optimal fronts) found by the parallel DEMO algorithm for various casting speeds. It can be seen that the two objectives can simultaneously be fulfilled to the highest degree at the regular casting speed of 1.8 m/min. On the other hand, the lower the speed, the more evident the conflicting nature of the two objectives: improving the coolant flow settings with respect to one objective makes them worse with respect to the other. In addition, a systematic analysis of the solutions confirms that the actual slab surface temperatures are in most cases higher than the target temperatures, while the core length is shorter than or equal to the target core length.

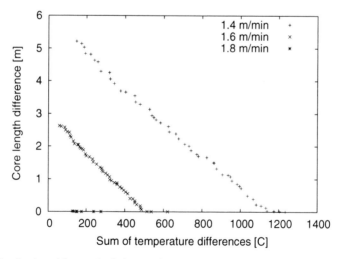

Fig. 7.6 Nondominated fronts of solutions to the two-objective steel casting optimization problem for various casting speeds.

In further experimentation, a detailed analysis of the framework speedup on various numbers of processors was carried out. To make the experimental results directly comparable, the framework parameters other than the number of processors did not vary between the tests. Because the framework is based on the master–slave parallelization type, the population size was first selected as the one that suits the problem while also being a multiple of the number of processors. As shown in previous work [6, 20], optimization of continuous casting with DE and DEMO seems to work best with population sizes between 20 and 40, which coincides well with the 34 available processors. Number 34 unfortunately has only four divisors (1, 2, 17 and 34). Having numerous divisors is important as it allows for numerous tests where population size is a multiple of the number of processors. Therefore, the population size of 32 was chosen, which has six divisors (1, 2, 4, 8, 16 and 32). With this population size, six tests with various number of processors and maximum efficiency (minimum processor idle time) were possible. In every test, the framework was run five times for each, the single- and multiobjective optimization.

Mean wall clock times of the tests were recorded and are summarized in Table 7.1. Two important observations can be made from the measured wall clock times alone. The first one is great variance of the results. The most likely cause of this is the variable fitness evaluation time, but we will explore this later. To simplify matters, we will only use mean values of the tests in further discussion. The second observation is that the multiobjective optimization appears slightly slower than the single-objective optimization. The single-sided paired t-test however returns the p value of 0.12, which means the difference in times is not statistically significant. Therefore, both algorithms can be considered equally fast and the following analysis can be generalized in terms of the algorithm choice. Multiobjective optimization will serve as the basis for all further speedup analyses with its differences towards single-objective optimization mentioned only when necessary.

Table 7.1 Mean wall clock times and their standard deviations for the tests with variable number of processors. All times are specified in seconds.

Number of processors	DE Mean	DE St. dev.	DEMO Mean	DEMO St. dev.
1	295735	1180	298502	1576
2	143661	945	145584	5646
4	79565	1018	79751	446
8	41412	370	41105	389
16	21123	93	21454	183
32	10925	122	11019	276

We can calculate the speedup directly from the mean wall clock times of the DEMO tests, but let us first try to estimate it with Eq. (7.11). First, we make a series of 100 test runs of fitness evaluations from which we estimate the fitness evaluation time to be distributed normally with $\mu = 32.2$ s and $\sigma = 1.5$ s. We estimate all other times in the equation to be in the order of milliseconds and therefore negligible compared to the fitness evaluation time. Now we can estimate the speedup for arbitrary number of processors, N_p, and compare it to the measured speedups. Fig. 7.7 shows the estimated and measured speedups, and the theoretical limit for the speedup on $N_p \in [1 \ldots 34]$.

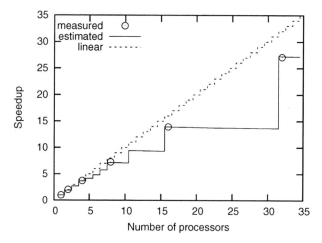

Fig. 7.7 Measured and analytically derived speedup for DEMO on the continuous casting problem, with population size 32, for various number of processors. Linear speedup as the theoretical limit of speedup for master–slave EAs is also shown for reference.

In addition to the total execution time, times of four mutually exclusive steps of the optimization procedure are measured. The first step, which should also be the most time-consuming, is fitness evaluation. The second step is the interprocess communication. This consists of sending the data on individuals from the master process to the slave processes, and sending the fitness evaluation results in the opposite direction. Waiting of the master process for the slave processes to start sending their results is also included in the communication, because in the source code the two are not separated. Next are the output operations, which consist of log keeping and storing the data on the individuals from each generation in a file. The last step is the application of the algorithm operators. The distribution of times among the steps described above for multiobjective optimization on 32 processors is shown in Table 7.2

Table 7.2 The distribution of total wall clock time among steps of the optimization procedure. All times are specified in seconds.

Algorithm stage	Mean	St. dev.
Total	11019	308
Evaluation	9911	376
Communication	1108	129
Input/output	0.307	0.001
EA operators	0.135	0.003

A quick scan over the times used by the algorithm steps reveals that the algorithm behaves as predicted. Fitness evaluation represents by far the largest part of execution time, while the times of input/output operations and the EA operators are negligible in comparison. On the other hand, the interprocess communication time, which should be negligible, represents a substantial proportion of the total algorithm wall clock time. But this view is misleading because the communication times are bundled together with the times of waiting for communication. The latter are a consequence of, and in Eq. (7.11) also a part of, varying fitness evaluation times. We can understand communication times better by analyzing them per generation.

Out of the four steps for which the times are recorded per generation, two – interprocess communication and fitness evaluation – are worth special attention. In addition to the interprocess communication time and the fitness evaluation time as measured on the master process, Fig. 7.8 also shows the maximum time of all fitness evaluations in a generation. It can be seen that the measured communication time roughly equals the difference between the longest fitness evaluation time and the fitness evaluation time on the master process. Measured communication time is therefore mostly spent waiting for the longest fitness evaluations. Pure communication time can be estimated as the sum of communication and the fitness evaluation times on the master process, from which the longest fitness evaluation time is subtracted. It sums up to 1.2 s for the shown optimization run, which can be translated to 4 ms per generation on average. Although this is only a rough estimate, it shows that communication times are an order of magnitude longer than the times of the input/output operations and the EA operators, but still negligible in comparison to the fitness evaluation time. In conclusion, the measured interprocess communication times are in good accordance with the estimates made before the experiments.

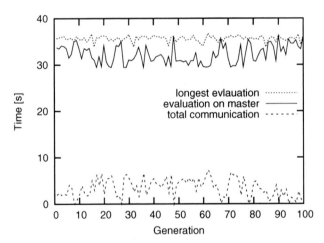

Fig. 7.8 Fitness evaluation and interprocess communication times per generation of multiobjective optimization on 32 processors, for the initial 100 generations. Fitness evaluation time of the master process is contrasted with the longest fitness evaluation time of all the processes.

7.8 Conclusion

In this chapter, a parallel evolutionary computation framework for solving numerical optimization problems with one or more objectives was presented. Master–slave parallel versions of the DE and DEMO algorithms were implemented for solving single- and multiobjective problems, respectively. The implementation was a straightforward one, parallelization was done only on the inherently parallel portion of the algorithms – the fitness evaluation – thus keeping the algorithm behavior independent of the number of processors. The interprocess communication was implemented in a simple manner, focusing on its robustness rather than speed.

The performance of the developed framework was empirically evaluated on an industrial optimization problem of tuning coolant flows in the continuous steel casting process. A single- and a two-objective fitness evaluation function were derived from a computer simulator, implementing a test case of the continuous casting procedure. The quality of the results and the achieved parallel speedups were evaluated separately. The results proved satisfactory and comparable to the results obtained previously on the same problem instances. The measured speedups were high (e.g., the speedup on 32 processors was 27) and matched the predictions.

The presented framework demonstrated that due to a relatively simple master–slave parallelization model, EAs can be extensively used on homogeneous parallel hardware. At the same time, it highlighted a weakness of the master–slave model – the sensitivity of the speedup to constant fitness evaluation time. In our case, we experienced variability in the execution time of fitness evaluation at the order of several percent. Similar effect would be expected from a constant-time fitness evaluation function executing on heterogeneous processors or even on homogeneous processors under some load, that is, executing other jobs. Therefore, our future work will focus on overcoming the demand for constant fitness evaluation time. This will be achieved by eliminating the synchronous nature of the master–slave parallelization type and thus maximizing the algorithm efficiency (minimizing processor idle time). In this way, we expect to increase the speedup and make the algorithms more usable on heterogeneous hardware architectures that are less suitable to ordinary master–slave EAs.

Acknowledgments The authors are grateful to Professor Erkki Laitinen from the Department of Mathematical Sciences, University of Oulu, Finland, for providing the mathematical model and technical details of the continuous casting process optimized in this study. The work was supported by the Slovenian Research Agency under research programmes P2-0095 *Parallel and Distributed Systems*, and P2-0209 *Artificial Intelligence and Intelligent Systems*.

References

1. A. E. Eiben, J. E. Smith, Introduction to Evolutionary Computing, Springer-Verlag, Berlin, (2003).
2. K. De Jong, Evolutionary Computation: A Unified Approach, The MIT Press, Cambridge, (2006).
3. G. B. Fogel, D. W. Corne (Eds.), Evolutionary Computation in Bioinformatics, Morgan Kaufmann Publishers, Amsterdam, (2003).
4. D. Dasgupta, Z. Michalewicz (Eds.), Evolutionary Algorithms in Engineering Applications, Springer-Verlag, Berlin, (1997).
5. J. Biethahn, V. Nissen (Eds.), Evolutionary Algorithms in Management Applications, Springer-Verlag, Berlin, (1995).
6. B. Filipič, E. Laitinen, Model-based tuning of process parameters for steady-state steel casting, Informatica 29 (4) (2005) 491–496.
7. R. Buyya, High Performance Cluster Computing: Architectures and Systems, Vol. 1, Prentice Hall, Upper Saddle River, (1999).
8. K. V. Price, R. Storn, Differential evolution: A simple evolution strategy for fast optimization, Dr. Dobb's Journal 22 (4) (1997) 18–24.
9. K. Price, R. M. Storn, J. A. Lampinen, Differential Evolution: A Practical Approach to Global Optimization (Natural Computing Series), Springer-Verlag, Berlin, (2005).
10. K. Deb, Multi-Objective Optimization using Evolutionary Algorithms, John Wiley & Sons, Chichester, (2001).
11. C. A. Coello Coello, D. A. Van Veldhuizen, G. B. Lamont, Evolutionary Algorithms for Solving Multi-Objective Problems, Kluwer Academic Publishers, New York, (2002).
12. A. Abraham, L. Jain, R. Goldberg (Eds.), Evolutionary Multiobjective Optimization, Springer-Verlag, London, (2005).
13. T. Robič, B. Filipič, Demo: Differential evolution for multiobjective optimization, in: C. A. Coello Coello, A. Hernández Aguirre, E. Zitzler (Eds.), Conference on Evolutionary Multi-Criterion Optimization, Vol. 3410 of Lecture Notes in Computer Science, Springer, Berlin, (2005), pp. 520–533.
14. K. Deb, A. Pratap, S. Agarwal, T. Meyarivan, A fast and elitist multiobjective genetic algorithm: NSGA-II, IEEE Transactions on Evolutionary Computation 6 (2) (2002) 182–197.
15. S. G. Akl, Parallel Computation: Models and Methods, Prentice Hall, Upper Saddle River, (1997).
16. E. Cantú-Paz, A survey of parallel genetic algorithms, Tech. rep., University of Illinois at Urbana-Champaign (1997).
17. D. A. van Veldhuizen, J. B. Zydallis, G. B. Lamont, Considerations in engineering parallel multiobjective evolutionary algorithms, IEEE Transactions on Evolutionary Computation 7 (2) (2003) 144–173.
18. M. Snir, S. Otto, S. Huss-Lederman, D. Walker, J. Dongarra, MPI – The Complete Reference, The MIT Press, Cambridge, (1996).
19. A. J. Nebro, F. Luna, E.-G. Talbi, E. Alba, Parallel multiobjective optimization, in: E. Alba (Ed.), Parallel Metaheuristics, John Wiley & Sons, New Jersey, (2005), pp. 371–394.
20. B. Filipič, T. Tušar, E. Laitinen, Preliminary numerical experiments in multiobjective optimization of a metallurgical production process, Informatica 31 (2) (2007) 233–240.

Chapter 8
WaLBerla: Exploiting Massively Parallel Systems for Lattice Boltzmann Simulations

Christian Feichtinger, Jan Götz, Stefan Donath, Klaus Iglberger and Ulrich Rüde

Abstract

In this chapter, a software concept for massively parallel computational fluid dynamics (CFD) applications is introduced. The focus thereby lies on the parallelization, which is based on a domain partitioning scheme named patch concept. This concept also enables a seamless specialization of the partitions to different application features as well as the possibility for further optimization such as memory reduction. It is discussed in detail how our design ensures an efficient and flexible implementation. The suitability and efficiency of this concept is demonstrated and evaluated with the waLBerla project, which aims at the development of an efficient massively parallel lattice Boltzmann framework providing the necessary features for several CFD applications. To discuss the suitability of the parallelization for massively parallel usage, various test scenarios have been investigated on different architectures. These tests include serial, weak and strong scaling experiments up to 810 cores and up to a domain size of 1530^3 lattice cells.

Christian Feichtinger
Chair for System Simulation, Friedrich-Alexander University Erlangen-Nuremberg, 91058 Erlangen, Germany, e-mail: christian.feichtinger@informatik.uni-erlangen.de

Jan Götz
Chair for System Simulation, Friedrich-Alexander University Erlangen-Nuremberg, 91058 Erlangen, Germany, e-mail: jan.goetz@informatik.uni-erlangen.de

Stefan Donath
Chair for System Simulation, Friedrich-Alexander University Erlangen-Nuremberg, 91058 Erlangen, Germany, e-mail: stefan.donath@informatik.uni-erlangen.de

Klaus Iglberger
Chair for System Simulation, Friedrich-Alexander University Erlangen-Nuremberg, 91058 Erlangen, Germany, e-mail: klaus.iglberger@informatik.uni-erlangen.de

Ulrich Rüde
Chair for System Simulation, Friedrich-Alexander University Erlangen-Nuremberg, 91058 Erlangen, Germany, e-mail: ulrich.ruede@informatik.uni-erlangen.de

R. Trobec et al. (eds.), *Parallel Computing*, DOI 10.1007/978-1-84882-409-6_8,
© Springer-Verlag London Limited 2009

8.1 Motivation

In computational fluid dynamics (CFD), many applications of scientific interest share physical and computational aspects. For research environments, the usual practice is one program for each application, leading to a reimplementation of the shared physics, the common data structures and also the parallelization, which often requires a considerable effort. Furthermore, this replicated functionality has to be validated for each application, again leading to unnecessary work. In addition to a design flexible enough to include further applications, a generic framework has to be suitable for the simulation of large domains, which cannot be simulated on a single CPU. Hence, the framework has to be adapted for parallel architectures. The waLBerla software library has been designed to provide such a framework. It will be used throughout this chapter to demonstrate the requirements and a possible solution for a parallel software library, i.e., flexible enough to support research for several physical applications that cannot be simulated by existing software packages. For a detailed description of the features of waLBerla, e.g., parallel simulation output or input descriptions see Feichtinger et al. [1].

Most of today's flow simulations are based on numerical schemes that solve the Navier-Stokes (NS) equations directly. However, there exists an alternative approach named lattice Boltzmann method (LBM). This method is based on solving an approximation of the Boltzmann equation and thus is a kinetic-based approach. For the waLBerla software library, the LBM has been chosen due to its advantages for the parallelization as well as its suitability for the scheduled applications. These applications cover moving charged colloids [2–4], fluid flow in blood vessels [5] and free surface flows [6,7] through micro porous media in fuel cells.

The requirement of large domains mentioned above results from, e.g., the simulation of a finely resolved representative volume (REV) of the gas diffusion layer (GDL) in a polymer electrolyte fuel cell [8]. Approximately, the size of the REV[1] is about $0.45\,\mathrm{mm} \times 0.45\,\mathrm{mm} \times 0.1$ mm. The volume of a lattice cell has to be $\delta x^3 = 0.1^3 \mu m^3$ due to accuracy reasons and the limitation of the LBM to small Knudsen Numbers. With a porosity of 10% this leads to $1.8 \cdot 10^{10}$ fluid cells, which results in a memory requirement of about 6.5 TiB for the LBM (for LBM resource requirements see Sect. 8.2). Such a simulation is not feasible on a single CPU. A possible system for solving the problem above is the (Höchstleistungsrechner in Bayern) HLRB II which is an SGI Altix 4700 [9] featuring 39 TiB main memory. A rough estimation shows the power of this machine: With the performance of the current implementation of waLBerla (see Sect. 8.5.1), the above example results in a theoretical computation time of about 3 hours per time step, given a single core CPU with enough memory. Assuming a parallel efficiency of 70%, a single time step would take about 1.5 seconds on the 4864 dual-core CPUs of the HLRB II. Thus running fifty thousand time steps would require about 20 hours, instead of 17 years. Hence, only with an efficient parallelization it is possible to simulate the fluid flow in a GDL.

---

[1] Minimum size of REV based on findings from internal projects. No publications yet.

The remainder of this chapter is organized as follows: In the subsequent paragraph, a brief overview of the LBM in three-dimensional is given, followed by the introduction of the waLBerla patch concept in Sect. 8.3. These patches are subdivisions of the fluid domain, which are the basic components for the parallelization, the optimization strategies, and the flexibility that is needed for the integration of further applications. In Sect. 8.4, the implementation of the process local and Message Passing Interface (MPI) communication is explained in detail. Performance results are given in Sect. 8.5, where the serial performance as well as the parallel performance for various architectures has been evaluated. This investigation discusses the suitability of the parallel concept for massively parallel usage in basic geometries. The article is concluded in Sect. 8.6 with a summary and outlook.

8.2 Introduction to the Lattice Boltzmann Method

The LBM is one approach to solve CFD problems numerically. It originates from the lattice gas cellular automata (LGCA), whereas McNamara and Zanetti were the first to introduce the Boltzmann collision operator to LGCA in 1988 [10]. Further work [11] has shown that the LBM can be directly derived from the continuous Boltzmann equation. Hence, it is independent of the LGCA and based on kinetic theory. It can also be shown that the LBM is equivalent to an explicit finite difference scheme of the NS equations with second-order spatial accuracy and first-order temporal accuracy [12]. Amongst others, the LBM has been successfully applied to free surface flows [7], multiphase flows [13], flows through porous media [14], fluid mixtures [15], blood flows [16] and metal foams [17]. Performance optimizations techniques for the LBM can be found in [18–20]. The advantages of the LBM are the explicit update rule, the fast mesh generation due to the Cartesian grids, and that many macroscopic and hydrodynamic effects result from mesoscopic quantities. A detailed description of the LBM can be found in [21–23]. In the remainder of this Section an overview of the governing equations of the LBM is provided.

For the waLBerla software library, the D3Q19 stencil [24] and the LBGK [22] model are used. With the D3Q19 stencil, the LBM is based on cubic cells with 19 unknowns, the particle distribution functions (PDF) $f_\alpha(x_i, t)$, which are defined as the expected amount of particles in the volume δx^3 located at the lattice position x_i with the lattice velocity $e_{\alpha,i}$. The lattice direction α points toward the neighboring cells (see Fig. 8.1 for an illustration). Discretized in time and space the LBGK model is given in tensor notation by:

$$f_\alpha(x_i + e_{\alpha,i}\delta t, t + \delta t) - f_\alpha(x_i, t) = -\frac{\delta t}{\tau}\left[f_\alpha(x_i, t) - f_\alpha^{(eq)}(\rho(x_i, t), u_i(x_i, t))\right]. \quad (8.1)$$

Due to simplicity, quantities depending on x_i and t will be written without their dependencies, e.g., $f_\alpha = f_\alpha(x_i, t)$. The relaxation time τ can be determined from the lattice viscosity defined by Eq. (8.8). Further, the equilibrium distribution

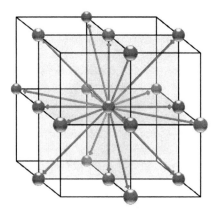

Fig. 8.1 The D3Q19 stencil.

$f_\alpha^{(eq)}(\rho, u_i)$, depending on the macroscopic velocity u_i, defined by Eq. (8.5), and the macroscopic density ρ, defined by Eq. (8.6) for the isothermal case, is given by the Maxwell-Boltzmann distribution function discretized for low mach numbers:

$$f_\alpha^{(eq)}(\rho, u_i) = \rho \cdot w_\alpha \cdot \left[1 + \frac{1}{c_s^2}(e_{\alpha,i} \cdot u_i) + \frac{1}{2c_s^4}(e_{\alpha,i} \cdot u_i)^2 - \frac{1}{2c_s^2}u_i^2 \right]. \qquad (8.2)$$

In the D3Q19 model, the thermodynamic speed of sound is given by $c_s = \frac{1}{\sqrt{3}}$ and the lattice velocities $e_{\alpha,i}$ and lattice weights w_α are:

$$e_{\alpha,i} = \begin{cases} (0,0,0), & \alpha = [0] \\ (\pm 1,0,0),(0,\pm 1,0),(0,0,\pm 1), & \alpha = [1,6] \\ (\pm 1,\pm 1,0),(0,\pm 1,\pm 1),(\pm 1,0,\pm 1), & \alpha = [7,18] \end{cases} \qquad (8.3)$$

$$w_\alpha = \begin{cases} 1/3, & \alpha = [0] \\ 1/18, & \alpha = [1,6] \\ 1/36, & \alpha = [7,18] \end{cases}. \qquad (8.4)$$

The macroscopic quantities of interest (ρ, p, u_i) can be determined from the moments of the distribution functions:

$$\rho u_i = \sum_{\alpha=0}^{18} e_{\alpha,i} \cdot f_\alpha = \sum_{\alpha=0}^{18} e_{\alpha,i} \cdot f^{(eq)}, \tag{8.5}$$

$$\rho = \sum_{\alpha=0}^{18} f_\alpha = \sum_{\alpha=0}^{18} f_\alpha^{(eq)}, \tag{8.6}$$

$$p = c_s^2 \rho, \tag{8.7}$$

$$\nu = (\tau - \tfrac{1}{2}) c_s^2 . \tag{8.8}$$

Due to the dependencies in Eq. (8.1), two grids are needed to store the PDFs. Hence, 2×19 *double* values are needed per lattice cell. Additionally, the update rule for each cell only depends on the neighboring cells. This locality can be exploited for an efficient parallelization. For further details on the actual implementation of the LBM in the waLBerla framework, see Feichtinger et al. [1].

8.3 Domain Partitioning Using Patches

To create a parallelization suitable for several thousand cores it is essential to subdivide the global simulation domain into small blocks that are independent of each other except for the necessary communication between the boundaries. In the waLBerla framework, these blocks are called patches and are always rectangular due to performance reasons. In addition to the sole purpose of communication, patches are an adequate tool to realize different simulation requirements, such as free surfaces and moving rigid objects, only in parts of the domain (see Fig. 8.2). This can

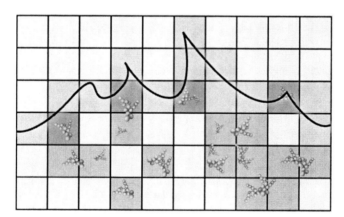

Fig. 8.2 Two-dimensional sketch of differently skilled patches interacting with each other. In the figure patches for pure fluid, free surface flows, particle laden flows and combined free surface / particle flows can be seen.

be exploited to increase the performance of the simulation: special treatment for free surfaces or rigid objects is only enabled in patches where it is needed, whereas a pure fluid patch can be optimized for performance. Furthermore, the patches can also be distributed on a hybrid computer architecture. For example, simple pure fluid patches could be calculated on the Cell processor and communicate to computationally more difficult patches via MPI, which are calculated on standard processors.

Since the waLBerla software library is written in C++, these thoughts can directly lead to the idea of a hierarchical patch design, which introduces specialized patches based on an abstract base class. Whereas the introduction of such a hierarchy seems to be the natural way in C++, a number of problems have been encountered with this hierarchy that challenged its application in the waLBerla framework (for more details see Feichtinger et al. [1]). The most important of the problems is that a hierarchical patch design would introduce multiple inheritance and therefore virtual inheritance to avoid duplicate data members. However, virtual inheritance introduces an additional indirection to the access of the data members of the common base class, which leads to a performance impact intolerable for a high performance implementation. Thus, for the waLBerla framework, we decided against a hierarchical patch design and use the approach illustrated in Fig. 8.3. Here, the single patch

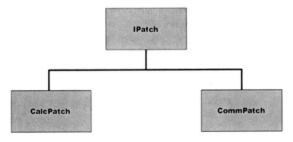

Fig. 8.3 Implementation possibility of choice for patch class hierarchy.

class *CalcPatch* handles all simulation aspects. Instead of creating different classes for the simulation features, the *CalcPatch* class uses different functions, which can be activated or deactivated, depending on the current situation. If these are again composed by basic inline functions a high code reuse among the different applications can be ensured; for example, in a pure fluid patch only the standard LBM functions are enabled, which allows a fast processing. For particle laden flows the same standard LBM functions as well as functions for the particle treatment are activated. The only other patch class besides the *CalcPatch* is the *CommPatch*, which handles the communication across process boundaries (see Sect. 8.4). With this design the flexibility needed for optimizations and integration of various applications is achieved.

8.3.1 Memory Reduction

Next to the primary purposes of building chunks of memory that can be distributed among several processes and to distinguish between different simulation requirements, the patch concept can also be used to reduce the overall amount of memory required for the simulation. The idea of this approach is described in detail in a previous work by Götz [25] who is dealing with LBM blood flow simulations. Due to the complex, arbitrary geometry of blood vessels and the Cartesian grid setup of the LBM, a large fraction of LBM cells ends up as solid nodes. To save memory, the domain is subdivided into patches and then all patches with no fluid cells are removed (see Fig. 8.4). This approach can also be used efficiently for porous media. However, the downside of this strategy is the additional communication across the boundaries of the patches, when the domain is split into several patches on one process.

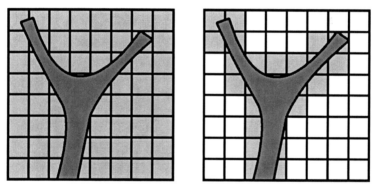

Fig. 8.4 Two-dimensional sketch of a vessel bifurcation and patches that can be omitted (*white*) in order to save memory.

8.4 Communication Concept

For a scalable simulation on a large number of processors, an efficient communication concept is inevitable. With the introduction of patches in Sect. 8.3, the communication is divided into local and MPI communication. Patches on the same process exchange their data via local communication, whereas patches on different processes communicate by using MPI. For the D3Q19 model, a patch has to communicate with at most 18 neighboring patches. Thereby, only the necessary PDFs have to be transferred: five for cells lying adjacent to boundary planes, one for edges and nothing for corners. Each patch stores a structure containing the neighborhood information, which is set up in the beginning of the simulation by first cutting the domain into *CalcPatch*es and then assigning the neighbors (either *CalcPatch*es or

*CommPatch*es) to each patch. For the parallel case, each process allocates a patch grid of the whole simulation domain including the neighborhood structures, but only allocates data fields in its own patches. The placement of the patches onto processes is either done by a Cartesian subdivision of the domain or done by distributing an equal number of patches to the processes. To treat local and MPI communication in a similar way, the patch class *IPatch* holds a common interface for the communication routines. Thus, a patch does not need to know if its neighbor is a *CalcPatch* or *CommPatch* and can just call the *Send()* routine of its neighboring patches. In the next two paragraphs, the difference in process local communication and communication across process boundaries is discussed.

8.4.1 Process Local Communication

For the process local communication a *CalcPatch* communicates with a neighboring

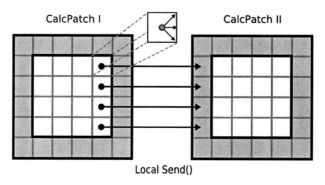

Fig. 8.5 Process local communication from *CalcPatch I* to *CalcPatch II*.

CalcPatch, which is depicted in Fig. 8.5. The sending patch (*CalcPatch I*) calls the *Send()* function of the receiving patch (*CalcPatch II*). This function directly copies the data from the source data fields into the ghost nodes (dark gray nodes, which are not part of the fluid domain) of the target data fields, without using any buffers in between.

8.4.2 MPI Communication

The MPI communication concept is illustrated in Fig. 8.6. Here, the *CalcPatch I* has to transfer its data to a second *CalcPatch II* on a different process using MPI. To make this action transparent to the *CalcPatch I*, it locally communicates with a *CommPatch I* as described in the previous section. Here, the *Send()* routine of the *CommPatch* copies the data into its *SendBuffer* and sends it with the MPI command *MPI_Isend* to *CommPatch II*. Afterwards it issues an appropriate *MPI_Irecv* to receive data from *CommPatch II*. After the same procedure has been executed on the other side, the data sent by *CalcPatch I* is located in the *RecvBuffer* of *CommPatch II*. With an additional call of the *CopyFromBuffer* routine the data is transferred to the ghost nodes of *CalcPatch II*. For each parallel communication, a compatible send–receive pair is needed. We use the function argument *tag* in MPI to match the messages on the processes by putting both IDs from the source and the target patch in *tag*. Unfortunately, the size of this identifier is limited and depends on the MPI implementation, which restricts the number of patches (see Sect. 8.5.2.1).

Since non blocking MPI commands are used for the MPI communication and all data transfer (process local and MPI) is issued after the patches finished their calculations, both communications overlap in time.

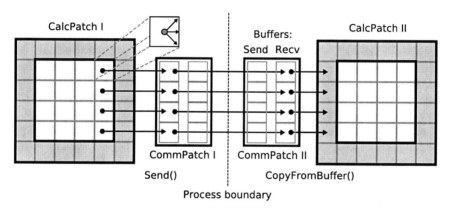

Fig. 8.6 MPI communication from *CalcPatch I* to *CalcPatch II* via two *CommPatches.*

8.5 Performance Studies

Before integrating more complex applications, the implementation of the parallelization has to be analyzed. Therefore, performance studies have to be performed in order to quantify whether the concept is suitable for the use on massively parallel systems. The performance evaluation of the current implementation presented in this paper consists of serial experiments, weak and strong scaling measurements for the parallel performance as well as an examination of a multi-core implementation on the Cell Broadband Engine. For all studies, a simple three-dimensional canal scenario with inflow and outflow conditions has been used. The results are given in terms of million fluid lattice updates per second (MFlups), which is an established performance measure in the lattice Boltzmann community since it allows for an estimation of the simulation runtime for a given problem size.

The later denoted parallel efficiency E is:

$$E(N,P) = \frac{S(N,P)}{P} = \frac{1}{P} \cdot \frac{\text{MFlups}(N,P)}{\text{MFlups}(N,1)} \cdot 100\%, \qquad (8.9)$$

where S is the speedup gained by the use of P cores or nodes and N is the problem size.

Two IA32-based clusters, namely the Woodcrest cluster at the Regional Computing Center of Erlangen (RRZE) and the Opteron cluster at the Chair for System Simulation (LSS) Erlangen, and an IA64-based supercomputer, the HLRB II [9] at the Leibnitz Computing Center (LRZ) in Munich, are chosen for the evaluation. The experiments of Cell performance have been performed on the JUICE Cell Cluster [26] at the Research Center Jülich.

The LSS cluster consists of 50 AMD Opteron processors resulting in a rough overall peak performance of about 220 GFlops. The nodes used for the benchmarks consist of four single-core CPUs with 4 GiB [2] dedicated memory each and are connected via Infiniband, providing a bandwidth of up to 10 GBit/s. On the Woodcrest cluster, there are 217 2-socket nodes (HP DL140G3) with dual-core 64-bit enabled Intel Xeon 5160 CPUs (codename Woodcrest) and Infiniband interconnection. The rough overall peak performance of the system is about 10.3 TFlops. The HLRB II features 4846 dual-core Itanium 2 CPUs of Montecito type, each of which capable to address the whole shared memory of 39 TiB capacity by non uniform memory access (NUMA). The CPUs are interconnected by a hierarchically organized NUMAlink 4 network with a nominal bandwidth of 6.4 GiB/s. This computer is listed as number 27 in the TOP500 list (June, 2008) [27] with an overall peak performance of 56.5 TFlops. The JUICE consists of 12 QS20 blades each equipped with 2 Cell processors and 2×512 MiB memory.

8.5.1 Serial Experiments

The serial performance of a parallel code is important for the quantification of the quality of parallel scaling results, since the parallel efficiency depends on the serial performance (see Eq. (8.9)). The presented results are restricted to one of the IA32 architectures and the IA64-based machine, namely the Woodcrest cluster and the HLRB II, whereby only for Woodcrest detailed benchmark comparisons are discussed. On the Woodcrest, the theoretical memory bandwidth of one node is 21.3 GiB/s. However, to estimate the upper limit of the possible memory throughput for an LBM solver Zeiser et al. [28] suggest to compare with the STREAM [29]

---

[2] 1 GiB/s = 1024^3 B/s, 1 TiB/s = 1024^4 B/s.

Table 8.1 Performance and memory transfer rates of waLBerla and STREAM benchmark on one node of Woodcrest cluster with different number of processes and different placements. STREAM benchmark values by courtesy of [30].

Configuration	1 Process		2 Processes 1 socket		2 Processes 2 sockets		4 Processes	
	MFlups	GB/s	MFlups	GB/s	MFlups	GB/s	MFlups	GB/s
waLBerla	4.4	2.31	6.4	3.35	8.6	4.51	11.4	5.97
STREAM triad		3.32		3.30		6.09		6.04

vector-triad benchmark. Table. 8.1 shows that the maximum achievable data transfer rate is around 6 GB/s only[3].

On architectures that perform a read for ownership before a write, waLBerla transfers 524 Bytes per cell update (for details see [1]). The single core performance of waLBerla obtains 4.4 MFlups which corresponds to a bandwidth usage of 2.3 GB/s. Compared to the STREAM triad, a single waLBerla process uses 70% of the available usable bandwidth.

Contrary to IA32-based architectures, achieving high performance on IA64 machines is generally more difficult. The in-order architecture requires explicit vectorization and thus the performance often relies on the capabilities of the compiler and the use of appropriate pragmas in the code. Performance optimization of LBM for Itanium 2 is a well-explored task in our group [20]. However, many findings that enhance the performance of simple kernels cannot easily be applied to complicated programs like waLBerla. To feature a suitable framework for the complex algorithms of the real-life applications to be implemented, the class design contains complex structures that are not comparable with flat loops of simple kernels. Currently, the single core performance of 1.68 MFlups on the HLRB II represents only 39% of the performance on the Woodcrest cluster.

8.5.1.1 Influence of the Process-Local Communication

For an estimation of the influence the local communication has on the performance, a domain of 100^3 lattice cells has been simulated with an increasing number of patches. While the size of the domain remains constant, the number of patches is varied from 1 to 1000. In each simulation run, all patches have the same size, resulting in 100^3–10^3 lattice cells per patch. If the domain does not fit exactly into the patches, the remaining cells in the patches are marked as obstacles, not influencing the MFlups rate. The results (Fig. 8.7) show that the loss in performance compared to the situation without local communication is architecture dependent. On the LSS cluster, the performance falls by about 13% for 64 patches and about 28% for 1000 patches, while on the Woodcrest cluster the loss is higher with 14%

---

[3] The STREAM benchmark uses a base of 1000 for the orders of magnitude, thus $1 \, GB/s = 10^9 \, B/s$.

and 35%, respectively. On the HLRB II, the performance losses are smoother with 7 and 23%, respectively. However, this results from the small fraction of time which the copy operations of the local communication consume compared to the slow computations. Due to the ghost layers of the patches, the memory requirement on all architectures increases for 64 patches by 19% and for 1000 patches by 73% compared to a single patch simulation.

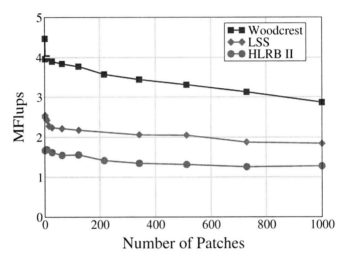

Fig. 8.7 Influence of the local communication on the performance.

8.5.2 Parallel Experiments

The parallel efficiency is measured by different parallel experiments: Weak scaling tests on the Woodcrest cluster and the HLRB II show the influence of both process-local and MPI communication. Strong scaling experiments on all three clusters test the proportion of MPI communication to computational time.

As Table 8.1 shows, the memory architecture of the Woodcrest cluster has a tremendous effect on concurrently running processes. Apparently, the chipset cannot provide the full achievable memory bandwidth to a single socket, but only nearly the half. Employing the second core on the same socket does not improve the bandwidth usage. In the case of waLBerla, one single process cannot utilize the full bandwidth one socket could achieve. Thus, employing the second core on the same socket can increase the usage by 45% until the machine's limit is reached. However, using two processes on separate sockets or two fully employed sockets does not double the performance and bandwidth usage, similarly to the STREAM benchmark. Using four cores results in a performance of 11.4 MFlups and a data transfer rate of 5.9 GB/s from memory, which equals the bandwidth usage of STREAM benchmark. The effect of getting not much more than half of the maximum achievable bandwidth on one socket is attributed to inherent limitations of the memory controller chipset. As a consequence, a comparison of the parallel scaling measurements based on fully employed sockets instead of single cores is preferable. However, running less than four processes without explicit process pinning can lead to different results depending on their distribution on the sockets (see columns 2 and 3 of Table 8.1). Therefore, the parallel performance experiments on Woodcrest cluster in this paper always have been conducted on fully employed nodes with four MPI processes, and the graphs are based on the number of nodes used. On the LSS cluster, the performance scales well with the number of cores because every core accesses its dedicated memory. For the HLRB II, the term "node" is not applicable. The smallest unit in the hierarchy is a blade, featuring one (high-bandwidth blades) or two (high-density blades) dual-core CPUs that share one connection to the NUMAlink 4 network. This connection is able to feed each core with the full bandwidth. Thus, the graphs of the performance experiments of the latter two architectures are based on the number of cores.

8.5.2.1 Weak Scaling

For the weak scaling test, the overall size of the problem is not fixed, but scaled by the number of processes. Thus the problem size on each process is constant for all simulation runs. This test is especially well-suited to measure the maximal overall parallel performance of the code and to determine the runtime to be expected for real-world applications. The weak scaling benchmark in Fig. 8.8 has been carried out on the Woodcrest cluster. From 1 to 203 nodes each core processes one patch containing 100^3 lattice cells such that the largest system size for this weak scaling experiment is $1000 \times 900 \times 900$ consuming approx. 300 GiB (measured value). Showing a nearly linear speed up, the parallel efficiency is 95.7% for 183 nodes and 93.3% for 203 nodes. Since CFD applications often need large simulation domains up to and beyond 1000^3 lattice cells, the parallel efficiency of the code was tested in a single experiment with large memory consumption. This test resulted in 94.4% parallel efficiency on 729 cores for a simulated system size of 1530^3 and 1.2 TiB. An investigation of the scaling behavior depending on the amount of process-local communication (see Figs. 8.9 and 8.10) reveals that the MPI communication is not as expensive as expected due to the low latencies and high bandwidths of the Infiniband and NUMAlink 4 interconnect on the one hand, and the overlapping of local and MPI communication in time on the other.

In this weak scaling experiments again, every process computes a domain with 100^3 lattice cells, but with different number of patches per process in each experiment. This does not only increase the effort of the local communication but also of the MPI communication, since every patch sends its data to its remote neighbors via a separate MPI call. Thus, a process with one patch that has neighbors to all sides

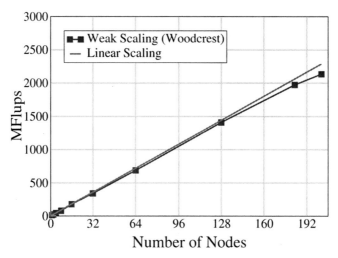

Fig. 8.8 Weak scaling experiment of waLBerla on the Woodcrest cluster with one single patch per process in comparison to the ideal scaling.

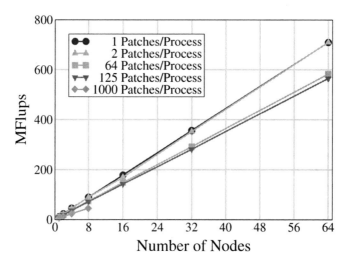

Fig. 8.9 Weak scaling experiment of waLBerla on the Woodcrest cluster with different numbers of patches per process in comparison with the single patch version.

sends 18 messages per time step. With two patches it sends 34 messages, with 125 patches 690 and with 1000 patches even 2880. The performance impact in comparison with the single patch version is for each scenario relatively constant over the increasing process number. With 64 patches per process the performance loss is 18%, with 125 patches 21% and with 1000 patches around 50%. The single core experiment above resulted in 14, 16 and 35%, respectively. Thus, the overhead of

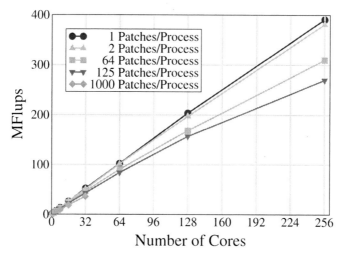

Fig. 8.10 Weak scaling experiment of waLBerla on high-bandwidth blades of the HLRB II with different numbers of patches per process in comparison with the single patch version.

MPI communication is low, which is attributed to the small latencies of the Infiniband network, but mainly to the fact that local and global communication overlap in time (as described in Sect. 8.4.2). In fact, a closer measurement with the Intel TraceAnalyzer shows that 5.6% of the total time are spent for MPI communication with 16 processes and 125 patches per process. Despite of the good results, further optimization of the MPI communication is inevitable: The weak scaling experiment with 1000 patches per process could not be performed with more than 8 nodes on the Woodcrest and 32 cores on the HLRB II, respectively, because of a limitation in the current implementation. The *tag* parameter for MPI communication is used to code the sending and receiving patch by using a globally unique ID for the patches. When using more than 32768 patches, the 32-bit signed integer value overflows. Although for real-world applications the use of 1000 patches per process is very unlikely, simulations using more than 9000 cores—as planned on the HLRB II—can quickly reach the limit. Therefore, a future implementation will combine the messages of the patches of the same process in order to reduce the message count and evade the patch number limitation.

The NUMAlink 4 interconnect of HLRB II, also having low latencies and a high bandwidth, has a low impact on the performance in this scaling experiment; however, the hierarchical structure influences the results (see Fig. 8.10). Up to 32 processes, the average performance losses of the parallel version of 10% for 64 patches and 30% for 1000 patches compare well to the 7 and 23% of the serial performance, respectively. From 64 to 128 cores as well as from 128 to 256 a decrease in parallel efficiency can be clearly determined. One compute partition of the HLRB II consists of 512 cores that are arranged in groups of 64 cores. Thus, messages between cores from different groups have to hop over routers of a higher hierarchy level. This effect

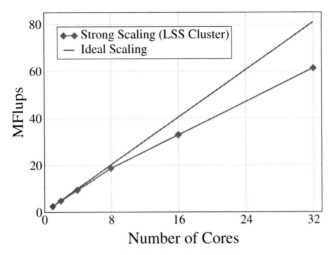

Fig. 8.11 Strong scaling experiment of waLBerla on the LSS Cluster.

can be neglected when the amount of communication is small, as in the case of one patch per process, while 64 and more patches per process experience a noticeable impact. This is another reason for changing the communication implementation in order to reduce the message count.

8.5.2.2 Strong Scaling

The strong scaling scenario has been performed with a size of 100^3 lattice cells. Here, the performance of the implementation is measured with an increasing number of cores, whereas the overall size of the problem remains fixed. This scenario enables to estimate the shortening of computation time when a higher number of processes is employed for solving the same problem. To neglect the effects of local communication one patch per MPI process has been used. The measurements for the LSS cluster are shown in Fig. 8.11. With 32 processes the parallel efficiency only drops to 75%.

For the strong scaling on the Woodcrest cluster (see Fig. 8.12) up to 16 nodes have been used, each node running four MPI processes. The resulting parallel efficiency on 16 nodes is about 84.9%. It has to be noted that for this result the patch size of 25^3 does not fit into the cache.

Since for the same problem size the Itanium 2 based implementation is slower, while having the same amount of communication, the ratio between work and communication is larger than on IA32-based architectures. According to Amdahl's Law

$$S = \frac{1}{(1-p) + p/P},$$

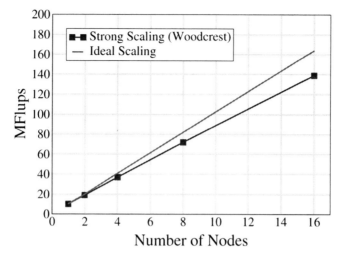

Fig. 8.12 Strong scaling experiment of waLBerla on the Woodcrest Cluster.

an increase of the parallelizable part p results in a higher scalability on the same number of cores P. Therefore, the scaling behavior in a strong-scaling experiment on HLRB II shows a good efficiency: With 32 cores the efficiency of up to 93.5% is much better than on Woodcrest cluster, where 8 nodes (32 cores) reach only 87.8%. Fig. 8.13 shows the scaling on high density and high bandwidth blades. From 8 to 16

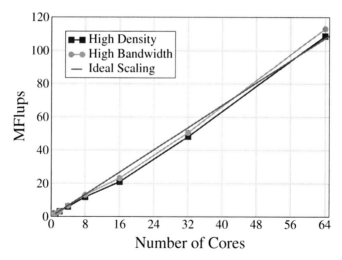

Fig. 8.13 Strong scaling experiment on both high density and high bandwidth blades of HLRB II.

cores one can clearly determine the additional overhead induced by the communication across the boundaries of so-called building blocks (high bandwidth building blocks consist of 8 cores), which represent a lower part of the sophisticated hierarchy in HLRB II. Due to the large caches, for 64 cores the system size per process is small enough to fit completely into the cache, which results in super-linear scaling. Since the communication via the NUMAlink 4 network shares the bus to the memory controller, it is obvious that the scaling on high density nodes is worse.

8.5.3 IBM Cell Processor

To investigate optimization techniques for special hardware and multi-core systems our group's research includes the IBM Cell processor, which is the heart of Sony's Playstation III gaming console. This hybrid multi-core processor combines one Power processor element (PPE), which is a PowerPC compliant general purpose core, and eight simple single instruction multiple data (SIMD) cores, so-called synergistic processor elements (SPEs). The PPE is mainly responsible to run the operating system and for program control, whereas the SPEs are optimized for efficient data processing. In the Playstation III only six of the SPEs are available to be used for programing. Whether with Playstations or as blades, this processor can be a valuable asset to a hybrid cluster enabling high performance for appropriate codes. Using the Cell, our group implemented a blood flow simulation [25] with similar concepts as in Sect. 8.3. With just compiling a straightforward implementation, one gains a meager performance of 2 MFlups on a single SPE, possibly summing up to 12 MFlups on a Playstation, after all. However, Stürmer et al. [5] showed that with architecture-related optimizations up to 95 MFlups are possible. Unfortunately, these results are based on single precision floating point operations. Since the code performance is limited by the memory bus, one can estimate that performance decreases by a factor of 2.5 for double precision, resulting in a sustained performance of around 40 MFlups.

8.6 Conclusion

In this chapter, the parallelization concept for the waLBerla framework has been presented, which aims at the integration of various CFD applications together with an efficient massively parallel implementation. The key component for our realization of these aims is the patch concept, which supports the specialization to different applications as well as domain decomposition needed for parallelization. To verify the suitability of the parallelization for massively parallel usage the serial and the parallel performance have been investigated on different architectures.

For the serial performance, it has been demonstrated that the sustained memory throughput is 70% of the maximum throughput (STREAM triad) on the Woodcrest cluster which gives an indication of the quality of the serial performance. Additionally, the influence of the process-local communication has been discussed and it has been shown that the performance with 64 patches only drops by 14% on the Woodcrest cluster, 7% on the HLRB II and 13% on the LSS cluster.

In the parallel case, weak scaling scenarios up to $1000 \times 900 \times 900$ lattice cells have been performed for process-local plus MPI communication and for pure MPI communication. The results show that the basic concept of the framework is suitable for massively parallel usage, as a parallel performance of about 94% has been achieved for $203(810)$ nodes (cores) on the Woodcrest cluster. Furthermore, the scenarios with several patches per process indicated that the MPI communication has a smaller impact on the performance than the local communication. The reason for that is the overlapping of local and MPI communication as well as the low latency of the interconnects. In addition to the weak scaling scenarios, strong scaling experiments have been performed. On the IA32-based architectures, these result with 32 cores in 87.8% parallel efficiency compared to 93.5% on the IA64-based HLRB II.

In future work, the efficiency of the framework will be tested on the HLRB II beyond 1000 cores. For this purpose, the communication will be redesigned to reduce the overall message count, as the connection network of the HLRB II has a strong influence on the efficiency when dealing with a high amount of messages. With further work the CFD applications, e.g., free surfaces and moving objects will be implemented together with a dynamic load balancing and the specialization of the patches.

References

1. C. Feichtinger, J. Götz, S. Donath, K. Iglberger, U. Rüde, Concepts of waLBerla prototype 0.1, Tech. Rep. 07–10, University of Erlangen-Nuremberg, Computer Science 10 – Systemsimulation (2007).
2. K. Iglberger, N. Thürey, U. Rüde, Simulation of moving particles in 3D with the Lattice Boltzmann method, Comp. Math. Appl. 55 (7) (2008) 1461–1468.
3. C. Binder, C. Feichtinger, H. Schmid, N. Thürey, W. Peukert, U. Rüde, Simulation of the hydrodynamic drag of aggregated particles, J. Colloid Interface Sci. 301 (2006) 155–167.
4. J. Horbach, D. Frenkel, Lattice-Boltzmann method for the simulation of transport phenomena in charged colloids, Phys. Rev. E 64 (6) (2001) 061507.
5. M. Stürmer, J. Götz, G. Richter, A. Dörfler, U. Rüde, Blood flow simulation on the Cell Broadband Engine using the lattice Boltzmann method, Tech. Rep. 07–9, University of Erlangen-Nuremberg, Computer Science 10 – Systemsimulation, submitted to the International Conference for Mesoscopic Methods in Engineering and Science, ICMMES (2007).
6. C. Körner, M. Thies, T. Hofmann, N. Thürey, U. Rüde, Lattice Boltzmann model for free surface flow for modeling foaming, J. Stat. Phys. 121(1-2) (2005) 179–196.
7. N. Thürey, T. Pohl, U. Rüde, M. Oechsner, C. Körner, Optimization and stabilization of LBM free surface flow simulations using adaptive parameterization, Comput. Fluid 35(8–9) (2006) 934–939.
8. Information on fuel cells, http://www.fuelcells.org (2008).
9. Information on the HLRB II, http://www.lrz-muenchen.de/services/compute/hlrb/ (2008).
10. G. McNamara, G. Zanetti, Use of the Boltzman equation to Simulate Lattice Gas Automata, Phys. Rev. Lett. 61 (20) (1988) 2332–2335.
11. X. He, L.-S. Luo, Theory of the lattice Boltzmann method: From the Boltzmann equation to the lattice Boltzmann equation, Phys. Rev. E 56 (6) (1997) 6811–6817.

12. M. Junk, A. Klar, L.-S. Luo, Asymptotic analysis of the lattice Boltzmann equation, J. Comput. Phys. 210 (2) (2005) 676–704.
13. X. Shan, H. Chen, Lattice Boltzmann model for simulating flows with multiple phases and components, Phys. Rev. E 47 (3) (1993) 1815–1819.
14. T. Zeiser, H.-J. Freund, J. Bernsdorf, P. Lammers, G. Brenner, F. Durst, Detailed Simulation of Transport Processes in Reacting Multi-Species Flows Through Complex Geometries by Means of the Lattice Boltzmann method, in: In High Performance Computing in Science and Engineering '01, Transactions of the High Performance Computing Center Stuttgart (HLRS), Springer (2002).
15. P. Asinari, Multiple-relaxation-time lattice boltzmann scheme for homogeneous mixture flows with external force, Phys. Rev. E (Statistical, Nonlinear, and Soft Matter Physics) 77 (5) (2008) 056706.
16. A. Artoli, A. Hoekstra, P. Sloot, Mesoscopic simulations of systolic flow in the human abdominal aorta, J. Biomech. 39 (5) (2006) 873–884.
17. C. Körner, T. Pohl, U. Rüde, N. Thürey, T. Hofmann, FreeWIHR: Lattice Boltzmann methods with free surfaces and their application in material technology, in: A. Bode, F. Durst (Eds.), High Performance Computing in Science and Engineering, Garching 2004, Springer (2005), pp. 225–236.
18. C. Körner, T. Pohl, U. Rüde, N. Thürey, T. Zeiser, Parallel Lattice Boltzmann Methods for CFD Applications, in: A. Bruaset, A. Tveito (Eds.), Numerical Solution of Partial Differential Equations on Parallel Computers, Vol. 51 of Lecture Notes for Computational Science and Engineering, Springer (2005) Ch. 5, pp. 439–465.
19. J. Wilke, T. Pohl, M. Kowarschik, U. Rüde, Cache Performance Optimizations for Parallel Lattice Boltzmann Codes, in: Proc. of the EuroPar-03 Conf., Vol. 2790 of Lecture Notes in Computer Science, Springer (2003), pp. 441–450.
20. G. Wellein, T. Zeiser, G. Hager, S. Donath, On the single processor performance of simple Lattice Boltzmann kernels, Comput. Fluid 35 (8–9) (2006) 910–919.
21. D. Hänel, Molekulare Gasdynamik, Springer (2004).
22. D. Wolf-Gladrow, Lattice-Gas Cellular Automata and Lattice Boltzmann Models, Springer (2000).
23. D. Yu, R. Mei, L.-S. Luo, W. Shyy, Viscous flow computation with the method of lattice Boltzmann equation, Prog. Aero. Sci. 39 (5) (2003) 329–367.
24. Y. H. Qian, D. D'HumiÄlres, P. Lallemand, Lattice BGK Models for Navier-Stokes equation, Europhys. Lett. 17 (6) (1992) 479–484.
25. J. Götz, Numerical Simulation of Blood Flow with Lattice Boltzmann Methods, Master's thesis, University of Erlangen-Nuremberg, Computer Science 10 – Systemsimulation (2006).
26. Information on the Juelicher Initiative Cell Cluster (JUICE), http://www.fz-juelich.de/jsc/service/juice (2008).
27. Top500, The top 500 supercomputer sites, http://www.top500.org (2008).
28. T. Zeiser, J. Götz, M. Stürmer, On performance and accuracy of lattice Boltzmann approaches for single phase flow in porous media: A toy became an accepted tool – How to maintain its features despite more and more complex (physical) models and changing trends in high performance computing!?On performance and accuracy of lattice Boltzmann approaches for single phase flow in porous media, in: Proceedings of 3rd Russian-German Workshop on High Performance Computing, Novosibirsk, Springer (2008).
29. J. D. McCalpin, STREAM: Sustainable memory bandwidth in high performance computers, http://www.cs.virginia.edu/stream/ (1991–2008).
30. T. Zeiser, Private correspondence with Thomas Zeiser, Regional Computing Center Erlangen (RRZE) (Aug. 2008).

Chapter 9
Parallel Pseudo-Spectral Methods for the Time-Dependent Schrödinger Equation

Tore Birkeland and Tor Sørevik

Abstract

Simulations in quantum mechanics can easily become extremely computationally demanding, making parallel computing a necessity. In this chapter we outline a computational technique of the time-dependent Schrödinger equation (TDSE) using pseudo-spectral methods. The split-step propagator method with dimensional splitting enables efficient parallelization; each fractional step can be perfectly parallelized, while redistribution is necessary between steps. It is showed that the scalability of the split-step method can be greatly increased by applying an improved data distribution scheme. The software framework PyProp is also introduced, implementing the methods described in this chapter. PyProp tries to combine the flexibility of object-oriented programming (C++), the convenience of high-level scripting language (Python) and high-performance computational libraries (blitz++, FFTW, LAPACK) to create a flexible framework for solving the TDSE.

9.1 Introduction

In quantum mechanics, a system of particles is completely described by the wavefunction $\psi(\mathbf{x},t)$ and the Hamiltonian operator H. H is a linear, but possibly time-dependent operator describing the setup of the system, and the wavefunction is a complex-valued function describing the state of the system at a given time t. A fundamental principle of quantum mechanics is that all measurable values can be calcu-

Tore Birkeland
Department of Mathematics, University of Bergen, 5008 Bergen, Norway,
e-mail: tore.birkeland@math.uib.no

Tor Sørevik
Department of Mathematics, University of Bergen, 5008 Bergen, Norway,
e-mail: tor.sorevik@math.uib.no

R. Trobec et al. (eds.), *Parallel Computing*, DOI 10.1007/978-1-84882-409-6_9,
© Springer-Verlag London Limited 2009

lated from the wavefunction. For instance, the absolute square of the wavefunction $|\psi(\mathbf{x},t)|^2$ yields the probability density, and an appropriate normalization factor is usually applied to make the integrated probability one. In general, physically observable values are obtained by calculating the expectation value of an operator. The Hamiltonian, for instance, corresponds to the total energy of the system, and the expectation value of energy can be calculated by

$$\langle E \rangle = \int_{\Omega} \psi^*(\mathbf{x},t) H \psi(\mathbf{x},t) dx. \tag{9.1}$$

The time evolution of a quantum mechanical system is described by the time-dependent Schrödinger equation (TDSE)

$$i\hbar \frac{\partial}{\partial t} \psi(\mathbf{x},t) = H \psi(\mathbf{x},t); \qquad \mathbf{x} \in C^s \tag{9.2}$$

Here, i is the imaginary unit and $\hbar$ is the reduced Planck's constant. The Hamiltonian operator, H, reads:

$$H = -\frac{\hbar^2}{2m} \nabla^2 + V. \tag{9.3}$$

Here we will consider the nontrivial time-dependent case where the potential operator, V, is a function not only of the spatial variables, $\mathbf{x}$, but of time as well.

The computational challenge of this equation is a consequence of the high dimensionality of interesting systems. The dimension, s, increases proportionally with the number of particles, p, in the system. Thus for direct discretization of Eq. (9.2), the amount of work scales exponentially with the number of dimensions, leaving many particle simulations out of reach for ab initio methods.

In cases where $p >> 1$ approximation models are the only possibility. Methods such as Hartree–Fock, Density Functional Theory (DFT) and others tackle the exponential growth in computational cost by decoupling the dimensionality, at the expense of accurately modeling the inter-particle exchange interaction. These methods work well when the exchange interaction is of less importance. However, for few particle problems where exchange interaction is of importance, none of these approximate models give satisfying results. In these cases, one is faced with the daunting task of computing accurate approximations to the full high-dimensional Schrödinger equation (9.2). Using current technology, realistic simulations beyond $s = 3$ challenge the computational limits. To face these challenges one needs to employ the most efficient algorithms for a specific problem and run the code on the most powerful computers available.

A typical situation is when the system has some sort of symmetry or only exhibits slow changes when formulated in the right coordinate system. Thus the first step to efficient computation will be to use the best possible problem formulation. A flexible code needs therefore to cater for different coordinate systems. We discuss some of these possibilities and problems in Sect. 9.3.

The next step is to choose a discretization scheme. In the spatial variables a spectral approximation appears to be the best choice for our problem, although alternatives do exist. In particular, spectral approximations are attractive in combination with split operator technique. By the right choice of spectral basis we may be able to diagonalize the operator, allowing for fast, stable and accurate time integration by an exponential integrator. In Sect. 9.2 we describe the pros and cons of this technique.

The exponential growth of single CPU performance has come to a halt; increasing parallelism appears to be the alternative provided by the hardware vendors for increasing performance. To take advantage of this technology trend, one needs highly scalable parallel algorithms to efficiently utilize tomorrow's computers. When parallelizing high-dimensional problems using spectral methods, the standard data decomposition technique has limited scalability. In Sect. 9.4 we explain how to overcome the scalability issue by taking advantage of the full dimensionality of the problem.

The primary requirement for a high-performance scientific code is certainly efficiency. To be able to study increasingly more complex systems, it is paramount to efficiently utilize modern supercomputers. The best way to get maximum performance from these machines is to write the computationally intensive parts of the program in a "close-to-the-hardware" language and reuse libraries which are optimized for the specific hardware. However, low-level languages such as Fortran have limited facilities for abstraction. It is therefore difficult to create software that is flexible enough to support different discretization and propagation methods, while at the same time keeping duplicate code to a minimum.

Object-oriented techniques can help encapsulate low-level calculation routines to more functionality oriented objects, and inheritance can be used to encapsulate the implementation of different discretization methods, while keeping a unified interface. Statically compiled languages such as C++ allow for a high level of abstraction and at the same time gives high performance (at least when combined with high-performance Fortran kernels). An issue with statically compiled languages is the necessity of recompiling the program every time a change is made. The overhead of recompilation is certainly insignificant for computationally intensive simulations taking several hours to complete, which is usually the case for the final simulations in a project. For the early stages of a project, however, exploration of different parameters and methods plays an important part, and the overhead of recompiling the project for every change can become tedious. On the other hand, a dynamic and interactive runtime combined with a rich plotting environment such as provided by MATLAB is efficient for exploration, but sacrifices performance for interactivity and ease of use.

A solution is to recognize that the performance critical parts of most scientific programs are only a small fraction of the program. The performance critical parts can then be implemented in a high-performance language, while the bulk of the program can then be written in an expressive high-level language. Python is an excellent example of such a language. It is well documented, readily extensible and has a number of extension modules for scientific computing [1].

We will now describe the split operator technique for solving the TDSE, specifically in the context of spectral approximations. In Sect. 9.5 we describe PyProp, a framework for solving the TDSE using the methods described in this chapter.

Finally, in Sect. 9.6.2, we illustrate the need for parallel computing in this field by a numerical experiment.

9.2 Time Stepping and Split Operator Technique

Let H be a linear differential operator over the spatial variables for a time-dependent system

$$i\frac{\partial}{\partial t}\psi(t) = H\psi(t). \tag{9.4}$$

It is convenient to split the operator into suboperators, $H = A + B$. Assuming H does not explicitly include any time dependence, the formal solution to the system can be written:

$$\psi(t) = e^{-iHt}\psi(0) = e^{-i(A+B)t}\psi(0). \tag{9.5}$$

If H is time dependent, the above equation can be amended by performing a sequence of smaller time steps Δt, such that H is essentially time independent on the interval $(t, t + \Delta t)$. As A and B do not in general commute, writing the above exponential as a product of two exponentials introduces a splitting error

$$\psi(t + \Delta t) = e^{-iA\Delta t}e^{-iB\Delta t}\psi(t) + \mathcal{O}(\Delta t^2 [A, B]), \tag{9.6}$$

where $[A, B] = AB - BA$ is the commutator between A and B. It can be shown that the error introduced by the splitting can be reduced by splitting in a symmetric manner, known as the Strang splitting,

$$\psi(t + \Delta t) = e^{-iA\Delta t/2}e^{-iB\Delta t}e^{-iA\Delta t/2}\psi(t) + \mathcal{O}(\Delta t^3), \tag{9.7}$$

and corresponds to first solving a half-time step of A, then a full-time step of B and finally another time step of A. The splitting procedure can be generalized further, in order to decrease the splitting error:

$$\psi(t + \Delta t) = e^{c_n A\Delta t}e^{d_{n-1}B\Delta t}\cdots e^{c_1 A\Delta t}e^{d_0 B\Delta t}e^{c_1 A\Delta t}\cdots e^{d_{n-1}B\Delta t}e^{c_n A\Delta t}\psi(t). \tag{9.8}$$

This corresponds to $2n$ time steps of various lengths with operator A and $2n - 1$ steps with operator B. Explicit values for the coefficients which give a global splitting error of $O(\Delta t^{2n})$ are known for $n \leq 4$ [2, 3].

Any numerical time-stepping scheme may be used to propagate the system to $4n - 1$ substeps. In particular, when the transformation to the eigenspace of the suboperator is known, the propagation may be executed efficiently and accurately. This strategy was, to our knowledge, first applied to the TDSE by Feit et. al [4] and later modified and used successfully by many others. See [5] for splitting when the problem is presented in spherical coordinates. For higher-dimensional hyperspherical coordinates, see [6] and for application to the nonlinear TDSE, [7].

The different operators generally have different eigenfunctions, and consequently the spectral representation of the wavefunction must be changed accordingly. Thus, for the strategy sketched above to be efficient, we need fast transformation routines mapping the data from one representation to another.

9.3 Variable Transformations and Spectral Bases

The choice of coordinate system for a given problem depends on many variables, such as the symmetries and near symmetries of the system, as well as which observable features of the system are of interest. An atom, for example, which in the single active electron model can be described by a spherically symmetric potential, clearly favors a spherical over a Cartesian representation. However, a spherical discretization requires more calculation per grid point than a Cartesian representation. This is due to the fact that the fast Fourier transform available for Cartesian coordinates is significantly faster than any known transformation for spherical harmonics. In this chapter, only Cartesian and spherical coordinate systems will be discussed. The methods, however, are readily extensible to other coordinate systems as well, such as cylindrical or prolate spheroidal coordinates.

9.3.1 Cartesian Coordinates and Fourier Basis

The eigenfunctions of the Laplacian operator on a hyper-rectangle are the Fourier functions, and for functions with a smooth periodic extension on this domain a Fourier expansion converges rapidly. Unfortunately, the wavefunction does not live on a finite domain, but is defined on the entire space. Nevertheless, for problems where the wavefunction decays rapidly away from origin, no great error is made if we confine the domain to a finite domain, say $[-R, R]^s$. In these cases, an s-dimensional Fourier expansion becomes the obvious choice of spectral basis when using the splitting indicated in Eq. (9.3). The algorithm for advancing one (partial) time step with the Laplacian operator becomes:

- represent $\psi(\mathbf{x}, t)$ by its Fourier expansion:

$$\psi(\mathbf{x}, t) = \sum_{||\mathbf{k}||_\infty < N} \widehat{\psi}(\mathbf{k}) e^{\pi i \mathbf{k}^T \mathbf{x}/R}, \tag{9.9}$$

- advance the solution by scaling each component by the corresponding eigenvalue:

$$\psi(\mathbf{x}, t + c_j \Delta t) = e^{c_j \Delta t \nabla^2} \psi(\mathbf{x}, t)$$

$$= c_j \Delta t \sum_{\|\mathbf{k}\|_\infty < N} e^{c_j \Delta t - \frac{(\pi \|\mathbf{k}\|)^2}{R^2}} \widehat{\psi}(\mathbf{k}) e^{\pi i \mathbf{k}^{\mathsf{T}} \mathbf{x}/R}.$$

The simplicity of this algorithm, the existence of the fast Fourier transform (FFT) algorithm to compute the expansion coefficients and the accuracy of the time stepping are the main reasons for choosing this method. One of the main issues with this method is the necessity to truncate the problem at a cutoff distance and assume a periodic extension of the wavefunction. Making the computational domain too small may introduce unacceptable errors, while making it too large implies wasting resources on storage and computation of insignificant grid points. It is very difficult to know in advance exactly where to truncate, as that in general requires a priori knowledge of the function we are computing. Furthermore, as the Fourier collocation method distributes its evaluation points equidistant, the extra cost of taking to large domain grows exponentially with dimension.

9.3.2 Spherical Coordinates

Configurations with angular symmetry or slow variations in angular directions are best expressed in spherical coordinates. In (hyper-) spherical coordinates the s-dimensional Laplacian takes the form:

$$\nabla^2 = \frac{\partial^2}{\partial r^2} + \frac{s-1}{r} \frac{\partial}{\partial r} - \frac{1}{r^2} \hat{\Lambda}^2(s). \tag{9.10}$$

Here, $\hat{\Lambda}^2(s)$ is the (hyper-) angular momentum operator. In the case of full angular symmetry, derivatives in angular directions become zero, the effect of the angular momentum operator vanishes and the problem effectively reduces to a one-dimensional problem. But also for problems without complete angular symmetry, spherical coordinates are of interest. One tractable computational feature is that the infinite boundaries are removed in the angular direction and replaced with accurate periodic boundary conditions. In the radial direction, however, there still is a semi-infinite boundary that must be dealt with.

The eigenfunctions of $\hat{\Lambda}^2(s)$ are the (hyper-)spherical functions which form an orthogonal basis in the angular directions. Thus, these are good candidates for basis functions.

How to deal with the radial direction is still a problem. Replacing $\psi(r)$ with the reduced wavefunction $\Phi(r) = r^{(1-s)/2} \psi(r)$ removes the first derivative term in Eq. (9.10), making the radial part similar to one of the coordinates in a Cartesian coordinate system. Any method used for Cartesian coordinates, such as the Fourier spectral method discussed in Sect. 9.3.1, can be used for the radial part, only on a slightly different domain, $r \in [0, \infty)$, with the boundary condition $\Phi(r = 0) = 0$.

Truncating the right boundary will in this case produce a periodic extension which is nowhere near a smooth function. Thus a Fourier spectral approximation will converge slowly. The simplest trick is to symmetrize the problem by defining the odd extension $\psi(-r, \Omega, t) = -\psi(r, \Omega, t)$ on $[-R, R]$ [5]. The success of this is dependent on the smoothness of the function at origin. Using the Fourier spectral approximation suffers from the standard dilemmas of interval truncation as described above. For many applications, it is better to use a basis of functions orthogonal on $[0, \infty)$, such as the Laguerre functions or the rational Chebyshev functions [8]. Another possibility is to apply a variable transformation which maps $[0, \infty)$ onto $[0, 1)$, and apply a standard Chebyshev basis [9] on that domain. As these are not eigenfunctions of the radial Laplacian in the new variable, an explicit diagonalization is needed in order to use an exponential propagator.

9.4 Parallelizing Many Dimensional FFTs

In order to propagate large systems, two restrictions occur: The processing time required to propagate the problem and the amount of memory required to hold the entire problem. Both these issues are addressed by large, distributed memory parallel systems.

Having split the spatial operator and applied a spectral eigendecomposition to each of the partial operators, the time propagation becomes trivial and "embarrassingly parallel". The difficulty is that each partial operator typically requires a different spectral representation, thus between each partial step there is a need for transforming the data, i.e., a multidimensional FFT. Parallelizing such transformations on huge dense data sets on distributed memory computers usually leads to a communication bottleneck.

The standard way of parallelizing FFT on a tensor product of data is to distribute one of the ranks across the processing nodes and perform calculations on the ranks which are local first [10–13]. Next, when it comes to processing the distributed rank, the data set is redistributed among the nodes such that one of the local ranks becomes distributed, while the distributed rank becomes local. This is a simple algorithm and is near optimal in terms of data elements that must be moved. However, it is clear that in the above parallelization scheme, the number of processors cannot exceed the number of data points in the distributed rank, and for modern supercomputers with a large number of cores, this is a severe limitation of scalability. Furthermore, the communication pattern is that of all processors participating in an all-to-all collective operation, transmitting almost the entire wavefunction. On most inter-processors network this operation does not scale well [14, 15]. In [14, 15] the authors describe how to perform a 3D FFT when the data were distributed across two of the ranks. In [16] it was shown how to extend this parallelization scheme by distributing r ranks of an s-dimensional array. Some of the details of this extension are given below.

Consider an s-dimensional data set of size $N_0 \times N_1 \times \cdots \times N_{s-1}$ which is mapped onto an r-dimensional processor array of size $P_0 \times P_1 \times \cdots \times P_{r-1}$, $1 \leq r < s$. The mapping is done by distributing the data set along r dimensions in equal pieces. We get different mappings depending on which dimensions we choose to distribute. There are $\binom{s}{r}$ possible mappings. In a computation, as exemplified by the s-dimensional FFT, the mapping will have to change during the computation. For convenience we here assume that $N_i \bmod P_j = 0$ for all $i = 0, \cdots, s-1$ and $j = 0, \cdots, r-1$. This requirement is not a practical limitation, but it simplifies notation and analysis. A straightforward way to deal with the $N_i \bmod P_j = 0$ requirement is to set $P_1 = P_2 = \cdots = P_r = P$ and pad the data array with zeros to satisfy $N_i \bmod P = 0$. Another way (which we have used in our implementation) is to modify the algorithm slightly so that it can work with different amounts of data residing on each processor. In a practical setting we may assume that we can configure our system, which means we have some control over r and the P_js, while on the other hand, s and N_i are defined by the problem. Let $S = \{i_0, i_1, ..., i_{r-1}\}$ be an index set where $0 \leq i_j < s$ for $j = 0, \cdots, r-1$. Then, S_{now} denotes the dimensions which are distributed among the r-dimensional processor array. A dimension can only be distributed over one set of processors, which gives $i_j \neq i_k$, if $j \neq k$. If we want to do computation on dimension k, where $k \in S_{now}$, a redistribution is required. Let S_{next} be a distribution where $k \notin S_{next}$. The dimensions $S_{now} \backslash S_{next}$ will be distributed, while the dimensions $S_{next} \backslash S_{now}$ will be gathered (Fig. 9.1).

Life on Proc(0,0)

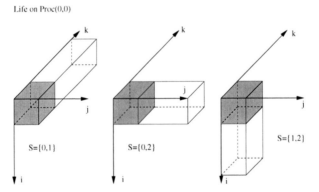

Fig. 9.1 This figure shows the three different slices of 3D data onto a 2D processor array. The slice of the data shown here is the local data to $P_{0,0}$. The shaded part is the portion of the local data that is invariant for all three different slices. Thus it does not have to be transmitted whenever a redistribution is needed.

Assuming that the difference between S_{now} and S_{next} is exactly one index, i.e., the operation to be performed is an all-to-all along one dimension. For such an operation, the processors can be organized in groups, where a processor only communicates with other processors in the same group. For redistribution along different dimensions in the processor grid, different groups will have to be formed. In general, one set of groups will be formed for each dimension in the process or array. A processor P_α, where $\alpha = (\alpha_0, \alpha_1, ..., \alpha_{r-1})$, will be a part of the groups $G^j_{\alpha_j}$, for $j = 0, 1, ..., r - 1$ (Fig. 9.2).

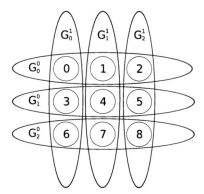

Fig. 9.2 A 3×3 processor array. The processors are organized into one group for each dimension in the processor array. For redistributing the pth dimension in the processor array, processors in the G^p groups will communicate internally.

For communication within one group, an algorithm similar to the standard implementation of all-to-all is used. Below is an implementation of this algorithm in simplified Python-like syntax. `inData` and `outData` are the input and output data arrays local to the current processor. `fullShape()` returns the shape of the global array and `shape(x)` returns the local size of the array x. `inDistrib` and `outDistrib` are the dimensions of the data set which is distributed at the beginning and end of the algorithm respectively. `groupSize` is the number of processors in the communication group.

Listing 9.1 Algorithm for redistributing one rank

```
sendSize = fullShape(inDistr)/groupSize
recvSize = fullShape(outDistr)/groupSize

for i in range(groupSize):
  sendProc = (groupRank + i) % groupSize
  recvProc = (groupRank + groupSize - i) % groupSize

  sendSlice = shape(inData)
  sendStart = sendProc*sendSize
  sendEnd = (sendProc+1)*sendSize
  sendSlice[inDistr] = sendStart:sendEnd
  sendBlock = inData[sendSlice]

  recvSlice = shape(outData)
```

```
recvStart = recvProc*recvSize
recvEnd = (recvProc+1)*recvSize
recvSlice[outDistr] = recvStart:recvEnd
recvBlock = outData[recvSlice]

irecv(fromProc, recvBlock)
isend(toProc, sendBlock)
wait()
```

The above algorithm is implemented in C++ using MPI and incorporated into PyProp (see Sect. 9.5). The processor groups are set up using the Cartesian topology routines and each group is assigned a unique communicator. This allows for optimized MPI implementations to exploit locality in the underlying network topology without user interaction. For handling multidimensional data in C++ we have used the excellent blitz++ library [17]. Using blitz++ and MPI data types, we have been able to hide the details of sending and receiving a strided hyper-slab, which has simplified the implementation of the redistribution considerably.

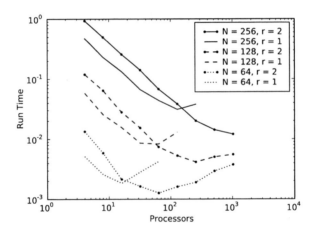

Fig. 9.3 Redistribution time as a function of number of processors plotted for different grid sizes on the Cray XT4.

We have compared the time it takes to redistribute the data set on a 2D vs. a 1D processor grid for a 3D data set of size $N \times N \times N = N^3$, on a processor grid of P^2 processors. The tests have been run for several values of P and N. In Fig. 9.3 the results for the main performance test are shown both for a 1D and a 2D processor grid. As expected, for few processors, the 1D processor grid is superior. However, the 2D configuration gives better scaling and eventually becomes faster than the 1D configuration. The crossover point appears to be $P^r \approx N/2$.

Assuming $N_0 = N_1 = \cdots = N_{s-1} = N$, $P_0 = P_1 = \cdots = P_{r-1} = P$ and $N \bmod P = 0$, the complexity of the two different algorithms becomes

$$W_1 \sim P^r t_s + t_w \frac{N^s}{P^r} \qquad (9.11)$$

for distributing along one dimension and

$$W_2 \sim r(P t_s + t_w \frac{N^s}{P^r}) \qquad (9.12)$$

for distributing along two dimensions. Here t_s is the latency and t_w the reciprocal bandwidth. Interestingly, we observe that for $P^r = N$ (which is the highest possible processor count for the 1D processor grid) the 2D processor grid performs best, even though twice the amount of data is being transferred. This means that not only does the 2D processor grid allow one to use more processors, it also enables more efficient utilization of the network. This is most likely due to larger blocks of data being sent at each step in the redistribution. The relative decrease in efficiency of the 2D processor grid seen for increasing values of N and P^r can be explained from the fact that the block size increases cubically with N and decreases linearly with P^r. As the block size increases, we expect the start-up effects for the 1D processor grid to decrease.

9.5 Creating a Framework for Combining Discretization Methods

For a given problem, one must carefully choose a coordinate system, a corresponding discrete representation and a time propagation scheme with the overall goal of minimizing discretization error and maximizing performance. The interesting problems in atomic physics are quite varied, and there does not exist a unique choice which is optimal in all cases. In fact, in many cases it is hard to guess upfront what the best choices are and thus there is a need for being able to experiment with different coordinate systems, discretization methods and time marching methods in a simple way, rather than depending on a priori knowledge about the system alone. This is usually an issue with existing codes, where either cylindrical, spherical or Cartesian coordinate systems have been chosen early in the development process, alongside discretization methods for the corresponding ranks. This can lead to a specific data structure being hard coded throughout the entire program, making it difficult to reconsider the choice of method later in the development process.

PyProp is an attempt to create a flexible yet efficient framework for solving the TDSE based on the ideas presented above. An overall design goal is to make it easy to mix-and-match the different modules such as coordinate system, spatial discretization and time marching, making it possible to test several methods before settling on the one best suited for the problem at hand. It uses the Python language for configuration and high-level program flow, while the computational intensive parts are written in C++ and Fortran90, utilizing high-efficiency libraries like BLAS/LAPACK, FFTW [11], blitz++ [17] and GNU Scientific Library (gsl) [18]. C++ is used to allow some object orientation in the computational routines, while some generated Fortran90 code is used for critical routines. The boost::python library [19] is used to "glue" C++ and Python together.

9.5.1 Wavefunction

The Wavefunction class is a self-contained model of the wavefunction $\psi(x,t)$ in Eq. (9.2) for a given t, and is the central class of PyProp. In Fig. 9.4, a simplified schematic of the Wavefunction class is shown.

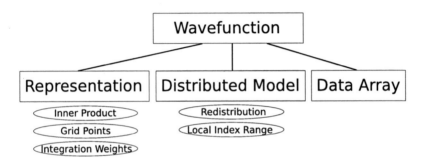

Fig. 9.4 Schematic showing the Wavefunction class and dependencies in PyProp. A wavefunction consists of a set of data buffers, a representation and a distributed model. The representation describes the data, and knows about grid points and quadrature rules, thus allowing for operations like inner products and integration. The distributed model knows how the wavefunction is distributed among the processors, and can redistribute the data. One of the data buffers is always active and contains the numerical data of the wavefunction, where the others serve as work data for out-of-place computations.

In addition to the numerical values, it has a representation object attached, which determines how the numerical values are interpreted. The representation also has information about integration weights, and can therefore perform inner products between two wavefunctions. In order to support changing representation in one dimension without changing the others, the representation object can either be attached directly to the wavefunction (and thus be responsible for all ranks), or combined with other representations through a Combined Representation, which delegates responsibility for each rank to a subrepresentation. Furthermore, the wavefunction has a Distributed Model, which describes the way the wavefunction is distributed across processors in an MPI environment. The distributed model knows how to change the distribution of the wavefunction through the algorithm described in Sect. 9.4.

9.5.2 Operators and Transforms

Given a spatial discretization of Eq. (9.2), there is not a unique method of propagation that will be superior in all cases. It is therefore important that the code is flexible enough to support different propagation methods. PyProp enables this by the concept of *Transforms* and *Operators*. Transforms are objects that transform one or more ranks in the wavefunction from one representation to another. Operators are objects that represent a part of the Hamiltonian (9.3), and can be applied to the wavefunction when one or more of the ranks are in a certain representation. Operators can be seen as tensors, which are either diagonal or nondiagonal in each rank, in the sense that a diagonal operator does not combine different grid points in a tensor product with the wavefunction. For a nondiagonal operator, the application of the operator onto the wavefunction can be written the following way

$$\Phi_{\mathbf{i}} = \sum_{i_0'} \sum_{i_1'} \cdots \sum_{i_s'} V_{\mathbf{i},\mathbf{i}'} \cdot \psi_{\mathbf{i}'}. \tag{9.13}$$

Here subscript $\mathbf{i}$ and $\mathbf{i}'$ are vector indices representing an index to an element in the s-dimensional arrays. If the operator is diagonal in all ranks, we call it a diagonal operator or a *potential* . In that case, it is a function of the coordinates in the given representation, scaling every grid point independently,

$$\Phi_{\mathbf{i}} = V_{\mathbf{i}} \psi_{\mathbf{i}}. \tag{9.14}$$

Because a diagonal operator V does not couple any grid points, any analytic function of the operator, $f(V)$, will have the same property. Applying $f(V)$ to the wavefunction can then be done elementwise,

$$\Phi_{\mathbf{i}} = f(V_{\mathbf{i}}) \psi_{\mathbf{i}}. \tag{9.15}$$

A Hamiltonian is always represented by a selfadjoint operator (or equivalently, a Hermitian matrix in the discrete case). This implies that it is possible to find a unitary similarity transform T which will make the operator V diagonal,

$$V = T V_T T^*. \tag{9.16}$$

Note that it is possible to split a selfadjoint operator into two suboperators that are not selfadjoint and use the split-step method on those suboperators. This is generally a bad idea, as it can lead to a break of unitarity in the propagation. It can be shown that any analytic function $f(V)$ will be unchanged under such a similarity transform, and can be applied directly to the transformed operator.

$$f(V) = T f(V_T) T^*. \tag{9.17}$$

The exponential of an operator $f(V) = \exp(-i\Delta t V)$ is a special case of the above, and is of special interest in split-step propagation schemes. Diagonal operators in

PyProp therefore have methods to apply both the operator itself and the exponentiated operator to a wavefunction.

9.5.3 Split-Step Propagator

Having transforms between representations that make the operators diagonal makes it possible to propagate the Schrödinger equation with the split-step method discussed in Sect. 9.2. Propagating one time step will be a series of exponentiated operators. Between two operators, transforms will be applied in order to get the wavefunction to a representation where the next operator is diagonal.

$$\psi(t + \Delta t) = P_n T_n \cdots P_1 T_1 P_0 T_0 \psi(t), \qquad (9.18)$$

where T_i are transforms and $P_i = exp[i\Delta t c_i V_i]$ are exponentiated operators.

In order to cater for different combinations of problems and discretizations, it is important to make the implementation of Eq. (9.18) as flexible as possible. In PyProp the sequence of operators and transforms is therefore implemented in Python, while the actual transforms and operators are implemented in C++ and Fortran.

Listing 9.2 Action wrappers for split-step propagation

```
class TransformAction(object):
  def __init__(self, transform):
    self.Transform = transform

  def Execute(self, psi, t, dt):
    self.Transform.Transform(psi)

class ExpOperator(object):
  def __init__(self, operator):
    self.Operator = operator

  def Execute(self, psi, t, dt):
    self.Operator.ApplyExponential(psi, t, dt)
```

Before starting propagation, a setup routine determines the transforms and operators to be used in this propagation. The setup routine wraps the transforms and operators in the action wrappers above, and returns a list of action wrappers. Propagating one time step with the split-step algorithm is performed with the code below.

Listing 9.3 Split-step propagation algorithm

```
def AdvanceStep(psi, t, dt, actionList):
  for action in actionList:
    action.Execute(psi, t, dt)
```

Parallelization of the split-step propagator is done more or less automatically with the parallelization scheme from Sect. 9.4, in the sense that neither the transforms nor the diagonal operators are concerned with the parallelization details. Through the Distributed Model object, operators get information about the local part of the wavefunction. Diagonal operators are applied to the wavefunction elementwise, and can therefore be applied to the local part of the wavefunction on each processor independently. A transform acting on one dimension is independent of the distribution in the other dimensions, and can be applied independently on each processor as long as the transform acts only on local dimensions. In order to make transforms local, a redistribution step is applied. The redistribution step can be seen as a kind of transform, which is well suited to the split-step scheme presented above.

Due to the dynamic nature of Python, it is feasible to create advanced systems on top of the basic scheme of transforms and operators. We are currently developing a system where the optimal (in terms of runtime) order of operators is determined at runtime and the transforms and redistributions needed are determined automatically.

9.5.4 Explicit Propagators

The scheme of separating operators and transforms works well for other propagators beside the split-step propagator. Explicit propagators, such as multistep methods, Runge–Kutta methods and Taylor propagators or Krylov subspace based propagators, are particularly well suited for this scheme. Explicit propagators require only the action of the Hamiltonian on the wavefunction, and this can be implemented by a small modification to the algorithm in listings 9.2 and 9.3.

Direct exponential propagators are only efficient when efficient transforms to the eigenspaces of the operators are available. If this is not the case, other time propagation schemes such as multistep methods, Runge–Kutta methods and Taylor propagators or Krylov subspace based propagators should be considered. These methods will typically need shorter time steps and more work for each time step, but on the other hand do not require the operator to be diagonalized, and can in some cases therefore be more efficiently parallelized. Again, there are no universal rules for choosing the optimal strategy. It is therefore of great advantage to have software that allows the user to test different time-marching strategies.

9.6 A Numerical Example

9.6.1 Physical Model

In order to demonstrate some of the flexibility and efficiency of PyProp, we will now consider a numerical example: The ionization of the molecular ion H_2^+. The molecule consists of two hydrogen nuclei and one electron. The nuclei are much heavier than the electron, and it is therefore a good approximation to assume that the center of mass of the system are on the axis between the nuclei. We can then write up the Hamiltonian for the field-free H_2^+ in atomic units:

$$H_0(\mathbf{R}, \mathbf{r}) = -\frac{1}{2M} \nabla_{\mathbf{R}}^2 - \frac{1}{2m} \nabla_{\mathbf{r}}^2 - \frac{1}{|r - \mathbf{R}|} - \frac{1}{|r + \mathbf{R}|} + \frac{1}{\mathbf{R}}, \quad (9.19)$$

where $\mathbf{R}$ is the nuclear coordinate and $\mathbf{r}$ is the electronic coordinate.

We will employ spherical coordinates for both coordinates. However, the mass of the nuclei makes the nuclear motion much slower than the motion of the electrons, which allows us to neglect the rotational motion of the nuclei and keep only the vibrational coordinate R.

The molecule is exposed to a dipole laser pulse polarized along the axis of the molecule. This can be modeled by the following Hamiltonian:

$$H_{laser} = E_0 f(t) \sin(\omega t) r \cos \theta, \quad (9.20)$$

where E_0 is the field strength of the laser field, ω is the laser frequency, and $f(t)$ is the convolution giving the shape of the pulse. As there is no explicit ϕ dependence in Eq. (9.19) or Eq. (9.20), the electronic motion can be fully described by r and θ. This gives us a total of three spatial coordinates: R, r and θ. Introducing the reduced wavefunction trick in both R and r, $\Phi(R, r, \theta) = Rr\psi(R, r, \theta)$, we can write the Hamiltonian for the molecule as

$$H = H_0 + H_1 + H_{laser}. \quad (9.21)$$

Here, H_0 is the kinetic energy part of the Hamiltonian, which is dependent on coordinate system and discretization, but independent on problem,

$$H_0 = -\frac{1}{2M} \frac{\partial^2}{\partial R^2} - \frac{1}{2m} \left(\frac{\partial^2}{\partial r^2} + \frac{\Lambda^2}{r^2} \right). \quad (9.22)$$

H_1 is the problem-specific part of the time-independent Hamiltonian, representing potential energy in the system,

$$H_1 = \frac{1}{\sqrt{(r\cos\theta \pm R)^2 + (r\sin\theta)^2}} + \frac{1}{R}. \quad (9.23)$$

9.6.2 Numerical Considerations

We now have to decide which discretization methods to use. In this example, we will use equispaced grids in both r and R, truncated at $r = r_{max}$ and $R = R_{max}$. The fast Fourier transform is used to map between grid space and Fourier space as described in Sect. 9.3.1. The eigenfunctions of the angular operator are the Legendre polynomials. Thus, we sample the function on the zeroes of the $N + 1$ degree Legendre polynomial and apply the associated transform.

PyProp can automatically set up, from a configuration file, the required transforms as well as the diagonal operators required for evaluating H_0. We only need to supply the problem-specific potentials (H_1 and H_{laser}) to PyProp, and the rest will be taken care of internally.

In order to study ionization mechanisms of the system, a good representation of both the bound and unbound states is needed. A high density of grid points near the origin of R and r is needed to have a good representation of the bound states. Correspondingly, for a good representation of the unbound states, r_{max} and R_{max} must be sufficiently large. In this example, we use $R_{max} = 15$ a.u., $r_{max} = 120$ a.u., $N_r = 512$, $N_R = 256$ and $N_l = 128$.

9.6.3 Scalability

Propagating the above-mentioned system on one processing unit takes ≈ 40 s/time step. In order to study effects under the influence of a laser with wavelength $\lambda = 800$ nm, the system must be propagated at least a few cycles of the laser, where each laser cycle corresponds to $t = 110$ a.u. Converged results are obtained for a time step of $\Delta t = 0.01$ a.u., which gives a minimum wall clock time of 5 days per laser cycle.

From the above estimate, it is clear that in order to study interesting physics of H_2^+ with nuclear motion, it is an absolute requirement to have efficient parallelization. We will therefore apply the parallelization technique described in Sect. 9.4. In Fig. 9.5, the speedup gained from adding processors to the job is shown for a Cray XT4 supercomputer.

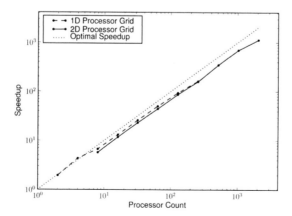

Fig. 9.5 Speedup in wall clock time $S = T_n/T_1$ for a realistic discretization of H_2^+. T_n is the wall clock time used for propagating the system a few time steps using n processors.

Using a 1D processor grid, we alternate between having the R dimension and the r dimension distributed. This limits the number of processors to $N_p = 256$. Using more than 256 grid points leads to some processors not having any data when the R dimension is distributed. The problem can be scaled further by switching to a 2D processor grid, as described in Sect. 9.4, as in this case the two distributed dimensions are distributed on $\sqrt{N_p}$ processors each. We observe that the 1D processor grid performs better than the 2D case, but the difference decreases towards $N_p = 256$. Increasing the number of processors further, the 2D processor grid continues to scale very well up to $N_p = 2048$, where a speedup of over 1000 has been achieved compared to running on a single processor.

9.7 Conclusion

In this brief chapter, we have tried to give an introduction to some of the fascinating aspects of simulating quantum systems. Scalability on modern supercomputers can be obtained by using the right distribution method for pseudo-spectral methods. PyProp demonstrates that it is possible to create flexible solvers without sacrificing significant performance. By recognizing the computational bottlenecks, a hierarchy of languages can be used to get both flexibility and high performance.

The computational example of Sect. 9.6.2 is intended as a practical demonstration of how problems beyond reach on desktop computers become readily available on modern HPC systems (provided adequate software is available). We stress that this is not the end of the story, only the beginning. In the future, we aim to do calculations on higher-dimensional systems, and expect to be able to do so by improving algorithms and implementation, with increasing scalability and system size. Parallel computing will be an indispensable tool in this effort.

Acknowledgments The numerical experiments were carried out on the Cray XT4 system operated by Bergen Center of Computational Science, UNIFOB. Computational resources were granted by NOTUR, the Norwegian infrastructure for high-performance computing and computational science.

References

1. H. P. Langtangen, Python Scripting for Computational Science (Texts in Computational Science and Engineering), 3rd Edition, Springer (2008).
2. H. Yoshida, Construction of higher order symplectic integrators, Phys. Lett. A 150 (5–7) (1990) 262–268.
3. R. I. McLachlan, R. W. Quispel, Splitting methods, Acta Numerica 11 (2002) 341–434.
4. M. D. Feit, A. Fleck jr., A. Steiger. Solution of the schrödinger equation by a spectral method, J. Comput. Phys. 47 (3) (1982) 412–433.
5. M. R. Hermann, A. Fleck jr., Split-operator spectral method for solving the time-dependent schrödinger equation in spherical coordinates, Phys. Rev. A 38 (12) (1988) 6000–6012.

6. T. Sørevik, L. B. Madsen, J. P. Hansen, A spectral method for integration of the time-dependent schrödinger equation in hyperspherical coordinates, J. Phys. A: Math. Gen. 38 (2005) 6977–6985.
7. G. Muslu, H. A. Erbay, Higher-order split-step fourier schemes for the generalized nonlinear schrödinger equation, Math. Comput. Simmul. 67 (2005) 581–595.
8. J. P. Boyd, C. Rangan, P. H. Bucksbaum, Pseudospectral method on a semi-infinite interval with application to the hydrogen atom: A comparison of the mapped fourier-sine method with laguerre series and rational chebyshev expansions, J. Comput. Phys. 188 (2003) 56–74.
9. T. Sørevik, T. Birkeland, G. Oksa, Numerical solution of the 3D time dependent Schrödinger equation in spherical coordinates: Spectral basis and effects of split operator technique, J. Comput. Appl. Math. 225(1) (2009) 56–67.
10. H. Q. Ding, R. D. Ferraro, D. B. Gennery, A portable 3d FFT package for distributed-memory parallel archite ctures, in: PPSC (1995) pp. 70–71.
11. M. Frigo, S. G. Johnson, Fftw: An adaptive software architecture for the fft (1998) 1381–1394.
12. C. E. Cramer, J. A. Board, The development and integration of a distributed 3d fft for a cluster of workstations, in: Proceedings of the 4th Annual Linux Showcase and Conference (2000) pp. 121–128.
13. P. D. Haynes, M. Cote, Parallel fast fourier transforms for electronic structure calculations, Comput. Phys. Commun. 130 (2000) 132–136.
14. M. Eleftheriou, B. G. Fitch, A. Rayshubskiy, T. J. C. Ward, R. S. Germain, Scalable framework for the 3d ffts on the blue gene/l supercomputer: Implementation and early performance measurements, IBM J. Res. Dev. 49 (2005) 457–464.
15. A. Dubey, D. Tessera, Redistribution strategies for portable parallel fft: a case study, Concurr. Comput. Pract. Exp. 13 (2001) 209–220.
16. T. Birkeland, T. Sørevik, Parallel redistribution of multidimensional data, in: C. Bischof, M. BÄijcker, P. Gibbon, G. R. Joubert, T. Lippert, B. Mohr, F. Peters (Eds.), Advances in Parallel Computing, Vol. 15, IOS Press, (2008).
17. T. L. Veldhuizen, Arrays in Blitz++.
18. M. Galassi, J. Theiler, J. Davies, GNU Scientific Library Reference Manual (2nd Ed.), Network Theory Limited (2003).
19. D. Abrahams, R. W. Grosse-Kunstleve, Building hybrid systems with boost.python, C/C++ Users Journal July.

Chapter 10
Parallel Approaches in Molecular Dynamics Simulations

Dušanka Janežič, Urban Borštnik and Matej Praprotnik

Abstract

In this contribution we will present the survey of our past and current endeavor on parallel approaches in molecular modeling algorithm development, for example, molecular dynamics (MD) simulation. In particular, we will describe the new split integration symplectic method for the numerical solution of molecular dynamics equations and methods for the determination of vibrational frequencies and normal modes of large systems, and the distributed diagonal force decomposition method, a parallel method for MD simulation.

Parallel computer programs are used to speed up the calculation of computationally demanding scientific problems such as MD simulations. Parallel MD methods distribute calculations to the processors of a parallel computer but the efficiency of parallel computation decreases due to inter processor communication. Calculating the interactions among all atoms of the simulated system is the most computationally demanding part of an MD simulation. Parallel methods differ in their distribution of these calculations among the processors, while the distribution dictates the method's communication requirements.

We have developed a computer program for molecular dynamics simulation that implements the split integration symplectic method and is designed to run on specialized parallel computers. The molecular dynamics integration is performed by the new integration method, which analytically treats high-frequency vibrational motion and thus enables the use of longer simulation time steps. The low-frequency motion

Dušanka Janežič
National Institute of Chemistry, Hajdrihova 19, 1000 Ljubljana, Slovenia,
e-mail: dusa@cmm.ki.si

Urban Borštnik
National Institute of Chemistry, Hajdrihova 19, 1000 Ljubljana, Slovenia,
e-mail: urban@cmm.ki.si

Matej Praprotnik
National Institute of Chemistry, Hajdrihova 19, 1000 Ljubljana, Slovenia,
e-mail: praprot@cmm.ki.si

R. Trobec et al. (eds.), *Parallel Computing*, DOI 10.1007/978-1-84882-409-6_10,
© Springer-Verlag London Limited 2009

is treated numerically on specially designed parallel computers, which decreases the computational time of each simulation time step. We study the computational performance of simulation on specialized computers and provide a comparison to standard personal computers. The combination of the new integration method with two specialized parallel computers is an effective way to significantly increase the speed of molecular dynamics simulations.

We have also developed a parallel method for MD simulation, the distributed-diagonal force decomposition method. Compared to other methods its communication requirements are lower and it features dynamic load balancing, which increase the parallel efficiency. We have designed a cluster of personal computers featuring a topology based on the new method. Its lower communication time in comparison to standard topologies enables an even greater parallel efficiency.

10.1 Split Integration Symplectic Method

The standard integrators for solving the classical equations of motion are the second-order symplectic leap-frog Verlet (LFV) algorithm [1] and its variants. Their power lies in their simplicity since the only required information about the studied physical system are its interacting potential and the timescale of the fastest motion in the system, which determines the integration time step size. Therefore they are employed for solving dynamics problems in a variety of scientific fields, for example, molecular dynamics (MD) simulation [2, 3], celestial mechanics [4–6], and accelerator physics [7]. However, in the case of MD integration, the integration time step size is severely limited due to the numerical treatment of the high-frequency molecular vibrations, which represent the fastest motion in the system [8]. Therefore, a huge number of integration steps is usually required to accurately sample the phase space composed of all the coordinates and momenta of all the particles. This is a time-consuming task and is often too demanding for the capabilities of contemporary computers.

One way of overcoming the limitation of the standard methods' integration time step size is to analytically treat high-frequency molecular vibrations. This requires the standard theory of molecular vibrations [9] to be built into the integration method. In this way the fast degrees of freedom are rigorously treated and not removed, as in case of rigid-body dynamics [10–12], where small molecules are treated as rigid bodies. Such semi-analytical second-order symplectic integrators were developed by combining MD integration and the standard theory of molecular vibrations [13–16]. The unique feature of these MD integrators is that the standard theory of molecular vibrations, which is a very efficient tool to analyze the dynamics of the studied system from computed trajectories [17–23], is used not to analyze, but to compute trajectories of molecular systems. Information about the energy distribution of normal modes and the energy transfer between them is obtained without additional calculations. The analytical description of coupled molecular vibrations can be employed only when using the normal coordinates [9, 13–15] and a translating and rotating internal coordinate system of each molecule [24,25]. The dynamics of an Eckart frame has to be adopted to be used within the second-order generalized leap-frog scheme [26, 27] for MD integration. This assures the time reversibility of the methods [13, 16]. In the following we shortly summarize technical details of the method.

In MD simulations for each atom of the system the Hamilton equations are solved

$$\frac{d\eta}{dt} = \{\eta, H\} = \hat{L}_H \eta \qquad (10.1)$$

where $\hat{L}_H$ is the Lie operator, $\{,\}$ is the Poisson bracket [28], and $\eta = (\mathbf{q}, \mathbf{p})$ is a vector of the coordinates of all the particles and their conjugate momenta.

The formal solution of the Hamiltonian system (10.1) can be written in terms of Lie operators as

$$\eta|_{t_k + \Delta t} = \exp(\Delta t \hat{L}_H) \eta|_{t_k} \qquad (10.2)$$

and represents the exact time evolution of a trajectory in phase space composed of coordinates and momenta of all the particles from t_k to $t_k + \Delta t$, where Δt is the integration time step [28].

The first step in the development of a new symplectic integration method is to split the Hamiltonian H of a system into two parts [29, 30]

$$H = H_0 + H_r, \qquad (10.3)$$

where H_0 is the part of the Hamiltonian that can be solved analytically and H_r is the remaining part.

Next, a second-order approximation for (10.2), known as the generalized leap-frog scheme [26, 27], is used

$$\eta|_{t_{k+1}} = \exp\left(\frac{\Delta t}{2} \hat{L}_{H_0}\right) \exp(\Delta t \hat{L}_{H_r}) \exp\left(\frac{\Delta t}{2} \hat{L}_{H_0}\right) \eta|_{t_k} + O(\Delta t^3), \qquad (10.4)$$

which defines the split integration symplectic method (SISM). The whole integration time step combines the analytical evolution of H_0 with a correction from the H_r resolved by numerical integration. The Eq. (10.4) on the operators level describes how to propagate from one point in phase space to another. First, the system is propagated for a half integration time step by H_0, then for a whole step by H_r, and finally for another half step by H_0. The whole integration time step thus combines the analytical evolution of H_0 with a correction arising from the H_r performed by numerical integration. This integration scheme was used as the basis for the development of the SISM.

The model Hamiltonian has the following form

$$H = \sum_i \frac{\mathbf{p}_i^2}{2m_i}$$

$$+ \frac{1}{2} \sum_{bonds} k_b (b - b_0)^2 + \frac{1}{2} \sum_{angles} k_\theta (\theta - \theta_0)^2 + \frac{1}{2} \sum_{torsions} V_0 (\cos \phi - \cos \phi_0)^2$$

$$+ \sum_{i>j} \frac{e_i e_j}{4 \pi \varepsilon_0 r_{ij}} + \sum_{i>j} 4 \varepsilon_{ij} \left[\left(\frac{\sigma_{ij}}{r_{ij}} \right)^{12} - \left(\frac{\sigma_{ij}}{r_{ij}} \right)^6 \right], \quad (10.5)$$

where i and j run over all atoms, m_i is the mass of the i-th atom, $\mathbf{p}_i$ is the linear momentum of the i-th atom, b_0 and θ_0 are reference values for bond lengths and angles, respectively, k_b and k_θ are corresponding force constants, ϕ_0 are the reference values for the torsion angles, and V_0 are the corresponding barrier heights; e_i denotes the charge on the i-th atom, ε_0 is the dielectric constant in vacuum, r_{ij} is the distance between the i-th and j-th atoms, and ε_{ij} and σ_{ij} are the corresponding constants of the Lennard–Jones potential.

The Hamiltonian (10.5) is a typical MD Hamiltonian that describes a system of molecules with only one equilibrium configuration and no internal rotation. We assume that the height of the barrier of the torsional potential is large enough that the motion of atoms in the vicinity of the minimum of the torsional potential can be treated as a harmonic vibration around the equilibrium configuration. The vibrational potential energy is therefore the sum of vibrational potential energies of all the molecules in the system

$$V_{vib} = \sum_{j'=1}^{m} V_{vib_{j'}} =$$

$$\frac{1}{2} \sum_{bonds} k_b (b - b_0)^2 + \frac{1}{2} \sum_{angles} k_\theta (\theta - \theta_0)^2 + \frac{1}{2} \sum_{torsions} V_0 (\cos \phi - \cos \phi_0)^2, \quad (10.6)$$

where $V_{vib_{j'}}$ is the vibrational potential energy of the j'-th molecule.

The pure harmonic Hamiltonian H_0 in the splitting (10.3) is defined as the sum of vibrational energies of all the molecules in the system

$$H_0 = T + V_{harm} = \sum_{j'=1}^{m} (T_{j'} + V_{harm_{j'}}), \quad (10.7)$$

where $T = \sum_i \mathbf{p}_i^2 / 2m_i$ is the kinetic energy of all the atoms in the systems, $T_{j'}$ is the kinetic energy of the j'-th molecule, V_{harm} is the harmonic vibrational potential energy, which is for an individual molecule defined by Eq. (10.11), $V_{harm_{j'}}$ is the corresponding harmonic vibrational potential energy of the j'-th molecule, and m is the number of all the molecules in the system.

The remaining part of the Hamiltonian

$$H_r = H - H_0 = V_{nb} + V_{ah} \quad (10.8)$$

is then equal to the sum of the nonbonded potential energy

$$V_{nb} = \sum_{i>j} \frac{e_i e_j}{4\pi\varepsilon_0 r_{ij}} + \sum_{i>j} 4\varepsilon_{ij} \left[\left(\frac{\sigma_{ij}}{r_{ij}}\right)^{12} - \left(\frac{\sigma_{ij}}{r_{ij}}\right)^6 \right] \tag{10.9}$$

and the anharmonic vibrational potential energy of higher terms (cubic, quartic, etc.) in terms of displacements of atoms from their equilibrium positions

$$V_{ah} = V_{vib} - V_{harm}. \tag{10.10}$$

The underlying principle to enable the SISM to permit longer integration time steps lies in the analytical treatment of high-frequency vibrations described by H_0. The propagation scheme (10.4) enables to treat the time evolution of the vibrational, rotational, and translational degrees of freedom of each molecule (described by $\exp\left((\Delta t/2)\hat{L}_{H_0}\right)$) independently of all other molecules in the system because the total intermolecular interactions are described by a separate term $\exp\left(\Delta t \hat{L}_{H_r}\right)$. Each molecule is treated as an isolated molecule when propagating by $\exp\left((\Delta t/2)\hat{L}_{H_0}\right)$. Propagation by $\exp\left((\Delta t/2)\hat{L}_{H_0}\right)$ can therefore be solved analytically using normal-mode analysis. In the latter, only quadratic terms are kept in the expansion of the vibrational potential energy V_{vib} and all higher terms are neglected [9]

$$V_{vib} \approx V_{harm} = \frac{1}{2} \sum_{i,j=1}^{3N} \left(\frac{\partial^2 V_{vib}}{\partial \Delta q_i \partial \Delta q_j}\right)_0 \Delta q_i \Delta q_j =$$

$$\frac{1}{2} \sum_{i,j=1}^{3N} \left(\frac{\partial^2 V_{harm}}{\partial \Delta q_i \partial \Delta q_j}\right)_0 \Delta q_i \Delta q_j$$

$$= \frac{1}{2} \sum_{i,j=1}^{3N} H_{ij} \Delta q_i \Delta q_j = \frac{1}{2} \Delta \mathbf{q} \cdot \mathbf{H} \cdot \Delta \mathbf{q}. \tag{10.11}$$

Here $\Delta \mathbf{q} = (\Delta x_1, \Delta y_1, \Delta z_1, \ldots, \Delta x_N, \Delta y_N, \Delta z_N)$ is a vector of the relative Cartesian displacement coordinates and their corresponding momenta are $\Delta \mathbf{p} = (m_1 \Delta v_{1_x}, m_1 \Delta v_{1_y}, m_1 \Delta v_{1_z}, \ldots, m_N \Delta v_{N_x}, m_N \Delta v_{N_y}, m_N \Delta v_{N_z})$, where subscripts x, y, x denote x, y, z components of the internal coordinate system, respectively (see Fig. 10.1).

The Hessian $\mathbf{H} \in \mathbb{R}^{3N \times 3N}$ is a symmetric matrix of the second derivatives of the vibrational potential energy with the elements

$$H_{ij} = H_{ji} = \left(\frac{\partial^2 V_{vib}}{\partial \Delta q_i \partial \Delta q_j}\right)_0 = \left(\frac{\partial^2 V_{harm}}{\partial \Delta q_i \partial \Delta q_j}\right)_0. \tag{10.12}$$

To determine the vibrational motions of the system, the eigenvalues and eigenvectors of the mass-weighted Hessian $\mathbf{M}^{-1/2} \cdot \mathbf{H} \cdot \mathbf{M}^{-1/2}$ have to be calculated [9, 20–22]. This leads to solving a secular equation

$$\det(\mathbf{M}^{-1/2} \cdot \mathbf{H} \cdot \mathbf{M}^{-1/2} - \lambda \mathbf{I}) = 0, \tag{10.13}$$

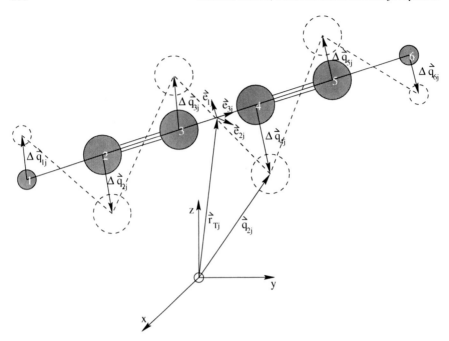

Fig. 10.1 Atom displacement in the Cartesian and the internal coordinate system.

where $\mathbf{M} \in \mathbb{R}^{3N \times 3N}$ is a diagonal mass matrix. The diagonal elements are $M_{11} = m_1$, $M_{22} = m_1$, $M_{33} = m_1, \ldots, M_{3N-2,3N-2} = m_N$, $M_{3N-1,3N-1} = m_N$ and $M_{3N,3N} = m_N$. For a nonlinear molecule composed of N atoms, Eq. (10.13) has $3N - 6$ nonzero eigenvalues $\omega_i = \sqrt{\lambda_i}$ describing molecular vibrations. The corresponding dynamics is described in the standard theory of molecular vibration by normal coordinates Q_i, $i = 1, 2, \ldots, 3N - 6$ [28]. Six of $3N$ roots in Eq. (10.13) are zero. They correspond to three translations and three rotations of a molecule as a whole while their dynamics is not described in terms of the normal coordinates [9, 13].

An alternative approach to standard theory's description of molecules' rotation and translation [9] is to describe rotation and translation of a molecule in terms of the normal coordinates. To do so the whole atom velocity needs to be expressed in terms of the relative Cartesian displacement coordinates. It has been shown in full detail that the dynamics of the internal coordinate system in this case differs from the dynamics of the Eckart frame, which is employed in the standard theory of molecular vibrations [13].

The equations of motion for the normal coordinates take the Hamiltonian form as [13]

$$\frac{d}{dt}P_i = -\omega_i^2 Q_i; \qquad \frac{d}{dt}Q_i = P_i, \qquad i = 1, 2, \ldots, 3N \qquad (10.14)$$

where P_i is the conjugate momentum to the normal coordinate Q_i [28].

The particular solution of the system (10.14) can be written as [13]

$$\begin{bmatrix} P_i(\frac{\Delta t}{2}) \\ Q_i(\frac{\Delta t}{2}) \end{bmatrix} = \begin{bmatrix} \cos(\omega_i \frac{\Delta t}{2}) & -\omega_i \sin(\omega_i \frac{\Delta t}{2}) \\ \frac{1}{\omega_i} \sin(\omega_i \frac{\Delta t}{2}) & \cos(\omega_i \frac{\Delta t}{2}) \end{bmatrix} \begin{bmatrix} P_i(0) \\ Q_i(0) \end{bmatrix}. \tag{10.15}$$

Equation (10.15) describes vibrational motion corresponding to the normal mode i with $\omega_i > 0$.

The equations of motion for the translation and rotation of a molecule in terms of the normal coordinates, obtained from Eq. (10.15) for the normal coordinates with $\omega_i = 0$ and using $\lim_{x \to 0} \frac{\sin x}{x} = 1$, are [13]

$$P_i\left(\frac{\Delta t}{2}\right) = P_i(0), \tag{10.16}$$

$$Q_i\left(\frac{\Delta t}{2}\right) = P_i(0)\frac{\Delta t}{2} + Q_i(0). \tag{10.17}$$

The expressions for the transformations between Cartesian, relative Cartesian displacement, and normal coordinates are obtained in a straightforward way [13]. The SISM then explicitly reads as follows:

- **Preparatory step**: at the outset of calculation, vibrational frequencies and normal modes of H_0, represented by the normal coordinates P, Q, are determined. The initial normal coordinates P_i^0, Q_i^0, $i = 1, ..., 3N$, are obtained from the initial atoms' velocities and the initial displacements of the atoms from their equilibrium positions by means of the transformational matrix $\mathbf{A}$. The columns of $\mathbf{A}$ are the eigenvectors of the root-mass-weighted second-derivative matrix $\mathbf{M}^{-1/2} \cdot \mathbf{H} \cdot \mathbf{M}^{-1/2}$ and N is the number of atoms in each molecule.
- **Analytical solution** $\exp\left(\frac{\Delta t}{2} \hat{L}_{H_0}\right)$: the normal coordinates, P_i^0, Q_i^0, are rotated in phase space by the corresponding vibrational frequency ω_i for $\frac{\Delta t}{2}$:

$$\begin{bmatrix} P_i' \\ Q_i' \end{bmatrix} = \mathbf{R} \begin{bmatrix} P_i^0 \\ Q_i^0 \end{bmatrix} \tag{10.18}$$

$$\mathbf{R} = \begin{bmatrix} \cos(\omega_i \frac{\Delta t}{2}) & -\omega_i \sin(\omega_i \frac{\Delta t}{2}) \\ (1/\omega_i) \sin(\omega_i \frac{\Delta t}{2}) & \cos(\omega_i \frac{\Delta t}{2}) \end{bmatrix} \tag{10.19}$$

$\omega_i \neq 0$ defines the vibrations of atoms in each molecule
$\omega_i = 0$ defines translations and rotations of molecules
The normal coordinates of the normal modes with frequency zero
($\lim_{x \to 0} \frac{\sin x}{x} = 1$ for $\omega_i = 0$) evolve as

$$P_i' = P_i^0 \tag{10.20}$$

$$Q_i' = P_i^0 \frac{\Delta t}{2} + Q_i^0 \tag{10.21}$$

Coordinate transformation: the normal coordinates P_k', Q_k' are transformed to the Cartesian displacement coordinates $\Delta p_i'$, $\Delta q_i'$ ($m_1 = m_2 = m_3, ..., m_{3N-2} = m_{3N-1} = m_{3N}$, where m_i, $i = 1, ..., 3N$ are the atoms' masses):

$$\Delta p'_i = \sqrt{m_i} \sum_k A_{ik} P'_k \qquad (10.22)$$

$$\Delta q'_i = \frac{1}{\sqrt{m_i}} \sum_k A_{ik} Q'_k \qquad (10.23)$$

- **Numerical solution**, $\exp(\Delta t \hat{L}_{H_r})$: momenta in the Cartesian coordinates are numerically integrated:

$$p''_i = p'_i - \Delta t \left(\frac{\partial H_r}{\partial q} \right) \qquad (10.24)$$

$$q''_i = q'_i + \Delta t \left(\frac{\partial H_r}{\partial p} \right) = q'_i \qquad (10.25)$$

Only one force calculation per integration step must be performed. Since $H_r = H_r(q)$ and $\left(\frac{\partial H_r}{\partial p} \right) = 0$, only momenta change at this step.

Back-transformation: the Cartesian displacement coordinates $\Delta p''_k$, $\Delta q''_k$ are back-transformed to the normal coordinates P''_i, Q''_i:

$$P''_i = \sum_k \frac{1}{\sqrt{m_k}} A^T_{ik} \Delta p''_k \qquad (10.26)$$

$$Q''_i = \sum_k \sqrt{m_k} A^T_{ik} \Delta q''_k \qquad (10.27)$$

- **Analytical solution**, $\exp\left(\frac{\Delta t}{2} \hat{L}_{H_0} \right)$: the normal coordinates are again rotated in phase space for $\frac{\Delta t}{2}$:

$$\begin{bmatrix} P_i \\ Q_i \end{bmatrix} = \mathbf{R} \begin{bmatrix} P''_i \\ Q''_i \end{bmatrix} \qquad (10.28)$$

This concludes one full SISM integration step, which is repeated until the desired number of integration steps is reached.

One time step of SISM is schematically presented in Fig. 10.2.

10.1.1 Calculation of Infrared Spectra

The vibrational and rotational motions of molecules are those which involve energies that produce the spectra in the infrared region. Therefore, the SISM is particularly suitable for computing the IR spectra because rotational, translational, and vibrational motions are resolved analytically, independently of the MD integration time step.

Figure 10.3(a) demonstrates that the IR spectra of bulk water at ambient conditions calculated by SISM and LFV using a 0.5 fs integration time step are in good agreement. These IR spectra were taken as a reference for comparison with calculated IR spectra using longer integration time steps. When using a 1.0 fs integration

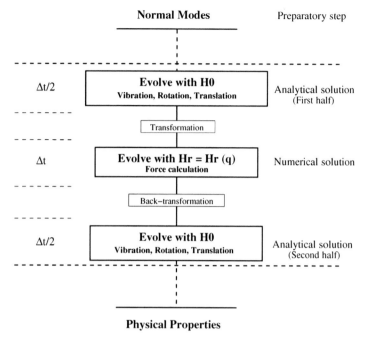

Fig. 10.2 Solution scheme for SISM.

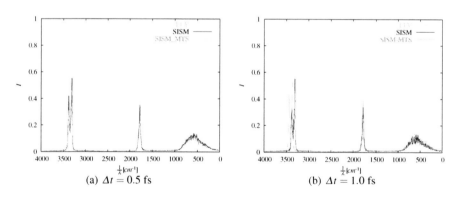

Fig. 10.3 Calculated (LFV, SISM) IR spectrum of bulk water for $\Delta t = 0.5$ fs and $\Delta t = 1.0$ fs.

time step, the high-frequency double peak at 3300 cm^{-1} in the IR spectrum calculated by the LFV already shifts to the higher frequencies as shown in Fig. 10.3(b). The observed blue shift suggests that when using a 1.0 fs integration time step, the LFV can no longer accurately describe the high-frequency vibrational motions of atoms in a water molecule. This phenomenon is even more evident in Fig. 10.4 for the cases of 1.5 fs and 2.0 fs integration time steps, where the peak at 1775 cm^{-1} also starts shifting toward higher frequencies. Peaks in corresponding IR spectra,

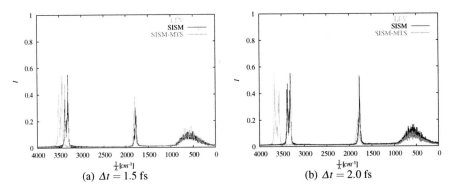

Fig. 10.4 Calculated (LFV, SISM) IR spectrum of bulk water for $\Delta t = 1.5$ fs and $\Delta t = 2.0$ fs.

which are calculated by the SISM, however remain at the same positions as corresponding peaks in the reference IR spectra calculated using the integration time step of 0.5 fs. This proves that owing to the analytical description of high-frequency molecular vibrations, the latter are accurately described by the SISM also using a 2.0 fs integration time step [15].

10.1.2 Enlarging the Integrational Time Step

The actual speedup of an integrational method is determined by measuring the required CPU time per integration step. Our results show that the computational cost per integration step is slightly larger for the SISM than the LFV for systems smaller than 1000 atoms. However, for larger systems consisting of more than 1000 atoms the computational cost per integration step becomes approximately the same for all of the methods due to the time-consuming $O(N^2)$ numerical calculation of non-bonded forces, which is performed by all three methods in the same way and prevails over the additional calculations in the iterative SISM, which scale linearly with N. Therefore, the speedup of the SISM over the LFV is determined mainly by the significant difference in the integration time step size owing to the analytical treatment of high-frequency motions by SISM [16, 31].

10.2 Parallel Computers

Computers are an essential tool used to solve computational problems in science today. The speed of computer processors is continually increasing, enabling its use to approach ever more complex computational problems [32, 33]. However, many existing problems would be well served by an increase in computational capacity today. For these problems, parallel computers provide a solution [34]. Many scientific problems can be effectively parallelized to run on a parallel computer.

10.2.1 Parallel Computing

In parallel computing, a problem is split into several subproblems that are solved concurrently on parallel processors in a shorter time. A parallel program is written to be executed on many processors at once and they must correctly share and exchange data to solve the problem. Generally, the processors must communicate throughout the computation since the results from one processor are needed by others. The manner in which the initial problem is divided among the processors – the data distribution and the distribution of computation – greatly affects how the parallel program is written and the time that is spent for communication. Generally, time that is spent for communication cannot be used for communication, since the processor is waiting for input for its next calculation.

10.2.1.1 Parallel Efficiency

Since it is the goal of parallel computing to reduce the total time required to solve a problem, the time spent for communication must be minimized. If it takes time T to solve the problem, an ideal parallel computation on P processors would take only T/p; however, due to the time lost to communication and other factors, the time T_p required by any processor is usually greater: $T_p > T/p$. We can now define the *speedup*

$$S = \frac{T}{T_p} \tag{10.29}$$

as the factor specifying how much faster the parallel computation is compared to a single processor computation. Ideally, the speedup S would equal the number of processors P, $S = P$, which is true if $T_p = T/P$. In several rare cases such a *linear* speedup is possible or even exceeded due to hardware effects [35]. We can define the *parallel efficiency*

$$E = \frac{S}{P} = \frac{T}{PT_p} \tag{10.30}$$

to measure the performance of the parallel computation relative to the ideal time. In optimizing a parallel program, we strive to obtain the highest parallel efficiency since it directly translates to increasing the speedup offered by the program. A higher parallel efficiency is obtained by bounding the communication time and by ensuring that all of the processors have an equal computational load. If processors have unequal computational loads, then whenever the processors communicate globally, the ones with the lowest load must wait for the most loaded processor to finish its computation and begin communication. Load balancing the computation attempts to keep an equal computional load among all processors, which minimizes waiting time and achieves a higher parallel efficiency.

10.2.2 Parallel Computer Types

Parallel computers may be divided into two broad categories depending on the way processors access memory. The type of memory access greatly influences the way in which a parallel program must be written.

Shared Memory

In shared memory computers, all processors may access all memory directly (i.e., a processor may read from or write to any memory location, as if the memory were local). The two common types of shared memory computers are symmetric multi-processing (SMP) computers, in which all memory is local to all of the processors. In effect, all memory accesses require the same access time. In nonuniform memory access computers (NUMA), processors have local memory, which provides the fastest access times; however, they can still directly access remote memory (i.e., another processor's local memory), albeit with a higher access time.

Distributed Memory

In distributed memory computers, processors can access only their local memory, but they cannot directly access remote memory. All data exchange between the processors must occur by explicit *message passing* that involves both processors exchanging messages over a processor interconnect, which provides the connection among the processors. Current interconnect technologies range from standard Ethernet to higher-performance Myrinet [36], Infiniband [37], and others.

Libraries, such as the Parallel Virtual Machine [38] (PVM) or the Message Passing Interface [39,40] (MPI) are used to abstract the implementation details of a given computer's message-passing hardware, providing a standard interface to the programmer. Since distributed memory computers are more specific than shared memory computers, parallel programs targeted for distributed memory computers can run on shared memory ones as well. Specific implementations of message-passing libraries on shared memory computers are often optimized to take advantage of the shared memory.

Modern parallel computers, such as clusters of personal computers, are increasingly hybrids of both shared memory and distributed memory computers: the parallel computer is composed of a number of shared-memory nodes (such as multiprocessor, multi-core personal computers), which are in turn connected by the interconnect. While the processors in one node share memory, the overall parallel computer is still characterized by its distributed memory. The programmer must still use a message-passing library as the overall data exchange mechanism.

10.2.2.1 Topologies of Clusters

Clusters are traditionally built using switching technologies. Indeed, the first clusters used the fastest Ethernet switches then available [41, 42]. However, switches often have limited number of connections, limiting the cluster size, and often have a limited amount of bandwidth that must be shared among all nodes connected to it, which is especially true if multiple levels of switches are used [43].

Many parallel computers have therefore been designed around point-to-point connections between individual processors. A point-to-point processor interconnect can be described by a mathematical graph. The vertices of a graph correspond to the processors while the edges correspond to the interconnect's connections between the processors. The topology of the interconnect is then described by the graph's topology. While it is virtually impossible to provide full direct connectivity among any processor pair for larger numbers of processors, the topology can be chosen to have desirable attributes from both a performance standpoint as well as from an ease of programming perspective. Generally, successful topologies used for MD simulation have been rings, meshes [44], and hypercubes [45].

10.2.3 Reducing Computational Complexity in Molecular Dynamics Simulations

The number of nonbonding interactions in a molecular system greatly outnumbers the number of bonding interactions. A system of N atoms has $O(N^2)$ nonbonding interactions arising from the $N^2/2$ atomic pairs. Since any atom can have at most a few bonds, the number of bonding interactions is $O(N)$. The calculation of the nonbonding interactions is the principal limiting factor in computer simulations, limiting not only the attainable simulation lengths but also the system sizes that can be feasibly simulated.

Several approaches are used to reduce the computational complexity of nonbonding interactions below $O(N^2)$. Among these are employing an interaction cutoff distance, the Barnes–Hut tree method [46], and the fast multipole methods [47, 48].

Cutoff Distance

Employing a cutoff distance is among the principal means of reducing the computational complexity of computing nonbonding interactions [49]. A characteristic of nonbonding interactions is their decreasing magnitude with increasing distance. Both commonly-employed potentials in classical MD simulations behave this way.

The Lennard–Jones potential used to describe van der Waals interactions between atomic pairs, decays as r^{-6} with increasing distance r and the Coulomb potential, which describes the electrostatic interaction between atomic pairs, decays as r^{-1} with increasing distance r. The limit at infinite distance for these interactions is 0. The potential can be changed or redefined to be 0 beyond a certain cutoff distance. Various methods are used to achieve this while retaining an accurate simulation despite the changed functional form [50, 51].

The gain is that only interactions with the cutoff distance need to be calculated. Since interactions among atoms farther apart than the cutoff distance is defined to be zero, their calculation can be ignored. Instead of calculating $O(N)$ interactions for each of the N atoms (yielding $O(N^2)$ interaction calculations), only a finite subset of interactions for each of the N atoms must be calculated. The size of the subset depends on the system density and the cutoff radius, but is independent of the system size. The computational complexity is therefore reduced to $O(N)$.

Tree and Fast Multipole Methods

Tree-based methods and fast multipole-based methods provide a means to account for all the pairwise interactions in a molecular system with a computational complexity less than $O(N^2)$. Both involve clustering spatially close atoms into clusters and using representative values of these clusters instead of individual atoms to calculate distant interactions. In the Barnes and Hut tree method, interactions are calculated individually for each atom. For close by atoms, the interaction is calculated directly. Beyond a certain distance, the interactions are calculated between the atom and the cluster. The computational complexity of the tree-based methods is $O(N \log N)$. In the fast multipole methods, several orders of multipoles are calculated for each atomic cluster. Atomic interactions are derived from the interactions of their representative multipoles. For most distant clusters, individual atoms are not even considered.

As seen later in Sect. 10.3.1, the tree- and multipole-based methods are especially well suited to parallelization techniques in which the presence of atoms on individual processors is limited.

10.3 Parallel Molecular Dynamics Computer Simulations

In parallel calculations of molecular dynamics simulations, processors are used in parallel to calculate the two parts of every MD integration step: the force calculation and the coordinate update [52]. MD simulation time steps are inherently sequential: the newest coordinates are needed to correctly calculate the forces and coordinates can be updated only when the latest forces have been calculated. While the force calculation and the coordinate update are calculated in parallel, the processors must exchange force and atomic coordinates between these two calculations in a global

Fig. 10.5 The parallel main loop in molecular dynamics. It consists of two computation phases indicated in white boxes (the force calculation and coordinate updates) and two communication phases indicated in grayed boxes (the force summation and coordinate broadcast). The global operations performed in the communication phases are detailed in Sect. 10.3.3.

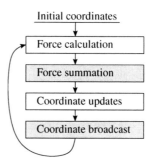

operation step. The parallel MD loop is shown in Fig. 10.5 and the global operations are detailed in Sect. 10.3.3.

10.3.1 Methods for Parallel Molecular Dynamics Simulations

Three main classes of parallel methods have been developed for MD simulations: replicated data [53, 54], spatial decomposition [55], and force decomposition [54, 56, 57]. Several advanced methods combine both the spatial and force decomposition approaches [58–61]. The methods differ in how interaction calculations are distributed among the processors. Since a processor needs coordinate data to calculate interactions, the distribution of interaction calculation determines the data distribution among the processors. The data distribution in turn governs the data that must be transferred among processors in each global operation. In addition, the atomic distribution maps atoms to processors for coordinate updates and other calculations that do not depend on interactions with other atoms.

Replicated Data

The replicated data method [53, 54] is the most straightforward parallelization method yet with the highest communication cost. As its name implies, all atomic data are replicated among all processors. As such, each global operation step entails the transfer of all N atomic data among all P processors, which has a higher communication cost than other methods. The global communication can easily be performed using a single collective operation routine. Any processor can calculate any interaction and perform any of the force updates, which simplifies load balancing. The atomic distribution is therefore very fluid.

Spatial Decomposition

In the spatial decomposition method, the space of the molecular system is divided into separate regions, nominally one per processor. The processors are then responsible for calculating the interactions among atoms in their region of space; for this, they need to communicate with at least their 27 neighboring processors resulting in a data transfer volume of $(N/P)^{2/3}$. The spatial decomposition method is well suited to simulations with a short cutoff distance. Since the transferred data volume is limited and the communication due to the global operations is also limited to nearby processors, it is straightforward to map processors onto common interconnect topologies such as a mesh. If no cutoff would be used, the communication would degenerate to data replication. If the molecular system does not have uniform density, the load balancing is nontrivial. The atomic distribution generally mirrors the spatial decomposition, that is, a processor updates coordinates of the atoms in its assigned spatial region.

Force Decomposition

The force decomposition [54,56,57] method divides the N^2 force matrix (representing the N^2 interactions among N atomic pairs) into P disjoint sets called blocks, where P is related to the number of processors employed for the calculation. Such a division of the force matrix implies that the set of N atoms is divided into $N/\sqrt{P}$ subsets. Each processor calculates the interactions in its region, that is, among the atoms in two blocks. Only $O(N/\sqrt{P})$ data is exchanged and a processor communicates only with $\sqrt{P}$ other processors that are in the same processor row or column. The atomic distribution is a refinement of the distribution of atoms into blocks. Atoms in a block are assigned to one of the $\sqrt{P}$ processors associated with the block for coordinate updates since its data are already present on the processor.

10.3.2 Specialized Processors

Specialized processors are processors that are designed for only a certain type of calculation. While they are much faster than general-purpose processors, they are more difficult to use. They are usually coprocessors, located in the host computer, and software must be specially written to effectively use them. A common example are the graphics processing units (GPU) found in modern personal computers. These processors optimized for calculating the linear algebra operations that are commonly used for computer graphics but are not as suited for other general purpose calculations as general-purpose processors [62,63].

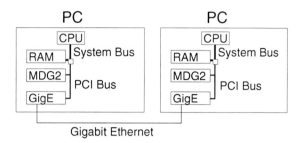

Fig. 10.6 The use of parallel MDGRAPE-II processors. Shown are two personal computers (PCs), each with one MDGRAPE-II processor (labeled MDG2). The PCs are directly connected with a gigabit Ethernet point-to-point connection.

Fig. 10.7 The calculation of forces by the MDGRAPE-II. The atomic position vectors **q** are input, and the MDGRAPE-II returns a vector of forces **f** exerted on the atoms.

MDGRAPE

The MDGRAPE (MD Gravity Pipeline) processor is a specialized processor for calculating MD simulations [64–67]. Specifically, it is used for the fast evaluation of pairwise interactions, which is precisely the most demanding part of MD simulations. Due to its specialization, it can be effectively used to calculate only the nonbonding interactions. Other calculations, including bonding interactions, are calculated on the general-purpose processor of the host computer. An example of two MDGRAPE-II processors placed in two PCs is shown in Fig. 10.6. Using the MDGRAPE-II processor achieves an eightfold speedup in the evaluation of pairwise interactions compared to standard contemporary processors [68].

In MD applications on specialized processors, the input data are the atomic co-ordinates and atomic types, while the output data are the interactions, for example the forces acting on the atoms or the energies of individual atoms. As an example for the MDGRAPE-II processor, the interaction to be calculated (i.e., the Coulomb and the Lennard–Jones potential) is defined as a function and uploaded to the processor. Coordinates are then sent to the processor in a vector, and the return value is the vector containing forces or the atomic energies. The process of calculating interactions is depicted in Fig. 10.7. The calculation on other specialized processors proceeds in a similar manner.

10.3.3 Global Communication in Parallel Molecular Dynamics Simulations

Global operations entail a communication operation in which all processors participate. A simple example is the broadcast of data by one processor to all others. The collective operations that are present in many message-passing libraries often include basic collective operations such as a broadcast-to-all and all-to-all data exchanges; however, more complex global operations must still be programmed by hand to be efficient [45, 69]. The two main operations found in parallel MD are the global sum and global broadcast [45, 69]. The role of these two global operations is illustrated in Fig. 10.5.

Global Sum

The global sum operation in MD is used after the calculation of interactions, for example, forces acting on atoms. After the calculation, many processors may have a partial force acting upon an atom, their sum being the total force, which is the same as if it were calculated by a single processor. The global sum operation therefore sums all of the partial forces to obtain the total forces. In addition, the force is needed only by the processor that updates coordinates. Therefore, an efficient implementation of the global sum operation leaves the total forces only on the processors performing the coordinate updates of the respective atoms. The global operation can be implemented using the `MPI_reduce_scatter` MPI routine in a parallel MD program using the replicated data parallelization method [69] in which any processor may have a force acting on any atom.

Global Broadcast

The global broadcast is used in MD simulations to broadcast updated coordinates to processors. After processors perform force updates for their respective atoms, other processors must receive the updated coordinates to correctly calculate the next interactions. The global broadcast operation performs this broadcast. In a replicated data parallel MD program, the `MPI_allgatherv` MPI routine may be used since every processor may need coordinates of any atom.

The global sum and broadcast operations for parallel MD not using the replicated data parallelization method tend to be more complex. In spatial decomposition, the global sum needs to sum interactions from neighboring processors only (assuming the cutoff distance is small enough) and the broadcast has a similarly small locality. In the force decomposition method, the communication in the global sum and global broadcast operations is limited to blocks. Only the processors that share a block communicate. Since data within a block is replicated, the processors within a block

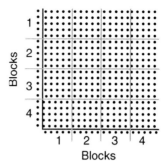

Fig. 10.8 The decomposition of the force matrix used for our parallel SISM MD program. An example for 20 atoms and 16 processors is shown. The atoms are divided into 4 blocks and one processor is assigned to calculate the interactions among each of the $4 \times 4 = 16$ block pairs.

perform a "block-limited" version of the global operation used in replicated data parallel MD.

10.4 Parallelization of SISM

Because the SISM method focuses on speeding up the calculation of bonding interactions and parallelization focuses on speeding up the calculation of non-bonding interactions, it is natural to complement the two approaches.

To showcase the complementarity of the SISM method, parallelization, and the use of specialized processors, we have developed a parallel program for MD simulation implementing the SISM method [68]. It supports the use of multiple MDGRAPE-II processors in many host computers. We opted to use the force decomposition approach to parallelization and do not rely on any special interconnect topology. The method is available for distributed memory parallel computers

The decomposition of the force matrix that we used in our program is depicted in Fig. 10.8. Molecules are never split into different blocks. The molecules in every block are also assigned to individual processors, forming an atomic distribution. A processor applies the SISM to the molecules assigned to it, including coordinate updates of its constituent atoms. The processor is also responsible for calculating interactions among the atoms in two of its associated blocks. If the MDGRAPE-II board is present, the calculations are performed on the board as shown in Fig. 10.9, otherwise the host processor calculates the interactions.

10.4.1 The Distributed Diagonal Force Decomposition Method

To enable calculations of the SISM method on larger, general parallel computers that do not rely on specialized processors, we have implemented the distributed diagonal force decomposition (DDFD) method [70,71]. The DDFD method is an extension of the general force decomposition method. It uses a minimal number of processors for

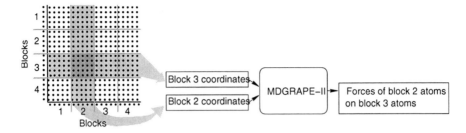

Fig. 10.9 Force calculation on the MDGRAPE-II processor using force decomposition. Shown is the force calculation of the interactions among the atoms in blocks 2 and 3, specifically the forces exerted by block 2 atoms on block 3 atoms. The blocks are highlighted with a light gray background; the dark gray square represents the interactions among the atoms of these two blocks. A separate calculation is used to calculate the equal but opposite forces of block 3 atoms on block 2 atoms.

the number of blocks used to decompose the force matrix. Since a larger number of blocks are smaller, the communication requirements are lower, resulting in a higher parallel efficiency.

In the DDFD method, the diagonal of the force matrix is distributed. As seen in Fig. 10.10(a), there are three types of interactions among the atomic blocks: a block product (interactions among two atomic blocks) lies either above, on, or below the diagonal. The interactions in the block products above the diagonal are opposite but equal to the interactions in the block products below the diagonal, so they do not have to be explicitly calculated. The interactions in block products on the diagonal are only among atoms in the same block. Any processor that has atomic data for these atoms can calculate any of the intra-block interactions for this block. As seen in Fig. 10.10(b), these interactions are distributed for calculation to processors below the diagonal; Fig. 10.10(c) shows the final state. The number of processors needed is equal to only the number of block products below the diagonal.

A side effect of the diagonal distribution process in the DDFD method is the straightforward implementation of load balancing. The distribution of interactions from a diagonal block product to processors holding the block data can easily be altered, assigning specific processors more or less interaction calculations. By altering the diagonal distribution in this way, the computational load of the processors is changed [70, 71]. Load balancing is especially crucial when using an interaction cutoff distance, since the computational load inherently varies among processors. In addition, due to atomic motion during the MD simulation, the atoms included with one atom's cutoff range varies throughout a simulation. Since the load balancing in the DDFD method is dynamic, the load balancing is dynamically tuned during the entire MD simulation, resulting in a higher parallel efficiency.

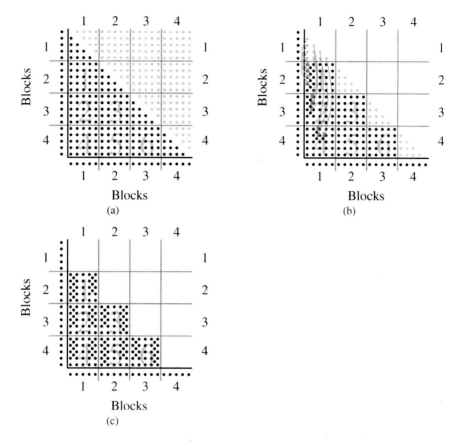

Fig. 10.10 The distributed diagonal force decomposition (DDFD) method. In **(a)** the interactions between the atoms are shown. The self-interactions (diagonal interactions) are 0 and not considered. The grayed interactions above the diagonal are equal but opposite to the ones below the diagonal and are therefore obtained from those. One processor is assigned to every block product of two different blocks. For example, processor 1 is assigned to the product of blocks 1 and 2, while no processor is assigned to the product of block 1 with itself; as shown in **(b)**, these interactions are rather assigned to processors calculating other interactions with block 1 (i.e., processors 1, 2, and 4). The final state when this diagonal distribution is performed for all 4 blocks is shown in **(c)**.

10.5 Conclusions

We have presented our research on parallel approaches to MD simulation. We have shown the complement between algorithmic approaches and parallelization in the quest to speed up the calculation of simulations.

The SISM, based on the standard theory of molecular vibrations, enables the use of much larger integration time steps than are possible with standard MD methods. Since the computational cost of an MD time step remains virtually constant, the computational time of an MD simulation is drastically reduced.

While the SISM allows larger integration time steps to be used, other methods must be used to reduce the computational time of the time steps themselves. Calculating nonbonding interactions dominates an MD time step, so focusing on reducing the time of calculating nonbonding forces is crucial. Specialized hardware can be effectively used to reduce the computational time of nonbonding interactions. We have shown the use of multiple MDGRAPE-II processors to speed up the calculation of nonbonding interactions.

As the algorithmic approaches and specialized hardware reduce the computational cost of individual MD time steps, efficient parallelization becomes even more important to achieving faster MD simulations, since the communication time increases relative to the computational time. The SISM is readily parallelized, including its implementation with multiple specialized processors. Used in combination with the force decomposition method, the communication between the distributed memory computers is guaranteed to be limited even for systems where no distance cutoff is employed. The DDFD method further reduces communication requirements among processors and enables a greater number of processors to be used. In addition, it intrinsically supports dynamic load balancing, which allows effective load balancing, which leads to higher parallel efficiencies and greater speedups of MD simulations.

Acknowledgments The authors would like to acknowledge the financial support of the Slovenian Research Agency under grant No. P1-0002.

References

1. L. Verlet, Computer "experiments" on classical fluids. I. Thermodynamical properties of Lennard-Jones molecules, Phys. Rev. 159 (1967) 98–103.
2. J. M. Sanz-Serna, M. P. Calvo, Numerical Hamiltonian Problems, Chapman & Hall, London (1994).
3. B. J. Leimkuhler, S. Reich, R. D. Skeel, Integration methods for molecular dynamics, IMA (1994) 1–26.
4. J. Wisdom, M. Holman, Symplectic maps for the N-body problem, Astron. J. 102 (1991) 1528–1538.
5. J. Wisdom, M. Holman, J. Touma, Symplectic correctors, Field Inst. Commun. 10 (1996) 217–244.
6. J. Laskar, P. Robutel, High order symplectic integrators for perturbed Hamiltonian systems, Celestial Mech. 80 (2001) 39–62.
7. L. Nadolski, J. Laskar, Application of a new class of symplectic integrators to accelator tracking, Proceedings of EPAC 2002 (2002) 1276–1278.
8. T. Schlick, E. Barth, M. Mandziuk, Biomolecular dynamics at long timesteps: Bridging the timescale gap between simulation and experimentation, Annu. Rev. Biophys. Biomol. Struct. 26 (1997) 181–222.
9. E. B. Wilson, J. C. Decius, P. C. Cross, Molecular Vibrations, McGraw-Hill Book Company, Inc., New York (1955).
10. N. Matubayasi, M. Nakahara, Reversible molecular dynamics for rigid bodies and hybrid Monte Carlo, J. Chem. Phys. 110 (1999) 3291–3301.

11. T. F. Miller III, M. Eleftheriou, P. Pattnaik, A. Ndirango, D. Newns, G. J. Martyna, Symplectic quaternion scheme for biophysical molecular dynamics, J. Chem. Phys. 116 (2002) 8649–8659.

12. M. Ikegutchi, Partial rigid-body dynamics in NPT, NPAT and NPγT ensembles for proteins and membranes, J. Comput. Chem. 25 (2004) 529–541.

13. D. Janežič, M. Praprotnik, F. Merzel, Molecular dynamics integration and molecular vibrational theory: I. New symplectic integrators, J. Chem. Phys. 122 (2005) 174101.

14. M. Praprotnik, D. Janežič, Molecular dynamics integration and molecular vibrational theory: II. Simulation of non-linear molecules, J. Chem. Phys. 122 (2005) 174102.

15. M. Praprotnik, D. Janežič, Molecular dynamics integration and molecular vibrational theory: III. The infrared spectrum of water, J. Chem. Phys. 122 (2005) 174103.

16. M. Praprotnik, D. Janežič, Molecular dynamics integration meets standard theory of molecular vibrations, J. Chem. Inf. Model 45 (2005) 1571–1579.

17. R. Rey, Vibrational energy of HOD in liquid D_2O, J. Chem. Phys. 104 (1996) 2356–2368.

18. R. Rey, Transformation from internal coordinates to Cartesian displacements in the Eckart frame for a triatomic molecule, Chem. Phys. 229 (1998) 217–222.

19. R. Rey, Vibrational phase and energy relaxation of CN^{-1} in water, J. Chem. Phys. 108 (1998) 142–153.

20. B. R. Brooks, D. Janežič, M. Karplus, Harmonic analysis of large systems: I. Methodology, J. Comput. Chem. 16 (12) (1995) 1522–1542.

21. D. Janežič, B. R. Brooks, Harmonic analysis of large systems: II. Comparison of different protein models, J. Comput. Chem. 16 (12) (1995) 1543–1553.

22. D. Janežič, R. M. Venable, B. R. Brooks, Harmonic analysis of large systems: III. Comparison with molecular dynamics, J. Comput. Chem. 16 (12) (1995) 1554–1566.

23. M. Praprotnik, D. Janežič, J. Mavri, Temperature dependence of water vibrational spectrum: a molecular dynamics simulation study, J. Phys. Chem. A 108 (2004) 11056–11062.

24. C. Eckart, Some studies concerning rotating axes and polyatomic molecules, Phys. Rev. 47 (1935) 552–558.

25. J. D. Louck, H. W. Galbraith, Eckart vectors, Eckart frames, and polyatomic molecules, Rev. Mod. Phys. 48 (1) (1976) 69–106.

26. H. F. Trotter, On the product of semi-groups of operators,, Proc. Am. Math. Soc. 10 (1959) 545–551.

27. G. Strang, On the construction and comparison of difference schemes, SIAM J. Numer. Anal. 5 (1968) 506–517.

28. H. Goldstein, Classical Mechanics, 2nd Edition, Addison-Wesley Publishing Company (1980).

29. D. Janežič, F. Merzel, An efficient symplectic integration algorithm for molecular dynamics simulations, J. Chem. Inf. Comput. Sci. 35 (1995) 321–326.

30. D. Janežič, F. Merzel, Split integration symplectic method for molecular dynamics integration, J. Chem. Inf. Comput. Sci. 37 (1997) 1048–1054.

31. D. Janežič, M. Praprotnik, Molecular dynamics integration time step dependence of the split integration symplectic method on system density, J. Chem. Inf. Comput. Sci. 43 (6) (2003) 1922–1927.

32. U. Borštnik, M. Hodošček, D. Janežič, Fast parallel molecular simulations, Croat. Chem. Acta 78 (2) (2005) 211–216.

33. W. F. van Gunsteren, H. J. C. Berendsen, Computer simulation of molecular dynamics: Methodology, applications, and perspectives in chemistry, Angew. Chem. Int. Ed 29 (9) (1990) 992–1023.

34. D. W. Heermann, A. N. Burkitt, Parallel Algorithms in Computational Science, Springer-Verlag, Berlin (1991).

35. R. Trobec, M. Šterk, M. Praprotnik, D. Janežič, Implementation and evaluation of MPI-based parallel MD program, Int. J. Quant. Chem. 84 (1) (2001) 23–31.

36. N. J. Boden, D. Cohen, R. E. Felderman, A. E. Kulawik, C. L. Seitz, J. N. Seizovic, W.-K. Su, Myrinet: A gigabit-per-second local area network, IEEE Micro 15 (1) (1995) 29–36.

37. J. Liu, J. Wu, D. K. Panda, High performance RDMA-based MPI implementation over Infini-Band, Int. J. Parallel Programm. 32 (3) (2004) 167–198.
38. V. S. Sunderam, PVM: A framework for parallel distributed computing, Concurr. Pract. Exper. 2 (4) (1990) 315–339.
39. G. Burns, R. Daoud, J. Vaigl, LAM: An open cluster environment for MPI, in: Proceedings of Supercomputing Symposium, Vol. 94 (1994) pp. 379–386.
    URL http://www.lam-mpi.org/download/files/lam-papers.tar.gz
40. W. Gropp, E. Lusk, N. Doss, A. Skjellum, A high-performance, portable implementation of the MPI message passing interface standard, Parallel Comput. 22 (6) (1996) 789–828.
41. T. Sterling, D. J. Becker, D. Savarese, Beowulf: A parallel workstation for scientific computa-tion, in: Proceedings, 24th International Conference on Parallel Processing, Vol. 1 (1995) pp. 11–14.
42. D. H. M. Spector, Building Linux Clusters: Scaling Linux for Scientific and Enterprise Appli-cations, O'Reilly & Associates, Sebastopol, CA (2000).
43. H. G. Dietz, T.I.Mattox, KLAT2's flat neighborhood network, in: Extreme Linux track of the 4th Annual Linux Showcase (2000).
44. R. Trobec, Two-dimensional regular d-meshes, Parallel Comput. 26 (13) (2000) 1945–1953.
45. U. Borštnik, M. Hodošček, D. Janežič, Improving the performance of molecular dynamics simulations on parallel clusters, J. Chem. Inf. Comput. Sci. 44 (2) (2004) 359–364.
46. J. Barnes, P. Hut, A hierarchical O(N log N) force-calculation algorithm, Nature 324 (4) (1986) 446–449.
47. J. A. Board, Jr., C. W. Humphres, C. G. Lambert, W. T. Rankin, A. Y. Toukmaji, Ewald and multipole methods for periodic N-body problems, in: P. Deuflhard, et al. (Eds.), Lecture Notes in Computational Science and Engineering, Springer-Verlag (1998).
48. J. Board, L. Schulten, The fast multipole algorithm, Comput. Sci. Eng. 2 (1) (2000) 76–79.
49. A. R. Leach, Molecular Modeling: Principles and Applications, Addison Wesley Longman Limited, Essex (1996).
50. R. Loncharich, B. Brooks, The effects of truncating long-range forces on protein dynamics, Proteins: Struct. Funct. Genet 6 (1989) 32–45.
51. S. Feller, R. Pastor, A. Rojnuckarin, S. Bogusz, B. Brooks, Effect of electrostatic force trunca-tion on interfacial and transport properties of water, J. Phys. Chem. 100 (1996) 17011–17020.
52. R. Trobec, I. Jerebic, D. Janežič, Parallel algorithm for molecular dynamics integration, Par-allel Comput. 19 (9) (1993) 1029–1039.
53. B. R. Brooks, M. Hodošček, Parallelization of CHARMm for MIMD machines, Chemical Design Auto. News 7 (1992) 16–22.
54. S. Plimpton, B. Hendrickson, Parallel molecular dynamics algorithms for simulation of molec-ular systems, in: T. G. Mattson (Ed.), Parallel Computing in Computational Chemistry, Amer-ican Chemical Society (1995) pp. 114–132.
55. T. G. Mattson (Ed.), Parallel Computing in Computational Chemistry, American Chemical Society (1995).
56. S. J. Plimpton, Fast parallel algorithms for short-range molecular dynamics, J. Chem. Phys. 117 (1) (1995) 1–19.
57. S. J. Plimpton, B. A. Hendrickson, A new parallel method for molecular-dynamics simulation of macromolecular systems, J. Comp. Chem. 17 (1996) 326–337.
58. M. Snir, A note on N-body computation with cutoffs, Tech. rep., IBM T. J. Watson Research Center (2001).
59. M. Snir, A note on n-body computations with cutoffs, Theory Comput. Systems 37 (2004) 295–318.
60. K. Bowers, R. Dror, D. Shaw, The midpoint method for parallelization of particle simulations, J. Chem. Phys. 124 (18) (2006) 184109–184109.
61. K. Bowers, R. Dror, D. Shaw, Overview of neutral territory methods for the parallel evaluation of pairwise particle interactions, J. Phys. Conf. Ser. 16 (2005) 300–304.
62. K. Moreland, E. Angel, The FFT on a GPU, in: Proceedings of the ACM SIGGRAPH/EURO-GRAPHICS conference on Graphics hardware, ACM (2003).

63. J. Krueger, R. Westermann, Linear algebra operators for GPU implementation of numerical algorithms, ACM Trans. Graphics 22 (3) (2003) 908–916.
64. T. Narumi, R. Susukita, T. Ebisuzaki, G. McNiven, B. Elmegreen, Molecular dynamics machine: Special-purpose computer for molecular dynamics simulations, Mol. Sim. 21 (1999) 401–415.
65. T. Narumi, Special-purpose computer for molecular dynamics simulations, Doctor's thesis, University of Tokyo (1998).
66. T. Narumi, A. Kawai, T. Koishi, An 8.61 Tflop/s molecular dynamics simulation for NaCl with a special-purpose computer: MDM, in: Proceedings of SuperComputing 2001, ACM, Denver (2001).
67. M. Taiji, T. Narumi, Y. Ohno, N. Futatsugi, A. Suenaga, N. Takada, A. Konagaya, Protein explorer: A Petaflops special-purpose computer system for molecular dynamics simulations, in: Proceedings of SuperComputing 2003, ACM, Phoenix (2003).
68. U. Borštnik, D. Janežič, Symplectic molecular dynamics simulations on specially designed parallel computers, J. Chem. Inf. Model. 45 (6) (2005) 1600–1604.
69. K. Kutnar, U. Borštnik, D. Marušič, D. Janežič, Interconnection networks for parallel molecular dynamics simulation based on hamiltonian cubic symmetric topology, J. Math. Chem. 45(2) (2009) 372–385.
70. U. Borštnik, Parallel computer simulations on clusters of personal computers, Ph.D. thesis, University of Ljubljana (2007).
71. U. Borštnik, B. R. Brooks, D. Janežič, The distributed diagonal force decomposition method. I. Description of the method, submitted for publication (2008).

Chapter 11
Parallel Computer Simulations of Heat Transfer in Biological Tissues

Roman Trobec

Abstract

Parallel computer simulation of heat transfer in parts of the human body is described. Realistic geometric models and tissues with different thermodynamic properties are analyzed. The principal steps of the computer simulations, including mathematical and geometric modeling, domain discretization, numerical solution, validation of simulated results, and visualization, are described. An explicit finite difference method for the inhomogeneous computational domain has been developed and tested on the diffusion equation. The bio-heat equation, which incorporates heat conduction, heat transfer between blood and tissues and heat production by metabolism, was used in our analysis. Because of significant calculation complexity, a parallel simulation code was also implemented.

Domain decomposition and communication with messages have been selected in the parallel implementation of the explicit finite difference method. Mapping of the computational domain on the parallel computer was addressed, followed by theoretical performance analysis of the proposed parallel algorithm. The implementation of all simulation steps is shown in detail for the simulation of the steady-state temperature and its evolution in time for a human knee exposed to external conditions and to topical cooling. The results have been validated by experimental measurements. Execution time was measured on a computing cluster with different numbers of processors and compared with theoretical expectations. It is shown that parallel computer simulations can be of great use in medicine, either for planning surgery or for evaluating doctrines of medical treatment. The chapter concludes with a summary of the results and a list of relevant references from the research field.

Roman Trobec
Department of Communication Systems, Jožef Stefan Institute, Jamova cesta 39, 1000 Ljubljana, Slovenia, e-mail: `roman.trobec@ijs.si`

R. Trobec et al. (eds.), *Parallel Computing*, DOI 10.1007/978-1-84882-409-6_11,
© Springer-Verlag London Limited 2009

11.1 Introduction

In recent decades, computer simulations have proved a great help in understanding and solving a variety of problems in science [1, 2], medicine [3, 4], engineering [5, 6], etc. Initial simulations and optimizations in the design of cars, aircrafts, or turbines [7, 8] are done on computers. Basic characteristics of the designed devices are obtained using computer simulations, which are significantly less expensive than real modeling and testing. Computer simulations are complementary methods for speeding up prototyping processes. Hazardous modifications on nuclear plants [9], natural phenomena like weather [10], and molecular dynamics [11] cannot be subject to experiment in reality, and computer simulations are the only viable option for obtaining some deeper insight into such phenomena.

Especially in medicine, experiments are often difficult to perform because human subjects are involved [12]. Measurements during clinical procedures are time consuming and often not as accurate as desired, because many parameters are difficult to control [13]. In many cases, measurements made during clinical procedures would be too invasive and limited to only a few test points at best. An example would be the search for the highest temperature or for its steepest gradient in a part of the human body. With the use of computer simulation, however, it is possible to calculate, analyze, and visualize the temperature changes that occur with time.

In most scientific computing applications, a physical system is represented by a mathematical model. Real problems cannot be solved analytically, that is, by a formula that provides the solution for any moment in time and any spatial point. Instead, the numerical solution must be obtained, for which the continuous physical domain has to be replaced by its discrete form.

The computational domain, which is a simplified presentation of the physical domain being simulated, cannot always be represented by a simple object for which there is an eventual analytical solution. For example, parts of the human body are irregularly shaped three-dimensional (3D) objects, which must be represented by a spatial geometric model [14]. Such models of the body organs can be created using the visible human dataset (VHD) [15] or a similar data source based on 2D slices.

The investigated phenomena are often described mathematically by partial differential equations (PDE). Usually, the physical domain is partitioned into many small subdomains and a simple algebraic equation is obtained for each. If the 3D geometric model is obtained by stacking the 2D slices, it is natural and simple to discretize the problem, using an orthogonal structured mesh of points and replacing derivatives in PDEs by finite differences in order to obtain a system of algebraic equations. Some other, more sophisticated, methods exist for the numerical solution of PDEs based on unstructured meshes, such as finite element methods (FEM) [16], or, more recently, meshless methods [17, 18] that construct the final linear system from small sets of nearest neighbors only. These methods are more efficacious for irregularly shaped domains, but conceptually and computationally more complex. Therefore, we have proceeded with the simplest finite difference methods (FDMs), which have given adequate results. Note that we will use the terms mesh and points for domain discretization and network and nodes for computer interconnection topology.

Solving the resulting system of equations over such a discretized domain gives the values for certain physical quantities at every subdomain. If the evolution in time is of interest, then time discretization also has to be implemented, by dividing the investigated interval into a set of short time-steps. For each time-step, an algebraic system of equations has to be solved [19]. For higher accuracy, the number of discretized subdomains and their equations can become too large to be managed with a single computer, and high performance computers are advantageous in such cases [20].

Computing performance can also be improved by numerous approaches, such as network computing, grid computing, cloud computing, and other variants of the more general term "distributed computing" [21]. This is characterized by heterogeneous, interconnected computers of different computing performance, different operating systems and, what is most important, different characteristics of the interconnecting channels. Such a heterogeneity constitutes a serious drawback for the application of distributed computing in computer simulations [22], particularly because of slow communication and the occasional unavailability of computing resources.

On the other hand, parallel computers are composed of fast, unified computers connected by fast, dedicated communication links. Today, low-cost parallel computers are available with computing clusters [23, 24]. The parallel speedup is defined as the ratio of the execution time on a single processor to that on a parallel computer. The time of running an application on a parallel computer consists of both computation and communication time. Increasing the number of processors usually increases the ratio of communication time to computation time; thus, both processor performance and communication time have to be improved in order to improve the overall performance of parallel algorithms. In some problems where a significant amount of global communication is needed, for example, molecular dynamics [25], optimal performance of intra-cluster communication is particularly important.

The temperature in human tissue is an important factor in many fields of physiology [26], surgery [27, 28], sport [29], cryotherapy [27, 30], etc. The temperature profiles are influenced by environmental conditions, by the temperatures of neighboring tissues, by the muscle metabolism, and by the circulating blood. Different tissues have different physical and thermodynamic properties and respond diversely to temperature change [31]. The temperature field varies in space and time in different parts of the investigated domain. In vivo measurements are usually invasive and often impossible, if deep tissue or vital organs are in question.

A substantial amount of work on analytical and numerical solutions of the bio-heat equation has been published [12, 32–34]. In this chapter a computational method for the solution of the bio-heat equation has been devised that incorporates heat diffusion, heat generation by tissue metabolism, and heat transfer between blood and tissues. Some important extensions have been introduced by our work, in particular an inhomogeneous spatial model composed of tissues with different characteristics, and modeling of the heat transfer and heat sources as functions of the surrounding tissue temperature. This solution of the bio-heat equation was evaluated in terms of stability and accuracy and solved numerically on single and parallel computers.

We implemented the parallel version of the proposed method, which runs efficiently on 16 or more connected computers – a computing cluster. In this way computation time can be shortened significantly. Such an approach enables the solution of several millions of equations for each time-step. Besides the results, the computer simulations also support the development of new ideas and theories because "unexpected" simulated results have to be explained. The proposed method is quite general and can be applied to investigations of a variety of living tissues.

The simulation procedure was previously tested using published measurements of the steady-state temperature fields of the human forearm, in particular from the well-known Pennes' paper published 60 years ago [32]. Other experimental data are available for the temperature of the forearm during immersion in water at various temperatures, either evolving in time [35] or near their steady state [36]. These measurements have since been elaborated by others [33, 37]. We have simulated most of these measurements with high accuracy, which has confirmed that the proposed simulation is stable and accurate. We have shown that the measured Pennes' results actually reflect, in detail, correct measured values.

We have also studied in detail other parts of the human body, like the human knee and heart. In paper [38] we have been interested in simulating possible modes of knee cooling following injury or surgery [26, 27]. Two different methods of topical cooling were compared; first, use of a gel pack filled with refrigerated gel [30], which is exposed to ambient temperature and therefore becomes less and less effective, and secondly, use of a cryo-cuff cooled by a liquid at constant temperature maintained by an external cooling device [29]. Lowering the tissue temperature reduces the need for pain medication and shortens the rehabilitation period. The aim of this work was to simulate topical cooling of the knee after injury or surgery and to calculate and display the development of temperature distribution in all tissues of the knee region.

A simulation of the cooling of the human heart during surgery has been described in [4]. We present results obtained using a more complex mathematical model that takes into account convection, diffusion and fluid flow, as well as providing a higher spatial resolution. The technique used can also be applied for the prediction of temperature elevation following coronary artery occlusion and many other medical situations. The human body and the heart have to be cooled appropriately in order to slow down their vital functions [39]. To lower metabolic requirements, the body and the heart have to be cooled, for example, by pumping a cold solution through coronary vessels (cardioplegia). For even better cardiac cooling a method of topical cooling is sometimes used [40], for example, submerging the heart in cooling liquid. In vivo temperature measurements are invasive and limited to a few test points, while computer simulation provides improved analysis of various cooling options. Some initial results on the simulation of heart cooling based on diffusion have been reported in [14, 41].

In this chapter the complete design of a parallel computer simulation procedure will be presented, from initial modeling and implementation to the final validation and analysis of the results. In the next section the principal steps of computer simulations are described and some basic terms from the field are introduced. Methods for the solution of the PDEs are then described. A time-dependent diffusion equation is taken as an example. The explicit FDM was selected as the most appropriate for our application, because of simple implementation of a numerical scheme for inhomogeneous tissue and a great potential for parallelization. In the second part of the chapter, the bio-heat equation, which was used in the simulation of heat transfer and temperature distribution in a human knee, is introduced. The equation incorporates heat conduction, heat transfer between blood and tissues, and heat production by metabolism. The results show the evolution of the temperature field and its steady state. The simulation results were evaluated by comparison with experimental measurements.

The solution method, based on the finite differences, was implemented on a parallel computing cluster. Opportunities for parallelization of the solution method are described and its calculation complexity analyzed. Domain decomposition and communication with messages were selected in the parallel program. Execution time was measured on a computing cluster with various numbers of computers and compared with theoretical expectation. The chapter concludes with a summary of results and a list of relevant references from the research field.

11.2 Principal Steps in Computer Simulation

Computer simulation of physical phenomena involves several steps. Some of them depend on user requirements and others on the results of previous steps. Some of them are not obligatory, for example, parallel implementation, if the simulated problem can be managed in a reasonable time by a single computer. The computer simulation process is iterative by its nature and can be implemented by loops for step refinement. The flowchart of a typical computer simulation process is shown in Fig. 11.1.

Goals of the simulation, with possible strategies and formulations of the mathematical model, have to be considered first. The designers have to locate a balance between requirements and limitations, using assumptions and approximations. Much previous knowledge, available from the mathematical modeling area, can be applicable to a different application area; however, models often need to be adapted in order to be useful for the specific application. The modeling step usually finishes with a system of PDEs, with initial and boundary conditions that depend on the simulated case. In our work we focus on heat transfer in biological tissue, modeled by a diffusion equation, with some additional terms for heat production and heat transport.

The simulated region, termed the computational domain, is usually bounded and placed in the environment that interacts with the simulated phenomenon through the boundary conditions. If the phenomenon is time dependent, an initial state must be specified, from which its evolution in time will be calculated. Geometric objects inside the domain can be analytically defined bodies from a single material, or bodies

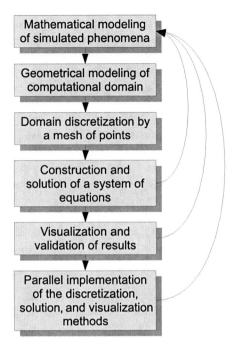

Fig. 11.1 Principal steps of computer simulation process loop.

of irregular shape, composed of several materials with different physical properties. Modeling the geometry is strongly related to the simulation scale. In molecular dynamics, atoms and molecules are modeled as interacting particles; at the other extreme, in the simulation of galaxies, stars are taken as a modeled unit. In our contribution, we will be modeling 3D organs or parts of a human body that are irregular and inhomogeneous. The geometric model must satisfy the user requirements and those of the mathematical model in order to provide a stable and accurate solution of the simulated system.

Usually the PDEs cannot be solved analytically, therefore they are converted into a set of equations, which is accomplished by approximating the solution in a set of points that are placed within simulated objects in the domain [42]. Points are also placed on the borders of objects and on boundary layers between different tissues, which enables the imposition of boundary conditions and continuous transitions between different materials in the domain. The points are traditionally "connected" by neighboring points to make a mesh, in order to be able to approximate derivatives in PDEs [43] or a local solution function [16]. Recently, meshless methods have been proposed that rely only on a set of neighboring points, weighted by a hat-shaped weight function [17, 18]. In both cases the density of points varies to an extent depending on the geometric detail of the model and the gradients in the solution. Usually, the initial mesh is not adequate and must be refined after validation of the first simulated results [44]. The mesh refinement may also introduce some changes in the geometrical and mathematical models.

Often, at the time of initial meshing, no knowledge is available about the regions with larger gradients or discontinuities in the simulated solution [45]. In order to improve the accuracy of the simulated solution, an adaptive domain discretization is needed [46].

A system of equations that provides a simulated solution is constructed using a discretized domain and different approximation principles. For example, using an FDM, derivatives in the mathematical model are replaced by finite differences and, for each internal point [43], an equation is formed using collocation, which states that the model equations should be satisfied exactly at all points, while by the FEM the weighted residual method [42] is used, which minimizes the difference between the solution and its approximation over the whole computational domain.

Only temporal derivatives, if present, remain, thus PDEs are transformed into a system of ordinary differential equations (ODEs) that can be solved using time discretization by explicit or implicit numerical methods [19]. The explicit methods are simple and require just matrix-vector multiplication at each time-step, while implicit methods require solution of a system of equations at each time-step.

The simulated solution has to be visualized in order to understand and analyze the simulated results. By FDM the solution values between points are obtained by interpolation, while those by FEM are obtained by evaluating the numerical solution at any desired point. The solution thus obtained must be validated by its application to some simplified problem, for example, a homogeneous cube with simple initial and boundary conditions for which the analytical solution is known [47]. Alternatively, previous measurements [32, 35, 48] of simulated variables from physical experiments can be used for validation.

The whole simulation process is iterative, with local loops for improving models or refining domain discretization, or for debugging the simulation program. Applications, as in the simulation of aircrafts, rocket engines, human organs, etc., require several millions of discretization points. Standard sequential computers may not be able to compute the simulation results in a reasonable time or to have enough memory for all the simulation steps. One of the possible solutions is reimplementation of the knowledge embedded in the sequential programming code by developing a new parallel program, which can run on a large number of parallel processors [49].

Parallel speedup is defined as the ratio of the execution time on a single processor to that on a parallel computer. Parallel efficiency is the speedup normalized by the number of processors [20]. Efficient parallel implementation of the simulation procedure must encompass the parallelization of most simulation steps including domain discretization, numerical solution, and visualization. None of them have been adequately solved today, so further research is needed in this area. We will focus in the rest of the chapter on the explicit FDM, which requires a simple structured mesh and is simple to parallelize. Other simulation approaches, for example, FEMs or meshless methods are based on similar steps, except that the domain discretization and consequent construction of the final system of equations could be more complex, however, also more stable and accurate.

11.3 Numerical Solution of Partial Differential Equations

The mathematical model of a system is a set of variables and equations that describe the system. If the equations involve derivatives of system variables, we talk about a system of ODEs with solutions that are functions of independent variables. Many of the basic laws in nature, for example, from the area of electromagnetic field (Maxwell's equations), fluid flow (Navier–Stokes equations), etc., can be expressed effectively by the use of partial derivatives of unknown functions with respect to independent system variables. Equations that involve partial derivatives of unknown functions with respect to more than one independent variable are termed PDEs.

In real cases the PDE solution cannot be written as an analytical expression, particularly in cases where the computational domain is not composed of simple geometric objects, as is the case in simulations of biological tissues in human organs. Numerical solution is a only viable option in such cases. Different approaches exist for obtaining a numerical solution.

A numerical solution of PDEs is based on spatial discretization of the global domain Ω and its boundary Γ, which convert a PDE of the form:

$$\mathscr{L}u(\mathbf{x}) = f(\mathbf{x}), \qquad \mathbf{x} \in \Omega \tag{11.1}$$

with boundary conditions

$$u = \bar{u} \qquad \mathbf{x} \in \Gamma \tag{11.2}$$

into a system of ordinary algebraic or differential equations. u is the unknown solution, $\mathscr{L}$ is a differential operator $u \mapsto \mathscr{L}u$, f is a linear form $f : \Omega \to \mathbb{R}$, and $\mathbf{x}$ is the vector of independent variables.

Replacing u with the unknown approximate solution $\hat{u}$ in the PDE, the residual of Eq. (11.1) can be defined as

$$r(\mathbf{x}) = \mathscr{L}\hat{u}(\mathbf{x}) - f(\mathbf{x}). \tag{11.3}$$

If a PDE depends only on spatial variables, a system of ordinary algebraic equations (OAE) is obtained after discretization.

A time-dependent PDE is converted into a system of ODEs with point parameters $u_i(t)$ as unknowns, which is then discretized in time to give the final system of OAE. The initial conditions $u(\mathbf{x}, t_0)$ are required in the time-stepping solution procedure. We will see that the relationship between the spatial and temporal discretization and the numerical methods used for the transformation of ODE into OAE are important issues in the stability of the numerical methodology.

The diffusion equation,

$$\rho c \frac{\partial T}{\partial t} = \nabla \cdot (\lambda \nabla T) \tag{11.4}$$

described in detail in the next section, is an example of PDE, which models the time-dependent heat transfer in 3D solid bodies. It has four independent variables, three space variables denoted by x, y, z and a time variable denoted by t. To solve the diffusion equation numerically, the domain Ω was discretized by a set of N points $\mathbf{x}_i$ on the equidistant orthogonal mesh of cubes in 3D. The approximate solution $\hat{T}$ is represented in each point by the corresponding time-dependent point parameters $T_i(t)$. With an explicit time-stepping method the approximate solution is obtained for the desired solution time and accuracy. The solution will provide a temperature field whose derivatives, with respect to the independent variables, boundary values, and initial state, satisfy the relationships defined by the diffusion equation (11.4).

We will first describe a solution process for a general PDE, then we will apply it to the solution of the bio-heat equation used in our simulations.

11.3.1 Finite Difference Method

The geometric domain has to be discretized in order to transform PDEs into a simpler system of equations. In the regular FDM, the domain is discretized in space with a rectangular mesh of points with four neighbors in 2D, and with six neighbors in 3D. The parameter of each point represents the value of $\hat{u}$ at the point $u_i = \hat{u}(\mathbf{x}_i)$ while, in between points, $\hat{u}$ is undefined. The values of spatial derivatives in PDEs are approximated using u_i, adjacent point parameters, and their finite differences. For each internal point $\mathbf{x}_i$, an equation is formed using collocation, which states that the residual should be zero at points, that is, that Eq. (11.1) should be satisfied exactly at $\mathbf{x}_i$

$$r(\mathbf{x}_i) = \mathcal{L}\hat{u}(\mathbf{x}_i) - f(\mathbf{x}_i) = 0, \qquad \mathbf{x}_i \in \Omega. \tag{11.5}$$

Similarly, equations for boundary points are formed from Eq. (11.2)

$$\hat{u}(\mathbf{x}_i) - \bar{u}(\mathbf{x}_i) = 0, \qquad \mathbf{x}_i \in \Gamma. \tag{11.6}$$

The spatial order of accuracy of FDM is the same as the lowest order of all the spatial derivative approximations used [43].

Only temporal derivatives remain, thus the PDE was transformed into a system of ODE with point parameters $u_i(t)$ as unknowns. The system will be solved numerically by approximating time derivatives with finite differences to obtain a system of OAE. Then, starting with initial condition $u(\mathbf{x}, t_0)$ and stepping in time for Δt, the solution of PDE u_i^{n+1} in time $t_n + \Delta t$ is obtained from the solution in the previous time-step u_i^n, either by explicit or implicit numerical methods, described in Sect. 11.3.3.

11.3.2 Finite Element Method

Another popular approach for obtaining the PDEs solution is by FEMs. The domain is discretized in space by an unstructured mesh of triangles or quadrilaterals in 2D, and tetrahedra or hexahedra in 3D, which is much more complicated than by FDM. The unknown solution $\hat{u}$ is approximated by shape functions ϕ_i and nodal parameters u_j as

$$\hat{u}(\mathbf{x}) = \sum_{j=1}^{N} u_j \phi_j(\mathbf{x}) = \mathbf{u}^{\mathrm{T}} \Phi(\mathbf{x}), \qquad \mathbf{u}^{\mathrm{T}} = [u_1, \ldots, u_N]. \tag{11.7}$$

where N is the number of points. The derivatives of $\hat{u}$ are obtained trivially, provided that the derivatives of shape functions are known. Simple bilinear or bicubic B-spline functions with local support are used for shape functions in 2D, and trilinear or tricubic in 3D, such that ϕ_i has its largest value at $\mathbf{x}_i$. It is said that the shape functions are C^k consistent if Eq. (11.7) can exactly reproduce any polynomial of degree up to k [16].

FEM equations for internal points can be constructed by stating that the residual (11.3) is orthogonal to a set of test functions W_i:

$$\int_{\Omega} W_i(\mathbf{x}) r(\mathbf{x}) \, d\Omega = \int_{\Omega} W_i(\mathbf{x}) \left[\mathscr{L} \hat{u}(\mathbf{x}_j) - f(\mathbf{x}_j) \right] d\Omega = 0, \tag{11.8}$$

which is termed the weighted residual method [42] and is one of the weak forms of Eq. (11.1). The test functions W_i can, in principle, be any nonzero functions that lead to a convenient formulation, although the choice affects the accuracy to some degree. If the test functions are equal to the shape functions we get the Galerkin method, which is the basis for the formulation of FEM.

For FEM to converge, i.e., for the approximate solution to approach the exact solution as nodal spacing approaches zero, the shape functions must at least be C^r consistent, where r is the order of the PDE being solved [16]. The shape functions are often simple enough, particularly in the case of triangular elements, so that the integrals can be calculated analytically. With more complicated functions or elements, numerical integration has to be applied. After integration of Eq. (11.8), PDE (11.1) is transformed into a system of ODEs with $u_i(t)$ as unknowns. From now on, the solution procedure is identical to that by FDM. Time discretization is applied for the transformation of the system of ODE to the system of OAE, which is then solved for each time-step.

11.3.3 Solution Methods of Time-Dependent PDEs

After spatial discretization of PDEs, only temporal derivatives remain, thus PDEs are transformed into a system of ODEs with time-dependent parameters $u_i(t)$ as unknowns. Higher-order time derivatives can always be transformed into the first-order derivatives by introducing new unknowns, which results in a larger system of first-order ODEs, written as

$$\mathbf{u}'(t) = \mathbf{f}(t, \mathbf{u}) \tag{11.9}$$

where $\mathbf{u}$ is the unknown solution, $\mathbf{u}' = d\mathbf{u}/dt$ is the first derivative with respect to the independent variable t, and $\mathbf{f}$ is a known function.

It is known that the solution $\mathbf{u}$ of a linear homogeneous system of ODEs with constant coefficients

$$\mathbf{u}' = A\mathbf{u} \tag{11.10}$$

can be expressed as the sum of exponential components [19] with eigenvalues of A in exponents. The solution of such a system is stable if, for every eigenvalue λ_i of A, $Re(\lambda_i) \leq 0$, which means that the error in the solution diminishes with time. For a general ODE $\mathbf{u}'(t) = \mathbf{f}(t, \mathbf{u})$ the stability is determined by the eigenvalue analysis of the Jacobian matrix J of derivatives of $\mathbf{f}$ with respect to $\mathbf{u}$

$$J_{ij} = \frac{\partial f_i(t, \mathbf{u})}{\partial u_j}, \tag{11.11}$$

which represents the locally linearized form of the original ODE. We see that the stability of the general ODE depends on a particular solution and is also limited to the local neighborhood of the analyzed point in the computational domain.

The system of ODEs (11.9) is solved numerically by discretizing the time to short time intervals, then approximating time derivatives with finite differences and transforming the system of ODEs into a system of OAE. We will be satisfied with an approximate solution in the discrete time instants obtained from the solutions in the consecutive time-steps, either by explicit or implicit numerical methods, as described in the following sections.

11.3.3.1 Explicit Numerical Methods

The system of ODEs (11.9) is solved numerically by first discretizing the time to short time intervals Δt, so that $t_n = t_0 + n\Delta t$ for $n = 0, 1, 2, \ldots$. Then, time derivatives in ODEs are approximated with finite differences, which transforms the system of ODEs into a system of OAE. Starting with the initial condition in t_0 with value

$$\mathbf{u}^0 = \mathbf{u}(t_0, \mathbf{x}) \tag{11.12}$$

and stepping in time for Δt, the solution of PDE in time $t_1 = t_0 + \Delta t$, denoted by $\mathbf{u}^1$, is obtained from the initial state. In general, the solution $\mathbf{u}^{n+1}$ in time $t_{n+1} = t_0 + (n+1)\Delta t$ is obtained from the solution in the previous time-step $\mathbf{u}^n$. Often, the time-step Δt can be adaptive, but we will suppose in the following that the time-step is constant through the whole simulation.

With the initial value, a particular solution of ODEs (11.9) is chosen. Additionally, the initial slope $\mathbf{u}'(t_0)$ can be determined by evaluating Eq. (11.9) at the given initial value. If we linearize the solution in the first time-step and apply finite difference approximation for the first derivative on time

$$\mathbf{u}'(t_0) = \frac{\mathbf{u}(t_1) - \mathbf{u}(t_0)}{\Delta t}, \tag{11.13}$$

we obtain, from Eq. (11.13), the predicted solution in time t_1 as

$$\mathbf{u}^1 = \mathbf{u}(t_1) = \mathbf{u}(t_0) + \Delta t \mathbf{u}'(t_0) = \mathbf{u}^0 + \Delta t \mathbf{u}'^0, \tag{11.14}$$

And, in an analogous way, for the predicted solution in time t_{n+1}, as

$$\mathbf{u}^{n+1} = \mathbf{u}^n + \Delta t \mathbf{u}'^n. \tag{11.15}$$

The term $\mathbf{u}'^n$ from Eq. (11.15) is obtained by evaluation of the right side of Eq. (11.9), using the already known solution $\mathbf{u}^n$ obtained in the previous time-step t_n

$$\mathbf{u}'^n = f(t_n, \mathbf{u}^n). \tag{11.16}$$

The method described here is known as explicit Euler's method, because the solution value in the next time-step is obtained from already known solutions in the previous time-step. It is a single-step method, because the next approximate solution value depends only on the solution value from the previous step.

The accuracy of this form of Euler's method, in each time-step, is proportional to $(\Delta t)^2$, because we take just the linear term from the Taylor series of the exact solution or, equivalently, we locally interpolate the solution by a linear function. The global error e_n of the Euler's method has an accuracy of the first order $O(\Delta t)$ because the global error is equal to the sum of all n step errors, however n is inversely proportional to Δt.

A numerical method is stable if small perturbations do not produce meaningless solutions. Assuming that our ODE has a form as in Eq. (11.10), we obtain from Eqs. (11.15) and (11.16) the recurrence

$$\mathbf{u}^{n+1} = (I + \Delta t A)\mathbf{u}^n \quad \text{and} \quad \mathbf{u}^n = (I + \Delta t A)^n \mathbf{u}^0, \tag{11.17}$$

where I is an identity matrix of the same dimension as the matrix A. The expression $(I + \Delta t A)$ is called the amplification factor. According to the exponential solution, components with eigenvalues of A in exponents, the errors will not grow if all the eigenvalues of $\Delta t A$ lie on the complex plane inside a circle of radius 1 and centered at -1, or equivalently, if the spectral radius $\rho(I + \Delta t A) \leq 1$. The spectral radius of a matrix is defined as the maximal absolute eigenvalue of the matrix. For the general ODE, a similar analysis can be carried out on the Jacobian matrix (11.11), however only a local estimate for the stability can be obtained.

From the above condition, we see that stable solutions can be obtained with a shorter time-step; however, a shorter time-step means more steps in the simulation procedure and thus greater computational complexity and longer simulation time. Because the elements of matrix A depend on the density of the spatial mesh, there is a relation between space and time discretization. This relation depends also on the type of initial PDEs and will be derived later for the diffusion equation. In practical cases, we can deduce the required time-step Δt that will guarantee a stable numerical solution, by applying spatial discretization. Intuitively, it is clear that a finer spatial mesh requires a shorter time-step.

Euler's method is simple, but inefficient for general problems, so that, in practice, more complicated methods are used, based on polynomial approximation of the solution values between several time-steps. Euler's method does not require the solution of a linear system. The simulated solution is obtained solely by a matrix-vector multiplication in each time-step, which can easily be implemented on parallel computers. Accuracy requirements for the whole simulated time interval often demand the use of such a short time-step that it falls in the same range as the time-step dictated by the stability criteria. In such cases, a short time-step, dictated by stability, is not an important disadvantage.

11.3.3.2 Implicit Numerical Methods

The stability of the numerical solution of a system of ODEs can be improved by implicit methods that will be described in brief. These methods are based also on the solution obtained in the current time-step t_{n+1}, consequently the solution of a linear system is required in each time-step [50].

The Crank–Nicolson method is a popular time-stepping scheme [19], implemented by using an approximation of the solution and its derivative in the intermediate points $(t + \frac{1}{2}\Delta t)$

$$\mathbf{u}\left(t + \frac{1}{2}\Delta t\right) = \frac{\mathbf{u}(t) + \mathbf{u}(t + \Delta t)}{2}, \tag{11.18}$$

$$\mathbf{u}'\left(t + \frac{1}{2}\Delta t\right) = \frac{\mathbf{u}(t + \Delta t) - \mathbf{u}(t)}{\Delta t}. \tag{11.19}$$

Substituting Eqs. (11.18) and (11.19) into Eq. (11.9) and rewriting the equation in the standard form gives

$$\mathbf{u}^{n+1} = \mathbf{u}^n + \Delta t \mathbf{f}\left(t + \frac{1}{2}\Delta t, \frac{\mathbf{u}^n + \mathbf{u}^{n+1}}{2}\right). \tag{11.20}$$

The function $\mathbf{f}$ cannot be evaluated directly because the solution $\mathbf{u}^{n+1}$ is not, at this stage, known. It can be obtained by solving the linear system resulting from Eq. (11.20). It has been shown that the amplification factor of the Crank–Nicolson scheme is always less than 1, and this scheme is therefore unconditionally stable. The method achieves second-order temporal accuracy, which is better than the simple Euler's method, and is convenient for the solution of the diffusion equation, because its spatial accuracy is also of the second order.

We have shown that implicit methods require the solution of a linear system, the number of equations being similar to that of the discretized points. The solution of a large linear system is one of the most computationally demanding steps in the solution procedure; however, the linear system is sparse and many efficient iterative methods are known for its solution [51, 52]. Parallel algorithms for solving an iterative linear system also exist [53, 54]. We will not go into further detail on this topic.

We have seen that, if evolution in time is needed, the advantageous stability of the implicit methods may not be so important, because the solution has to be known in the predefined time-steps, which are usually more dense than that required for the implicit solution. Additionally, accuracy requirements can be so strong that the implicit time-steps must be shortened in order to meet them. For these reasons, despite its general inefficiency, the simple Euler's method has been implemented in our simulation software, described in detail in Sect. 11.4.

11.3.4 Computational and Memory Complexity

The computational complexity consists of two main parts: the construction of the global system of linear equations and its solution.

11.3.4.1 System Construction

The N points of the regular FDM are placed onto an equidistant orthogonal mesh within the domain. The point parameters represent the solution values at points. When the spatial derivatives are replaced by their numerical approximations, a PDE is transformed to a system of ODE. The solution values of boundary points are always equal, because they are prescribed by the essential boundary condition and can therefore be eliminated from the system. For example, by the 2D diffusion equation, the final linear system, obtained by an implicit numerical method, can be written in the matrix form as

$$A\mathbf{u}^{n+1} = B\mathbf{u}^n, \tag{11.21}$$

where n is the index of the current time-step, the system matrix A is symmetrical, and both A and B contain five nonzero elements in each row. For N points, FDM systems can be generated trivially in time $O(N)$, incorporating all relevant neighboring contributions for each point, or can even be solved without being generated explicitly.

For comparison, we have shown [55] that the asymptotic computational complexity for the construction of the global FEM linear system is $O(eN)$, where e is the number of points for each element, for example, $e = 3$ for triangular elements.

11.3.4.2 System Solution

Direct methods for the solution of a general linear system are not appropriate because they have the calculation complexity of $O(N^3)$ for matrix factorization and $O(N^2)$ for the triangular system solution. Even if the system matrix is constant, the calculation complexity $O(N^2)$ remains in each time-step. Therefore, iterative solvers have to be used.

In the case of explicit time integration the system is not needed and the solution value in each step is obtained by a simple matrix-vector multiplication. The matrix is sparse and possibly banded, so that the asymptotic computational complexity of such multiplication is equal to $O(bN)$, where b is the width of the nonzero band, for example, $b = 7$ in 3D FDM and $b \approx 15$ in 3D FEM.

Using implicit methods, the linear system is solved by an iterative method in each time-step. If good preconditioners are known, the number of iterations of efficient iterative solvers for sparse matrices is a small constant, for example, less than 5 and independent of N. The calculation of a temperature field as it evolves with time offers the possibility of using the information from the solution in the previous time-step for constructing efficient preconditioners. The asymptotic complexity of a single iteration is near the number of nonzero elements which is $O(bN)$ [51, 53].

The performances of different iterative solvers: Gauss–Seidel (GS), conjugate gradient with incomplete Cholesky preconditioning (PCG), and full multimesh (often termed as multigrid MG) are shown in Fig. 11.2, where the number of iterations is given for a single step on a 3D cubic domain with $N = k^3$ points. MG(i, j) stands for a full multimesh method with i Gauss–Seidel iterations at each mesh level and j or more mesh points at the coarsest level. The iteration stopping criterion was $||\mathbf{err}||_\infty \leq 10^{-6}$.

The Gauss–Seidel and PCG methods require approximately $1.5k^{2.1}$ and $1.4k$ iterations respectively, while the full MG method outperforms both, the number of iterations being independent of k. MG(5,8) requires only 4 iterations in larger systems also.

11.3.4.3 Memory Requirements

We have seen in the previous description that, for FDM and FEM, the sizes of all the data structures are, at most, proportional to bN, because all the matrices involved are sparse and no lists are longer than N. Therefore, the asymptotic memory requirement of both methods is equal to $O(bN)$.

11.4 Diffusion Equation

Heat transfer, that is, energy transport as a result of a temperature gradient, is important in homeothermic organisms, because maintenance of a specific temperature

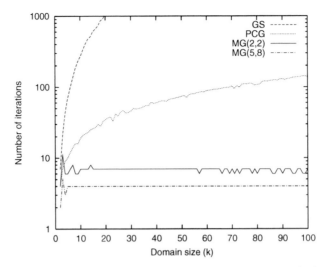

Fig. 11.2 The number of iterations in a single step as a function of the number of points for different iterative solvers.

is crucial for their functioning [33]. Besides real experiments, computer simulations based on theoretical physical models [56] can be very useful for understanding heat transfer processes in such systems. There are three fundamental heat transfer mechanisms: conduction, convection and radiation. Conduction, described by the heat conduction equation, is usually the most important for thermal energy transport within a solid substance. In order to study the temperature distribution in a system such as a living organ, efficient numerical schemes for solving the heat conduction equation are required [19]. The heat conduction model describes conduction on the macroscopic scale and the numerical schemes are usually based on the finite difference approximation [57], as opposed, for example, to molecular dynamics simulations [58], where the properties of the physical system are computed on the microscopic scale, and efficient algorithms for solving the Hamilton's equations for each atom in the system have to be developed [59].

The basic equation that describes heat transfer is known as the heat conduction equation [42], often referred to as the diffusion equation

$$\nabla \cdot (\boldsymbol{\lambda} \nabla T) = \rho c \left(\frac{\partial T}{\partial t} + (\mathbf{v} \cdot \nabla) T \right), \tag{11.22}$$

where $\nabla = (\frac{\partial}{\partial x}, \frac{\partial}{\partial y}, \frac{\partial}{\partial z})$ is a differential operator in terms of Cartesian coordinates, $\mathbf{v}$ is the velocity of a part of the substance volume, $\rho = \rho(\mathbf{r})$ is the mass density, and $c = c(\mathbf{r})$ is the specific heat at constant pressure. $\boldsymbol{\lambda} = \boldsymbol{\lambda}(\mathbf{r}, T)$ is the heat conductivity of a substance, which could also be temperature dependent.

Although Eq. (11.22) is in general nonlinear, we focus here only on linear cases, with $\boldsymbol{\lambda}$ independent of temperature. However, $\boldsymbol{\lambda} = \boldsymbol{\lambda}(\mathbf{r})$ because we are simulating inhomogeneous objects composed of several tissues with different heat conductivities. $\boldsymbol{\lambda}$ is in general a 3×3 tensor. For isotropic substances $\boldsymbol{\lambda} = \lambda I$, where I is the 3×3 identity matrix. It holds now that $\nabla \boldsymbol{\lambda} = \nabla \lambda$, and λ can therefore be treated as a scalar. $T = T(\mathbf{r},t)$ is the temperature as a function of the position $\mathbf{r} = (x,y,z)$ and time t.

For a 1D system that is motionless, Eq. (11.22) is now written as

$$\rho c \frac{\partial T}{\partial t} = \lambda(x) \frac{\partial^2 T}{\partial x^2} + \frac{\partial \lambda(x)}{\partial x} \cdot \frac{\partial T}{\partial x}, \tag{11.23}$$

or equivalently

$$\rho c \frac{\partial T}{\partial t} = \frac{\partial}{\partial x} \left[\lambda(x) \frac{\partial T}{\partial x} \right]. \tag{11.24}$$

Equation (11.23) can be used for determining the stationary temperature distribution in a 1D heat conductor of length l shown in Fig. 11.3.

$$\lambda_1, \rho_1, c_1 \qquad \lambda_2, \rho_2, c_2$$

T_L $\qquad$ T_R

Fig. 11.3 A 1D heat conductor composed of two materials with the same thickness but different λ, ρ, and c.

The conductor is composed of two different materials of the same thickness. The boundary conditions at the left and right ends are the temperatures T_L and T_R. The left part of the conductor is composed of a substance with λ_1, ρ_1, and c_1, and the substance of the right part is characterized by λ_2, ρ_2, and c_2. This simple system provides a test case for the analysis of different numerical methods because its analytical solution can easily be obtained.

Different tissues have different thermodynamic characteristics and any part of a living organ is composed of several different tissues. Therefore, the numerical methods have to support such inhomogeneities if we want to obtain accurate results. After the analytical solution of our simple test case, it will be shown in detail, how to develop an FDM scheme that is stable and still correct for interfaces between two different layers. In other types of numerical methods, for example, FEM or meshless methods, similar approaches have to be applied for the development of appropriate schemes.

11.4.1 Analytical Solution

The analytical solution of Eq. (11.23), which is used as the reference to check the accuracy of the numerical solution, can be derived as follows. In the stationary state,

$$0 = \frac{\partial T}{\partial t} = D\frac{\partial^2 T}{\partial x^2} \tag{11.25}$$

holds for each separate part of the conductor, where λ is constant and its derivative equal to zero. D is the thermal diffusivity of the substance, defined as $D = \lambda/(\rho c)$. The boundary conditions are $T(0) = T_L$ and $T(l) = T_R$ and the transient conditions between the left and right parts of the conductor are

$$T_1(l/2) = T_2(l/2) \tag{11.26}$$

and

$$\lambda_1 \frac{\partial T_1}{\partial x}|_{l/2} = \lambda_2 \frac{\partial T_2}{\partial x}|_{l/2}, \tag{11.27}$$

which state that the temperature at the point of contact is the same for both parts and that the heat flux density, determined by the Fourier law of heat transfer [47], is conserved. The solution of Eq. (11.25) comprises the linear functions

$$T_1 = A_1 x + T_L, \tag{11.28}$$

$$T_2 = A_2(x - l) + T_R. \tag{11.29}$$

Inserting Eqs. (11.28) and (11.29) into (11.26) and (11.27), the constants A_1 and A_2 are

$$A_1 = \frac{2(T_R - T_L)}{l(\frac{\lambda_1}{\lambda_2} + 1)}, \qquad A_2 = \frac{2(T_R - T_L)}{l(\frac{\lambda_2}{\lambda_1} + 1)}, \tag{11.30}$$

and

$$T_2(l/2) = \frac{\lambda_1 T_L + \lambda_2 T_R}{\lambda_1 + \lambda_2}. \tag{11.31}$$

The temperature profile of our test case has the shape of a linear function that is broken at the contact point between the two parts of the conductor.

11.4.2 Finite Differences with Gradient Term

For the numerical solution of Eq. (11.23), the explicit finite difference scheme can be used by replacing the spatial and temporal derivatives with finite differences

$$\rho_i c_i \frac{T_i^{n+1} - T_i^n}{\Delta t} = \lambda_i \frac{T_{i+1}^n - 2T_i^n + T_{i-1}^n}{(\Delta x)^2} + \frac{(\lambda_{i+1} - \lambda_{i-1})(T_{i+1}^n - T_{i-1}^n)}{4(\Delta x)^2}. \tag{11.32}$$

Index i denotes the spatial discretization and Δx is the step size. Index n refers to the time discretization and Δt is the length of the time-step. Suppose first that the gradient term in Eq. (11.32)

$$\frac{\partial \lambda}{\partial x} \cdot \frac{\partial T}{\partial x} = \frac{(\lambda_{i+1} - \lambda_{i-1})(T_{i+1}^n - T_{i-1}^n)}{4(\Delta x)^2} = 0 \qquad (11.33)$$

is zero. The rest of Eq. (11.32) can be written in a matrix form

$$T^{n+1} = \left(I + \frac{D\Delta t}{(\Delta x)^2} \begin{bmatrix} -2 & 1 & & \\ 1 & -2 & 1 & \\ & 1 & -2 & 1 \\ & & \cdots & \cdots & \cdots \end{bmatrix} \right) T^n \qquad (11.34)$$

where I is the identity matrix. Denoting the tridiagonal matrix by A, the eigenvalues of the Jacobian matrix $\frac{D}{(\Delta x)^2} A$ are between $-4 \frac{D}{(\Delta x)^2}$ and 0. Regarding stability criteria for the explicit Euler's method, described in Sect. 11.3.3.1, we obtain the stability condition for this scheme

$$\left| 1 - \frac{4D\Delta t}{(\Delta x)^2} \right| < 1 \qquad \text{or} \qquad \Delta t < \frac{(\Delta x)^2}{2D}. \qquad (11.35)$$

If λ is not constant, the maximal value of thermal diffusivity in the system D_{max} has to be inserted in the condition (11.35).

In the case of significant changes in λ, for example because of different substances with different thermal properties, the difference $(\lambda_{i+1} - \lambda_{i-1})$ can be large and so too the gradient, which can significantly increase the spectral radius of the Jacobian matrix of Eq. (11.32). This will introduce instability into the numerical method [60] and force Δt to be smaller. A direct consequence is the increased computational complexity of the simulation.

11.4.3 Explicit Finite Difference Scheme

In developing the explicit finite difference scheme an approach similar to that in the analytical solution is used. The heat conduction equation (11.32), without the troublesome gradient term (11.33), is solved separately for each mesh element of the conductor because they are homogeneous. The separate solutions are then smoothly matched by transient conditions at the contact. The transient condition for the equality of the temperatures at the contact point is fulfilled by introducing a new intermediate mesh point $T_{i+1/2}$ that lies on the boundary between the mesh elements as shown in Fig. 11.4.

The temperature at the boundary $T_{i+1/2}$ is derived from the transient condition for the heat flux density, using Eq. (11.27)

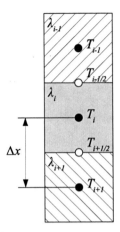

Fig. 11.4 Mesh elements
with different λ, ρ, c, denoted
by different patterns.

$$\lambda_i \frac{\partial T_i}{\partial x}|_{\Gamma} = \lambda_{i+1} \frac{\partial T_{i+1}}{\partial x}|_{\Gamma}, \qquad (11.36)$$

where Γ denotes the contact boundary and index i the mesh point where the sudden
change in thermal conductivity occurs. Replacing derivatives by finite differences

$$\lambda_i \frac{T_{i+1/2} - T_i}{\frac{\Delta x}{2}} = \lambda_{i+1} \frac{T_{i+1} - T_{i+1/2}}{\frac{\Delta x}{2}} \qquad (11.37)$$

we obtain the temperature at the contact between the two materials

$$T_{i+1/2} = \frac{\lambda_i T_i + \lambda_{i+1} T_{i+1}}{\lambda_i + \lambda_{i+1}}. \qquad (11.38)$$

The second derivative of the temperature at the mesh points not touching the contact
boundary is computed as

$$\frac{\partial^2 T_i}{\partial x^2} = \frac{T_{i+1} - 2T_i + T_{i-1}}{(\Delta x)^2}, \qquad (11.39)$$

and at the mesh points from the boundary as

$$\frac{\partial^2 T_i}{\partial x^2} = \frac{\frac{T_{i+1/2} - T_i}{\frac{\Delta x}{2}} - \frac{T_i - T_{i-1/2}}{\frac{\Delta x}{2}}}{\Delta x} = \frac{2(T_{i+1/2} - 2T_i + T_{i-1/2})}{(\Delta x)^2}. \qquad (11.40)$$

In Eq. (11.40) the left and right differences were used for the first derivatives in
the mesh points $T_{i+1/2}$ and $T_{i-1/2}$, respectively. Thus only the temperatures from the
same mesh element are used in order to avoid the troublesome gradient term (11.33).
If $\lambda_i = \lambda_{i-1} = \lambda_{i+1}$ then Eq. (11.40) simplifies to (11.39). $T_{i-1/2}$ is, analogously to
Eq. (11.38), defined as

$$T_{i-1/2} = \frac{\lambda_i T_i + \lambda_{i-1} T_{i-1}}{\lambda_i + \lambda_{i-1}}. \tag{11.41}$$

The devised explicit FD scheme is now

$$\rho_i c_i \frac{T_i^{n+1} - T_i^n}{\Delta t} = \lambda_i \frac{2(T_{i+1/2}^n - 2T_i^n + T_{i-1/2}^n)}{(\Delta x)^2}. \tag{11.42}$$

The iterative solution Algorithm using this scheme is as follows:

Algorithm 11.1 Iterative_Explicit_FD_scheme

1: **for** each time-step n **do**
2:     **for** each point i in the domain **do**
3:         Compute $T_{i+1/2}$ or $T_{i-1/2}$ from the temperatures from the previous time-step by Eqs. (11.38) and (11.41).

        - Note that $T_{i+1/2}$ is equal to $T_{i-1/2}$ of the neighboring mesh element and need not be calculated separately.

        - Computing the second derivatives from Eqs. (11.39) and (11.40) the separate solutions for each mesh element, which itself is homogeneous, are smoothly matched together. The gradient term (11.33) is thus omitted in the heat conduction equation (11.23). One does not have to consider how or to what extent the heat conductivity varies with position. The transient conditions are also considered implicitly.
        - Some unnecessary extra computation is performed for calculating $T_{i+1/2}$ or $T_{i-1/2}$ in parts where the substance is homogeneous. It is interesting that Eq. (11.40) for an inhomogeneous substance is similar to Eq. (11.39) for a homogeneous substance. The only difference is the use of weighted temperature averages in Eq. (11.40), as expressed in Eqs. (11.41) and (11.38).

4:         Calculate new temperature by applying the explicit FD scheme devised from Eq. (11.42)

$$T_i^{n+1} = T_i^n + \frac{\lambda_i \Delta t}{\rho_i c_i} \frac{2(T_{i+1/2}^n - 2T_i^n + T_{i-1/2}^n)}{(\Delta x)^2}.$$

        - The temperature field is obtained effectively, in lD, by multiplication of a tridiagonal matrix and vector.
        - Note that the scheme (11.42) requires the computation of $T_{i+1/2}$ or $T_{i-1/2}$, which can be used advantageously in increasing the resolution of the simulated solution (see [61] for further details).

5:     **end for**
6: **end for**

Formulas for 3D Solution

For the 3D example, Eq. (11.23) is generalized to:

$$\rho c \frac{\partial T}{\partial t} = \lambda \left(\frac{\partial^2 T}{\partial x^2} + \frac{\partial^2 T}{\partial y^2} + \frac{\partial^2 T}{\partial z^2} \right) + \frac{\partial \lambda}{\partial x} \cdot \frac{\partial T}{\partial x} + \frac{\partial \lambda}{\partial y} \cdot \frac{\partial T}{\partial y} + \frac{\partial \lambda}{\partial z} \cdot \frac{\partial T}{\partial z}, \quad (11.43)$$

where $\lambda = \lambda(x,y,z)$ and $T = T(x,y,z)$. Equation (11.43) is a 3D analog of (11.23) and can be rearranged to

$$\rho c \frac{\partial T}{\partial t} = \left(\lambda \frac{\partial^2 T}{\partial x^2} + \frac{\partial \lambda}{\partial x} \cdot \frac{\partial T}{\partial x} \right) + \left(\lambda \frac{\partial^2 T}{\partial y^2} + \frac{\partial \lambda}{\partial y} \cdot \frac{\partial T}{\partial y} \right) + \left(\lambda \frac{\partial^2 T}{\partial z^2} + \frac{\partial \lambda}{\partial z} \cdot \frac{\partial T}{\partial z} \right). \quad (11.44)$$

By comparing Eq. (11.44) with (11.23) and by virtue of (11.42), the described explicit finite difference scheme for an inhomogeneous 3D example yields the form

$$\rho_{i,j,k} c_{i,j,k} \frac{T_{i,j,k}^{n+1} - T_{i,j,k}^{n}}{\Delta t} = 2\lambda_{i,j,k} \left(\frac{T_{i+1/2,j,k}^{n} - 2T_{i,j,k}^{n} + T_{i-1/2,j,k}^{n}}{(\Delta x)^2} + \frac{T_{i,j+1/2,k}^{n} - 2T_{i,j,k}^{n} + T_{i,j-1/2,k}^{n}}{(\Delta y)^2} + \frac{T_{i,j,k+1/2}^{n} - 2T_{i,j,k}^{n} + T_{i,j,k-1/2}^{n}}{(\Delta z)^2} \right). \quad (11.45)$$

Indices i, j, and k denote the spatial discretization in the x, y, and z directions, respectively, $\Delta x, \Delta y$, and Δz are the corresponding spatial step sizes, and

$$T_{i\pm1/2,j,k} = \frac{\lambda_{i,j,k} T_{i,j,k} + \lambda_{i\pm1,j,k} T_{i\pm1,j,k}}{\lambda_{i,j,k} + \lambda_{i\pm1,j,k}}. \quad (11.46)$$

$T_{i,j\pm1/2,k}$ and $T_{i,j,k\pm1/2}$ can be calculated in the same way as in Eq. (11.46), just by varying the other indices.

11.4.4 Comparison of Results

The analytical solution was compared with the two numerical approaches, (11.32) and (11.42), presented for solving the heat conduction equation on the ld heat conductor from Fig. 11.3. The boundary conditions were $T_L = 273.2$ K, $T_R = 295$ K. The temperature is dimensionless and is measured in units of the temperature T_L at the left-hand boundary of the conductor. For $\Delta x = 10^{-3}$ m, the time-step was $\Delta t = (\Delta x)^2/(2D_{air}) = 0.026$ s. The thermodynamic constants used in our test case are given in Table 11.1.

We found that, using the similar thermal conductivities λ_1 and λ_2, both schemes, (11.32) and (11.42), provide correct solutions. The temperature evolution and the steady-state profiles were computed also for a ld heat conductor composed of water and air that have significantly different D. The left half was assumed to be water and the right half air. The resulting steady-state temperature profiles are shown in

Table 11.1 Thermodynamic constants of water and air used in the test case.

	λ	c	ρ	D
	$\mathrm{W\,(m\,K)^{-1}}$	$\mathrm{J\,(kg\,K)^{-1}}$	$\mathrm{kg\,m^{-3}}$	$\mathrm{m^2\,s^{-1}}$
Water	0.58	4204	1000	$1.38 \cdot 10^{-7}$
Air	0.025	1012	1.29	$1.92 \cdot 10^{-5}$

Fig. 11.5. T_{anal} is the analytical solution and T_{EFD} is the numerical solution obtained by the described explicit FD scheme (11.42).

We see that solutions obtained by the explicit FD scheme agree with the analytical solution even in the case with very different diffusivities. The numerical solution obtained by (11.32) is not shown because it was not stable using the time-step calculated above, because of the large value of the gradient term at the contact between water and air.

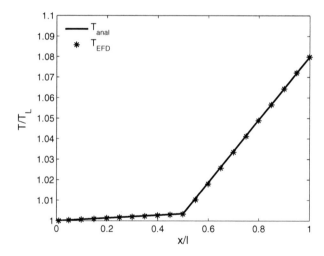

Fig. 11.5 The steady-state analytical solution and near steady-state (after 2 hours of simulation) numerical solution for a 1d heat conductor composed of water and air. The numerical solution was obtained by explicit FD scheme (11.42) and time-step 0.026 s. It is shown with asterisks for each fifth point only, for better visibility. The temperature is normalized by T_L and the position by the length of the conductor l.

11.5 Bio-Heat Equation

Because the human knee is an irregularly shaped, 3D object it cannot be represented as a combination of simple geometric objects. The temperature distribution cannot be obtained analytically. We selected the explicit finite difference scheme presented in Sect. 11.4.3 for the numerical solution procedure, because it can be used in inhomogeneous domains. The meshing is inherited from geometric modeling. The method has low calculation complexity, it provides time evolution, and it is sufficiently accurate and simple to implement on parallel computers.

Conduction, in and between tissues, was modeled by the diffusion equation, described in Sect. 11.4. Convection and radiation were approximated by increased conduction in order to keep the mathematical model simpler. Besides conduction, additional heat sources have to be modeled in the living tissue; for example, heat transfer from arterial blood in large and small vessels to the tissue, and heat production from the tissue metabolism. Heating from the arterial blood depends on the amount of blood perfusion, V, and the difference between the arterial, T_a, and local tissue temperatures, T. Metabolism itself depends on the metabolic heat production h_m of the local tissue, which is a function of T. Some additional terms could also be added that would model the heat sink, like breathing and sweating; however, in the simulation of knee cooling we neglected their impact. The model PDE used in our simulation is the well-known bio-heat equation [12, 32, 33] that can be written as

$$\rho c \frac{\partial T}{\partial t} = \nabla \cdot (\lambda \nabla T) + (1-k)\rho_b c_b V(T)(T_a - T) + \rho h_m(T) \qquad (11.47)$$

where k is the coefficient of thermal equilibrium between blood and tissue, ρ_b and c_b are the density and specific heat of blood, and all other notations are as defined previously. Note that temperature changes with time and space and that all constants depend on the specific tissue. The dependence of perfusion and metabolism on the temperature could be arbitrary functions.

Published measurements of the blood flow V show that it increases with increasing skin temperature and temperature of the surrounding tissue. The function $V(T)$ has been approximated by an exponential function, using published experimental data [35, 62]. For the range of temperatures in our simulation from 10 to 40°C, we obtained

$$V(T) = (5.142 \cdot 10^{-5} e^{0.322T} + 0.705) \text{ ml} (100 \text{ ml})^{-1} \text{ min}^{-1} \qquad (11.48)$$

which results, for example, in $V(33\,°C) = 2.82$ ml $(100\,\text{ml})^{-1}$ min^{-1} or in SI units $4.71 \cdot 10^{-4}$ s^{-1}. The blood flow in bones was assumed to be much smaller, so we set it at 10 times smaller than in other tissues. In Fig. 11.6 the fitted measurements [35, 62] of blood flow are shown as a function of tissue temperature $V(T)$.

Metabolic heat production in the human can be separated from unregulated heat production from voluntary muscle contraction and normal metabolic pathways, and from regulated heat production for maintaining temperature homeostasis at lower ambient temperatures [31]. The rate of metabolic heat production per unit mass, h_m,

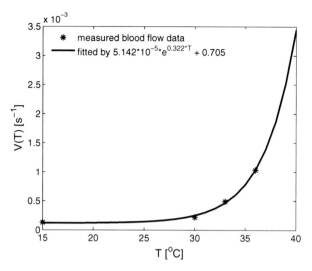

Fig. 11.6 Measured (*asterisks*) and fitted blood flow (*solid*) as a function of tissue temperature.

was assumed to obey the Q_{10} rule [63], and is expressed as a function of the tissue temperature

$$h_m(T) = h_r 2^{(T-T_r)/10} \tag{11.49}$$

where h_r is the reference metabolic heat production of a tissue at the reference temperature $T_r = 35\,°C$. For resting muscles, h_r was taken to be $0.58\ \mathrm{J\,kg^{-1}\,s^{-1}}$, which is about half the human basal metabolic rate. As in the case of blood flow, the metabolism of bones was assumed to be negligible.

The modeled nonlinear contribution of heat production in the muscle tissue from both factors from the right side of the bio-heat equation (11.47) is shown in Fig. 11.7. Graphs are obtained by evaluating the model Eqs. (11.48) and (11.49). We see that, at lower temperatures, the impact of blood flow and metabolism is minimal – they both act as small heat sources. However, if the tissue temperature rises above $37\,°C$, the arterial blood rapidly cools the tissue. Such behavior is in good agreement with measured data [62].

Moving air was was not simulated because of its significant contribution to calculation complexity. We simulated convection and radiation with the same simulation program as that used for conduction. A layer of "still" air was inserted between the model surface and the remaining external still air at constant room temperature. Preliminary simulations at room temperature of $28\,°C$, and without any other cooling, were made to find the appropriate air layer thickness. We found that a 5 mm thick air layer results in a skin temperature of $33\,°C$, which was also the measured value [35]. In this way, the convection and radiation were approximated by increased conduction. Conduction in the inter-layer air was simulated by a time-step 20 times smaller than in the rest of the model. Alternative approaches have been proposed using approximate boundary conditions based on the continuity of the heat flux

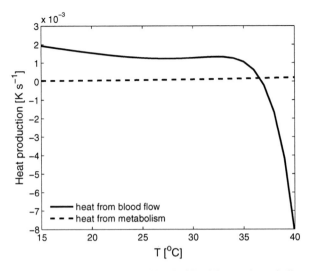

Fig. 11.7 The heat production rate contributed by the blood flow and metabolism as a function of muscle temperature.

perpendicular to the surface of the body [12], which however requires knowledge of the convection coefficient.

Applying the explicit FDM scheme, developed in Sect. 11.4.3, after time and space discretization of Eq. (11.47), we obtain one equation for each voxel (i, j, k). $T_{i,j,k}^n$ is the approximation of the temperature in voxels (i, j, k) in the n-th time-step Δt, which has to be selected so that the numerical solution remains stable and accurate. The time-step Δt is limited by stability with Eq. (11.35) as

$$\Delta t < \frac{min\{(\Delta x)^2, (\Delta y)^2, (\Delta z)^2\}}{6 D_{max}}. \tag{11.50}$$

Inserting the parameter values used in our simulation (see Table 11.2) we can calculate the minimal step length for air voxels as $\Delta t_a = 10^{-6}/(6 \cdot 1.9 \cdot 10^{-5}) = 0.0088\,\text{s}$ and for bone voxels as $\Delta t_b = 0.157\,\text{s}$. Bone voxels require the shortest time-step of the remaining substances, which is still 18 times longer than that for air. We could substantially shorten the simulation time if the air and all other substances were simulated with the finest and coarsest time-steps, respectively.

A further limiting factor in the choice of the time-step length is the solution accuracy. We made preliminary simulations of knee temperatures after 300 s of cooling with a gel pack and constant air temperature. The shortest time-step was dictated by bones. We obtained the reference solution with Δt 30 times shorter than required by Δt_b. In Fig. 11.8 the absolute solution errors are shown on the cross section from the central part of the knee. The errors were calculated for time-steps Δt_b and $\Delta t_b/4$ as the differences from the reference solution. The maximal errors, which appear in the regions with maximal temperature gradients, were 0.012 and 0.003 °C for Δt_b

and $\Delta t_b/4$, respectively. Because we want to simulate with an accuracy higher than 1%, all subsequent simulations have been run with four times shorter time-step as dictated by the stability condition $\Delta t = \Delta t_b/4 = 0.039\,\text{s}$.

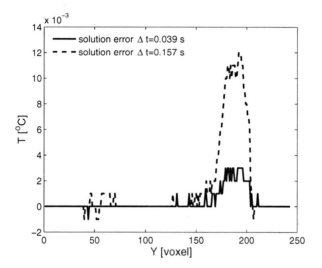

Fig. 11.8 Solution errors by maximal time-step Δt_b dictated by stability condition and by $\Delta t_b/4$.

We have shown that, starting with the initial state T^0, it is possible to compute T^n for every $t > 0$, stepping through time with the time-step Δt, just by banded matrix-vector multiplication and a matrix-vector addition. The calculation complexity of a single iteration step remains proportional to the number of voxels multiplied by the number of diagonals or, equivalently, the number of neighboring voxels.

If greater accuracy is needed, the FEM could be applied; however, meshing in 3D, a greater calculation and memory complexity, and a potential requirement for evolutionary temperature data could diminish the advantages of FEM.

11.6 Geometric Modeling of a Knee

The basic anatomical data for a computer model of the knee was derived from the colored digital photographs of the human body cross-sectional slices that are available in the VHD [64]. A similar whole body model at a resolution of 1 mm^3 has also been developed within VHD project, but is not always available for public use. Additionally, for our model, manual segmentation was applied on slices with the full resolution of $\frac{1}{3}$ mm^2. Cross sections of the human male lower limb were used that contain tissues from the central region of the knee joint. The knee area was cropped from the original photographs as a rectangle of 550 (width) × 610 (height) pixels. Cropped VHD slice 2301 is shown in Fig. 11.9.

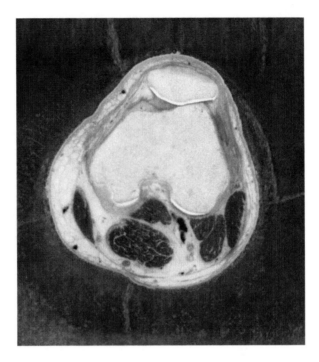

Fig. 11.9 Knee area of 550 × 610 pixels, cropped from the human male VHD slice 2301 (model slice 102).

Different tissues were identified manually by an expert anatomist who assisted the segmentation. The total number of pixels was too large for our simulation. Because the distance between neighboring cross-sections in the Z dimension was 1 mm we decided to reduce the resolution in the X and Y dimensions to 1 mm, using averaging of neighboring pixels. A model slice $z = 102$ from the central knee region, at the level of the femoral intercondylar notch, with the resolution used in the simulation, is shown in Fig. 11.10. The picture was taken from the custom program *ResultsViewer* developed for the visualization of simulated results. This program is used also for the extraction of the desired plane from the simulated results, in the selected dimensions and time for figures generated in the subsequent sections.

A 3D geometric model was built simply by stacking 191 consecutive VHD slices from the knee region, that is, VHD slice 2200 (model slice 1 on the top) to VHD slice 2390 (model slice 191 on the bottom). The 3D model is shown in Fig. 11.11. Skin, joint liquid, and subcutaneous tissues are not shown; the artero-lateral quadrant is removed to see into the inside of the knee.

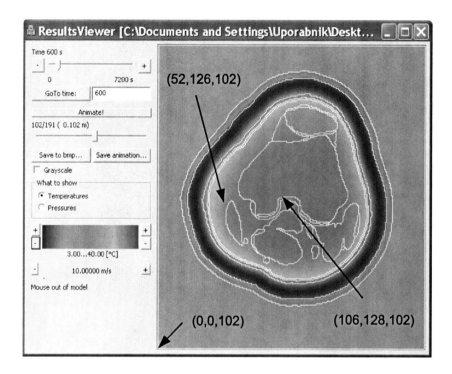

Fig. 11.10 Model slice $z = 102$ from the central region of the knee at the resolution used in the simulation. The knee is surrounded by an isolating blanket, cooling layer (*black*), and a protective bandage. Skin, subcutaneous tissue, bones, knee joint liquid, muscles, cartilage, nerves, and vessels are separated by contours. Points of measurements are marked with arrows and denoted by (x, y, z) coordinates.

Some surrounding space was added around the 3D knee model for the protective bandage, gel pack, blanket, and for ambient air. The simulation environment was imitated by an isolated cube composed of $x \times y \times z = 223 \times 243 \times 191 = 10,350,099$ small voxels characterized independently by thermodynamic properties and initial temperatures, each with a volume of $1\,\mathrm{mm}^3$. The boundary layers of the simulated box were held at constant initial temperatures to mimic the effect of the ambient air. The heat flux from the first and last slices was kept constant in order to imitate the influence of the leg not exposed to cooling. The knee was covered by a 2 mm thick protective bandage and embraced by a 12 mm thick cooling layer (gel pack or cryo-cuff). An additional 5 mm thick isolating blanket was used to cover the cooling layer in order to reduce convection and slow down its warming from the outside. The protective bandage, cooling layer, and isolating blanket were inserted into the model automatically by a computer program.

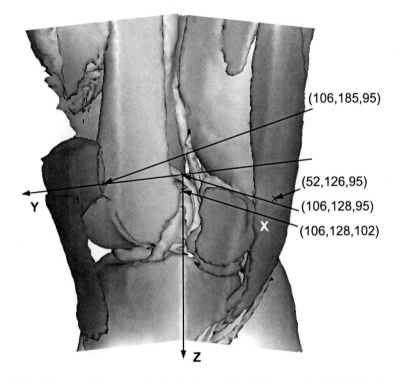

Fig. 11.11 3D geometric knee model obtained by stacking 191 consecutive VHD slices from top to bottom. The frontal quadrant is removed to see into the central knee region.

11.7 Simulation Methods and Parameters

The simulation environment was imitated by an isolated cube composed of voxels from different substances and characterized independently by thermodynamic properties and initial temperatures. Boundary layers of the simulated box were held at constant initial temperatures to mimic the ambient air. The influence of the rest of the leg, which was not simulated, was managed by setting the temperature flux at boundaries. The simulation time can be set as desired. Intermediate results can be written in an output file for later visualization and analysis. The simulation can be carried out in 2D with the temperature flux in the axial direction set to zero. In this way an infinitely long "knee", with homogeneous structure in the axial dimension, can be simulated. Simulation in 3D was performed on the described model with all boundaries set to values that are similar to those of the measuring conditions.

We first simulated the steady-state temperatures of the resting knee by the thermo-neutral conditions [36] at the ambient air temperature of $27\,^\circ$C, which is equivalent to the ambient water, for example, a water bath of temperature $33\,^\circ$C. A steady-state was reached after three hours of simulation, the maximal change in temperature near the end of the third hour being less than $0.01\,^\circ$C.

The simulated temperatures can now be recorded at arbitrary positions. For example, the temperature field over the model slice $z = 102$ is shown with a 2D surface in Fig. 11.12. Note that point $(0,0)$ is on the right and the frontal part of the knee with the patella on the left, for better visibility of the temperature field. The white curve represents the temperature along the transverse axis Y by $x = 106$, which is used in a later analysis. The steady state obtained was used as the initial condition in all our subsequent simulations. The tissues nearer the patella, toward the surface, are seen to be colder than the internal part. The impact of the main knee artery, with its constant blood temperature, is visible as a peak in the temperature field. The location of the bones is also evident as shallow depressions in the temperature field, mainly the result of lower blood flow in bones.

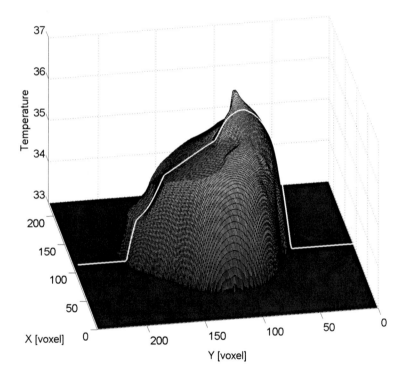

Fig. 11.12 Steady-state temperature field over model slice $z = 102$. Point $(0,0)$ is in the right corner for better visibility.

Initial and boundary conditions, thermal constants, rates of blood flow and metabolism, knee dimensions, and positions of measuring points can be varied in order to test the stability of the simulation method and to analyze the results obtained with different simulated conditions. In order to compare the simulated results with the measurements and to analyze them in more detail, the simulated temperatures were recorded from the points of interest marked in Figs. 11.10 and 11.11.

Numerical values of tissue thermal parameters are not known precisely because they depend on measurement conditions and state of the tissues. We determined their constant values from published data for thermal constants [33], for blood flow [35, 62], and for metabolic heat production [31]. The thermal constants used in the simulation are listed in Table 11.2.

Table 11.2 Thermal parameters of simulated substances with initial temperatures.

Substance	λ $W(m\,K)^{-1}$	c $J(kg\,K)^{-1}$	ρ $kg\,m^{-3}$	T $°C$
Ambient air	0.025	1012	1.29	25
Blanket	0.04	1200	150	25
Cryo-cuff water	0.58	4204	1000	15*
Gel pack	0.1	4000	990	0
Bandage	0.04	1200	150	30
Skin	0.51	3431	1200	35
Subcutaneous Tissue	0.55	2241	812	35.6
Muscle	1.03	4668	1179	36
Bone	2.28	1260	1700	36
Ligament	0.33	3966	1250	36
Cartilage	1.5	2275	1160	36
Joint liquid	0.58	4204	1000	22
Nerve	0.5	3277	1190	36
Venous blood	0.67	3890	1057	36
Arterial blood	0.67	3890	1057	36.8*

* Constant temperatures are denoted by asterisks.

11.8 Variation of Simulation Parameters

The steady-state temperature profiles of the resting knee under the TN conditions and along the transverse knee axis Y by $z = 102$ and $x = 106$ are analyzed in this section. In Fig. 11.13 the solid curve shows the simulated steady-state temperature obtained with the 3D simulation. The dotted curve was obtained under the same conditions with a 2D simulation and zero flux in the third dimension (infinite knee). The difference between the two simulations is in the range of 0.5 °C, and can be explained by underestimating the amount of muscle tissue in the whole knee region. Slice 102 is taken, namely, from the central part of the knee with less muscle that could contribute additional heat through its blood flow. However, the shapes of the profiles are similar, so we used 2D simulation for the initial analysis, to reduce the simulation time.

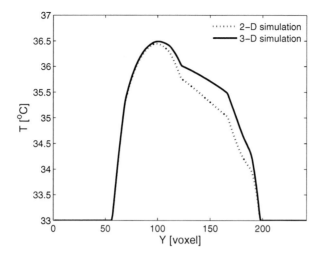

Fig. 11.13 Simulated steady-state temperature profiles by thermo-neutral conditions at the slice $z = 102$ along the transverse knee axis Y by $x = 106$ for 3D simulation and 2D simulation.

We can observe and analyze arbitrary simulated points or regions of the domain. For example, in Fig. 11.14 the steady-state temperature profiles of the knee are shown for temperatures on the transverse knee axis Y and model slice $z = 102$ by $x = 96$ (dotted), $x = 106$ (solid), and $x = 116$ (dashed). It is clear that there can be significant differences in the temperature, even for analyzed points as close as 10 mm. The effect of altering the observed position in other directions is similar to that seen in Fig. 11.14.

The stability and conditioning of the simulated steady-state solution was evaluated by varying the simulation parameters. Input parameters were varied within selected ranges and the variations in the solution were analyzed. In Fig. 11.15 the simulated steady-state temperatures from model slice 102 on the transversal axis Y at $x = 106$ are shown for various knee dimensions, diffusion constants, blood flow, and metabolism, all differing within for +20% (dotted) and −20% (dashed) of their nominal values. In Fig. 11.15b, c, and d the nominal knee dimension, described in Sect. 11.6, was used. Fig. 11.15c shows the impact of the blood flow by equating the coefficient of thermal equilibrium to $k = 0.2$ (blood flow 20 % below the nominal value), $k = 0$ (nominal blood flow) and $k = -0.2$ (blood flow 20 % above the nominal value).

The most important impact on the temperature profiles seen in Fig. 11.15 arises from varying the knee dimension (Fig. 11.15a). Larger knees result in a temperature plateau slightly above 36.7 °C, while smaller knees are cooled more intensively to the central temperature of about 36 °C, with larger temperature gradients in the superficial regions of the knee.

Changes in thermal diffusivity (Fig. 11.15b) have a smaller impact on the temperature profiles than changes in the dimensions, and in the opposite direction, that

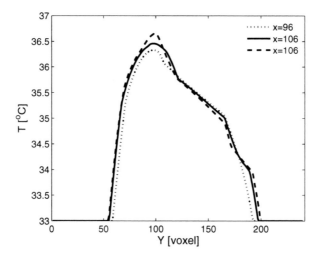

Fig. 11.14 Simulated steady-state temperature profiles by thermo-neutral conditions on the transverse knee axis Y and model slice $z = 102$ by $x = 96$, $x = 106$ and $x = 116$.

is, larger diffusivity constants result in lower central temperatures. This result is in accordance with Eq. (11.34), from which it follows that the changes in temperature are proportional to the diffusivity and inversely proportional to the square of the changes in dimensions.

The impact of blood flow (Fig. 11.15c) is similar to that of dimensions, with higher internal temperatures arising from greater blood flow, however with the important difference that the shapes of the temperature profiles remain unchanged with variations of the blood flow.

The impact of metabolism (Fig. 11.15d) is analogous to that of the blood flow – greater metabolism results in higher internal temperatures, but its impact is so small that it can be considered in our experiments as negligible.

11.9 Simulation Results

Simulated steady-state temperatures of the resting knee under TN conditions have been used as the initial condition for all further simulations. We simulated first a naked knee in the steady state at ambient air temperature of $25\,^\circ$C. We then simulated a two-hour period of arthroscopic operation, during which the knee joint was washed out by sterilized water at $22\,^\circ$C, and therefore cooled. During the following two-hour period we simulated the temperature evolution in the operated knee while it was resting and covered by a blanket, and therefore warming. Finally, we simulated the subsequent two-hour postoperative topical cooling.

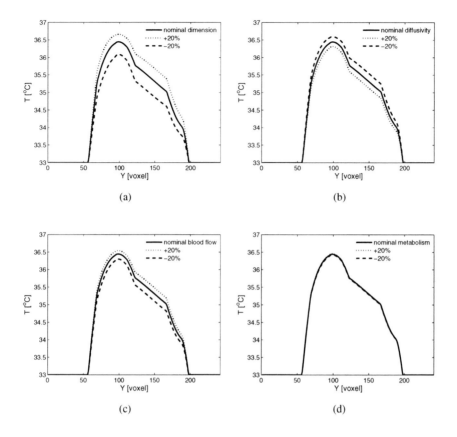

Fig. 11.15 Simulated steady-state temperatures from model slice $z = 102$ and the transversal axis Y at $x = 106$ for various (**a**) knee dimension, (**b**) thermal diffusivity, (**c**) blood flow, and (**d**) metabolism. In all figure-panels nominal values are solid, 20 % above nominal values are dotted and 20 % below nominal values are dashed.

11.9.1 Washing Out During Arthroscopy

During arthroscopic reconstruction of ligaments, the central part of the knee is washed by sterilized water at 22 °C. The water is circulated in the space around the femoral intercondylar notch, normally filled by the joint liquid. The initial temperatures were taken from the steady state of a naked knee at an ambient temperature of 25 °C. The temperature of the joint liquid was fixed at 22 °C. The temperature profile after two hours of washing out are shown in Fig. 11.16 for the same plane as before, that is, along the Y axis by $x = 106$ but for the 7 mm higher slice at $z = 95$, because this was nearer the actual position of our measuring probes for validating the simulated results. As a consequence, the internal knee temperature decreased significantly, maximally by more than 14 °C in places with direct contact with the washing water, as can be seen from the temperature profiles shown.

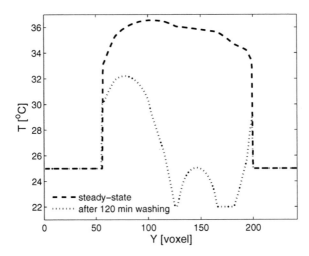

Fig. 11.16 Steady-state temperature profiles for naked knee and after a two hour washing out during surgery, from the model slice $z = 95$ and the transversal axis Y at $x = 106$.

11.9.2 Resting After Arthroscopy

We next simulated the two-hour period immediately after surgery. The knee was resting and covered with a blanket in a room temperature of 25 °C. The evolution of the temperatures in voxel $(106, 128, 95)$, which is at the level of the femoral inter-condylar notch in the central part of the knee, and in voxel $(52, 126, 95)$, which is nearer to the knee surface, 1 cm below the skin in the subcutaneous tissue, are shown in Fig. 11.17. Positions of both voxels are denoted in Figs. 11.10 and 11.11. Note, that arbitrary voxels could be selected for the analysis. The knee was initially colder in the central region because of the previous washing out with cold water. During resting, its temperature increases and approaches the steady state, with colder regions nearer the skin.

11.9.3 Postoperative Topical Cooling

Finally, we simulated postoperative topical cooling, by two different cooling methods, gel pack and cryo-cuff. In both cases the knee was bound with a protective blanket surrounded by ambient air at 25 °C as shown in Fig. 11.10. The protective bandage, remaining from the surgery, was also simulated.

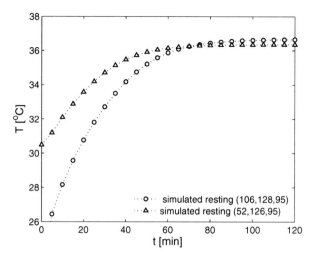

Fig. 11.17 Simulated temperature evolution in two resting hours after surgery for voxels $(106, 128, 95)$ (the central part of the knee) and $(52, 126, 95)$ (in the subcutaneous tissue).

11.9.3.1 Temperature Evolution During Two Hours of Simulated Cooling

In Fig. 11.18 the simulated temperatures of voxels $(106, 128, 95)$ and $(52, 126, 95)$ are shown for the two-hour simulated period and for the cooling with a gel pack (initial temperature $0\,^{\circ}C$) and a cryo-cuff (water with constant temperature of $15\,^{\circ}C$). The effects of the two methods and the knee temperatures are quite different.

On cooling with the gel pack, the temperature of voxel $(106, 128, 95)$ in the central knee region initially increases slightly, on account of the arterial blood perfusion and metabolism, and because of the weak influence of the initial cooling. After 5 minutes the voxel temperature starts to decrease but after 40 minutes the gel pack has received enough heat, from the knee surface and the ambient air, to allow the inner knee temperature to increase during the second part of the cooling period. For voxel $(52, 126, 95)$, in the subcutaneous tissue 10 mm below the skin, the temperature first decreases sharply; the effectiveness of the gel pack then becomes weaker and the voxel starts warming and after 120 minutes reaches almost $36.0\,^{\circ}C$.

Cooling with a cryo-cuff was found to be more effective, in that it induces lower tissue temperatures, even if the temperature of the cooling liquid is as high as $15\,^{\circ}C$. In the initial phase, both voxels experience the same cooling rate as that with the gel pack; however, there is no subsequent increase in temperature, because the cryo-cuff is a constant heat sink, which gradually cools the knee. After two hours of cooling, the near surface voxel reaches a temperature of $27\,^{\circ}C$. In the same way, but with smaller intensity, the inner voxel is cooled to $33\,^{\circ}C$.

In the case of topical cooling with an ice pack the heat of fusion should also be simulated, which is necessary for the transition between aggregate states. Obviously, for a crushed ice pack, a significant part of the heat is needed for such a transition,

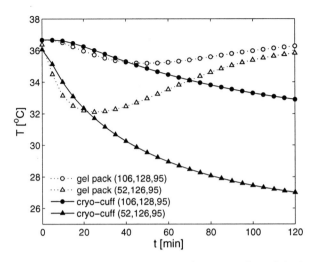

Fig. 11.18 Simulated temperature evolution in voxels $(106, 128, 95)$ and $(52, 126, 95)$, as a function of time for cooling with a gel pack and a cryo-cuff.

which would prolong the effective cooling time. This phenomenon has not been incorporated in our mathematical model. Instead, we increased the heat capacity c of the ice pack in order to recognize and account for this behavior.

From the simulated results it follows that the topical cooling with a cryo-cuff provides more constant lowering of the temperatures in the whole region of the knee. Cooling with gel packs is less stable; consequently, they should be changed every half hour in order to be effective.

11.9.3.2 Temperature Profiles After One Hour of Simulated Cooling

In Fig. 11.19 temperature profiles after one hour of simulated cooling by the gel pack and cryo-cuff are shown for a cross section from the patella to the lateral side of the knee on our standard axis, that is, along the Y (anteroposterior) axis at $x = 106$ on the model slice $z = 95$.

After one hour of simulated cooling the gradients in the temperature profiles were much more pronounced than in the initial state. The temperature of the outer knee layers at the skin level remained cooled to $32\,°C$ and in the centre of the knee to $36\,°C$ by cooling with a gel pack. Peaks in tissue temperatures around $y = 100$ result from the simulated heat conduction from the middle popliteal artery. Significantly lower temperatures are observed by cooling with a cryo-cuff, even though its constant temperature was as high as $15\,°C$.

Given the above results, it would be interesting to test how effective simple cooling with ambient air would be. The knee would remain uncovered and exposed to the ambient air temperature of approximately $20\,°C$. We expect from the results

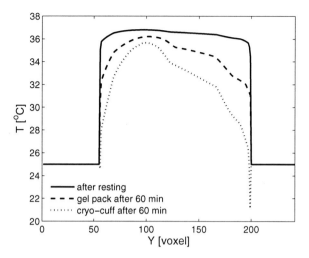

Fig. 11.19 Simulated initial knee temperature profile after resting and temperatures after one-hour period of cooling with the gel pack and cryo-cuff along the Y axis at $x = 106$ and $z = 95$.

under TN conditions that, in such a case, the skin temperature would be about $26\,^{\circ}\mathrm{C}$, which could lower the temperatures inside the knee. This could be tested preliminarily using the proposed simulation method.

11.10 Validation of Results

To evaluate the simulated results we made two control measurements of the knee temperature following surgery in a room with constant ambient temperature $25\,^{\circ}\mathrm{C}$. We measured the temperature of the knee covered with a blanket during the two-hour resting period immediately after surgery, then during the next two-hour period in which topical cooling with a gel pack was applied.

Two small thermistors were placed into the knee in thin sterile tubes (Foley-catheter with temperature sensor, 3 mm, Ch8-thermistor; Curity, Degania Silicone Ltd., Degania Bet, Israel). Similar tubes, without thermistors, are ordinarily inserted for wound drainage following surgery. The thermistors were connected to a registration device for continuous measurement with a sample rate of 0.1 Hz and resolution of $0.01\,^{\circ}\mathrm{C}$.

The first thermistor was placed in the centre of the knee near voxel $(106, 128, 95)$ and the second approximately 1 cm below the skin in the subcutaneous tissue, near voxel $(52, 128, 95)$. The measurements were approved by the Slovenian State Medical Ethics Committee and the patient gave written informed consent prior to participation.

In Fig. 11.20 the simulated and measured temperatures are shown for the two-hour resting period after washing out during the arthroscopic surgery. The knee was wrapped in a protective blanket at the ambient temperature of 25 °C under the same conditions as in the simulation.

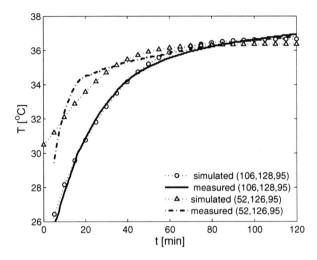

Fig. 11.20 Measured and simulated temperature evolution of voxels $(106, 128, 95)$ and $(52, 126, 95)$ in a two hour resting period after arthroscopic surgery.

The simulated temperature evolution for the resting period shows very good agreement with measured values for the test point $(106, 128, 95)$ in the central part of the knee. However, the simulated rate of cooling in point $(52, 126, 95)$ in the subcutaneous tissue was much smaller in the initial phase than those obtained by measurements. One of the possible reasons is the fact that we did not simulate the cold washing out inlet that also cooled the surrounding tissue from the skin to the central part of the knee. The subcutaneous thermistor was placed in such a cooled environment which could then exhibit faster warming than in our simulation.

In Fig. 11.21 the simulated and measured temperatures are shown for two hours of cooling with a gel pack, which follows immediately after the resting period. The initial temperature of the gel pack was 0 °C and the ambient temperature 25 °C. The knee was bound with elastic bandages approximately 2 mm thick, surrounded with fixed gel packs and wrapped in a protective blanket as shown in Fig. 11.10.

The simulated evolution of temperature for the cooling period shows good agreement with the measured values for both test points. The simulated rates of warming in the second hour are slightly greater than by measurement. One of the possible reasons lies in the incomplete mathematical model that does not include the heat of fusion for the gel pack. In fact, all the thermodynamic characteristics of the gel were not available and we just took some approximate values provided by the supplier. Another possible reason is inaccurate measurement, because we did not collect

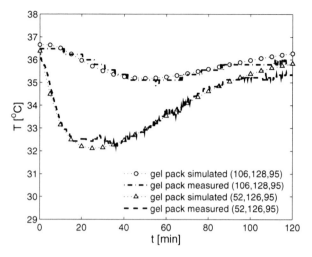

Fig. 11.21 Measured and simulated temperature evolution of voxels $(106, 128, 95)$ and $(52, 126, 95)$ as a function of time for cooling with a gel pack.

detailed data during measurements, for example, the wetness of the protective bandage, which could have a significant impact on the cooling intensity.

At this point we could start another loop in the simulation process shown in Fig. 11.1 by improving the mathematical model and recalculating the simulation results. We will leave such investigations for further work.

11.11 Parallel Implementation

The number of equations to be solved was several millions, therefore the execution time on a single processor could become unacceptably long. We will describe only the parallel implementation of the discretization and the solution method. Other steps of the computer simulation process from Fig. 11.1 are not so demanding, at least for the research purposes, and therefore they were not parallelized. It is well known [20] that any parallel algorithm has to avoid serial parts because even small serial tasks, like synchronization of all processors, global communication or some individual action of the master processing node can significantly degrade the speedup. The computational domain of our application is static and has a simple form because we close the simulated object in a box with boundary conditions. Consequently, the domain decomposition and load distribution can be simplified.

11.11.1 Opportunities for Parallelization

The time of running the application on a parallel computer t_{ex} is the sum of the times for computation t_{cp} and communication t_{cm}. The associated idle time is neglected in this idealized model. By increasing the number of processors the t_{cp} decreases, in the ideal case with the number of processors P. However, if t_{cm} does not scale with the same rate, the ratio t_{cm}/t_{cp} increases and slows down the execution. Thus both, processor performance and communication speed, have to be improved in order to improve the overall performance of parallel algorithms.

The computation time t_{cp} depends on many different factors, such as the CPU clock, cache system, floating point units, internal processor architecture, etc. In our simple model we take into account only the time needed for floating point calculation t_{FP}. We neglect instruction decoding, integer operations, etc., because we suppose that all these activities run concurrently with the floating point unit, so that t_{cp} is proportional to t_{FP}.

The simplest communication model was used where t_{cm} of a message transfer is modeled as a sum of the setup time, t_s, needed for preparation of the actual data to be communicated (delays of operating system, memory buffers, etc.) and the time, t_w, needed for the transmission of a single datum. Each processing node can have a single connection to the main communication switch or, alternatively, some additional connections that implement, for example, an interconnection topology with four nearest neighboring nodes. Such topologies can support, for example, a ring of computing nodes which can be quite efficient in some demanding applications. In problems where a significant amount of global communication is needed, such as molecular dynamics [25, 59], optimal performance of intra-cluster communication is particularly important.

With reference to the numerical methods described, calculations of each time-step are independent and can thus be parallelized for each time-step. Data in the computational domain can be distributed among processing nodes using data domain decomposition [20]. We implemented the simplest ld domain decomposition which divides the whole 3D geometric knee model into P subdomains with the shape of flat squares and allocates a single square to each processor. We assume that processing nodes are connected into a ring topology, that the communication is bidirectional, and that the two communication channels, such as those to the left and right neighbors on the ring topology, run concurrently with no degradation in speed.

If a similar calculation is needed in the whole domain, then the calculation is automatically balanced by using the same number of voxels in each subdomain. If some processors are dealing mostly with an empty space around the knee model, or if others have to calculate more voxels with a shorter time-step, then some further load-balancing principle should be used for better results.

The explicit FDM requires values from neighboring points; hence, in each time-step, the new border values of squares have to be exchanged between two neighboring processors, using communication channels. The communication of ld domain decomposition is shown schematically by arrows in Fig. 11.22. To calculate new step temperatures, processors 1 and 2 have to exchange the last and first slice, respectively, which are indicated by the lowest arrow. In the same way all other processor pairs have to exchange their first and last slices. First and last processors, 1 and P, will not exchange their first and last slice, because the temperatures of these slices are fixed by boundary conditions.

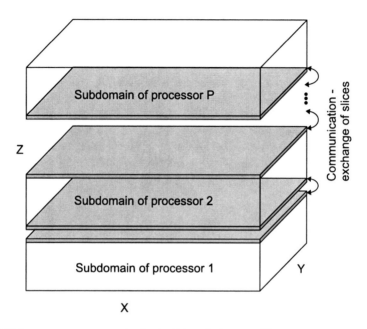

Fig. 11.22 Inter processor communication in ld domain decomposition.

11.11.2 Computation and Communication Complexity

From Eqs. (11.45) and (11.46) given in Sect. 11.4, we see that in a 3D simulation the new temperature of each voxel $T_{i,j,k}^{n+1}$ is obtained from its temperature in the previous time-step $T_{i,j,k}^{n}$ and the previous temperatures of its six neighbors $T_{i\pm1,j,k}^{n}$, $T_{i,j\pm1,k}^{n}$, $T_{i,j,k\pm1}^{n}$. Approximately 50 floating point (FP) multiplications or additions are needed at each time-step. Regarding the bio-heat equation (11.47), two additional evaluations of exponential functions are needed and about 10 additional FP operations. Because of the internal parallelism built into the processing unit, we reckon that in each time-step the computation time for each voxel is about $60t_{FP}$.

For the geometric knee model, at the resolution used in our simulation, we have $N = N_x \times N_y \times N_z$ voxels in a box-like domain being simulated on P parallel processors. If a balanced load distribution is supposed, then the parallel computation time is

$$t_{cp} = 60 N t_{FP}/P. \tag{11.51}$$

The simulation was parallelized using ld domain decomposition in order to preserve simplicity and also to enable effective execution on all clusters that can embed at least the ring topology. Each processor accepts the decomposed data of its subdomain and runs a simulation on them, for example, on a $\lceil N_z/P \rceil$ number of consecutive slices. If P cannot divide N_z then the last processor will run only the remaining slices, which will produce a small load imbalance and prolongation of the execution time.

In order to implement explicit FDM calculations at every point of the subdomain, including the first and last slice of the subdomain, the values of T from neighboring slices are needed. Edge slices of subdomains have their neighboring slices on neighboring processors. Because T is updated at every time-step, point-to-point communication is required with the neighboring processors that exchange $N_x \times N_y$ values of T from the subdomain's edge slices. Assuming that the computation and communication of any time-step will not run concurrently and that all data to be communicated are packed into a single message, then the communication time of each time-step is equal to

$$t_{cm} = t_s + (N_x \times N_y) t_w. \tag{11.52}$$

With $N_x \times N_y = N^{2/3}$, the parallel execution time is

$$t_{ex} = 60 N t_{FP}/P + t_s + N^{2/3} t_w. \tag{11.53}$$

In calculating the theoretical speedup, we suppose that one FP number is coded by 40 bits. The typical parameter values in the above equation are $t_w = 40 \cdot 7 \cdot 10^{-8}\,\text{s} = 2.8 \cdot 10^{-6}\,\text{s}$, $t_s = 10^{-4}\,\text{s}$, $t_{FP} = 10^{-8}\,\text{s}$, and $N = 5 \cdot 10^6$ because half of the voxels is the ambient air at constant temperature that are not included in the calculation. According to Eq. (11.51), the expected single processor execution time for a single time-step is $60 \cdot 5 \cdot 10^6 \cdot 10^{-8} = 3\,\text{s}$, which results by $\Delta t_b/4 = 0.039\,\text{s}$ in 77 hours of computation for every simulated hour of tissue temperatures.

The theoretical speedup of our parallel simulation program can be written as

$$S = 60 N t_{FP}/t_{ex} \tag{11.54}$$

because no communication is needed on a single processor. The theoretical speedup of the parallel simulation code for ld domain decomposition and ring topology is plotted in Fig. 11.23 as a function of the number of processors P. The ideal speedup is also shown for comparison.

The theoretical efficiency of our parallel program is $E = S/P$ and, in an ideal case, is at most unity. E is shown on the same graph as speedup in Fig. 11.23, but its value is multiplied by a factor of 40 for better visibility. We see that the efficiency is again a decreasing function of the number of processors, because of the prevailing communication time. Therefore, it is very important to think about the minimization of the t_{cm}, either by parallel algorithm design, by interconnection topology, or by advanced transmission technology.

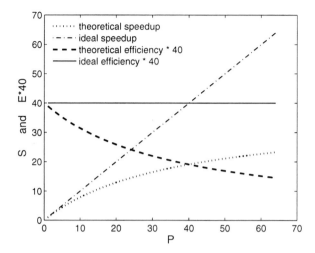

Fig. 11.23 Theoretical speedup and efficiency of the parallel simulation code for ld domain decomposition and ring topology. Ideal speedup, and efficiency, multiplied by a factor of 40, are also shown.

Because the communication time does not decrease with larger P, and due to unbalanced computation, we cannot expect any significant speedup above a certain number of processors. It is quite complicated to balance the computation load for general domains, because solid as well as air voxels would have to be evenly distributed among processors. The computational complexity for air voxels is approximately 18 times greater than for solid voxels. Automatic load balancing was not implemented in this stage, therefore an even smaller speedup is expected, as shown in Fig. 11.23.

We neglected the global communication which is needed at the beginning of the simulation for the implementation of the domain mapping on the parallel computing nodes. Additionally, the simulated temperatures have to be transferred, either on the host node or on a common disk, for later visualization and analysis, because we did not implement parallel programs for these two tasks. The global communication can be a bottleneck, particularly if the computing nodes are connected just through a common data switch.

On a parallel computer with a greater number of processors connected in 2D or 3D network topologies, with 4 or 6 neighboring nodes, respectively, the natural domain decomposition would also be in 2D or 3D. In this case, more communication channels work in parallel, which decreases the communication time and increases overall performance. However, with only a moderate number of processors the benefit of such a complicated communication scheme is minimal.

11.11.3 Measured Speedup

The parallel simulation program was run on a computing cluster comprising 16 computing nodes connected in a toroidal four-neighbors network. Each node contains two 64-bit processors (AMD Opteron 244), 1024 MB RAM, and six Gigabit Ethernet ports (2 × Broadcom BCM5704C + 4 × Intel Pro/1000 MT). Besides the four neighbors, all nodes are also directly connected to a Gigabit switch (Level One GSW-2451T).

The software used includes Fedora Core 2 Linux with a kernel 2.6.8-1.521smp, LAM/MPI v3.0.7. communications library [65], and gcc 3.3.3. The communication speed within MPI was measured preliminarily, then optimized [22] to achieve the fastest communication rates. The parallel cluster topology used in our simulation is shown in Fig. 11.24.

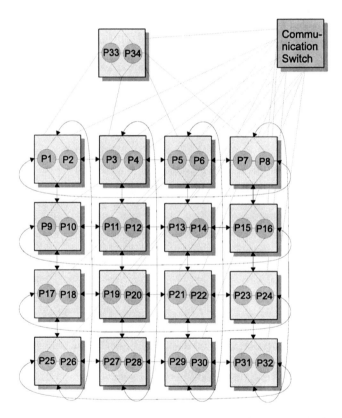

Fig. 11.24 Parallel computing cluster with 16 two-processor nodes in a toroidal four-neighbors network. Each node is also connected to the communication switch.

The parallel execution time was measured with the parallel program on the 3D geometric model and for different numbers of processors. The execution time of a single processor for simulating two cooling hours with the gel pack was 114 hours, that is, 57 times slower than real time, which is slightly less as expected, probably because of overestimating the number of FP operations. The speedup of the parallel code is shown in Fig. 11.25. An almost linear speedup is achieved with up to 8 processors, after which the effect of increasing communication time starts to be visible. The measured speedup with 16 and 32 processors was 10.9 and 16.2, respectively. The simulation on 32 processors remains 3.5 times slower than the real time. If greater simulation errors are acceptable, the simulation time can be shortened for a factor of four on the account of larger time-step (see Eq. (11.50) and explanation in Sect. 11.5).

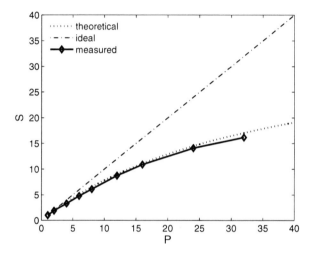

Fig. 11.25 Measured speedup of a parallel 3D simulation on a computing cluster. Theoretical and ideal speedup are also shown for comparison.

As expected, the speedup is close to linear for a number of processors less than 8, and departs significantly from linearity as the number of processors is increased. As noted, this behavior is a consequence of the increasing proportion of time spent for communication, and of load unbalance. The first problem could be partially alleviated by using 2D or 3D domain decomposition. The second problem could also be solved to some extent by a more detailed analysis of computational domain decomposition, which should distribute appropriate proportions of air voxels (shorter time-step) and the remaining tissue voxels among processors.

11.12 Conclusions

A method for computer simulation of heat transfer in inhomogeneous tissues is described. An explicit finite difference scheme has been developed and optimized for this purpose. A general method for the development of 3D models, based on digitalized slice data, is described and applied for generating a spatial geometric model of a knee. Heat transfer in the nonhomogeneous knee tissue was modeled with the bio-heat equation. We modeled and simulated the diffusive heat transfer, the heat exchange from blood to tissue and the heat production by tissue metabolism. The effects of blood flow and metabolism on the distribution of temperature in the knee are functions of the local tissue properties and temperatures. A corresponding mathematical model and a computer simulation program have been implemented.

We have confirmed the stability of the method by varying thermal parameters, initial and boundary values, and model dimensions. We demonstrated a practical application of the simulation program on the topical postoperative cooling of a knee. The results have been validated by experimental measurements of knee temperatures. The simulation results confirm that the model and methodology used are appropriate for the thermal simulation of bio-tissues. Computation time has been studied on a parallel computer for different numbers of processors, in order to achieve execution times comparable to those in real time. Similar procedures could be used for other parts of the human body, such as heart, forearm, eye, brain, etc., using an analogous approach.

We have shown that blood flow has a significant and complex impact on stationary-state temperatures, and on the gradient of temperature change in subcutaneous tissue and in the tissues nearer the central part of the knee. Lower blood flow results in linear temperature profiles with larger gradients, for example, in bones. The dimensions of the knee are a very important factor influencing its temperature distribution and gradient. Temperatures in smaller knees will differ more from arterial blood temperature than those in larger knees. At the same time the temperature gradient will be much greater in smaller knees. Thermal constants and metabolism have a relatively minor impact on the temperature field.

We show that fluctuations in temperature profiles are natural results and a consequence of the complex interplay between the positions of the measuring probes, the dimensions of the investigated body part, the anatomical positions of the main arteries and bones and the environmental temperature. The fluctuations and augmented peaks in the temperature profiles can be explained by our simulation results, since the temperature changes in the center of the knee are influenced by the nearest artery with constant blood temperature. The steepest, almost linear, gradients in the area of subcutaneous tissues are observed in the outer parts of the simulated temperature profiles, particularly in cases with lower cooling temperatures. The temperature plateaus, measured earlier by other investigators, in the inner knee region with increased dimensions of the knee were also demonstrated by our simulation.

We have assumed that blood flow and metabolism change as exponential functions of the temperature of the neighboring tissues. No firm justification exists for this except for some measurements of average blood flow in the human forearm.

In [35], blood flow was measured for two hours and remained practically constant if skin temperatures were lower than that of the arterial blood. Therefore the local blood flow depends on the local tissue temperature in the same way as the average blood flow depends on the bath temperature. Such an assumption needs to be supported by further physiological research. However, an arbitrary function for blood flow and metabolism could be incorporated in the simulation tool.

We simulated topical cooling of a knee with a gel pack. The inner knee tissues reached their lowest temperature in 40 minutes. For continuing effective cooling the gel pack has to be replaced. The topical cooling with a cryo-cuff was more effective. We simulated situations with relatively small cooling rates in order to be able to analyze the small influences of the blood flow and metabolism. The thickness of a protective bandage and isolating blanket, or their thermal conductivity, together with the cooling temperature, can be used to regulate cooling intensity.

Simulated results were visualized and compared with measured values. Good agreement was obtained, leading to the conclusion that the model and method used in our simulation are appropriate for such medical simulations. Although there are not many studies of knee temperatures measured in vivo, and the measuring conditions are often not described in sufficient detail, we have also run our simulation software with the initial conditions described in some published measurements. In [27] the temperature in the lateral gutter of the knee decreased by $4\,°C$, one hour after knee arthroscopy. Similar values have been obtained by our simulation program. More detailed control studies should be done to compare experimentally measured and simulated results in order to fine-tune the simulation method. Computer simulations could help in the study of various medical applications. It has been proved that the technique described can be used to predict the temperature distribution inside a knee, at any point and time of interest, for particular cooling methods.

For one hour of 3D simulation with a model resolution of 1 mm, the simulation program runs on a single processor for approximately 57 hours, which is impracticable. The parallel execution time exhibited a speedup of 10.9 on 16 computers and 16.2 on 32 computers. The 3D simulation time on 32 processors is comparable with the real time if a moderate accuracy suffices. We have shown that a 2D simulation with a constant temperature flux in the axial knee direction led to results that did not differ significantly from those obtained by 3D simulation. Such a simulation can be achieved two orders of magnitude faster and is appropriate also for simulation on a single computer, at least for the initial analysis.

The method described has several limitations. The 3D knee model used in our simulation is not complete, as only a small part of the leg above and below the knee was included. The remaining part of the leg that was not included was compensated for by a constant flux in boundary conditions. Spatial models differ with different persons and with time, and consequently the simulated results can differ. Minor errors in tissue segmentation and inaccurate thermodynamic constants could also produce small errors in the simulated results. Moving air was compensated by an artificial thin layer of air with nonconstant temperature. Incorporation of a fluid-flow model would be needed for even more accurate results.

The possible influence of blood perfusion by different regulatory mechanisms has not been simulated. Personal regulatory mechanisms have not been included in the simulation model but could easily be incorporated. All such limitations could have some impact on the simulated temperatures, but the essential findings are in remarkably good agreement with those from experimental in vivo measurements. The spatial knee model and simulation program are available from the author for research purposes and for the continuation of the work.

Acknowledgments The author acknowledges many colleagues for their previous cooperation or contributions, either by discussion, measurements or testing of heat transfer in biological tissues, in particular Prof. Dr. Borut Geršak, Dr. Marjan Šterk, and Prof. Dr. Matjaž Veselko. We acknowledge financial support from the state budget by the Slovenian Research Agency under grant P2-0095.

References

1. R. L. Martino, C. A. Johnson, E. B. Suh, et al., Parallel computing in biomedical-research, Science 265 (1994) 902–908.
2. R. Rook, S. Dost, The use of smoothed particle hydrodynamics for simulating crystal growth from solution, Int J Eng Sci 45 (2007) 75–93.
3. Y. Aggarwal, B. M. Karan, B. N. Das, R. K. Sinha, Computer simulation of heat transfer in different tissue layers of body extremities under heat stress in deep anesthetic condition, J Med Syst 32 (2008) 283–90.
4. M. Šterk, R. Trobec, Biomedical simulation of heat transfer in a human heart, J Chem Inf Mod 45 (2005) 1558–1563.
5. T. Liszka, J. Orkisz, The finite difference method at arbitrary irregular grids and its application in applied mechanics, Comput Struct 11 (1980) 83–95.
6. K. T. Danielson, R. A. Uras, M. D. Adley, S. Li, Large-scale application of some modern CSM methodologies by parallel computation, Adv Eng Softw 31 (2000) 501–509.
7. C. Hirsch, Numerical Computation of Internal and External Flows: Fundamentals of Computational Fluid Dynamics, Butterworth-Heinemann (2007).
8. A. Lipej, Optimization method for the design of axial hydraulic turbines, Proc Inst Mech Eng A – J Power Energy 218 (2004) 43–50.
9. A. Horvat, M. Leskovar, B. Mavko, Comparison of heat transfer conditions in tube bundle cross-flow for different tube shapes, Int J Heat Mass Transfer 49 (2007) 1027–1038.
10. L. F. Richardson, Weather Prediction by Numerical Process, Dover Publications, New York (1965).
11. B. Urban, D. Janežič, Symplectic molecular dynamics simulations on specially designed parallel computers, J Chem Inf Modell 45 (2005) 1600–1604.
12. P. Bernardi, M. Cavagnaro, S. Pisa, E. Piuzzi, Specific absorption rate and temperature elevation in a subject exposed in the far-field of radio-frequency sources operating in the 10-900-MHz range, IEEE Trans Biomed Eng 50 (2003) 295–304.
13. M. Depolli, V. Avbelj, R. Trobec, Computer-simulated alternative models of U-wave genesis, J Cardiovasc Electrophysiol 19 (2008) 84–89.
14. R. Trobec, B. Slivnik, B. Gersak, T. Gabrijelčič, Computer simulation and spatial modelling in heart surgery, Comput Biol Med 28 (1998) 393–403.
15. M. J. Ackerman, The visible human project, Proc IEEE 86 (1998) 504–511.
16. O. C. Zienkiewicz, R. L. Taylor, J. Z. Zhu, The Finite Element Method: Its Basis and Fundamentals, Elsevier Butterworth-Heinemann (2005).

17. M. Šterk, R. Trobec, Meshless solution of a diffusion equation with parameter optimization and error analysis, Eng Anal Bound Elem 32 (2007) 567–577.
18. V. Nguyen, T. Rabczuk, S. Bordas, M. Duflot, Meshless methods: a review and computer implementation aspects, Math Comput Simul 79 (2008) 763–813.
19. M. T. Heath, Scientific Computing: An Introductory Survey, 2nd Ed., McGraw-Hill (2002).
20. S. G. Akl, Parallel Computation: Models and Methods, Prentice Hall, New Jersey (1997).
21. A. Sulistio, U. Čibej, S. Venugopal, B. Robič, R. Buyya, A toolkit for modelling and simulating data Grids: an extension to GridSim, Concurr Comput Pract Exp 20 (2008) 1591–1609.
22. I. Rozman, M. Šterk, J. Močnik, B. Robič, R. Trobec, Performance measurements of computing networks, Scalable Comput Pract Exp 9 (2008) 143Ű–150.
23. A. A. C. Braga, Technical aspects of beowulf cluster construction, Quimica Nova 26 (2003) 401–406.
24. U. Borštnik, M. Hodošček, D. Janežič, Improving the performance of molecular dynamics simulations on parallel clusters, J Chem Inf Comput Sci 44 (2004) 359–364.
25. R. Trobec, U. Borštnik, D. Janežič, Communication performance of d-meshes in molecular dynamics simulation, J Math Chem DOI 10.1007/s10910-008-9423-2.
26. K. L. Knight, Cryotherapy: Theory, Technique and Physiology, Chatanooga Corporation, Chattanooga (1985).
27. S. S. Martin, K. P. Spindler, J. W. Tarter, K. Detwiler, H. A. Petersen, Cryotherapy: an effective modality for decreasing intraarticular temperature after knee arthroscopy, Am J Sports Med 29 (2001) 288–291.
28. S. S. Martin, K. P. Spindler, J. W. Tarter, K. Detwiler, H. A. Petersen, Accelerated rehabilitation after anterior cruciate ligament reconstruction, Am J Sports Med 18 (1990) 292–299.
29. W. Grana, Cold modalities, in: J. C. DeLee and D. Drez (Eds.), Orthopaedic Sports Medicine, Principles and Practice, WB Saunders, Philadelphia (1994).
30. W. C. McMaster, S. Liddle, T. R. Waugh, Laboratory evaluation of various cold therapy modalities, Am J Sports Med 6 (1978) 291–294.
31. D. H.Silverthorn, Human Physiology, An Integrated Approach, Prentice-Hall, New Jersey (2001).
32. H. H. Pennes, Analysis of tissue and arterial blood temperature in the resting human forearm, J Appl Physiol 1 (1948) 93–122.
33. H. F. Bowman, E. G. Cravalho, M. Woods, Theory, measurement, and application of thermal properties of biomaterials, Annu Rev Biophys Bioeng 4 (1975) 43–80.
34. C. K. Charny, Mathematical models of bioheat transfer, in: Y.I. Cho (Ed.), Advances in Heat Transfer, Academic Press, New York (1992).
35. H. Barcroft, O. G. Edholm, Temperature and blood flow in the human forearm, J Physiol 104 (1946) 366–376.
36. M. B. Ducharme, W. P. VanHelder, M. W. Radomski, Tissue temperature profile in the human forearm during thermal stress at thermal stability, J Appl Physiol 71 (1991) 1973–1978.
37. E. H. Wissler, Pennes' 1948 paper revisited, J Appl Physiol 85 (1998) 35–41.
38. R. Trobec, M. Šterk, S. AlMawed, M. Veselko, Computer simulation of topical knee cooling, Comput Biol Med 38 (2008) 1076–1083.
39. S. Karthik, A. D. Grayson, A. Y. Oo, et al., A survey of current myocardial protection practices during coronary artery bypass grafting, Ann Roy Coll Surg 86 (2004) 413–415.
40. C. L. Olin, I. E. Huljebrant, Topical cooling of the heart – a valuable adjunct to cold cardioplegia, Scand J Thorac Card 41 (1993) 55–58.
41. P. Trunk, B. Gersak, R. Trobec, Topical cardiac cooling – computer simulation of myocardial temperature changes, Comput Biol Med 33 (2003) 203–214.
42. G. B. Pollard, Lectures on Partial Differential Equations, Wiley, New York (1964).
43. M. N. Özisik, Finite Difference Methods in Heat Transfer, CRC Press, Boca Raton (1994).
44. S. J. Owen, A survey of unstructured mesh generation technology, in: Proceedings of 7th International Meshing Roundtable, Sandia National Laboratories (1998), pp. 239–267.
45. T. Rabczuk, S. Bordas, G. Zi, A three-dimensional meshfree method for continuous crack initiation, nucleation and propagation in statics and dynamics, Comput Mech 40 (3) (2007) 473–495.

46. M. Thuné, Straightforward partitioning of composite grids for explicit difference methods, Parallel Comput 17 (1991) 665–672.
47. H. S. Carslaw, J. C. Jaeger, Conduction of Heat in Solids, Oxford University Press, London (1959).
48. P. Trunk, R. Trobec, B. Gersak, Measurement of porcine heart temperatures, Pflügers Arch 440 (2000) R132–R133.
49. G. Golub, J. M. Ortega, Scientific Computing – An Introduction with Parallel Computing, Academic Press Inc., Boston (1993).
50. M. Vajteršic, Algorithms for Elliptic Problems, Efficient Sequential and Parallel Solvers, Kluwer Academic Publishers (1993).
51. H. A. van der Vorst, BI-CGSTAB: A fast and smoothly converging variant of BI-CG for the solution of nonsymmetric linear systems, SIAM J Sci Stat Comput 13 (1992) 631–644.
52. R. Barrett, M. Berry, T. F. Chan, et al., Templates for the Solution of Linear Systems: Building Blocks for Iterative Methods, SIAM, Philadelphia (1994).
53. M. Šterk, R. Trobec, Parallel performance of a multigrid poisson solver, in: Proceedings of Second International Symposium on Parallel and Distributed Computing, IEEE Computer Soc (2003), pp. 238–243.
54. C. Shen, J. Zhang, Parallel two level block ILU preconditioning techniques for solving large sparse linear systems, Parallel Comput 28 (2002) 1451–1475.
55. R. Trobec, M. Šterk, B. Robič, Computational complexity and parallelization of the meshless local Petrov-Galerkin method, Comput Struct 87 (2009) 81–90.
56. J. W. Mitchell, G. E. Myers, An analytical model of the counter-current heat exchange phenomena, Biophys J 8 (1968) 897–911.
57. W. H. Press, B. P. Flannery, S. A. Teukolsky, W. T. Vetterling, Numerical Recipes, The Art of Scientific Computing, Cambridge University Press, Cambridge (1986).
58. M. P. Allen, D. J. Tildesley, Computer Simulation of Liquids, Clarendon Press, Oxford (1987).
59. D. Janežič, M. Praprotnik, Molecular dynamics integration time step dependence of the split integration symplectic method on system density, J Chem Inf Comput Sci 43 (2003) 1922–1927.
60. I. Kuščer, A. Kodre, Mathematik in Physik und Technik, Springer Verlag, Berlin (1993).
61. M. Praprotnik, M. Šterk, R. Trobec, Inhomogeneous heat-conduction problems solved by a new explicit finite difference scheme, Int J Pure Appl Math 13 (2004) 275–291.
62. M. B. Ducharme, P. Tikuisis, In vivo thermal conductivity of the human forearm tissues, J Appl Physiol 70 (1991) 2682–2690.
63. P. Tikuisis, M. B. Ducharme, Finite-element solution of thermal conductivity of muscle during cold water immersion, J Appl Physiol 70 (1991) 2673–2681.
64. The visible human project, United States National Library of Medicine, http://www.nlm.nih.gov/research/visible/getting_data.html.
65. M. Snir, S. Otto, S. Huss-Lederman, D. Walker, J. Dongarra, MPI – The Complete Reference, The MIT Press, Cambridge (1996).

Chapter 12
Parallel SVD Computing in the Latent Semantic Indexing Applications for Data Retrieval

Gabriel Okša and Marián Vajteršic

Abstract

One of the main sources of information in our society is a written word. Since times of Sumerians, a written document became the main tool to inform, to teach, to entertain and to archive the knowledge. Today, some 6000 years after Sumerians, nothing has changed with respect to the importance of a written text. To become widely available, the knowledge must be manipulated in an easy and reliable way, and some type of text encoding on a computer is needed.

The Latent Semantic Indexing (LSI) is a concept-based automatic indexing method for overcoming the two fundamental problems which exist in the traditional lexical-matching retrieval schemes: synonymy and polysemy. It is based on the modeling of a term – document relationship using the reduced-dimension representation of a term-document matrix computed by its partial Singular Value Decomposition (SVD). We describe main principles of the LSI in the form of a mathematical model and discuss its implementation on a parallel computer with distributed memory.

12.1 Introduction

Latent Semantic Indexing (LSI) is a concept-based automatic indexing method for overcoming the two fundamental problems which exist in the traditional lexical-matching retrieval schemes: synonymy and polysemy [1]. With respect to the

Gabriel Okša
Mathematical Institute, Department of Informatics, Slovak Academy of Sciences, Bratislava, Slovakia, e-mail: Gabriel.Oksa@savba.sk.

Marián Vajteršic
Department of Computer Sciences, University of Salzburg, Austria, and Mathematical Institute, Department of Informatics, Slovak Academy of Sciences, Bratislava, Slovakia, e-mail: marian@cosy.sbg.ac.at

R. Trobec et al. (eds.), *Parallel Computing*, DOI 10.1007/978-1-84882-409-6_12,

synonymy, several different words can be used to express a concept and the key-words in a user's query may not match those in the relevant documents. On the other hand, polysemy means that certain words can have multiple meanings and the user's words may match those in the irrelevant documents. LSI is an extension of the vector space model for information retrieval [1, 2]. In the vector space model, the collection of text documents is represented by a *term-document* matrix $A = (a_{ij}) \in \mathbb{R}^{m \times n}$, where a_{ij} is based on the number of times the term i appears in the document j, m is the number of terms, and n is the number of documents in the collection. Hence, a document becomes a column vector, and a user's query can also be represented as a vector of the same dimension. The similarity between a query vector and a document vector is usually measured by the cosine of the angle between them, and for each query a list of documents ranked in a decreasing order of similarity is returned to the user.

LSI modifies this vector space model by modeling the term – document relationship using a *reduced-dimension representation* (RDR) of term-document matrix A computed by its singular value decomposition (SVD). Let

$$A = P \Sigma Q^T, \quad \Sigma = \text{diag}(\sigma_1, \sigma_2, \ldots, \sigma_{\min\{m,n\}}), \quad \sigma_1 \geq \sigma_2 \geq \ldots \geq \sigma_{\min\{m,n\}},$$

be the SVD of A. Then the RDR is given by the best rank-k approximation $A_k = P_k \Sigma_k Q_k^T$, $k < \min\{m,n\}$, where P_k and Q_k consist of the first k columns of P and Q, respectively, and Σ_k is the kth leading principal submatrix of Σ. Each of the k reduced dimensions represents a so-called *pseudo-concept* [2], which may not have any explicit semantic content but helps to discriminate documents [2,3].

Typically, a term-document matrix A is a large, sparse matrix with, say, $O(10^6)$ rows and/or columns. Its RDR is of order, say, $O(10^1 - 10^2)$, so that only a relatively small part of largest singular triplets are needed for representing a given, static, homogeneous database of documents. Hence, a computational method used for the reduced SVD of a large, sparse matrix is usually iterative. One can choose, for example, the Arnoldi method [4], subspace iteration [5,6] or trace minimization [7]. Parallel versions of these algorithms are discussed in detail in [8].

In rapidly changing environments such as the World Wide Web, the document collection is frequently updated with new documents and terms constantly being added (or deleted). Hence, the task arises to efficiently update the old LSI-generated RDR after an addition of new documents and terms. As shown in next sections, such updating (or downdating) requires the SVD computation of structured matrices that are upper or lower triangular and dense, so that the sparsity of an original term-document matrix is lost. Although the above mentioned SVD methods can also be used in these cases, they are not so efficient anymore. Therefore, other parallel SVD algorithms can be used.

We present some new ideas with respect to the 'old' Jacobi SVD method that are able to accelerate the computation. The emphasis is on the block approach with an efficient use of the hierarchy of memory modules in modern (parallel) computers via BLAS-3 matrix operations, on the new type of parallel ordering (so-called *dynamic ordering* for the parallel two-sided block-Jacobi SVD algorithm), and on a possible preservation of the matrix triangular structure (the parallel block-Kogbetliantz variant of the Jacobi SVD method).

The chapter is organized as follows. In Sect. 12.2, the mathematical model of updating is briefly presented, which is based on algorithms derived in [9]. It turns out that the computationally most intensive task in the correct updating is the SVD computation of some upper or lower triangular matrix. The same is true for two downdating problems (i.e., deleting the documents and/or terms) that are discussed in Sect. 12.3. In Sect. 12.4 we design a parallel SVD algorithm for solving these updating/downdating problems, which is based on the Kogbetliantz method with a special cyclic ordering that preserves the (block) triangular structure of a matrix. In Sect. 12.5, the parallel two-sided block-Jacobi method with dynamic ordering is introduced. Using the dynamic ordering of subproblems leads to much faster convergence, but the (block) triangular structure of a matrix is lost after first parallel iteration.

In the second part of the chapter, two models for the encoding of a distributed database are analyzed using the paradigm of the LSI. The first model comprises a grid of individual nodes that possess only a restricted inter-communication, so that only local level of approximation for individual nodes is possible. In the second model, the individual processors are connected by a communication network that enables to build, besides the local approximation, also the global one with respect to the whole database.

12.2 Two Updating Problems in LSI

12.2.1 Updating Documents

Let us suppose that the RDR of order k was already computed and stored for some term-document matrix A, and the original matrix was discarded (e.g., for the memory reasons), so that only $A_k = P_k \Sigma_k Q_k^T$ is available in the factored form. Let $D \in \mathbb{R}^{m \times r}$ be r new documents. The task is to compute the best rank-k approximation of the column partitioned matrix

$$B \equiv (A_k, D) \ .$$

Using the factorization of A_k, the matrix B can be written as

$$B = \left(P_k \Sigma_k Q_k^T, D \right)$$
$$= \left(P_k, (I_m - P_k P_k^T) D \right) \cdot \begin{pmatrix} \Sigma_k & P_k^T D \\ 0 & I_r \end{pmatrix} \cdot \begin{pmatrix} Q_k^T & 0 \\ 0 & I_r \end{pmatrix} \ .$$

Note that $I_m - P_k P_k^T$ is the matrix representation of the orthogonal projection, which maps the columns of matrix D into the subspace $\mathscr{P}_k^\perp$ that is orthogonal to the column range of matrix P_k. Let $(I_m - P_k P_k^T) D = \hat{P}_r R$ be the QR decomposition of the matrix $(I_m - P_k P_k^T)D$. Then

$$B = (P_k, \hat{P}_r) \cdot \begin{pmatrix} \Sigma_k & P_k^T D \\ 0 & R \end{pmatrix} \cdot \begin{pmatrix} Q_k^T & 0 \\ 0 & I_r \end{pmatrix}. \tag{12.1}$$

The crucial point in the above derivation is the observation that the r-orthonormal columns of matrix $\hat{P}_r$ are mutually orthogonal to the k-orthonormal columns of matrix P_k because the columns of $\hat{P}_r$ constitute the orthonormal basis of the subspace $\mathscr{P}_k^\perp$. Note that two exterior matrices on the right hand side of Eq. (12.1) are orthogonal, but the inner matrix is not diagonal. Hence, from the computational point of view, the updating problem is reduced to the SVD of the inner matrix in Eq. (12.1).

Based on these facts, Zha and Simon [9] have derived a method for solving the problem of updating documents. Their approach is summarized in Algorithm 12.1. Notice that step 4 in Algorithm 12.1 requires the SVD of structured matrix $\hat{B}$, which

---

Algorithm 12.1 Algorithm for updating documents

1: *Input:* k, $P_k \in \mathbb{R}^{m \times k}$, $\Sigma_k \in \mathbb{R}^{k \times k}$, $Q_k \in \mathbb{R}^{n \times k}$, $D \in \mathbb{R}^{m \times r}$.
2: Compute the projection: $\hat{D} = (I_m - P_k P_k^T)D$.
3: Compute the QR decomposition: $\hat{D} = \hat{P}_r R$, where $\hat{P}_r \in \mathbb{R}^{m \times r}$, $R \in \mathbb{R}^{r \times r}$.
4: Compute the SVD of matrix

$$\hat{B} \equiv \begin{pmatrix} \Sigma_k & P_k^T D \\ 0 & R \end{pmatrix} \in \mathbb{R}^{(k+r) \times (k+r)}$$

    in the form:

$$\hat{B} = (U_k, U_k^\perp) \cdot \mathrm{diag}(\hat{\Sigma}_k, \hat{\Sigma}_r) \cdot (V_k, V_k^\perp)^T,$$

    where $U_k, V_k \in \mathbb{R}^{(k+r) \times k}$ and $\hat{\Sigma}_k \in \mathbb{R}^{k \times k}$.
5: *Output:* The best rank-k approximation of $B = (A_k, D)$ is given by:

$$B_k \equiv \left[(P_k, \hat{P}_r) U_k \right] \cdot \hat{\Sigma}_k \cdot \left[\begin{pmatrix} Q_k & 0 \\ 0 & I_r \end{pmatrix} V_k \right]^T.$$

---

is upper triangular with the diagonal left upper block of order $k \times k$. At the same time, this step represents the most intensive computation in Algorithm 12.1.

12.2.2 Updating Terms

In this case, let $T \in \mathbb{R}^{q \times n}$ be the q new term vectors that should be added to the existing terms at the bottom of the old term-document matrix. The task is to compute the best rank-k approximation of the row partitioned matrix

$$C \equiv \begin{pmatrix} A_k \\ T \end{pmatrix}.$$

Using steps similar to those in the previous paragraph (see [9]), one gets the Algorithm 12.2 for the correct updating of terms. Similarly to the problem of updating

---

Algorithm 12.2 Algorithm for updating terms

1: *Input: k, $P_k \in \mathbb{R}^{m \times k}$, $\Sigma_k \in \mathbb{R}^{k \times k}$, $Q_k \in \mathbb{R}^{n \times k}$, $T \in \mathbb{R}^{q \times n}$.*
2: Compute the projection: $\hat{T} = (I_n - Q_k Q_k^T) T^T \in \mathbb{R}^{n \times q}$.
3: Compute the QR decomposition: $\hat{T} = \hat{Q}_q L^T$, where $\hat{Q}_q \in \mathbb{R}^{n \times q}$, $L \in \mathbb{R}^{q \times q}$.
4: Compute the SVD of matrix

$$\hat{C} \equiv \begin{pmatrix} \Sigma_k & 0 \\ TQ_k & L \end{pmatrix} \in \mathbb{R}^{(k+q) \times (k+q)}$$

in the form:

$$\hat{C} = (U_k, U_k^\perp) \cdot \text{diag}(\hat{\Sigma}_k, \hat{\Sigma}_q) \cdot (V_k, V_k^\perp)^T,$$

where $U_k, V_k \in \mathbb{R}^{(k+q) \times k}$ and $\hat{\Sigma}_k \in \mathbb{R}^{k \times k}$.

5: *Output:* The best rank-k approximation of $C = \begin{pmatrix} A_k \\ T \end{pmatrix}$ is given by:

$$C_k \equiv \left[\begin{pmatrix} P_k & 0 \\ 0 & I_q \end{pmatrix} U_k \right] \cdot \hat{\Sigma}_k \cdot \left[(Q_k, \hat{Q}_q) V_k \right]^T.$$

---

documents, the computationally most intensive step is the SVD of the lower triangular matrix $\hat{C}$ with the upper left diagonal block. Since the upper and lower triangular matrices are related by the matrix transposition that affects the SVD only by interchanging the left and right singular vectors, in the following we focus on the upper triangular matrix $\hat{B}$ in Algorithm 12.1. The conclusions with respect to the efficiency of the SVD computation will be valid for both updating problems.

12.3 Two Downdating Problems in LSI

In downdating problems, there exists the k-dimensional approximation of the original term-document matrix A in the form $A_k = P_k \Sigma_k Q_k^T$. As above we assume that only factors P_k, Σ_k and Q_k are available. In contrast with updating problems, our task is now either to delete r documents, i.e., the matrix D of order $m \times r$ from the representation $A_k = [D, \tilde{A}_k]$, or to delete q terms, i.e., the matrix T of order $q \times n$ from the representation $A_k = \begin{pmatrix} T \\ \hat{A}_k \end{pmatrix}$. Our new database is represented by matrix $\tilde{A}_k$ or $\hat{A}_k$, and we must end with the k-dimensional SVD representation of a reduced matrix in either case. Next we describe effective serial algorithms published in [10], which solve both downdating problems. Since both algorithms are very similar, we will describe in detail only the algorithm for downdating the documents and comment on differences when downdating the terms.

12.3.1 Downdating Documents

Let us start with the representation $A_k = [D, \tilde{A}_k] = P_k \Sigma_k Q_k^T$. Let the matrix $I_n^{1:r}$ denote the first r rows of the identity matrix of order n (we assume $n > r$, which is natural – not all n documents are deleted from a database). Let us define the matrix $\hat{Q}$ as an $n \times (k+r)$ orthogonal matrix of the form

$$\hat{Q} = (Q_k, S),$$

where S contains r orthogonal columns of length n, which are orthogonal also to the columns of Q_k – i.e., $Q_k^T S = 0$ (to be sure that such S exists, we assume that $k + r \leq n$). Then one can check by direct computation that the following decomposition is valid:

$$\begin{pmatrix} I_r & 0 \\ 0 & P_k^T \end{pmatrix} \cdot \begin{pmatrix} I_n^{1:r} \\ A_k \end{pmatrix} \cdot \hat{Q} = \begin{pmatrix} Q_k^{1:r} & S^{1:r} \\ \Sigma_k & 0 \end{pmatrix} \equiv W.$$

Thus the right-hand side matrix above, referred to as W, is composed of the first r rows of Q_k followed by the first r rows of S.

The key step now is to reduce W^T by orthogonal transformations into a special form, which will contain the identity I_r as the left upper diagonal block. Since Σ_k is diagonal, we can write

$$W^T = \begin{pmatrix} (Q_k^{1:r})^T & \Sigma_k \\ (S^{1:r})^T & 0 \end{pmatrix}.$$

Notice the special structure of W^T. First r columns are dense (in fact, these are the first r orthogonal rows of the orthogonal matrix $\hat{Q}$), but next k columns are very sparse, because Σ_k is diagonal and the bottom diagonal block is zero. It is this special structure which allows to use left and right Givens rotations in a so-called *non-zero chasing scheme* [11, pp. 145–149] to obtain:

$$G_L W^T G_R = G_L \begin{pmatrix} Q_k^{1:r} & S^{1:r} \\ \Sigma_k & 0 \end{pmatrix}^T G_R = \begin{pmatrix} I_r & 0 \\ Y & \tilde{B} \end{pmatrix}^T,$$

where $\tilde{B}$ is the lower triangular matrix. Here G_L and G_R are orthogonal matrices of order $(k+r)$ constructed as products of individual Givens rotations. It can be shown that G_R^T does not act on first r rows of the composed matrix. Therefore

$$G_R^T \begin{pmatrix} I_r & 0 \\ 0 & P_k^T \end{pmatrix} = \begin{pmatrix} I_r & 0 \\ 0 & \tilde{P}_k^T \end{pmatrix}.$$

However, G_L^T reduces exactly r first columns of $\hat{Q}$ to $(I_r, 0)^T$, so that

$$\hat{Q} G_L^T = \begin{pmatrix} I_r & 0 \\ 0 & \bar{Q}_k \end{pmatrix}$$

(since the columns of $\hat{Q}$ are orthogonal and remain so also after the orthogonal transformation, the block 12 must be zero). Then it follows that

$$\begin{pmatrix} I_r & 0 \\ 0 & \bar{P}_k^T \end{pmatrix} \cdot \begin{pmatrix} I_r & 0 \\ D & \tilde{A}_k \end{pmatrix} \cdot \begin{pmatrix} I_r & 0 \\ 0 & \bar{Q}_k \end{pmatrix} = \begin{pmatrix} I_r & 0 \\ Y & \tilde{B} \end{pmatrix},$$

and the second row yields the downdated $\tilde{A}_k$ given by

$$\bar{P}_k^T \tilde{A}_k \bar{Q}_k = \tilde{B},$$

where $\tilde{B}$ is the lower triangular, banded matrix of order k. If the full SVD of $\tilde{B}$ is

$$\tilde{B} = P_B \Sigma_B Q_B^T,$$

then the SVD of $\tilde{A}_k$ is given by

$$\tilde{A}_k = (\bar{P}_k P_B) \cdot \Sigma_B \cdot (\bar{Q}_k Q_B)^T \equiv \tilde{P}_k \tilde{\Sigma}_k \tilde{Q}_k^T,$$

with $\tilde{P}_k = \bar{P}_k P_B$, $\tilde{\Sigma}_k = \Sigma_B$ and $\tilde{Q}_k^T = (\bar{Q}_k Q_B)^T$.

All steps required for deleting a block of documents are summarized in Algorithm 12.3. The most computationally demanding task in Algorithm 12.3 is the SVD

Algorithm 12.3 Algorithm for downdating documents

1: *Input:* k, $P_k \in \mathbb{R}^{m \times k}$, $\Sigma_k \in \mathbb{R}^{k \times k}$, $Q_k \in \mathbb{R}^{n \times k}$, $D \in \mathbb{R}^{m \times r}$.
2: Complete Q_k into the orthonormal matrix $\hat{Q} = (Q_k, S)$ of order $n \times (k + r)$ by taking r random vectors and orthogonalizing them by the modified Gram-Schmidt process.
3: Form the matrix $W = \begin{pmatrix} Q_k^{1:r} & S^{1:r} \\ \Sigma_k & 0 \end{pmatrix}$ and find orthogonal matrices G_L and G_R so that

$$G_L W^T G_R = G_L \begin{pmatrix} Q_k^{1:r} & S^{1:r} \\ \Sigma_k & 0 \end{pmatrix}^T G_R = \begin{pmatrix} I_r & 0 \\ Y & \tilde{B} \end{pmatrix}^T,$$

where $\tilde{B}$ is the lower triangular matrix of order k.
4: Compute $\bar{P}_k$ and $\bar{Q}_k$ by:

$$G_R^T \begin{pmatrix} I_r & 0 \\ 0 & P_k^T \end{pmatrix} = \begin{pmatrix} I_r & 0 \\ 0 & \bar{P}_k^T \end{pmatrix}, \quad \hat{Q} G_L^T = \begin{pmatrix} I_r & 0 \\ 0 & \bar{Q}_k \end{pmatrix}.$$

5: Compute the SVD of $\tilde{B}$, $\tilde{B} = P_B \Sigma_B Q_B^T$. All matrices are square of order k.
6: *Output:* The best rank-k approximation of $\tilde{A}_k$ is given by $\tilde{P}_k = \bar{P}_k P_B$, $\tilde{\Sigma}_k = \Sigma_B$ and $\tilde{Q}_k^T = (\bar{Q}_k Q_B)^T$.

---

of the lower triangular matrix $\tilde{B}$ in step 5.

12.3.2 Downdating Terms

This case is indeed very similar to the above one for deleting documents. Let $A_k = \begin{pmatrix} T \\ \tilde{A}_k \end{pmatrix} = P_k \Sigma_k Q_k^T$ be our original database of order $m \times n$, from which q terms should be removed. These terms are placed on the top and are present in all documents, so that their influence is defined by the matrix T of order $q \times n$. Notice that the matrix A_k^T has the structure identical to the case of deleting documents. Hence, to delete a block of terms, one can work with the representation of A_k^T and use the algorithm from previous subsection.

However, similar steps as in the previous subsection, applied directly to the SVD factors of A_k, lead to Algorithm 12.4. Again, the most computationally expensive

---

Algorithm 12.4 Algorithm for downdating terms

1: *Input:* $k, P_k \in \mathbb{R}^{m \times k}, \Sigma_k \in \mathbb{R}^{k \times k}, Q_k \in \mathbb{R}^{n \times k}, T \in \mathbb{R}^{q \times n}$.
2: Complete P_k into the orthonormal matrix $\hat{P} = (P_k, Z)$ of order $m \times (k+q)$ by taking q random vectors and orthogonalizing them by the modified Gram-Schmidt process.
3: Form the matrix $H = \begin{pmatrix} P_{1:q,k}^T & \Sigma_k \\ Z_{1:q}^T & 0 \end{pmatrix}$ and find orthogonal matrices G_L and G_R so that

$$G_L H G_R = \begin{pmatrix} I_q & Y \\ 0 & \tilde{B} \end{pmatrix},$$

where $\tilde{B}$ is the upper triangular matrix of order k. Here $P_{1:q,k}^T$ denotes first q columns of P_k^T; similarly for $Z_{1:q}^T$.
4: Compute $\bar{P}_k$ and $\bar{Q}_k$ by:

$$G_L \hat{P}^T = \begin{pmatrix} I_q & 0 \\ 0 & \bar{P}_k^T \end{pmatrix}, \quad \begin{pmatrix} I_q & 0 \\ 0 & Q_k \end{pmatrix} G_R = \begin{pmatrix} I_q & 0 \\ 0 & \bar{Q}_k \end{pmatrix}.$$

5: Compute the SVD of $\tilde{B}$, $\tilde{B} = P_B \Sigma_B Q_B^T$. All matrices are square of order k.
6: *Output:* The best rank-k approximation of $\tilde{A}_k$ is given by $\tilde{P}_k = \bar{P}_k P_B$, $\tilde{\Sigma}_k = \Sigma_B$ and $\tilde{Q}_k^T = (\bar{Q}_k Q_B)^T$.

---

task is the SVD of the upper triangular matrix $\tilde{B}$.

When comparing together Algorithms 12.1, 12.2, 12.3 and 12.4, the computational pattern is similar – each task requires the SVD of a lower or upper triangular matrix, which can have some interesting additional structure. Next we describe the parallel Kogbetliantz variant of the Jacobi method for doing this.

12.4 Kogbetliantz Method for Triangular Matrices

Special form of the Jacobi method for obtaining the SVD of (upper or lower) triangular matrices was proposed by Kogbetliantz; see [12, 13]. However, in his original proposal the method was used for the solution of a system of linear equations, where the coefficient matrix was first transformed to a triangular form by the QR decomposition; then the R-factor was diagonalized by the two-sided unitary (orthogonal, in real case) transformations. However, today his method is mainly used for the SVD computation of triangular matrices.

From the numerical point of view, the Kogbetliantz algorithm is relatively stable [14], i.e., the tiniest singular values are computed with high relative accuracy. This property is similar to the one-sided Jacobi method. The convergence criterion can be checked without any extra cost, whereas the one-sided Jacobi method requires approximately $n^2/2$ dot products to do this. However, the main weakness of the Kogbetliantz method is its need to update *both* matrix columns and rows, which means twice as many matrix multiplications as compared with a one-sided method.

We start with the serial approach and describe a special, so-called *butterfly* form of a triangular matrix. The second 'brick' of the method is the modulus pivot strategy, which essentially preserves the butterfly form during the whole iterative process. Although the scalar algorithm which deals with individual elements of a matrix can be parallelized, better efficiency is achieved when working with matrix blocks because the BLAS-3 algorithms for matrix multiplication can be used. Therefore, the last subsection describes the approach when a matrix is divided into blocks and possible parallelization of the Kogbetliantz algorithm is also discussed.

12.4.1 Butterfly Form of Triangular Matrices

For $n = 6$ and $n = 7$, the butterfly form of a square matrix A of order n has the following form:

$$A = \begin{pmatrix} x\,0\,0\,0\,0\,0 \\ x\,x\,0\,0\,0\,x \\ x\,x\,x\,0\,x\,x \\ x\,x\,x\,x\,x\,x \\ x\,x\,0\,0\,x\,x \\ x\,0\,0\,0\,0\,x \end{pmatrix} \quad \text{and} \quad \begin{pmatrix} x\,0\,0\,0\,0\,0\,0 \\ x\,x\,0\,0\,0\,0\,x \\ x\,x\,x\,0\,0\,x\,x \\ x\,x\,x\,x\,x\,x\,x \\ x\,x\,x\,0\,x\,x\,x \\ x\,x\,0\,0\,0\,x\,x \\ x\,0\,0\,0\,0\,0\,x \end{pmatrix}.$$

It is shown in [15] that each dense square (even rectangular) matrix A can be reduced to the butterfly form by a series of Householder reflections and Givens rotations applied from the left.

In our updating/downdating problems, a matrix under interest is upper or lower triangular. Let us concentrate to the upper triangular form (the lower triangular case is similar and by transposition can be brought to the upper triangular one). If T is a general, upper triangular matrix, then an example in [15] shows how T can be transformed into B, which is in the butterfly form, using a very cheap similarity transformation by a permutation matrix: $B = P^T T P$. The permutation matrix P is composed of the product of simple transposition matrices:

$$P = \begin{cases} I_{12}I_{13}(I_{14}I_{23})(I_{15}I_{24})(I_{16}I_{25}I_{34}) \cdot (I_{1,n}I_{2,n-1}\cdots I_{k,k+1}) & \text{if } n = 2k, \\ I_{12}I_{13}(I_{14}I_{23})(I_{15}I_{24})(I_{16}I_{25}I_{34}) \cdot (I_{1,n}I_{2,n-1}\cdots I_{k,k+2}) & \text{if } n = 2k+1. \end{cases}$$

Here, $I_{pq} = (e_1,\ldots,e_q,\ldots,e_p,\ldots,e_n)$, $p < q$, is the transposition of columns p and q, where e_i is the ith column of the identity matrix I_n. The parentheses emphasize those transpositions that can be performed in parallel, because the corresponding pairs of indices are mutually disjunct. For example, for $n = 6$ we have $k = n/2 = 3$, and the transformation can be depicted as follows:

$$\begin{pmatrix} x \star x\,x\,x\,x \\ 0\,x\,x\,x\,x\,x \\ 0\,0\,x\,x\,x\,x \\ 0\,0\,0\,x\,x\,x \\ 0\,0\,0\,0\,x\,x \\ 0\,0\,0\,0\,0\,x \end{pmatrix} \mapsto \begin{pmatrix} x\,0 \star x\,x\,x \\ x\,x\,x\,x\,x\,x \\ 0\,0\,x\,x\,x\,x \\ 0\,0\,0\,x\,x\,x \\ 0\,0\,0\,0\,x\,x \\ 0\,0\,0\,0\,0\,x \end{pmatrix} \mapsto \begin{pmatrix} x\,0\,0 \star x\,x \\ x\,x \star x\,x\,x \\ x\,0\,x\,x\,x\,x \\ 0\,0\,0\,x\,x\,x \\ 0\,0\,0\,0\,x\,x \\ 0\,0\,0\,0\,0\,x \end{pmatrix}$$

$$\mapsto \begin{pmatrix} x\,0\,0\,0 \star x \\ x\,x\,0 \star x\,x \\ x\,0\,x\,x\,x\,x \\ x\,0\,0\,x\,x\,x \\ 0\,0\,0\,0\,x\,x \\ 0\,0\,0\,0\,0\,x \end{pmatrix} \mapsto \begin{pmatrix} x\,0\,0\,0\,0 \star \\ x\,x\,0\,0 \star x \\ x\,0\,x \star x\,x \\ x\,x\,0\,x\,x\,x \\ x\,0\,0\,0\,x\,x \\ 0\,0\,0\,0\,0\,x \end{pmatrix} \mapsto \begin{pmatrix} x\,0\,0\,0\,0\,0 \\ x\,x\,0\,0\,0\,x \\ x\,x\,x\,0\,x\,x \\ x\,x\,x\,x\,x\,x \\ x\,x\,0\,0\,x\,x \\ x\,0\,0\,0\,x\,x \end{pmatrix}.$$

The symbol $\star$ denotes the position of one *pivot* element, the subscripts of which define the rows and columns which are to be swapped.

It can be easily seen that the whole transformation can be performed in $n - 1$ parallel steps on $n/2$ processors by exchanging appropriate columns and rows. Note that when processors contain whole matrix columns, only matrix columns are sent/received, whereas the matrix rows can be exchanged locally in the processors. This is an example of a 'fine-grained' parallelism, because the number of processors increases linearly with the matrix order n, which is not feasible for very large n.

12.4.2 Modulus Pivot Strategy

In general, the *pivot strategy* is a fixed list containing the order in which the off-diagonal matrix elements of matrix A are nullified. For triangular matrices, the special, so-called *modulus* strategy was proposed in [16]. It is defined by the *modulus ordering* of the set $\mathscr{P}_n = \{(p,q) : 1 \le p < q \le n\}$ and is illustrated below for $n = 7$.

$$\begin{pmatrix} . & 6 & 9 & 11 & 14 & 16 & 19 \\ & . & 12 & 15 & 17 & 20 & 1 \\ & & . & 18 & 21 & 2 & 4 \\ & & & . & 3 & 5 & 7 \\ & & & & . & 8 & 10 \\ & & & & & . & 13 \\ & & & & & & . \end{pmatrix} \quad \begin{aligned} \mathscr{S}_1 &= \{(2,7),(3,6),(4,5)\} \\ \mathscr{S}_2 &= \{(3,7),(4,6),(1,2)\} \\ \mathscr{S}_3 &= \{(4,7),(5,6),(1,3)\} \\ \mathscr{S}_4 &= \{(5,7),(1,4),(2,3)\} \\ \mathscr{S}_5 &= \{(6,7),(1,5),(2,4)\} \\ \mathscr{S}_6 &= \{(1,6),(2,5),(3,4)\} \\ \mathscr{S}_7 &= \{(1,7),(2,6),(3,5)\} \end{aligned} \quad \begin{pmatrix} . & 2 & 3 & 4 & 5 & 6 & 7 \\ & . & 4 & 5 & 6 & 7 & 1 \\ & & . & 6 & 7 & 1 & 2 \\ & & & . & 1 & 2 & 3 \\ & & & & . & 3 & 4 \\ & & & & & . & 5 \\ & & & & & & . \end{pmatrix} .$$

The leftmost matrix represents the ordering in which the pivot elements are annihilated within one sweep. By $\mathscr{S}_t$, $1 \leq t \leq 7$, we denote the so-called *rotation sets* containing index pairs of matrix elements which can be annihilated *simultaneously* because all index pairs are mutually disjoint (or commuting). Finally, the rightmost matrix depicts the pivot positions according to rotation sets which can be zeroed in parallel.

Hence, the modulus pivoting for triangular matrices enables to introduce a parallel algorithm based on rotation sets. At parallel step t, the rotation set $\mathscr{S}_t$ determines which elements will be nullified. Since the Kogbetliantz method is iterative, the algorithm goes through a sequence

$$\mathscr{S}_1, \mathscr{S}_2, \ldots, \mathscr{S}_n, \mathscr{S}_1, \mathscr{S}_2, \ldots, \mathscr{S}_n, \ldots,$$

until convergence is achieved.

Let $\mathrm{Piv}(t)$ denote the *pivot set* that is currently used as a rotation set. We start with $B^{[1]} = B$, and at the beginning of time step t, $t \geq 1$, all rotation matrices (i.e., all rotation angles) $U_{ij}^{[t]}, V_{ij}^{[t]}, (i,j) \in \mathrm{Piv}(t)$ are computed using the elements of the same matrix $B^{[t]}$. Then the transformation

$$B^{[t+1]} = U^{[t]^T} B^{[t]} V^{[t]}, \quad U^{[t]} = \prod_{(i,j) \in \mathrm{Piv}(t)} U_{ij}^{[t]}, \quad V^{[t]} = \prod_{(i,j) \in \mathrm{Piv}(t)} V_{ij}^{[t]} \tag{12.2}$$

is performed. Here $U^{[t]}$ and $V^{[t]}$ are not computed explicitly; only all $V_{ij}^{[t]}, (i,j) \in \mathrm{Piv}(t)$ are applied simultaneously, and afterwards the same is done with $U_{ij}^{[t]}, (i,j) \in \mathrm{Piv}(t)$. If the right and/or left singular vectors are needed, then the right transformation $V^{[t]}$ can be accumulated into the orthogonal matrix V during iterations, and then U can be computed a posteriori from the equation $BV = U\Sigma$. Alternatively, one can accumulate $U^{[t]}$ into U and then compute V a posteriori.

The advantages of using the butterfly form together with the modulus pivot strategy in the Kogbetliantz method is discussed in detail in [15]. If B is in the butterfly norm then it is permutationally similar to the upper triangular matrix (B is PST). Therefore, it is also *essentially triangular* (ET) since it holds: $b_{pq} b_{qp} = 0$ for $p < q$. Moreover, it can be shown that if one starts with a triangular matrix in the butterfly form, then all matrices generated by the Kogbetliantz method using the modulus strategy are PST. In particular, when $B^{[t]}$ denotes the iterated matrix in the time step t with $t > n$ (n is the size of B), then $B^{[t]}$ and $B^{[t-n]}$ have zero structures which are transposed to each other. Each matrix $B^{[t]}$ is PST, therefore it is ET, and can be compactly stored in the upper triangle of square array. Hence, the upper triangular matrix $G^{[t]}$ can be defined by prescription

$$G^{[t]} + G^{[t]^T} = B^{[t]} + B^{[t]^T}.$$

Then the Kogbetliantz method with modulus strategy (KMMS) can be formulated in terms of matrices $G^{[t]}$ (see [15]). The result is a sequential KMMS algorithm which works with the upper triangular matrices and in each step applies approximately $n/2$ non-commuting rotations.

These rotations can be applied in parallel, but the disadvantage of such 'direct' parallelization of the KMMS is its low efficiency. We need approximately $n/2$ processors to exploit fully the inherent parallelism of the algorithm, which is certainly not efficient for large n. The parallelization strategy, which uses the number of processors as a (linear) function of the matrix size, belongs to a 'fine-grained' approach and can be very inefficient with respect to the cost of inter-processor communication for large n. Much better way is to work with matrix blocks, whereby the size of a block is given by the size of the matrix divided by a given number of processors. Next we describe a parallelization approach for the block KMMS.

12.4.3 Block Version and Parallelism

When working with matrix blocks, the numerical algorithms become much more efficient in general, because the memory hierarchy of modern computers can be used. However, the size of matrix blocks should be tuned according to the size of fast cache memory of a processor. Ideally, the whole matrix block should fit into the cache so that no additional calls for data will be made when working with that matrix block. In this way the algorithm can use the advantage of the so-called *BLAS-3 matrix multiplications* which are very fast. They are implemented also in modern linear algebra libraries, for example, LAPACK and ScaLAPACK. Moreover, working with matrix blocks leads to the 'coarse-grained' parallelism, in which the number of processors is given beforehand and can be quite small.

We start with the block upper triangular matrix T of order n in the form

$$T = \begin{pmatrix} T_{11} & T_{12} & \cdots & T_{1m} \\ 0 & T_{22} & \cdots & T_{2m} \\ \vdots & \vdots & \ddots & \vdots \\ 0 & 0 & \cdots & T_{mm} \end{pmatrix}.$$

Each diagonal block T_{ii} is of order $n_i \geq 1$, so that $\mathscr{M} = \{n_1, n_2, \ldots, n_m\}$ is the partition of n. We can assume $n_1 = n_2 = \cdots = n_m = n/m$.

To reduce T to the *block-butterfly form*, we can use permutations similar to the scalar case. However, I_{ij} is now a product of simple transpositions. The effect of $I_{ij}^T T$

is to swap block rows i and j of T; similarly, $T I_{ij}$ means swapping the appropriate block columns.

12.4.3.1 Parallel Step Zero

Let $B^{[0]} = B$ be a matrix in the block-butterfly form. Before starting the iteration process, some matrix *preprocessing* is needed which is called the *parallel step zero*. It can be described as follows:

$$B^{[1]} = U^{[0]^T} B^{[0]} V^{[0]}, \quad U^{[0]} = \prod_{(i,j)\in \text{piv}(m)} U_{ij}^{[0]}, \quad V^{[0]} = \prod_{(i,j)\in \text{piv}(m)} V_{ij}^{[0]},$$

where $\text{piv}(m) = \{(1,m), (2,m-1), \ldots, (m/2, m/2+1)\}$ is the mth *pivot set* associated with the block algorithm (notice that it is defined with respect to the block index).

The result of this zero step can be summarized as follows:

- The matrix blocks $B_{1,m}, B_{2,m-1}, \ldots B_{m/2,m/2+1}$ on the upper half of the main block anti-diagonal are nullified. Recall that B is in the block-butterfly form so that the lower part of the main block anti-diagonal is zero.
- All diagonal blocks B_{ii} are diagonalized by computing their SVDs. For this, any numerically reliable serial SVD algorithm can be used.

Moreover, $B^{[1]}$ and all subsequent iteration matrices $B^{[2]}, B^{[3]}, \ldots$ are represented in the *factored form*

$$B^{[t]} = E^{[t]^T} C^{[t]} F^{[t]},$$

where $E^{[t]}$ and $F^{[t]}$ are block diagonal and orthogonal. The main idea behind the factorization is to work with relatively small matrix blocks in updating matrix iterates, so that all matrix multiplications can be done in the fast cache memory by calling the appropriate data only once [14]. In addition, the diagonal elements of the current iterate $B^{[t]}$ are kept separately in the vector $\gamma^{[t]}$.

Consequently, the parallel step zero must provide the formulae for computing $E^{[1]}, C^{[1]}, F^{[1]}$ and $\gamma^{[1]}$. Let $B^{[0]} = (B_1^{[0]}, B_2^{[0]}, \ldots, B_m^{[0]})$ be the block column partition of $B^{[0]}$. Then the algorithmic description of the parallel step zero is depicted below as Algorithm 12.5.

Several remarks are in order to better understand the various tricks 'behind the scene':

1. The trick with the factorization of $B^{[0]}$ into the product of three matrices, $B^{[t]} = E^{[t]^T} C^{[t]} F^{[t]}$, is taken from [17]. The main idea here is to arrive at small enough matrices with good numerical properties (e.g., orthogonality) which can be handled in the cache memory of a processor. It is well known that the cache memory is up to $6-8$ times faster than the main memory. Therefore, even when the number of flops using the three-term recursion is larger than the direct approach to updating, the exclusive use of the cache memory can overcome this shortage with respect to the time complexity of the whole algorithm.

2. Consequently, all matrix multiplications in subsequent step for updating B, B' and $\bar{B}'$ are made in the cache memory – hence, they are fast.

3. The cosine–sine (CS) decomposition of an orthogonal matrix has a special structure and special properties; see [17]. It is still an open question how to compute it in a numerically reliable way for this class of matrices. Another interesting problem is its efficient parallelization.

4. The logical variable `left` controls the set of singular vectors which are computed during iterations. Only one set of singular vectors is computed in step zero (and in the iterative process below). The other set is computed a posteriori after finishing the process by solving the linear systems of equations

$$BV = U\,\Sigma \quad \text{or} \quad B^T U = V\,\Sigma.$$

This approach almost halves the number of matrix multiplications in each iteration step as compared to the iterative computation of both sets of singular vectors.

Algorithm 12.5 Algorithm for the parallel step zero

1: **for** $i = 1$ to $m/2$ in parallel **do**

2:     Set $j = m + 1 - i$.

3:     Compute the SVD:

$$\begin{pmatrix} B_{ii}^{[0]} & B_{ij}^{[0]} \\ 0 & B_{jj}^{[0]} \end{pmatrix} = \mathbf{U}_{ij}^{[0]}\, \mathbf{G}_i\, \mathbf{V}_{ij}^{[0]T}.$$

4:     Compute the CS decomposition of $\mathbf{U}_{ij}^{[0]}$ and $\mathbf{V}_{ij}^{[0]}$:

$$\mathbf{U}_{ij}^{[0]} = \begin{pmatrix} \acute{U}_{ii} & 0 \\ 0 & \acute{U}_{jj} \end{pmatrix} \Theta_{ij} \begin{pmatrix} \grave{U}_{ii} & 0 \\ 0 & \grave{U}_{jj} \end{pmatrix}, \ \mathbf{V}_{ij}^{[0]} = \begin{pmatrix} \acute{V}_{ii} & 0 \\ 0 & \acute{V}_{jj} \end{pmatrix} \Phi_{ij} \begin{pmatrix} \grave{V}_{ii} & 0 \\ 0 & \grave{V}_{jj} \end{pmatrix}.$$

5:     Apply: $B_i' = B_i \grave{V}_{ii},\, B_{ij}' = B_j \grave{V}_{jj}$.

6:     Apply: $(B_i'', B_j'') = (B_i', B_j')\, \Phi_{ij}$.

7:     Transpose: $\bar{B} = (B'')^T$ and let $\bar{B} = (\bar{B}_1, \bar{B}_2, \ldots, \bar{B}_m)$ be the block column partition of $\bar{B}$.

8:     Apply: $\bar{B}_i' = \bar{B}_i \grave{U}_{ii},\, \bar{B}_j' = \bar{B}_j \grave{U}_{jj}$.

9:     Apply: $(\bar{B}_i'', \bar{B}_j'') = (\bar{B}_i', \bar{B}_j')\, \Theta_{ij}$.

10:    Transpose: $C^{[1]} = (\bar{B}'')^T$.

11:    Copy: $E_{ii}^{[1]} = \acute{U}_{ii},\, E_{jj}^{[1]} = \acute{U}_{jj},\, F_{ii}^{[1]} = \acute{V}_{ii},\, F_{jj}^{[1]} = \acute{V}_{jj}$.

12:    Copy the first n_i and last n_j diagonal elements of $\mathbf{G}_i$ into the appropriate parts of the vector $\gamma^{[1]}$.

13:    **if** (`left`) **then**

14:        $U^{[1]} = E^{[1]T}$

15:    **else**

16:        $V^{[1]} = F^{[1]T}$

17:    **end if**

18: **end for**

12.4.3.2 Iterative Process

Recall that after the parallel step zero the matrix $B^{[1]}$ is in the block-butterfly form. In the iterative process, the block-modulus pivot strategy is applied in each parallel step until convergence. Thus, the algorithm in the parallel step t proceeds by annihilating the off-diagonal pivot submatrices $B_{ij}^{[t]}$, $(i,j) \in \text{piv}(t)$ and by diagonalizing the diagonal blocks $B_{ii}^{[t]}$, $B_{jj}^{[t]}$. Hence, at the beginning of the parallel step $t+1$ the new Frobenius off-norm is given by

$$\|\Omega(B^{[t+1]})\|^2 = \|\Omega(B^{[t]})\|^2 - \sum_{(i,j)\in\text{piv}(t)} \|B_{ij}^{[t]}\|^2.$$

Since $B^{[t]}$ is kept in the factored form of a matrix triple $E^{[t]}, C^{[t]}, F^{[t]}$, one has to derive the recursions for updating these matrices together with vector $\gamma^{[t]}$.

The main equation of the Kogbetliantz method is given by (12.2). For each pair $(i,j) \in \text{piv}(t)$ this orthogonal transformation can be written as

$$\mathbf{B}_{ij}^{[t]} = \begin{pmatrix} B_{ii}^{[t]} & B_{ij}^{[t]} \\ 0 & B_{jj}^{[t]} \end{pmatrix} = \mathbf{U}_{ij}^{[t]} \, \mathbf{G}_i \, \mathbf{V}_{ij}^{[t]^T}, \quad \mathbf{G}_i \text{ is diagonal.} \tag{12.3}$$

Notice that this equation is the SVD of $\mathbf{B}_{ij}^{[t]}$. We assume that the diagonal blocks $B_{ii}^{[t]}$ and $B_{jj}^{[t]}$ are diagonal matrices which is certainly true, by construction, for the initial matrix $B^{[1]}$. Here, $\mathbf{U}_{ij}^{[t]}$ and $\mathbf{V}_{ij}^{[t]}$ are orthogonal matrices of order $(n_i + n_j) \times (n_i + n_j)$; they are called *block rotations* in [17].

However, $B^{[t]}$ is given in its factored form, so that the upper-triangular matrix $\mathbf{B}_{ij}^{[t]}$ can be computed as follows:

$$\begin{pmatrix} B_{ii}^{[t]} & B_{ij}^{[t]} \\ 0 & B_{jj}^{[t]} \end{pmatrix} = \begin{pmatrix} E_{ii}^{[t]} & 0 \\ 0 & E_{jj}^{[t]} \end{pmatrix}^T \begin{pmatrix} C_{ii}^{[t]} & C_{ij}^{[t]} \\ 0 & C_{jj}^{[t]} \end{pmatrix} \begin{pmatrix} F_{ii}^{[t]} & 0 \\ 0 & F_{jj}^{[t]} \end{pmatrix}$$

$$= \begin{pmatrix} E_{ii}^{[t]^T} C_{ii}^{[t]} F_{ii}^{[t]} & E_{ii}^{[t]^T} C_{ij}^{[t]} F_{jj}^{[t]} \\ 0 & E_{jj}^{[t]^T} C_{jj}^{[t]} F_{jj}^{[t]} \end{pmatrix}.$$

Since $B_{ii}^{[t]}$ and $B_{jj}^{[t]}$ are diagonal, we can fill them by zeros and then copy appropriate diagonal elements from the vector $\gamma^{[t]}$ onto the diagonal of $B_{ii}^{[t]}$ and $B_{jj}^{[t]}$ (hence, the diagonal blocks of $\mathbf{B}_{ij}^{[t]}$ are *not* computed explicitly). After that we need to compute $E_{ii}^{[t]^T} C_{ij}^{[t]} F_{jj}^{[t]}$ on the processor which is associated with the pair $(i,j) \in \text{piv}(t)$ using the fast BLAS-3 LAPACK routine *GEMM.

Next, the SVD of $\mathbf{B}_{ij}^{[t]}$ is computed according to (12.3). Since $\mathbf{B}_{ij}^{[t]}$ is upper triangular with diagonal blocks being diagonal matrices, one can here choose among several fast and accurate serial methods, for example, the one-sided Jacobi or (cyclic or modulus) Kogbetliantz algorithm. This SVD is computed serially for one $\mathbf{B}_{ij}^{[t]}$, but, of course, $m/2$ processors compute in parallel for $m/2$ pairs of indices (i,j), $1 \le i < j \le n$.

The next step is the CS decomposition of orthogonal matrices $\mathbf{U}_{ij}^{[t]}$ and $\mathbf{V}_{ij}^{[t]}$, which can be written in the form (see [17]):

$$
\begin{aligned}
\mathbf{U}_{ij}^{[t]} &= \begin{pmatrix} \dot{U}_{ii}^{[t]} & 0 \\ 0 & \dot{U}_{jj}^{[t]} \end{pmatrix} \Theta_{ij}^{[t]} \begin{pmatrix} \ddot{U}_{ii}^{[t]} & 0 \\ 0 & \ddot{U}_{jj}^{[t]} \end{pmatrix}, \\
\mathbf{V}_{ij}^{[t]} &= \begin{pmatrix} \dot{V}_{ii}^{[t]} & 0 \\ 0 & \dot{V}_{jj}^{[t]} \end{pmatrix} \Phi_{ij}^{[t]} \begin{pmatrix} \ddot{V}_{ii}^{[t]} & 0 \\ 0 & \ddot{V}_{jj}^{[t]} \end{pmatrix}.
\end{aligned}
\tag{12.4}
$$

The matrices $\Theta_{ij}^{[t]}$ and $\Phi_{ij}^{[t]}$ are orthogonal products of at most $\min\{n_i, n_j\}$ commuting plane rotations (see [17]).

Now comes the parallel computation of the next iteration matrix, $B^{[t+1]}$, using all available pairs (i,j). It is computed as

$$
B^{[t+1]} = U^{[t]^T} (B^{[t]} V^{[t]}),
$$

where $U^{[t]}$ and $V^{[t]}$ are composed from all available matrices $\mathbf{U}_{ij}^{[t]}$ and $\mathbf{V}_{ij}^{[t]}$, respectively. However, $B^{[t+1]}$ is never computed explicitly. Recall that we have it in the factored form, so actually we need recursions how to compute $E^{[t+1]}$, $C^{[t+1]}$ and $F^{[t+1]}$. To this end, let us introduce the matrix $J_{ij} = (J_i, J_j)$ where $I_n = (J_1, J_2, \ldots, J_m)$ is the block-column partition of the identity. Then

$$
B^{[t]} J_{ij} = (B_i^{[t]}, B_j^{[t]}).
$$

and

$$
B^{[t+1]} \equiv E^{[t+1]^T} C^{[t+1]} F^{[t+1]} = U^{[t]^T} (E^{[t]^T} C^{[t]} F^{[t]}) V^{[t]}.
$$

Post-multiplying the last equality above by J_{ij} and writing the identity between C and F on both sides as $J_{ij}^T J_{ij}$, we obtain:

$$
\begin{aligned}
E^{[t+1]^T} &(C_i^{[t+1]}, C_j^{[t+1]}) \begin{pmatrix} F_{ii}^{[t+1]} & 0 \\ 0 & F_{jj}^{[t+1]} \end{pmatrix} \\
&= U^{[t]^T} E^{[t]^T} (C_i^{[t]}, C_j^{[t]}) \begin{pmatrix} F_{ii}^{[t]} & 0 \\ 0 & F_{jj}^{[t]} \end{pmatrix} \mathbf{V}_{ij}^{[t]} \\
&= U^{[t]^T} E^{[t]^T} (C_i^{[t]}, C_j^{[t]}) \left[\left[\begin{pmatrix} F_{ii}^{[t]} & 0 \\ 0 & F_{jj}^{[t]} \end{pmatrix} \begin{pmatrix} \dot{V}_{ii}^{[t]} & 0 \\ 0 & \dot{V}_{jj}^{[t]} \end{pmatrix} \right] \Phi_{ij}^{[t]} \right] \begin{pmatrix} \ddot{V}_{ii}^{[t]} & 0 \\ 0 & \ddot{V}_{jj}^{[t]} \end{pmatrix}.
\end{aligned}
$$

Hence, we have immediately the first set of updates:

$$F_{ii}^{[t+1]} = \dot{V}_{ii}^{[t]}, \quad F_{jj}^{[t+1]} = \dot{V}_{jj}^{[t]},$$

$$(\bar{C}_i^{[t]}, \bar{C}_j^{[t]}) = (C_i^{[t]}, C_j^{[t]}) \begin{pmatrix} F_{ii}^{[t]} \dot{V}_{ii}^{[t]} & 0 \\ 0 & F_{jj}^{[t]} \dot{V}_{jj}^{[t]} \end{pmatrix} \Phi_{ij}^{[t]}. \tag{12.5}$$

These updates can be performed in parallel for all $(i, j) \in \mathrm{piv}(t)$. This results in the matrix $F^{[t+1]}$ and auxiliary matrix $\bar{C}$.

The second set of updates starts with the equation

$$E^{[t+1]^T} C^{[t+1]} = U^{[t]^T} E^{[t]^T} \bar{C}^{[t]}.$$

After pre-multiplying it by J_{ij}^T ($J_i^T X$ is the ith block-row of X) and using the decomposition of identity $I = J_{ij} J_{ij}^T$ between E and C on both sides, we obtain:

$$\begin{pmatrix} E_{ii}^{[t+1]} & 0 \\ 0 & E_{jj}^{[t+1]} \end{pmatrix}^T \begin{pmatrix} J_i^T C^{[t+1]} \\ J_j^T C^{[t+1]} \end{pmatrix} = \mathbf{U}_{ij}^{[t]^T} \begin{pmatrix} E_{ii}^{[t]} & 0 \\ 0 & E_{jj}^{[t]} \end{pmatrix}^T \begin{pmatrix} J_i^T \bar{C}^{[t]} \\ J_j^T \bar{C}^{[t]} \end{pmatrix}$$

$$= \begin{pmatrix} \dot{U}_{ii}^{[t]} & 0 \\ 0 & \dot{U}_{jj}^{[t]} \end{pmatrix}^T \left[\Theta_{ij}^{[t]^T} \begin{pmatrix} \dot{U}_{ii}^{[t]} & 0 \\ 0 & \dot{U}_{jj}^{[t]} \end{pmatrix}^T \begin{pmatrix} E_{ii}^{[t]} & 0 \\ 0 & E_{jj}^{[t]} \end{pmatrix}^T \begin{pmatrix} J_i^T \bar{C}^{[t]} \\ J_j^T \bar{C}^{[t]} \end{pmatrix} \right],$$

and we have the second set of updates:

$$E_{ii}^{[t+1]} = \dot{U}_{ii}^{[t]}, \quad E_{jj}^{[t+1]} = \dot{U}_{jj}^{[t]},$$

$$(C_i^{[t+1]^T}, C_j^{[t+1]^T}) = (\bar{C}_i^{[t]^T}, \bar{C}_j^{[t]^T}) \begin{pmatrix} E_{ii}^{[t]} \dot{U}_{ii}^{[t]} & 0 \\ 0 & E_{jj}^{[t]} \dot{U}_{jj}^{[t]} \end{pmatrix} \Theta_{ij}^{[t]}. \tag{12.6}$$

Again, these updates can be performed in parallel for all $(i, j) \in \mathrm{piv}(t)$. Notice, that the auxiliary matrix $\bar{C}^{[t]}$, which was computed in (12.5), is to be transposed in (12.6). Then it is updated from the right hand side and $C^{[t+1]^T}$ is obtained. Finally, the diagonal elements of $\mathbf{G}_i$ are copied to the appropriate positions of the vector $\gamma^{[t+1]}$.

Equations (12.3), (12.4), (12.5) and (12.6) constitute one iterative step of the parallel block-Kogbetliantz algorithm. The parallelism is achieved by computing all updates for pivot indices $(i, j) \in \mathrm{piv}(t)$ simultaneously. This means that there are $p = m/2$ processors (recall that m is the blocking factor) whereby each processor works over two block columns of matrix data.

We write the iterative part in the form of Algorithm $\sim$ 12.6. The iteration index $[t]$ is omitted. The array C is $n \times n$, arrays E and F are $nb \times nb$ where $nb = \max_i\{n_i\}$. The block-column partition of C is given by $C = (C_1, \ldots, C_m)$. We denote $E_{ii}^{[t]}$ by E_i and similarly for F_i. The vector g is for $\gamma^{[t]}$, and several arrays U, B, V are square of size $2nb$. The matrices of left or right singular vectors are updated in arrays VECL or VECR, respectively, according to the logical variables left and right. Similarly

Algorithm 12.6 Algorithm for an iteration step

1: **for** $(i, j) \in \text{piv}(t)$ in parallel **do**
2:     Compute: $B_{12} = E_i^T C_{ij} F_j$.
3:     Copy the appropriate elements from g to $\text{diag}(B_{11})$ and $\text{diag}(B_{22})$.
4:     Form: $B = \begin{pmatrix} B_{11} & B_{12} \\ 0 & B_{22} \end{pmatrix}$ where diagonal blocks are diagonal matrices.
5:     Compute the SVD: $B = U \Gamma V^T$.
6:     Update: $g \leftarrow \Gamma$ (copy to appropriate positions).
7:     Compute the CS decompositions

$$U = \begin{pmatrix} U1 & 0 \\ 0 & U2 \end{pmatrix} H \begin{pmatrix} U3 & 0 \\ 0 & U4 \end{pmatrix}, \quad V = \begin{pmatrix} V1 & 0 \\ 0 & V2 \end{pmatrix} K \begin{pmatrix} V3 & 0 \\ 0 & V4 \end{pmatrix}.$$

8:     Compute: $X = F_i V1, \quad Y = F_j V2$.
9:     Update block columns of C: $C_i \leftarrow C_i X, \quad C_j \leftarrow C_j Y$.
10:    If (**right**) update: $\text{VECR}_i \leftarrow \text{VECR}_i X, \quad \text{VECR}_j \leftarrow \text{VECR}_j Y$.
11:    Update: $(C_i, C_j) \leftarrow (C_i, C_j) K$.
12:    If (**right**) update: $(\text{VECR}_i, \text{VECR}_j) \leftarrow (\text{VECR}_i, \text{VECR}_j) K$.
13:    Update: $F_i \leftarrow V3, \quad F_j \leftarrow V4$.
14:    Transpose: $C \leftarrow C^T$.
15:    Compute: $X = E_i U1, \quad Y = E_j U2$.
16:    Update block columns of C^T (i.e., block rows of C): $C_i \leftarrow C_i X, \quad C_j \leftarrow C_j Y$.
17:    If (**left**) update: $\text{VECL}_i \leftarrow \text{VECL}_i X, \quad \text{VECL}_j \leftarrow \text{VECL}_j Y$.
18:    Update: $(C_i, C_j) \leftarrow (C_i, C_j) H$.
19:    If (**left**) update: $(\text{VECL}_i, \text{VECL}_j) \leftarrow (\text{VECL}_i, \text{VECL}_j) H$.
20:    Update: $E_i \leftarrow U3, \quad E_j \leftarrow U4$.
21:    Transpose: $C \leftarrow C^T$ (back to the original form of C).
22: **end for**

to the parallel step zero, several comments are in order also for the iterative part of the algorithm:

1. The recursions have one important purpose – namely, to arrive at small enough matrices (or matrix blocks) which can be stored *at once* in the fast cache memory. These are the diagonal blocks of E, F and all factors of the CS decompositions. These all are square matrices of size nb, which is substantially smaller than the size n of the original upper triangular matrix provided that the blocking factor m is large enough. Perhaps more importantly, given n, the number of processors p and the blocking factor m ($p = m/2$) can be chosen in such way that all small blocks will indeed be stored in the cache memory at once. This means great time savings in computing updates by matrix multiplications.

2. The only matrix without any structure used in recursions is C. We see that C is updated in two steps, whereby the second update works with C^T. The reason is that using the transposition one can update *both* block columns and block rows by matrix multiplications *from the right*, whereby the updating matrices X and Y are small and should fit in the cache. Such updating will be very fast. If each processor contains two full block columns of C, then all updates can be computed *locally* in processors and there is no need to use the distributed matrix

multiplication. The price paid for this 'comfort' is the need of two transpositions of C which can be slow on distributed parallel architectures (e.g., on a cluster of PCs). In other words, one needs some fast, parallel (distributed) algorithm for the matrix transposition.

12.5 Parallel Two-sided Block-Jacobi SVD Algorithm with Dynamic Ordering

The modulus pivot strategy described above for the parallel version of the Kogbetliantz algorithm is an example of a so-called *static cyclic* ordering, in which there exists a prescribed list of matrix blocks that are nullified in each parallel iteration step. The SVD computation then consists of so-called *sweeps*; in one sweep, each off-diagonal matrix block is nullified exactly once. The value of the off-diagonal matrix norm is usually checked at the end of a sweep using the appropriate convergence criterion.

However, all proposed variations of the static ordering in the parallel two-sided block-Jacobi algorithm do not take into account the actual status of matrix A. Another possibility that was designed, implemented and tested in [18, 19], tries to maximize the decrease of the off-diagonal Frobenius norm in each parallel iteration step of the parallel two-sided block-Jacobi SVD algorithm. The idea is based on finding a set of the off-diagonal blocks $\{A_{ij}\}$, $i \neq j$, at the beginning of each parallel iteration step that, when zeroed, would decrease the off-diagonal Frobenius norm as much as possible and, at the same time, can be nullified concurrently. Having p processors and the blocking factor $\ell = 2p$ (so that matrix A is cut in ℓ block rows and ℓ block columns), such task is equivalent to the solution of the so-called *maximum-weight perfect matching* problem on a complete graph with ℓ vertices.

A *perfect matching* in an edge-weighted graph $G(\mathcal{V}, \mathcal{E})$, where $\mathcal{V}$ is the set of nodes and $\mathcal{E}$ is the set of edges, is a subset of edges such that each node of G is met by exactly one edge in the subset. If w is a real weighting function on the edges of G, then the weight of a matching is the sum of the weights of its edges. A matching is called a *maximum-weight matching* if its weight is the highest of all possible matchings. It is shown in [20] that the maximum-weight matching can be found in time $O(|\mathcal{V}| \cdot |\mathcal{E}| + |\mathcal{V}|^2 \cdot \log |\mathcal{V}|)$. The historical overview of problem solutions can be found in [21, 22].

Consider a weighted complete graph $G = (\mathcal{V}, \mathcal{E}) = K_\ell$, where the nodes are numbered from 0 to $\ell - 1$, $\mathcal{E} = \{(i, j) \mid i < j\}$ and the edge (i, j) has the weight $w_{ij} = \|A_{ij}\|_F^2 + \|A_{ji}\|_F^2$. The maximum-weight perfect matching of this graph can be computed in time $O(p^3)$ (recall that $\ell = 2p$). When an approximate solution is good enough, a simple greedy approach can be used. Let us sort the edges with respect to their weights in non-increasing order. Then let us scan this ordered sequence from left to right and add an edge to the matching only if neither of its endpoints is an endpoint of any previously selected edge. The complexity of this greedy algorithm is $O(p^2 \cdot \log p)$ due to the complexity of sorting. An example of perfect matching for $p = 3$, $\ell = 6$ is depicted on Fig. 12.1.

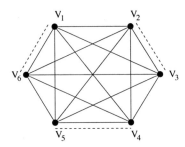

Fig. 12.1 An example of maximum-weight perfect matching for $p = 3$, $\ell = 6$. Edges $(1,6)$, $(2,3)$ and $(4,5)$ were chosen (*dashed*), which correspond to three parallel SVDs of three 2×2 block subproblems.

Next we describe basic constituents of the parallel two-sided block-Jacobi SVD algorithm with dynamic ordering; details can be found in [18]. The parallel algorithm for processor me, $me = 0, 1, \ldots, p - 1$, can be written in the form of Algorithm 12.7. For $\ell = 2p$, each processor contains exactly two block columns of dimensions $m \times n/\ell$ so that $\ell/2$ SVD subproblems of block size 2×2 are solved in parallel in each iteration step.

The procedure ReOrderingComp (Algorithm 12.7, step 6) computes the optimal reordering destinations of all block columns residing in a given processor (*dest*1 and *dest*2) and their locations at new position (*tag*1 and *tag*2). The reordering is based on the maximum-weight perfect matching that operates on the $\ell \times \ell$ updated weight matrix W using the elements of $W + W^T$, where $(W + W^T)_{ij} = \|A_{ij}\|_F^2 + \|A_{ji}\|_F^2$ [18]. The argument *tag* provides the matching between the corresponding send and receive calls.

The kernel operation is the SVD of 2×2 block subproblems

$$S_{ij} = \begin{pmatrix} A_{ii} & A_{ij} \\ A_{ji} & A_{jj} \end{pmatrix}, \tag{12.7}$$

where, for a given pair (i, j), $i, j = 0, 1, \ldots, \ell - 1$, $i \neq j$, the unitary matrices X_{ij} and Y_{ij} are generated such that the product

$$X_{ij}^H S_{ij} Y_{ij} = D_{ij}$$

is a block diagonal matrix of the form

$$D_{ij} = \begin{pmatrix} \hat{D}_{ii} & 0 \\ 0 & \hat{D}_{jj} \end{pmatrix},$$

where $\hat{D}_{ii}$ and $\hat{D}_{jj}$ are diagonal.

Algorithm 12.7 Parallel block-Jacobi SVD algorithm with dynamic ordering

1: $U = I_m$
2: $V = I_n$
3: $(i,j) = (2me, 2me + 1)$
4: **while** $F(A, \ell) \geq \varepsilon$ **do**
5:     update(W)
6:     ReOrderingComp$(i, j, W, me) \rightarrow dest1, dest2, tag1, tag2$
7:     copy$(A_i, U_i, V_i, i) \rightarrow A_r, U_r, V_r, r$
8:     copy$(A_j, U_j, V_j, j) \rightarrow A_s, U_s, V_s, s$
9:     send$(A_r, U_r, V_r, r, dest1, tag1)$
10:    send$(A_s, U_s, V_s, s, dest2, tag2)$
11:    receive$(A_i, U_i, V_i, i, 1)$
12:    receive$(A_j, U_j, V_j, j, 2)$
13:    **if** $F(S_{ij}, \ell) \geq \delta$ **then**
14:       ▷ *computation of X_{ij} and Y_{ij} by SVD of S_{ij}*
15:       SVD$(S_{ij}) \rightarrow X_{ij}, Y_{ij}$
16:       ▷ *update of block columns*
17:       $(A_i, A_j) = (A_i, A_j) \cdot Y_{ij}$
18:       $(U_i, U_j) = (U_i, U_j) \cdot X_{ij}$
19:       $(V_i, V_j) = (V_i, V_j) \cdot Y_{ij}$
20:    **else**
21:       $X_{ij} = I_{(m/p)}$
22:    **end if**
23:    AllGather$(X_{ij}, i, j) \rightarrow XX(t) = (X_{rs}, r, s), t = 0, 1, \ldots, p-1$
24:    ▷ *update of block rows*
25:    **for** $t = 0$ to $p - 1$ **do**
26:       $\begin{pmatrix} A_{ri} & A_{rj} \\ A_{si} & A_{sj} \end{pmatrix} = X_{rs,t}^H \cdot \begin{pmatrix} A_{ri} & A_{rj} \\ A_{si} & A_{sj} \end{pmatrix}$
27:    **end for**
28: **end while**

The termination criterion of the entire process is

$$F(A, \ell) = \sqrt{\sum_{i,j=0, i \neq j}^{l-1} \|A_{ij}\|_F^2} < \varepsilon, \tag{12.8}$$

where $\varepsilon = prec \cdot \|A\|_F$ is the required accuracy (measured relatively to the Frobenius norm of the original matrix A), and *prec* is a suitably chosen small constant, $0 < prec < 1$.

A subproblem (12.7) is solved only if

$$F(S_{ij}, \ell) = \sqrt{\|A_{ij}\|_F^2 + \|A_{ji}\|_F^2} \geq \delta, \tag{12.9}$$

where $\delta = 2\varepsilon / [\ell(\ell - 1)]$ is a given subproblem accuracy.

After the embedded SVD is computed (step 15), the matrices X_{ij} and Y_{ij} of local left and right singular vectors, respectively, are used for the local update of block columns (steps 16–22). In the procedure AllGather (step 23), each processor sends its matrix X_{ij} to all other processors, so that each processor maintains an array (denoted by XX) of p matrices. These matrices are needed in the orthogonal updates of block rows (steps 24–27).

From the implementation point of view, the embedded SVD is computed using the procedure *GESVD from the LAPACK library [23] while the matrix multiplications are performed by the procedure *GEMM from the BLAS (Basic Linear Algebra Subroutines). The point-to-point (steps 9–12) as well as collective (step 23) communications are realized by the MPI.

Numerical experiments with random full matrices of order from 2000 to 10000 show that the dynamic ordering needs in average about 40% less parallel iteration steps for convergence than the static cyclic method (cf. [18, 19]). When applied to updating/downdating problems in the LSI, the efficiency is sometimes even greater; however, the dynamic ordering does *not* preserve a triangular structure of matrices.

We now report and discuss results of numerical experiments performed with the parallel two-sided block-Jacobi algorithm with the dynamic ordering. The above parallel algorithm was implemented in Fortran on an SGI – Cray Origin 2000 parallel computer using the Message Passing Interface (MPI) library. The number of used processors covered the range $p = 2, 5, 10$ and 15. The constant $prec = 10^{-10}$ was chosen for the computation of ε and δ (see Eqs. (12.8) and (12.9)). All computations were made using the IEEE standard double precision floating point arithmetic with the machine precision $\varepsilon_M \approx 1.11 \times 10^{-16}$.

We have computed the SVD of matrix $\hat{B}$ in step 4 of Algorithm 12.1. Recall that $\hat{B}$ is the upper triangular, square matrix of order $v = k + r$ with the diagonal upper left block of order k. In practice, the parameter k depends on the text collection and covers the range from 100 to 300 (cf. [1–3, 9]). It is clear from the output of Algorithm 12.1 that only k largest singular triplets are needed for the construction of B_k so that some iterative method for the *partial* SVD can be considered in this case. On the other hand, the Jacobi SVD algorithm computes the *complete* SVD. However, when $k \gg r$ (or $k \gg q$) – i.e., the number of added documents (or added terms) is small as compared to k – and when the parallel computation is performed with a sufficient speedup, the usage of the parallel two-sided block-Jacobi SVD algorithm can be justified in the updating problems of LSI.

In our experiments, the order of the square upper triangular matrix $\hat{B}$ together with the size of its upper left diagonal block Σ_k were fixed: $v = k + r = 500$ and $k = 150$. The elements of $\hat{B}$ were generated randomly in two steps using two positive constants α and β. First, k values uniformly distributed in the interval $[0, 1]$ were obtained and multiplied by α; they constituted the diagonal of Σ_k. Next, the remaining elements of matrix blocks $P_k^T D$ and R uniformly distributed in the interval $[-1, 1]$ were generated and multiplied by β. The adopted approach enabled us to modify the ratio between the Frobenius norm of Σ_k and that of the rest of matrix $\hat{B}$. This is equivalent to the modeling of the relative weight that the new documents brought to the document collection. In our experiments, the value of $\alpha = 100$ was fixed and $\beta = 1, 2, 5, 10, 20$ and 50.

The experimental results are presented in the following tables. For a given number of processors p, the parallel computational time in seconds (first column) and the corresponding number of parallel iteration steps (second column) are shown for all values of parameter β mentioned above. In Table 12.1, the performance of the parallel algorithm is documented using the sweep technique with the static cyclic odd–even ordering ($CO(0)$, see [24, 25]). Table 12.2 contains the results of the dynamic ordering method. Let us discuss these results in more detail.

Table 12.1 Total parallel execution time in seconds and number of parallel iteration steps for the parallel SVD with the *static cyclic ordering CO(0)* using $v = 500$, $k = 150$, $\alpha = 100$, variable number of processors p and parameter β (see details in text).

β / p	2		5		10		15	
1	43.5	21	32.5	90	22.7	228	17.1	348
2	43.9	21	29.1	99	23.4	247	18.2	348
5	64.5	27	42.5	117	35.4	266	29.8	406
10	71.3	30	62.7	153	53.0	285	46.2	425
20	81.8	36	65.0	153	55.3	342	49.7	522
50	72.3	30	52.4	153	48.4	342	45.4	580

Table 12.2 Total parallel execution time in seconds and number of parallel iteration steps for the parallel SVD with the *greedy dynamic ordering* using $v = 500$, $k = 150$, $\alpha = 100$, variable number of processors p and parameter β (see details in text).

β / p	2		5		10		15	
1	22.4	10	17.1	47	13.5	107	10.0	173
2	24.2	10	17.4	48	14.7	111	11.4	183
5	27.2	11	21.2	49	15.2	112	12.1	187
10	29.7	11	20.8	51	17.5	115	14.9	195
20	29.7	12	19.9	53	18.1	121	15.0	198
50	29.4	12	22.1	55	20.3	126	17.7	211

For both orderings, the number of parallel iteration steps increases with an increase of the blocking factor $\ell = 2p$, i.e., with an increase of the number of processors p. This is in accordance with the statistical analysis given in [18] and documents the rather low scalability of the parallel two-sided block-Jacobi SVD algorithm. In other words, it has no sense to use a large number of processors for rather small matrices. This 'toy' example serves only for the comparison of two types of parallel ordering.

The number of parallel iterations steps needed for the convergence is an objective, machine-independent measure of the algorithm's performance. (Note that the computational time depends very much on the organizational details of computations adopted in a parallel computer, and the user has usually no direct access to influence these rules.) Comparing Tables 12.1 and 12.2, the greedy dynamic ordering clearly outperforms the cyclic one for all combinations of β and p. The ratio of the number of parallel iteration steps between the old method and the new one lies in the range 1.9–3.0 (average is 2.4). For a given number of processors p, this ratio has a tendency to grow with an increase of β, i.e., the greedy dynamic ordering is more efficient in reducing the relatively larger off-diagonal norms of matrix blocks than the cyclic ordering. This observation can be explained by the inherent property of the greedy dynamic ordering to pair the matrix blocks with the maximal sum of Frobenius norms.

Figure 12.2 depicts (in the logarithmic scale) the decrease of Frobenius norm of the off-diagonal blocks for $\beta = 50$ and $p = 10$. Similar behavior can be observed

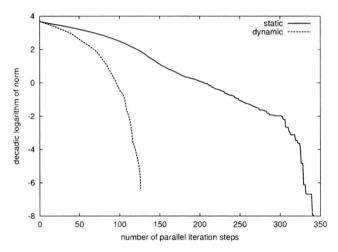

Fig. 12.2 Decrease of Frobenius norm of the off-diagonal blocks for $\beta = 50$ and $p = 10$.

also for other combinations of β and p. The different final norms for the dynamic and static cyclic ordering result from the fact that, in the case of cyclic ordering, the whole sweep must end before the convergence criterion is checked. Notice that for the static cyclic ordering there are many 'empty' parallel iteration steps that do not change the Frobenius norm of the off-diagonal blocks at all due to the pre-scribed combinations of non-diagonal blocks that do not fulfill the criterion given by Eq. (12.9). These steps correspond to the horizontal segments on the curve for the cyclic ordering in Fig. 12.2. In other words, the static cyclic ordering of sub-problems is, so to say, 'blind', because it does not take into account the actual status of the matrix, i.e., how the overall Frobenius norm is spread over the individual off-diagonal matrix blocks. Since the dynamic ordering combines the non-diagonal blocks with maximal Frobenius norms, no such effect is observed in this case, and the Frobenius norm of the off-diagonal blocks decreases strictly monotonically.

12.6 LSI Implemented on a Grid

We have shown that the updating/downdating problems in the LSI can be reduced to the computation of SVDs of upper or lower triangular matrices. For this purpose, the parallel block-Kogbetliantz algorithm was described and analyzed from the point of view of its implementation on a parallel distributed architecture. Another approach is the use of the two-sided block-Jacobi SVD method with the dynamic ordering, which is much more efficient than any prescribed cyclic ordering. However, the dynamic ordering does not preserve the triangular matrix structure.

Using the paradigm of the LSI, we now describe two models for the encoding of a distributed database. In the first model, the computational system is organized into a grid of individual nodes, which possess only a restricted inter-communication. The distributed database consists of mutually uncoupled local databases, which are updated for new documents/terms, and scanned independently for documents matching a given query. There is no easy way of comparing the accuracy of retrieval from individual nodes, because there exists no global approximation of the whole database. In the second model (which can describe, e.g., one node of a grid), the individual processors are connected by some sort of the communication network, so that the distributed computation and mutual communication are possible. We develop a two-stage model – the local and global levels of approximation – based on the LSI of documents for such a distributed system. Some interesting computational issues are discussed including the efficiency of a distributed SVD. Finally, it is possible to analyze the relationship between the local and global approximations with respect to the accuracy of retrieval of documents in this case.

Grid computing is based on a distributed computing model that provides the ability to perform high performance computing by using resources of many separate computers which are networked to a virtual computer architecture. We adopt for our model the term *computational grid* as defined in [26]. It is a hardware and software infrastructure that provides dependable, consistent, pervasive and inexpensive access to high-end computational capabilities.

Let us assume that the computational grid consists of p nodes. In context of the LSI, each node can store and maintain its individual database, which is represented by its own term-document matrix $A^{(i)}$, $i = 1, 2, \ldots, p$.

In the following two subsections we will describe the algorithms, which can be used for updating of individual databases and the retrieval of documents in individual nodes.

12.6.1 Storage

Let the index i denote one of the nodes of the grid under consideration, $1 \leq i \leq p$. This node first accumulates its term-document matrix $A^{(i)}$ of order $m \times n_i$, then computes its RDR of order k_i and, finally, stores the matrices P_{k_i}, Q_{k_i} and Σ_{k_i}. Although the individual nodes of the grid work independently, we assume that their data-term matrices $A^{(i)}$ describe information from the same (or very close) areas of interests, so that the terms are the same for all nodes – hence, the number of rows m is the same across the nodes.

12.6.2 Updating Documents

Since each node of the grid works independently, it can receive new documents from the outside world. Therefore, it must be capable to update its term-document matrix $A^{(i)}$. However, the original $A^{(i)}$ is not at our disposal anymore – all that is left is only its k_i-dimensional approximation in the factored form. Therefore, a natural question arises, how to compute a new approximation, which will incorporate a new information from new documents.

Let $D^{(i)} \in \mathbb{R}^{m \times r_i}$ be the r_i new documents vectors that should be added to the existing documents at the right-end of the old term-document matrix on node i. The algorithm for updating documents on node i is identical to the Algorithm 12.1, in which the index i is used to distinguish the node's number. Note that all nodes in a grid use the same algorithm. For the sake of clarity, the algorithm is summarized as Algorithm 12.8. Notice that step 4 in Algorithm 12.8 requires the SVD of the

---

Algorithm 12.8 Algorithm for updating documents on a node

1: *Input:* $k_i, P_{k_i} \in \mathbb{R}^{m \times k_i}, \Sigma_{k_i} \in \mathbb{R}^{k_i \times k_i}, Q_{k_i} \in \mathbb{R}^{n \times k_i}, D^{(i)} \in \mathbb{R}^{m \times r_i}$.
2: Compute the projection: $\hat{D}^{(i)} = (I_m - P_{k_i} P_{k_i}^T) D^{(i)}$.
3: Compute the QR decomposition: $\hat{D}^{(i)} = \hat{P}_{r_i} R_i$, where $\hat{P}_{r_i} \in \mathbb{R}^{m \times r_i}, R_i \in \mathbb{R}^{r_i \times r_i}$.
4: Compute the SVD of matrix

$$\hat{B}^{(i)} \equiv \begin{pmatrix} \Sigma_{k_i} & P_{k_i}^T D^{(i)} \\ 0 & R_i \end{pmatrix} \in \mathbb{R}^{(k_i + r_i) \times (k_i + r_i)}$$

in the form:

$$\hat{B}^{(i)} = (U_{k_i}, U_{k_i}^{\perp}) \cdot \mathrm{diag}(\hat{\Sigma}_{k_i}, \hat{\Sigma}_{r_i}) \cdot (V_{k_i}, V_{k_i}^{\perp})^T,$$

where $U_{k_i}, V_{k_i} \in \mathbb{R}^{(k_i + r_i) \times k_i}$ and $\hat{\Sigma}_{k_i} \in \mathbb{R}^{k_i \times k_i}$.
5: *Output:* The best rank-k_i approximation of $B^{(i)} = (A_{k_i}, D^{(i)})$ is given by:

$$B_{k_i}^{(i)} \equiv \left[(P_{k_i}, \hat{P}_{r_i}) U_{k_i} \right] \cdot \hat{\Sigma}_{k_i} \cdot \left[\begin{pmatrix} Q_{k_i} & 0 \\ 0 & I_{r_i} \end{pmatrix} V_{k_i} \right]^T.$$

---

structured matrix $\hat{B}^{(i)}$, which is upper triangular with the diagonal left upper block of order $k_i \times k_i$. At the same time, this step represents the most intensive computation in Algorithm 12.8. For that purpose one can use the parallel version of the Kogbetliantz algorithm, which was described in Sect. 12.4.

12.6.3 Updating Terms

Similarly to the previous case of updating the documents, the algorithm for updating terms on node i is identical to Algorithm 12.2. Again, the index i is used to denote local matrices on node i. Therefore, let $T^{(i)} \in \mathbb{R}^{q_i \times n_i}$ be the q_i new term vectors that should be added to the existing terms at the bottom of the old term-document matrix. The task is to compute the best rank-k_i approximation of the row partitioned matrix

$$C^{(i)} \equiv \begin{pmatrix} A_{k_i}^{(i)} \\ T \end{pmatrix}.$$

Writing Algorithm 12.2 for node i, one gets the Algorithm 12.9 summarized below. As in the case of updating documents, all nodes of a grid use the same algorithm for updating the terms. Similarly to the problem of updating documents, the computa-

---

Algorithm 12.9 Algorithm for updating terms on a node

1: *Input:* k_i, $P_{k_i} \in \mathbb{R}^{m \times k_i}$, $\Sigma_{k_i} \in \mathbb{R}^{k_i \times k_i}$, $Q_{k_i} \in \mathbb{R}^{n_i \times k_i}$, $T^{(i)} \in \mathbb{R}^{q_i \times n_i}$.
2: Compute the projection: $\hat{T}^{(i)} = (I_{n_i} - Q_{k_i} Q_{k_i}^T) T^{(i)T} \in \mathbb{R}^{n_i \times q_i}$.
3: Compute the QR decomposition: $\hat{T}^{(i)} = \hat{Q}_{q_i} L_i^T$, where $\hat{Q}_{q_i} \in \mathbb{R}^{n_i \times q_i}$, $L_i \in \mathbb{R}^{q_i \times q_i}$.
4: Compute the SVD of matrix

$$\hat{C}^{(i)} \equiv \begin{pmatrix} \Sigma_{k_i} & 0 \\ T^{(i)} Q_{k_i} & L_i \end{pmatrix} \in \mathbb{R}^{(k_i + q_i) \times (k_i + q_i)}$$

 in the form:
$$\hat{C}^{(i)} = (U_{k_i}, U_{k_i}^\perp) \cdot \mathrm{diag}(\hat{\Sigma}_{k_i}, \hat{\Sigma}_{q_i}) \cdot (V_{k_i}, V_{k_i}^\perp)^T,$$

 where $U_{k_i}, V_{k_i} \in \mathbb{R}^{(k_i + q_i) \times k_i}$ and $\hat{\Sigma}_{k_i} \in \mathbb{R}^{k_i \times k_i}$.
5: *Output:* The best rank-k_i approximation of $C^{(i)} = \begin{pmatrix} A_{k_i}^{(i)} \\ T^{(i)} \end{pmatrix}$ is given by:

$$C_{k_i}^{(i)} \equiv \left[\begin{pmatrix} P_{k_i} & 0 \\ 0 & I_{q_i} \end{pmatrix} U_{k_i} \right] \cdot \hat{\Sigma}_{k_i} \cdot \left[(Q_{k_i}, \hat{Q}_{q_i}) V_{k_i} \right]^T.$$

---

tionally most intensive step is the SVD of the lower triangular matrix $\hat{C}^{(i)}$ with the upper left diagonal block. Again, for that purpose one can use the parallel version of the Kogbetliantz algorithm described in Sect. 12.4.

12.6.4 Downdating Documents

When r_i documents are to be deleted on node i, Algorithm 12.3 can be easily modified to get Algorithm 12.10 below. The most computationally demanding task in Algorithm 12.10 is the SVD of the lower triangular matrix $\tilde{B}$ in step 5.

Algorithm 12.10 Algorithm for downdating documents on a node

1: *Input:* $k_i, P_{k_i} \in \mathbb{R}^{m \times k_i}, \Sigma_{k_i} \in \mathbb{R}^{k_i \times k_i}, Q_{k_i} \in \mathbb{R}^{n \times k_i}, D \in \mathbb{R}^{m \times r_i}$.
2: Complete Q_{k_i} into the orthonormal matrix $\hat{Q} = (Q_{k_i}, S)$ of order $n \times (k_i + r)$ by taking r random vectors and orthogonalizing them by the modified Gram-Schmidt process.
3: Form the matrix $W = \begin{pmatrix} Q_k^{1:r_i} & S^{1:r_i} \\ \Sigma_{k_i} & 0 \end{pmatrix}$ and find orthogonal matrices G_L and G_R so that

$$G_L W^T G_R = G_L \begin{pmatrix} Q_k^{1:r_i} & S^{1:r_i} \\ \Sigma_{k_i} & 0 \end{pmatrix}^T G_R = \begin{pmatrix} I_r & 0 \\ Y & \tilde{B} \end{pmatrix}^T,$$

where $\tilde{B}$ is the lower triangular matrix of order k_i.
4: Compute $\bar{P}_{k_i}$ and $\bar{Q}_{k_i}$ by:

$$G_R^T \begin{pmatrix} I_{r_i} & 0 \\ 0 & P_{k_i}^T \end{pmatrix} = \begin{pmatrix} I_{r_i} & 0 \\ 0 & \bar{P}_{k_i}^T \end{pmatrix}, \quad \hat{Q} G_L^T = \begin{pmatrix} I_{r_i} & 0 \\ 0 & \bar{Q}_{k_i} \end{pmatrix}.$$

5: Compute the SVD of $\tilde{B}$, $\tilde{B} = P_B \Sigma_B Q_B^T$. All matrices are square of order k_i.
6: *Output:* The best rank-k_i approximation of $\tilde{A}_{k_i}$ is given by $\tilde{P}_{k_i} = \bar{P}_{k_i} P_B$, $\tilde{\Sigma}_{k_i} = \Sigma_B$ and $\tilde{Q}_{k_i}^T = (\bar{Q}_{k_i} Q_B)^T$.

12.6.5 Downdating Terms

Similarly to downdating the documents, when q_i terms are to be deleted on node i, Algorithm 12.4 can be easily modified to get Algorithm 12.11. Again, the most

---

Algorithm 12.11 Algorithm for downdating terms on a node

1: *Input:* $k_i, P_{k_i} \in \mathbb{R}^{m \times k_i}, \Sigma_{k_i} \in \mathbb{R}^{k_i \times k_i}, Q_{k_i} \in \mathbb{R}^{n \times k_i}, T \in \mathbb{R}^{q_i \times n}$.
2: Complete P_{k_i} into the orthonormal matrix $\hat{P} = (P_{k_i}, Z)$ of order $m \times (k_i + q_i)$ by taking q_i random vectors and orthogonalizing them by the modified Gram-Schmidt process.
3: Form the matrix $H = \begin{pmatrix} P_{1:q_i,k_i}^T & \Sigma_{k_i} \\ Z_{1:q_i}^T & 0 \end{pmatrix}$ and find orthogonal matrices G_L and G_R so that

$$G_L H G_R = \begin{pmatrix} I_{q_i} & Y \\ 0 & \tilde{B} \end{pmatrix},$$

where $\tilde{B}$ is the upper triangular matrix of order k_i. Here $P_{1:q_i,k}^T$ denotes first q_i columns of $P_{k_i}^T$; similarly for $Z_{1:q_i}^T$.
4: Compute $\bar{P}_{k_i}$ and $\bar{Q}_{k_i}$ by:

$$G_L \hat{P}^T = \begin{pmatrix} I_{q_i} & 0 \\ 0 & \bar{P}_{k_i}^T \end{pmatrix}, \quad \begin{pmatrix} I_{q_i} & 0 \\ 0 & Q_{k_i} \end{pmatrix} G_R = \begin{pmatrix} I_{q_i} & 0 \\ 0 & \bar{Q}_{k_i} \end{pmatrix}.$$

5: Compute the SVD of $\tilde{B}$, $\tilde{B} = P_B \Sigma_B Q_B^T$. All matrices are square of order k_i.
6: *Output:* The best rank-k_i approximation of $\tilde{A}_{k_i}$ is given by $\tilde{P}_{k_i} = \bar{P}_{k_i} P_B$, $\tilde{\Sigma}_{k_i} = \Sigma_B$ and $\tilde{Q}_{k_i}^T = (\bar{Q}_{k_i} Q_B)^T$.

---

computationally expensive task is the SVD of the upper triangular matrix $\tilde{B}$.

12.6.6 Retrieval of Documents

A retrieval of relevant documents is based on the notion of a *query*, which is the m-dimensional binary vector q with ones at positions matching the terms that should be found and retrieved from the database. Notice that the dimension of a query is equal to the number of rows of $A_{k_i}^{(i)}$, i.e., to the size of the set of terms used for coding the documents into a database.

We assume that, despite the fact that the nodes of a grid do not communicate regularly, it is possible to send the same query through the connecting network to individual nodes. These nodes are capable to receive the query and search independently for relevant documents in their individual databases $A_{k_i}^{(i)}$, $i = 1, 2, \ldots, p$. Recall, however, that the low-rank approximation is stored in each node in its factored form given by matrices $P_{k_i} \Sigma_{k_i}$ and Q_{k_i}. Therefore, one has to work with these matrices and not with the matrix $A_{k_i}^{(i)}$, which is never computed explicitly.

The *query matching* is based on the comparison of a query vector q to the columns of the approximation $A_{k_i}^{(i)}$ by means of the acute angle $\theta_j^{(i)}$ between them; i.e., for $j = 1, 2, \ldots, n_i$, one should compute

$$\cos \theta_j^{(i)} = \frac{(A_{k_i}^{(i)} e_j)^T q}{\|A_{k_i}^{(i)} e_j\|_2 \cdot \|q\|_2} = \frac{e_j^T P_{k_i} \Sigma_{k_i} (Q_{k_i}^T q)}{\|\Sigma_{k_i} Q_{k_i}^T e_j\|_2 \cdot \|q\|_2}.$$

For a fixed low-dimensional approximations, this computation can be made more efficient by pre-computing n_i values:

$$s_j^{(i)} = \Sigma_{k_i} V_{k_i}^T e_j.$$

Then

$$\cos \theta_j^{(i)} = \frac{s_j^{(i)T} (P_{k_i}^T q)}{\|s_j^{(i)}\|_2 \cdot \|q\|_2}. \tag{12.10}$$

Producing a list of relevant documents is based on the geometric insight about alignment of two vectors in the k_i-dimensional Euclidean space: Two vectors are the more aligned (more 'identical') the less is the acute angle between them. Since cosine is the decreasing function in the interval $[0, \pi/2]$, this allows for the *ordering* of retrieved documents by listing a non-increasing sequence of their cosines. Usually, some sort of *thresholding* is applied for retrieved documents – retrieved are only documents for which

$$\cos \theta_j^{(i)} \geq \alpha^{(i)},$$

where $\alpha^{(i)}$ is the constant, which can be specific for each node of the grid.

12.6.6.1 Comparison of Retrieval Between Individual Nodes

Since the individual nodes do not communicate, one can *not* build the 'global' database that would represent the compound matrix from individual nodes $A = (A^{(1)}, A^{(2)}, \ldots, A^{(p)})$. From the mathematical point of view, there is no way to represent the matrix A by some lower, k-dimensional approximation, because there is no way to get the individual low-dimensional approximations of matrices A_i together and build upon them the approximation of A.

Therefore, the retrieval of documents is possible only on a local level of individual nodes. Moreover, because the global approximation is missing, we can *not*, strictly speaking, directly compare the results from individual retrievals of documents on individual nodes. In other words, since the individual low-dimensional approximations of A_i were built independently, there is no way how to compare the accuracy of retrievals coming from two different nodes.

Nevertheless, some general conclusions can be made in a special case. For the local approximation in each node, the most important parameters are (i) the number of encoded documents n_i and (ii) the dimension of the approximation vector space k_i. If the number of encoded documents is approximately the same in each node, and if the dimensions of approximations are also the same, than one can expect the same quality of encoding – i.e., when the same terms are used in each node, the structure of individual low-dimensional vector spaces will be very similar. In this case, one can use the same threshold in each node and merge and sort p individual lists of matched documents into one list according to, for example, non-increasing cosines. In other words, we can expect approximately the same accuracy of retrieval over the nodes in this special case.

The problem is, of course, how to manage the first requirement above during the updates. If the nodes of a grid do not communicate at all, there is no way how to ensure that the local databases will be built from the approximately same amount of documents. Therefore, next discussion is devoted to the second model of a distributed system where the inter-processor communication is available.

12.7 LSI Implemented on a Distributed System

We now consider the second possible paradigm with respect to the storage of documents and their retrieval. In contrast to the grid, in a *distributed* system the individual processors can communicate and mutually exchange data. Therefore, the term-document matrix A of order $m \times n$ can be distributed column-wise among, say, p processors in the form $A = (A_1, A_2, \ldots, A_p)$ where A_i is of order $m \times n_i$. This distributed system may even correspond to one node of a grid analyzed above.

Each processor builds its own k_i-dimensional approximation of the SVD of its block A_i as described above. Notice that these computations can be computed in parallel without any communication between processors. This means that all computations are perfectly local to processors and can be realized by some serial numerical library, for example, using the LAPACK.

After this initial computation, however, comes the main difference between a grid and a distributed system. In contrast to a grid, a distributed system *can* build another 'global' approximation of the original matrix A atop of individual approximations which were computed in individual processors. This is something completely new as compared to a grid. This *global approximation* is then used in the retrieval of documents.

Next we will describe how such a global approximation can be computed and updated from individual approximations stored in individual processors.

12.7.1 Building a Global Approximation

For the sake of simplicity of exposition, let us first consider the case of two processors, PE1 and PE2. All following derivations can be easily extended to the case of p processors with $p > 2$.

Let us assume that PE1 has computed its $m \times k_1$ approximation of A_1 and stored the corresponding partial factors of SVD P_{11}, Σ_{11} and Q_{11}. Similarly, PE2 has the local approximation P_{21}, Σ_{21} and Q_{21} of A_2 at its disposal. Notice that the first index in these local approximations can be interpreted as the processor index while the second one denotes a local level of approximation. Since the original A_i has n_i columns, we must have $k_i \leq n_i$, $i = 1, 2$.

To build a global approximation of local factors, we must first choose the order k of that approximation. For that purpose, let us organize two local approximations into a global matrix G,

$$G = (P_{11}\Sigma_{11}Q_{11}^T, P_{21}\Sigma_{21}Q_{21}^T),$$

which is the matrix of order $m \times n$ (notice that the first matrix is of order $m \times n_1$, the second one is of order $m \times n_2$ and $n = n_1 + n_2$).

To compute the k-dimensional global approximation of this matrix, we must clearly have $k \leq k_1 + k_2$. Notice that

$$(P_{11}\Sigma_{11}Q_{11}^T, P_{21}\Sigma_{21}Q_{21}^T) = (P_{11}\Sigma_{11}, P_{21}\Sigma_{21}) \begin{pmatrix} Q_{11} & 0 \\ 0 & Q_{21} \end{pmatrix}^T,$$

where the rightmost matrix is of order $(k_1 + k_2) \times n$ with orthonormal rows (after transposition).

Now the k-dimensional global approximation of G is computed in two steps by Algorithm 12.12. We have just shown that in the special case, when $k = k_1 + k_2$, the global approximation can be computed quite efficiently. Notice that the formation of G requires the scaling of the local columns of left singular vectors by local singular values. This scaling can be performed in parallel without any communication

Algorithm 12.12 Algorithm for global approximation

1: Compute the full SVD of the $m \times (k_1 + k_2)$ matrix

$$(P_{11}\Sigma_{11}, P_{21}\Sigma_{21}) = (\tilde{U}_k, \tilde{U}_k^\perp) \begin{pmatrix} \tilde{\Sigma}_k & 0 \\ 0 & \tilde{\Sigma}' \end{pmatrix} (\tilde{V}_k, \tilde{V}_k^\perp)^T,$$

where $\tilde{\Sigma}_k$ contains k largest singular values in a non-increasing order. Here, k denotes the dimension of a global approximation, which must be chosen (but see next step).

2: Now consider the matrix product

$$\tilde{U}_k \tilde{\Sigma}_k \left[\begin{pmatrix} Q_{11} & 0 \\ 0 & Q_{21} \end{pmatrix} \tilde{V}_k \right]^T \equiv \tilde{U}_k \tilde{\Sigma}_k \tilde{W}_k^T.$$

Notice that $\tilde{W}_k$ can be computed if and only if $k = k_1 + k_2$. In this special case it has orthonormal columns so that the above matrix product *is* the truncated k-dimensional SVD of G.

between processors. Then, the SVD of a distributed matrix G has to be computed. This can be achieved using the ScaLAPACK library, or using some new parallel block-Jacobi algorithm – see [18, 19]. Finally, a distributed matrix multiplication (e.g., by the ScaLAPACK routine PDGEMM) has to be performed for the computation of new global right singular vectors $\tilde{W}_k$.

In the case of p processors with $p > 2$, the above matrix G consists of p blocks with n_i columns, $i = 1, 2, \ldots p$. If we choose the special value of global approximation by $k = k_1 + k_2 + \ldots k_p$, then the local approximations are again not needed in full – only the locally scaled left vectors are needed in each processor. Hence, one has to compute explicitly only the SVD of matrix G of order $m \times k$ instead of a 'full' matrix of order $m \times n$. Therefore, when $k \ll n$ the substantial saving in computation time can be achieved.

At this moment, the new global approximation is available in the form of a triple $\tilde{U}_k, \tilde{\Sigma}_k, \tilde{W}_k$, whereby each matrix is distributed through p processors. We can either collect this global approximation into one (or each) processor by using the procedure GATHER (or ALLGATHER), or leave the computed global k-dimensional approximation in the distributed form. The latter approach is more advantageous from the point of view of storage requirements, since each processor stores the $m \times k_i$ submatrix of the global left singular vectors, one vector of k_i global singular values and the $n \times k_i$ submatrix of the global right singular vectors. This means that no processor has to store the complete k-dimensional factors; since $k = \sum_{k=1}^{p} k_i$, this means a substantial saving in storage space per processor.

With respect to the global approximation, the requirements for each processor are the same if $k_1 = k_2 = \cdots = k_p$. The local approximation in each processor can require different amount of storage if k_i's differ across the processors. However, if we consider a set of processors which process the qualitatively *same* database (i.e., documents from very similar areas of interest, e.g., mathematics, physics and astronomy), then there is no reason why the local orders of approximation k_i should differ too much. This is true provided that the individual k_i-dimensional approximations of local databases in individual processors are based on the roughly *same* amount of information, i.e., the starting column dimensions n_i of local databases are roughly the same. Hence, we require that $n_i \approx n/p$ where n is the initial number of documents in the whole (huge) database, which should be distributed among p processors. To provide an initial 'portion' of information equally to each processor, we can randomly choose n/p items from the initial set of documents and send them to a given processor. This starting phase ensures that local approximations can be computed with the same dimension across the processors (i.e., $k_1 = k_2 = \cdots = k_p$), and the accuracy in approximating the original local databases will be approximately the same in each processor.

12.7.2 Updating and Downdating

When new documents are to be inserted into an existing database, it is necessary to decide which processor(s) should receive all documents and which a portion of documents. From the computational point of view, it is not advisable to add documents into database one by one. Instead, new documents should be added in a batch of, say, d items with $d \gg 1$. When necessary, a new batch of documents can be divided among, say, p_1 processors with $p_1 \leq p$, where p is the total number of processors. Let us call these p_1 processors *locally active*.

After receiving new documents, locally active processors modify (update) their local databases in parallel using Algorithm 12.8. Notice that no inter-processor communication is needed at this stage of computation. After finishing local updates in locally active processors, however, the update on the global level is needed. All processors must participate in the global update and perform Algorithm 12.12. After finishing the global update, a new global, k-dimensional approximation of the distributed database is available, whereby this global approximation is itself distributed.

The same procedure is applied when some documents are to be deleted from a database using Algorithm 12.10 and Algorithm 12.12.

Updating terms is a bit different. Here, even on the local level, all processors must be active because all processors build their local, k_i-dimensional approximations over the same set of terms (vocabulary). Hence, after receiving new terms, all processors perform Algorithm 12.9 in parallel. Therefore, there are no inactive processors as opposed to the local updating of documents, where some processors may be idle (if they do not receive a batch of new documents). After finishing local updates of terms, all processors are involved in the computation of a new global k-dimensional approximation by performing Algorithm 12.12.

The same approach is required when some terms are to be deleted from a database using Algorithms 12.11 and 12.12.

12.7.3 Retrieval of Documents

In Sect. 12.6.6, the retrieval of documents was described for the set of non-communicating processors in a grid. In the view of our two-stage procedure for building the database of documents for the distributed LSI, we can call this approach a *local retrieval*. Since the processors in a distributed system build their local as well as global representation of a database, there is also the possibility of a two-stage retrieval of documents in this case. This possibility opens a new, interesting approach to the estimation of the quality of retrieval.

The first possibility is – as in the case of a grid – the *local* retrieval of documents. The query q is sent to all p processors and all of them go through their k_i-dimensional approximations of local databases computing the cosines according to the Eq. (12.10). Each processor provides its own list of relevant documents according to the algorithm described in Sect. 12.6.6, and there is no need for the inter-processor communication in this stage of retrieval from the local databases. When assuming that the local databases were built by encoding approximately the same amount of documents in each processor, then the thresholds α_i for cosines can be chosen the same (say, 0.5) in all processors, i.e., $\alpha_i = \beta$ for all i. Individual lists can be sent to a marked processor, which can then sort all matched documents into a final list $\mathscr{L}_1$ according to the local cosines obtained in processors.

However, since also the global k-dimensional approximation has been built in the case of a distributed system, one can try also the different retrieval of documents by using the distributed factors $\tilde{U}_k$ (order $m \times k$), $\tilde{\Sigma}_k$ (order $k \times k$) and $\tilde{W}_k$ (order $n \times k$). This is the *global* retrieval of documents. It proceeds by computing the cosines

$$\cos \tilde{\theta}_j = \frac{\tilde{s}_j^T (\tilde{U}_k^T q)}{\|\tilde{s}_j\|_2 \cdot \|q\|_2}, \quad j = 1, 2, \ldots, n, \tag{12.11}$$

where $\tilde{s}_j$ are n pre-computed values given by

$$\tilde{s}_j = \tilde{\Sigma}_k \tilde{W}_k^T e_j. \tag{12.12}$$

Both the above equations require a clever data organization in computing the required matrix-vector products with distributed factors and final scalar products. Let us suppose that the matrices $\tilde{U}_k$ and $\tilde{W}_k^T$ are distributed column-wise, so that processor i contains the respective blocks of dimension $m \times k_i$ and $k \times n_i$. Then, according to Eq. (12.12), $\tilde{s}_j$ is computed in two steps: (i) take the jth column of $\tilde{W}_k^T$ residing as a whole in some processor, and (ii) scale its ℓth component by $\tilde{\sigma}_\ell, \ell = 1, 2, \ldots, k$. These two steps are most easily performed locally if each processor contains all k global singular values from $\tilde{\Sigma}_k$. Then, for the computation of $\tilde{\theta}_j$ according to Eq. (12.11), one has (i) to compute the distributed matrix-vector product $\tilde{y} = (\tilde{U}_k^T q)$ with the query q residing in each processor, (ii) to compute the distributed scalar product $\tilde{s}_j^T \tilde{y}$, and, finally, (iii) to scale the scalar product by $1/(\|\tilde{s}_j\|_2 \cdot \|q\|_2)$. All these computations can be performed by appropriate functions from the ScaLA-PACK library.

The global retrieval ends with sorting of the set $\{\cos \tilde{\theta}_j\}$ and thresholding them by some threshold α. Assume that the same threshold is used as in the case of the local retrieval, i.e., $\alpha = \beta$. Thus, a list of matched documents $\mathscr{L}_2$ is produced.

Now comes an interesting part of the retrieval process – the comparison of lists $\mathscr{L}_1$ and $\mathscr{L}_2$. This comparison enables to make some conclusions with respect to the *accuracy* of local and global retrieval. It is assumed that the global and local threshold for retrieving documents are the same. Furthermore, it is assumed that all retrieved documents are indeed relevant, so that one can compare both lists without caring about wrongly matched documents.

In general, the following scenarios with respect to the number of items $|\mathscr{L}_1|$ and $|\mathscr{L}_2|$ and their contents are thinkable:

1. $|\mathscr{L}_1| = |\mathscr{L}_2|$ and both lists contain the same documents. This is the ideal situation, which says that both approximations on the local as well as global level are equally accurate.
2. $|\mathscr{L}_1| = |\mathscr{L}_2|$, but the lists do *not* contain the same documents. Hence, there are at least two different documents $\mathscr{D}_1$ and $\mathscr{D}_2$ such that $\mathscr{D}_1 \in \mathscr{L}_1$, $\mathscr{D}_1 \notin \mathscr{L}_2$ and $\mathscr{D}_2 \in \mathscr{L}_2$, $\mathscr{D}_2 \notin \mathscr{L}_1$. Since we assume $k = k_1 + k_2 + \cdots + k_p$ for the dimension of global approximation, it is unlikely that this discrepancy is based on the fact of wrong dimensions in the approximations. It is more likely that some documents with low ranking in the global list will not be found in the local list because of falling just below the local threshold (and vice versa). Lowering the threshold should help to achieve the matching of the same documents on both levels (both lists can be then larger than the original ones).
3. $|\mathscr{L}_2| > |\mathscr{L}_1|$. This situation tells us that the global approximation is more accurate than the local one. This can happen, for example, when one (or more) local databases differ substantially in their column dimensions n_i, i.e., they locally encode widely differing numbers of documents. Then the use of the same dimensionality of approximation at the local level (i.e., $k_1 = k_2 = \cdots = k_p$) leads to the *under-estimate* of optimal dimension because this value must be derived from the portion A_{n_i} having the *least* number of documents (columns). It is best to prevent such a situation by keeping the number of documents encoded in individual processors approximately the same. In other words, at the beginning, n documents should be divided evenly among processors, and, at the updating, a new batch of documents of large enough size should be processed so that, again, each processor receives approximately the same number of documents for its local update. Consequently, the new global update will be computed from locally *balanced* updates.
4. $|\mathscr{L}_1| > |\mathscr{L}_2|$. Can the retrieval at local level yield more documents than that at global level? Since $k = k_1 + k_2 + \cdots + k_p$, each locally approximating LSI space is the subspace of the globally approximating LSI space. Hence, all the latent couplings between terms and documents, which exist in the local databases, exist automatically also in the global database. (Notice that the reverse is not true.) Therefore, this situation should not occur in practice.

12.8 Conclusions

This chapter has been devoted to the computational issues arising in the field of LSI. We have shown that the updating/downdating of large databases, which are encoded by the low-rank approximation of term-document matrices via the SVD, leads to the algorithms where the most complex part is the computation of the SVD of large matrices with special structure (upper triangular). For that purpose, the parallel block Kogbetliantz method with a special ordering seems to be well suited, since it preserves the triangular (or block triangular) structure of a matrix. Another possibility is to use the parallel two-sided block-Jacobi algorithm with dynamic ordering that leads to faster convergence than any cyclic ordering of subproblems; however, the (block) triangular structure is lost after first parallel iteration.

Next, we have described two models of a distributed database of documents that are encoded using the paradigm of the LSI. The first model is devoted to the grid, when the individual nodes have a very limited possibility of mutual communication. In this case, the documents are encoded on individual processors without any connection between them. All updates of documents/terms are also performed independently as well as the retrieval of documents. If the grid consists of p nodes, the retrieval of documents for a given query yields p lists, which have no mutual relations. In particular, one can not compare the accuracy of retrieval from individual nodes. In general, it is possible to say only that the accuracy of individual nodes will be approximately the same when each node encodes approximately the same number of relevant documents (provided that the dimensions in approximation are the same in each node).

The more interesting situation arises in the case of a distributed database, when p processors are connected with some sort of inter-processor network (this can be, e.g., one node of a large grid). In this case, we have developed a two-stage compression of the latent semantic information. In the first step, the local approximation of the whole database is constructed by dividing the whole database evenly among the processors. Since the processors can communicate, the global approximation of the whole database can be built in the second step. We have shown that the SVD computation at the global level can be made very efficient when the dimension of global approximation is equal to the sum of dimensions of local approximations. We have briefly discussed the implementation issues and shown that the retrieval of documents will require the use of distributed numerical libraries like ScaLAPACK. Finally, it is now possible to compare the accuracy of the locally and globally encoded database by considering the lists produced when answering the same query. We have discussed some interesting scenarios that can arise in real life.

Acknowledgments The authors were supported by the VEGA grant no. 2/7143/27 from the Scientific Grant Agency of the Ministry of Education and Slovak Academy of Sciences, Slovakia.

References

1. M. W. Berry and M. Browne, Understanding Search Engines: Mathematical Modeling and Text Retrieval, First ed., SIAM, Philadelphia, PA (1999).
2. M. W. Berry, Z. Drmač and E. R. Jessup, Matrices, vector spaces, and information retrieval, SIAM Rev. 41 (1999) 335–362.
3. H. Zha, A subspace-based model for information retrieval with applications in latent semantic indexing, in: Proc. Irregular '98, LNCS 1457, Springer Verlag, New York, NY (1998) 29–42.
4. R. B. Lehoucq and D. C. Sorensen, Deflation techniques for an implicitly restarted Arnoldi iteration, SIAM J. Matrix Anal. Appl. 17 (1996) 789–821.
5. B. Parlett, The symmetric eigenvalue problem, First ed., SIAM, Philadelphia, PA (1996).
6. H. Rutishauser, Simultaneuos iteration method for symmetric matrices, Num. Math. 16 (1970) 205–223.
7. A. H. Sameh and J. A. Wasniewski, A trace minimization algorithm for the generalized eigenvalue problem, SIAM J. Num. Anal. 19 (1982) 1243–1259.
8. M. W. Berry, Large scale sparse singular value computations, J. Supercomp. Appl. 6 (1992) 13–49.
9. H. Zha and H. D. Simon, On updating problems in latent semantic indexing, SIAM J. Sci. Comput. 21 (1999) 782–791.
10. D. I. Witter and M. W. Berry, Downdating the latent semantic indexing model for conceptual information retrieval, Comput. J. 41 (1998) 589–601.
11. A. Björck, Numerical Methods for Least Squares Problems, First ed., SIAM, Philadelphia, PA (1996).
12. E. Kogbetliantz, Diagonalization of general complex matrices as a new method for solution of linear equations, Proc. Intern. Congr. Math. Amsterdam 2 (1954) 356–357.
13. E. Kogbetliantz, Solutions of linear equations by diagonalization of coefficient matrices, Quart. Appl. Math. 13 (1955) 123–132.
14. V. Hari and J. Matejaš, Accuracy of the Kogbetliantz method, preprint, University of Zagreb (2005).
15. V. Hari and V. Zadelj-Martič, Parallelizing Kogbetliantz method: A first attempt, J. Num. Anal. Industr. Appl. Math. 2 (2007), 49–66.
16. F. T. Luk and H. Park, On parallel Jacobi orderings, SIAM J. Sci. Statist. Comput. 10 (1989) 18–26.
17. V. Hari, Accelerating the SVD block-Jacobi method, Computing 75 (2005) 27–53.
18. M. Bečka, G. Okša and M. Vajteršic, Dynamic ordering for a parallel block-Jacobi SVD algorithm, Parallel Comput. 28 (2002) 243–262.
19. M. Bečka and G. Okša, On variable blocking factor in a parallel dynamic block-Jacobi SVD algorithm, Parallel Comput. 29 (2003) 1153-1174.
20. H. N. Gabov, Data structures for weighted matching and nearest common ancestors with linkings, in: Proceedings of the First Annual ACM-SIAM Symposium on Discrete Algorithms, ACM, New York (1990) 434–443.
21. W. J. Cook and A. Rohe, Computing minimum-weight perfect matchings, INFORMS J. Comput. 11 (1999) 138–148.
22. J. Van Leeuwen, ed., Handbook of Theoretical Computer Science. Volume A: Algorithms and Complexity, Elsevier, Amsterdam (1990) 587.
23. A. Anderson, Z. Bai, C. Bischof, J. Demmel, J. Dongarra, J. Du Croz, A. Greenbaum, S. Hammarling, A. McKenney, S. Ostrouchov and D. Sorensen, LAPACK Users' Guide, Second ed., SIAM, Philadelphia (1999).
24. M. Bečka and M. Vajteršic, Block-Jacobi SVD algorithms for distributed memory systems: I. Hypercubes and rings, Parallel Alg. Appl. 13 (1999) 265–287.
25. M. Bečka and M. Vajteršic, Block-Jacobi SVD algorithms for distributed memory systems: II. Meshes, Parallel Alg. Appl. 14 (1999) 37–56.
26. I. Foster and C. Kasselman, Computational Grids. In: The Grid: Blueprint for a Future Computing Infrastructure, I.Foster and C.Kasselman (Eds.), Morgan and Kaufmann Publishers (1998).

Chapter 13
Short-Vector SIMD Parallelization in Signal Processing

Rade Kutil

Abstract

Short-vector Single-instruction-multiple-data (SIMD) units have become common in signal processors. Moreover, almost all modern general-purpose processors include SIMD extensions, which makes SIMD also important in high performance computing. This chapter gives an overview of approaches to the vectorization of signal processing algorithms. Despite their complexity, these algorithms have a relatively regular data flow. This regularity makes them good candidates for SIMD vectorization. They fall in two categories: filter banks that operate on streaming signal data, and Fourier-like transforms that operate on blocks of data. For the first category, simple FIR filters, IIR filters and more complicated filter banks from the field of wavelet transforms are investigated to develop and present general vectorization strategies. Well-known loop transformations as well as novel vectorization approaches are combined and evaluated. For the second category, basic approaches for the fast Fourier transform (FFT) are shown and the workings of automatic vectorizing performance tuning systems are explained. The presented solutions are tested on Intel processors with SIMD extensions and the results are compared. Wherever possible, the reasons for performance gains or losses are uncovered so that good vectorization strategies can be derived for arbitrary signal processing algorithms.

13.1 Introduction

The trend in parallelization goes toward multi-level parallelism. In addition to the combination of clusters, shared-memory architectures, and multi-core processors, CPU cores exploit more and more internal parallelity. Among methods such as ex-

Rade Kutil
Department of Computer Sciences, University of Salzburg, J.-Haringer-Strasse 2,
5020 Salzburg, Austria, e-mail: rkutil@cosy.sbg.ac.at

R. Trobec et al. (eds.), *Parallel Computing*, DOI 10.1007/978-1-84882-409-6_13,

cessive pipelining, specialized units, as used in signal processors, and VLIW (very large instruction word), SIMD (single instruction multiple data) plays an important role. One reason for its popularity is the availability of short-vector SIMD extensions in all modern general-purpose processors.

These processors are very cost-effective and, thus, heavily used in high performance computing (HPC). As a consequence, their SIMD extensions are exploited in most HPC software. SIMD always benefits from regularity in algorithms. Fortunately, this is exactly what makes the difference between signal processing and other applications. In signal processing, large amounts of data are processed in a continuous way, which makes the use of SIMD techniques promising.

13.1.1 Signal Processing Algorithms

Most signal processing algorithms fall into two categories: filter banks and Fourier-like transforms. Other algorithms are usually quite similar to one of the two, or include at least one of the two as an essential ingredient.

There are differences between the two categories. The most important one is that Fourier-type transforms operate on blocks of signal data, while filters operate on streams of data. Another difference is that filters have the simple algorithmic form of a convolution, whereas fast Fourier-type transforms employ more complicated butterfly-like schemes. Note also that it is possible to implement convolutions and, thus, filters via Fourier transforms by applying the convolution theorem. This method is feasible whenever the filters are long. Yet another difference is that Fourier-type algorithms usually operate on complex numbers, whereas filter banks are almost always real-valued.

Let us look at the basic algorithms in more detail. The simplest form of a finite impulse response (FIR) filter is

$$y(n) = \sum_k x(n-k)h(k), \qquad (13.1)$$

where x is the discrete input signal, y the output signal, and h the (finite) filter. For causal filters, k is non-negative. In any case, k has finite limits. The general case can have more than one input and output signals. This leads to the form

$$y_i(n) = \sum_j \sum_k x_j(n-k)h_{i,j}(k). \qquad (13.2)$$

Additionally, input and output signals can be down-sampled, i.e., only every m-th value has to be calculated in the output signal, or is non-zero in the input signal. While this reduces the computational demand by omitting zero products, as well as memory demands by omitting zero values from arrays, it complicates the algorithms. Moreover, some values of $h_{i,j}(k)$ may be equal, or just have opposite signs. This happens for symmetric filters and quadrature mirror filter pairs, for instance.

Depending on the position of the filter coefficients and down-sampling factors, this may lead to redundant products, which means further potential for computational reduction at the price of higher algorithmic irregularity. Finally, the filters may have "holes," i.e., inner zero coefficients. All this renders a general-purpose implementation highly inefficient. Each filter bank has to be handled individually, or automatic compilation techniques must be used.

Infinite impulse response (IIR) filters are an extension of FIR filters, where the output signal is reused as input signal.

$$y(n) = \sum_l y(n-l)a(l) + \sum_k x(n-k)b(k), \tag{13.3}$$

where, of course, $l > 0$. The main difficulty in implementing this scheme is the recursive data flow that introduces loop dependencies and, thus, complicates parallelization and makes algebraic reformulations of the filter algorithm necessary.

On the other hand, Fourier-type algorithms are relatively irregular to start with. Despite the easy definition of the discrete Fourier transform

$$y(n) = \mathop{\mathscr{F}}_{N} x(n) = \sum_{k=0}^{N-1} x(k)e^{-i\frac{2\pi}{N}kn}, \tag{13.4}$$

where N is the size of the input signal block $(x(0),\ldots,x(N-1))$, and $0 \le n < N$, fast versions of the Fourier transform employ more complicated recursive reformulations such as

$$\mathop{\mathscr{F}}_{N} x = (\hat{x}_0 + \hat{x}_1, \hat{x}_0 - \hat{x}_1), \quad \hat{x}_0 = \mathop{\mathscr{F}}_{N/2} x_0, \quad \hat{x}_1(n) = \mathop{\mathscr{F}}_{N/2} x_1(n)e^{-i\frac{2\pi}{N}n}, \tag{13.5}$$

where x is split into even samples $x_0 = (x(0),x(2),\ldots,x(N-2))$, and odd samples $x_1 = (x(1),x(3),\ldots,x(N-1))$. This scheme is due to Cooley and Tukey [1]. In this version, N has to be even for one recursion level and a power of two for full recursion (radix 2). Similar schemes can be found for other radices. Further schemes include the split-radix algorithm [2] and the Rader algorithm [3] for prime sizes N. All these schemes may be mixed and lead to different memory access patterns with different computational performances which depend also on machine properties. Automatic tuning systems have been developed [4, 5] which recursively search the space of possible implementations, starting from abstract formulations of the algorithms to rewriting schemes in dedicated signal processing languages such as SPL [6].

13.1.2 Short-Vector SIMD

In SIMD architectures, data is organized in registers containing vectors of several values. These registers can be used in operations such as multiplication and addition just as normal registers. The difference is that the values in the vectors are operated on independently in parallel. Since it is common that a vector consists of $p = 4$ values, we will use this for demonstration throughout this chapter. A vector is written as $a = (a_0, a_1, a_2, a_3)$. Vector operators are displayed with circles:

$$a \odot b = (a_0 \cdot b_0, a_1 \cdot b_1, a_2 \cdot b_2, a_3 \cdot b_3), \quad a \oplus b = (a_0 + b_0, \dots, a_3 + b_3). \quad (13.6)$$

SIMD computers have been popular in the 1980s and early 1990s, mainly due to MasPar and the Connection Machines. Modern SIMD extensions of general purpose CPUs are different from those in that the vectors are much shorter, i.e., $p = 2$, 4, or 8, hence the name "short-vector SIMD." All these architectures have different constraints in accessing and arranging data in vector registers. While traditional vector computers only offered certain shift or rotation operations, new SIMD extensions include almost general variations of values in vector registers, written as

$$a_{(p,q,r,s)} = (a_p, a_q, a_r, a_s), \quad (13.7)$$

or, in the more common form with two operands,

$$(a,b)_{(p,q,r,s)} = (c_p, c_q, c_r, c_s), \quad (13.8)$$

where $c = (a,b) = (a_0, a_1, a_2, a_3, b_0, b_1, b_2, b_3)$, and $0 \le p, q, r, s < 8$. Not all of these so-called *shuffle operations* are available as single instruction on all architectures. As an important example, in Intel MMX and SSE, the shuffle operation has the restriction that the first two values of the destination vector have to be from the first operand and the last two from the second operand, i.e., $0 \le p, q < 4 \le r, s < 8$ in Eq. (13.8). Additionally, there are two operations called "unpack operations" which interleave the values of the first or second halves of the source operands, i.e., $(a,b)_{(0,4,1,5)}$ and $(a,b)_{(2,6,3,7)}$. The maximum number of necessary instructions for an arbitrary shuffle operation is two. On the other hand, the Motorola AltiVec architecture provides instructions for arbitrary shuffle operations.

Architectures can also differ in the allowed numerical precisions, and in the vector size depending on the precision. The common configuration, though, is that vector registers have 128 bit, so they support 4-fold SIMD for single precision (i.e., 32 bit) and 2-fold SIMD for double precision floating point numbers (i.e., 64 bit). Integer numbers are also possible, but we will concentrate on floating point numbers in this chapter.

Another restriction of most SIMD architectures is that they require aligned data access to memory. This means that p consecutive values that are read from memory into a vector register must have a starting address that is a multiple of the vector size. As a consequence, the programmer has to take care that arrays are properly aligned when they are allocated, and that they are read and written in non-overlapping blocks of p values. Although some processors allow unaligned reads and writes, these are usually much slower than aligned accesses.

13.2 General Vectorization Approaches

Most compilers today include options to automatically vectorize the code in order to utilize SIMD extensions. Although these vectorizations rarely lead to optimal code, it is advisable to look at vectorization strategies that might also help in manual vectorization of our signal processing algorithms.

13.2.1 Loop Unrolling

If the inner loop of the algorithm contains only a small number of operations, as is the case for the filter algorithm, then a simple approach is to unroll p iterations of the inner loop, where p is the vector length. The corresponding p operations, one from each iteration, are scheduled to be executed in parallel in a vector instruction.

This approach has only one advantage and many disadvantages. The advantage is that the data to be processed probably lies consecutively in memory and can simply be read into a vector register. However, this is mostly not true for both, input and output data simultaneously. Moreover, the data is unlikely to be aligned. For instance, in a simple filter algorithm the data to be read is shifted by one for every outer loop iteration. Therefore, it is aligned only every p-th time.

If iterations depend on previous iterations, the method is hardly usable at all. This is partly so for the filter algorithm. The multiplication of source data with filter coefficients can be done in parallel, but the summation of the products is inherently serial. Some SIMD architectures provide instructions for horizontal sums which could be used in this situation. However, this reintroduces scalars in the algorithm and, therefore, is suboptimal.

Nevertheless, unrolling a larger number of iterations, or even the whole inner loop, may allow good vectorization through clever shuffling of data in registers. This is, however, a complex problem to solve, and is treated next.

13.2.2 Straight Line Code Vectorization

Algorithms may contain blocks of code with no loops at all. If not, such blocks can be produced by loop unrolling. Reference [7] presents a basic approach to automatic vectorization of such a block. It starts with the speculative aggregation of destination variables into vector variables, followed by a depth-first search for appropriately aggregated operations and source variables. If no feasible solution can be found, backtracking is used to explore other combinations of variables into vectors. Because a full search may be too expensive, heuristics are used for choosing good candidates for aggregation.

This optimizing compiler technique is used and is especially important in automatically tuned FFT packages [8, 9], where small FFTs are recursively expanded into straight line codelets which are then included in larger FFTs.

13.2.3 Loop Fusion

If an algorithm consists of several passes that process the same arrays of data, where each pass reads the data that a previous pass has written, these data accesses degrade the performance and make the algorithm dependent on large cache sizes. Often, it is possible to fuse these passes into a single one. This is done by interleaving the loop iterations of different passes. Of course, one has to make sure that data is not read by an iteration of a later pass before it is written by an iteration of an earlier pass. In other words, a proper rescheduling of all passes' loop iterations has to be applied through a reformulation of the algorithm that respects data dependencies.

As a consequence, intermediate data is likely to be read immediately after it is written. Therefore, it is better to remove these writes and reads in the first place and keep the data in registers, local variables, or local buffers instead. The resulting algorithm consists of a single fused loop containing a larger loop body. In addition to the improved performance due to decreased cache dependency, the larger loop body may be vectorized more easily using techniques for straight line code vectorization.

13.2.4 Loop Transposition

Most algorithms contain nested loops. The inner loop is likely to have dependencies between iterations, which makes vectorization difficult. On the other hand, the outer loop very often has independent iterations. This is the case, for instance, if the outer loop iterates the output index, and the output values are calculated independently from each other, or if the outer loop iterates rows of a row-wise transform.

It should then be possible to transpose the outer and inner loop in order to eliminate dependencies in the new inner loop. This corresponds to the commutation of sum operators if the algorithm is formulated as double sum. Temporary variables that pass data between iterations, such as running sums, have to be avoided or taken care of by storing one value for each outer iteration.

Of course, this introduces new memory accesses and reduces the parallel efficiency. Therefore, it may be better to transpose only blocks of the outer loop, ideally blocks of exactly p iterations. This leads to an algorithm that is basically a copy of the original algorithm, but operates on vectors instead of scalars. Temporary variables are kept in vectors as well and do not have to be saved.

This approach is a simple example of iteration rescheduling. It may have benefits even if the outer loop has dependencies. However, a disadvantage is that data access may not be contiguous any more. This can make shuffle operations or even redundant data accesses necessary. In many cases, a simple $p \times p$ block transposition can solve the problem. Such a transposition can be implemented by

$$
\begin{array}{ll}
b^{(0)} = (a^{(0)}, a^{(1)})_{(0,2,4,6)}, & b^{(1)} = (a^{(2)}, a^{(3)})_{(0,2,4,6)}, \\
b^{(2)} = (a^{(0)}, a^{(1)})_{(1,3,5,7)}, & b^{(3)} = (a^{(2)}, a^{(3)})_{(1,3,5,7)}, \\
c^{(0)} = (b^{(0)}, b^{(1)})_{(0,2,4,6)}, & c^{(1)} = (b^{(2)}, b^{(3)})_{(0,2,4,6)}, \\
c^{(2)} = (b^{(0)}, b^{(1)})_{(1,3,5,7)}, & c^{(3)} = (b^{(2)}, b^{(3)})_{(1,3,5,7)}.
\end{array}
\tag{13.9}
$$

This scheme uses a minimum of eight shuffle instructions and can also be used on Intel SSE architectures. It arranges non-consecutive data $(a_i^{(0)}, a_i^{(1)}, a_i^{(2)}, a_i^{(3)})$ into the vectors $c^{(i)}$. On the other hand, it distributes the consecutive data in vectors $a^{(j)}$ to corresponding slots of different vectors $(c_j^{(0)}, c_j^{(1)}, c_j^{(2)}, c_j^{(3)})$. Very often, algorithms can operate more easily on transposed vectors $c^{(i)}$.

13.2.5 Algebraic Transforms

If it is possible to reformulate an algorithm algebraically, it is worth checking whether the reformulation is more suitable for vectorization. Reformulations can be as simple as applying associative and distributive laws to addition and multiplication. The associative law can, for instance, reverse the dependencies of summing loops.

Moreover, it is important to distinguish between dynamic and static data. In our algorithms dynamic data is mainly signal data that keeps changing. Static data consists of filter or transform coefficients that are constant over loops and, in most cases, available at compile-time. By applying the distributive law, it can be possible to shift operations on dynamic data to operations on static data.

An example would be $a(x + y) + by$, where x and y represent dynamic signal data and a and b are static coefficients. This expression can be transformed into $ax + (a + b)y$, where $a + b$ can be calculated outside of the signal data loop, thus saving one addition per iteration.

This approach can also reduce shuffle operations if applied cleverly. Combined with loop unrolling and vector aggregation, the space of possible reformulations is usually large. Therefore, algorithm specific approaches have to be found, or automatic optimizers with heuristics have to be applied.

Exploring the space of reformulations is even more important for Fourier-type transforms. This is already done in optimized sequential algorithms [4,5], as stated in Sect. 13.1.1. Vectorization of automatically generated straight-line code blocks (codelets) increases the necessity for testing different possible code blocks since some may be vectorized more efficiently than others. Inside the code blocks, the above method of algebraic reformulation could be applied if simple rescheduling, i.e., the aggregation strategy [8], is not sufficient. However, sequential optimization is usually the only algebraic reformulation step within code blocks.

13.3 Convolution Type Algorithms

The most common type of algorithm in signal processing is filtering. Filtering is basically a convolution of signal data $x(t)$ with the filter impulse response $h(t)$. If the impulse response is finite, the convolution can be implemented directly. If it is infinite or too large, a recursive formulation has to be found that is equal to, or approximates the filter. The latter will be treated in the next section.

In this section we will examine simple filters as well as more complex filter banks in order to develop and evaluate the most important vectorization approaches. As examples of filter banks, filter pairs which are common in wavelet transforms (see Sect. 13.6.1) are used. Automatic vectorization so far has not produced any performance increase for wavelet transforms [10, 11]. Also, approaches on old SIMD arrays [12–14] cannot be adapted directly. Therefore, good manual vectorization strategies [15, 16] are important.

Experimental results will also be presented, which were conducted on an Intel Pentium 4 CPU with 3.2 GHz and 2 MB cache size using the SSE extension with vectors of 4 single precision numbers. All implementations use the same amount of code optimization, i.e., memory access through incremented pointers instead of indexed arrays, and compilation with gcc 3.3.5 with the -O3 option. SIMD operations are implemented using gcc's built-in functions for vector extensions and the -msse option. Note that, in order to have full control over generated code, no automatic vectorization is applied.

13.3.1 Simple FIR Filter

The simplest case of an FIR filter has one input signal x and one output signal y, and does not apply any down- or upsampling. It is defined by

$$y(n) = \sum_k x(n-k)h(k). \tag{13.10}$$

There are two loops, the inner one for k and the outer for n. The loop iteration dependencies are shown in Fig. 13.1. We will now vectorize this expression by various methods and evaluate their advantages and disadvantages. The first method to try is simple loop vectorization. It is depicted as method A in Fig. 13.1. Four consecutive iterations shall be combined into one vectorized iteration. However, as the sum operation imposes dependencies between iterations, we have to break the parallelity. We get

$$y(n) = \sum_k S(x(n-4k-m,\ldots,n-4k-m+3) \odot h(4k+m,\ldots,4k+m-3)). \tag{13.11}$$

The operator $S()$ calculates the scalar sum of a vector's elements. On some architectures there is an instruction that implements the S-operator. If there is no such

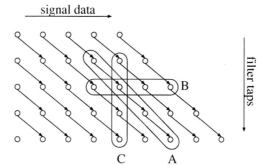

signal data

filter taps

Fig. 13.1 Loop iteration dependencies and vectorization strategies for simple FIR filtering.

instruction, a sequence of shuffle and add operations followed by an element extraction must be used, which is costly and may degrade the performance.

The dislocation parameter m does not have an influence on the result. It has, however, an influence on the range of k. If indices of $h(\cdot)$ lie outside of its finite support, h has to be padded with zeros, which introduces redundant calculations and degrades the parallel efficiency, especially for short filters. For causal filters, where indices have a minimum of 0, $m = 3$ avoids zero padding at least at the lower end of indices. m also determines the alignment of vectorized data access. To make the read operations on x aligned, m should depend on n such that $n - m$ is a multiple of the vector size p, i.e., four in our examples. The alignment of read operations on h cannot be set independently, but this could be solved by preparing p copies of h with different alignments.

The application of the S operator already makes mild use of the associative law. It can be further exploited to vectorize most of the summing operation by commutating the sum and S operator:

$$y(n) = S\left(\sum_k x(n-4k-m,\ldots,n-4k-m+3) \odot h(4k+m,\ldots,4k+m-3)\right).$$
$$(13.12)$$

There are still scalar operations in this algorithm such as the S operator and also the store operation on y. To make the entire process parallel, we have to look for a different approach. Therefore, we make use of the loop transposition method described in Sect. 13.2.4 by introducing another index l that shall be used to vectorize blocks of n-indices. It turns out that we have two options to reformulate Eq. (13.10), namely

$$B: y(n+l) = \sum_k x(n+l-k)h(k), \quad \text{and} \quad (13.13)$$

$$C: y(n+l) = \sum_k x(n-k)h(k+l). \quad (13.14)$$

Let us look at method C first. The resulting vectorization strategy is depicted in Fig. 13.1 as C, and can be formulated as

$$y(n,\ldots,n+3) = \sum_k x(n-k)_{(0,0,0,0)} \odot h(k,k+1,k+2,k+3), \qquad (13.15)$$

where the so-called *splat* operator $a_{(0,0,0,0)} = (a,a,a,a)$ on a scalar a creates a vector filled with the value a. We see that this method is still not completely vectorized because it reads the x array sequentially before applying the splat operator. However, this may be circumvented by vectorized reads followed by four simple shuffle operations for each read, i.e., $x(n-k,\ldots,n-k+3)_{(i,i,i,i)}$ for $0 \le i < 4$.

Note that the range of the index k has to be extended to generate all products. For causal filters, k has to start at $k = -3$. This introduces the need of additional zero-padding of h and, as a consequence, redundant operations. Moreover, the access of the h array is entirely non-aligned.

Therefore, our hope lies in method B. Its vectorization strategy is depicted in Fig. 13.1 as B, and can be formulated as

$$y(n,\ldots,n+3) = \sum_k x(n-k,\ldots,n-k+3) \odot h(k)_{(0,0,0,0)}. \qquad (13.16)$$

This method has the big advantage that no zero-padding of h is necessary. Therefore, there are no redundant calculations. Two disadvantages are the non-aligned access of x and the sequential access of h. The latter problem can be reduced by preparing vectors $h(k)_{(0,0,0,0)}$ in advance, which is favorable especially for short filters.

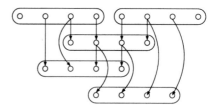

Fig. 13.2 Shuffle operations for all vector realignments on Intel architecture.

The non-aligned access of x implies one shuffle operation per non-aligned read, i.e., $p-1 = 3$ shuffles for $p = 4$ reads. However, these shuffle operations may not be available as single instructions on certain architectures. Unfortunately, this is the case for Intel SSE. However, as all possible realignments are necessary, shuffled vectors can be reused in other shuffle operations to also achieve a rate of one shuffle per non-aligned read. The method is depicted in Fig. 13.2 and can be written as

$$\begin{aligned}
a &= (x(n,\ldots,n+3),x(n+4,\ldots,n+7))_{(2,3,4,5)}, \\
x(n+1,\ldots,n+4) &= (x(n,\ldots,n+3),a)_{(1,2,5,6)}, \\
x(n+2,\ldots,n+5) &= a, \\
x(n+3,\ldots,n+6) &= (a,x(n+4,\ldots,n+7))_{(1,2,5,6)}.
\end{aligned} \qquad (13.17)$$

To summarize, we have applied the associative law and the loop transposition method to reschedule and reformulate loop iterations in order to vectorize the simple FIR filter algorithm. Method B turns out to be the most efficient due to the lack of redundant calculations. This is confirmed by experiments. We will now apply these insights in the vectorization of some exemplary and more complicated filter banks.

13.3.2 The Haar Filter

The Haar filter is the simplest orthogonal wavelet filter. It is a 2-tap filter. The coefficients are $(a,a) = (\frac{\sqrt{2}}{2}, \frac{\sqrt{2}}{2})$ in the low-pass form and $(a,-a) = (\frac{\sqrt{2}}{2}, -\frac{\sqrt{2}}{2})$ in the high-pass form, where the low- and high-pass filters form a filter bank. Together with down-sampling by a factor of 2, the following assignments define the filtering algorithm of the Haar wavelet transform.

$$\text{for all } i : L(i) \leftarrow ax(2i) + ax(2i+1), \; H(i) \leftarrow ax(2i) - ax(2i+1) \qquad (13.18)$$

L and H are the low-pass and the high-pass subbands, respectively. As a first sequential improvement we can reuse already computed products, which leads to

$$\text{for all } i : p \leftarrow ax(2i), \; q \leftarrow ax(2i+1), \; L(i) \leftarrow p+q, \; H(i) \leftarrow p-q . \qquad (13.19)$$

We see that for each pair $L(i), H(i)$ of output values we have to read two input values $x(2i), x(2i+1)$. Since we want to read and write only full vectors when using SIMD, we consequently have to read two vectors in each iteration. We find the vectorization of the Haar filter as

for all i :
$$\begin{aligned} &p \leftarrow x(8i,\ldots,8i+3) \odot a_{(0,0,0,0)}, \; q \leftarrow x(8i+4,\ldots,8i+7) \odot a_{(0,0,0,0)}, \\ &r \leftarrow (p,q)_{(0,2,4,6)}, \qquad\qquad\qquad s \leftarrow (p,q)_{(1,3,5,7)}, \\ &L(4i,\ldots,4i+3) \leftarrow r \oplus s, \qquad H(4i,\ldots,4i+3) \leftarrow r \ominus s . \end{aligned} \qquad (13.20)$$

In the first line two perfectly aligned vectors are read and each element is immediately multiplied by the coefficient a. In the second line the elements are rearranged into one vector containing all even elements and one containing all uneven elements using shuffle operations. To calculate the sum and difference of every two neighboring elements, we just have to add and subtract the two vectors, which is done in the third line.

While the sequential algorithm requires two multiplies and two additions (or subtractions) for every two input values, the SIMD version requires two packed multiplies and two packed additions for every eight input values. This gives a theoretical speedup of 4. However, since the shuffle operations also require some execution time and memory access can be a bottleneck, the speedup is reduced and we get an actual speedup of 2.7.

13.3.3 Biorthogonal 7/9 Without Lifting

In the following sections we will discuss the more complicated example of the biorthogonal 7/9-tap filter which is used in many multimedia applications such as the JPEG2000 standard [17]. Note that all algorithms will show the same phases: memory read, coefficient multiplication, data rearrangement, summation and memory write. Some will have a different order of execution, though. Especially coefficient multiplication and data rearrangement will be interchanged.

13.3.3.1 Sequential Algorithm

The biorthogonal 7/9 filter is an example of an uneven, symmetrical filter. It has 9 low-pass (a,b,c,d,e,d,c,b,a) and 7 high-pass coefficients (p,q,r,s,r,q,p). The sequential algorithm is

for all i :
$$\begin{aligned}
L(i) \leftarrow\ & ax(2i-4)+bx(2i-3)+cx(2i-2)+dx(2i-1)+ex(2i)\\
& +dx(2i+1)+cx(2i+2)+bx(2i+3)+ax(2i+4), \qquad (13.21)\\
H(i) \leftarrow\ & px(2i-2)+qx(2i-1)+rx(2i)+sx(2i+1)\\
& +rx(2i+2)+qx(2i+3)+px(2i+4).
\end{aligned}$$

However, this algorithm can be optimized in terms of the number of required multiplications due to the symmetry of the filters. Samples that have to be multiplied by the same coefficient and added afterwards can be added before multiplication instead, saving one multiply.

for all i :
$$\begin{aligned}
L(i) \leftarrow\ & a(x(2i-4)+x(2i+4))+b(x(2i-3)+x(2i+3))\\
& +c(x(2i-2)+x(2i+2))+d(x(2i-1)+x(2i+1))+ex(2i), \quad (13.22)\\
H(i) \leftarrow\ & p(x(2i-2)+x(2i+4))+q(x(2i-1)+x(2i+3))\\
& +r(x(2i)+x(2i+2))+sx(2i+1).
\end{aligned}$$

Thus, 14 adds and only 9 multiplies (instead of 16) are required in each iteration. To see the gain in performance of the optimized sequential algorithm, look at Fig. 13.3. This plot shows the execution times in ns/sample over the size of transformed data. The algorithm has been performed several times on the same data in order to unveil the influence of cache on the execution time. However, the fact that execution times per sample do not vary significantly with the data size shows that accessing cached data has little impact on the performance. This shows that memory access is not a bottleneck and the speedups shown in this and the following sections represent algorithmic improvements. The improved algorithm gains a sequential speedup of 1.18. All parallel speedups in this section will be measured against the improved algorithm.

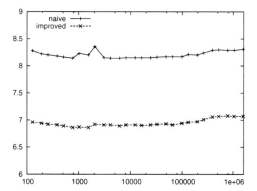

Fig. 13.3 Execution time of naive and improved sequential algorithm in ns/sample. The *horizontal axis* shows the size of the repeatedly transformed data set in number of single precision values.

13.3.3.2 SIMD Parallelization – Variant 1

There are many possibilities to parallelize the above algorithm. The main difference between these variants is when to apply the phase of shuffle operations – before or after multiplying with filter coefficients. The first variant performs this multiplication directly after source data is read from memory.

As with the Haar filter, two vectors have to be read to calculate one new low-pass vector and one new high-pass vector. However, since the filter is now longer than the two taps, the contents of more than two vectors are actually needed. This can be overcome by reusing intermediate results from previous iterations, which amounts to passing values from iteration to iteration.

In this first variant, the values of each of the two recently read vectors are immediately multiplied by all necessary filter coefficients. Then appropriate shuffles of the products have to be added, leading to the following algorithm:

for all i :
$$
\begin{aligned}
&Y \leftarrow x(8i+4,\ldots,8i+7), Z \leftarrow x(8i+8,\ldots,8i+11) \\
&A \leftarrow C, \ B \leftarrow D, \ C \leftarrow Y \odot (a,b,a,b), \ D \leftarrow Z \odot (a,b,a,b), \\
&E \leftarrow G, F \leftarrow I, \ G \leftarrow Y \odot (c,d,c,d), \ I \leftarrow Z \odot (c,d,c,d), \\
&J \leftarrow M, K \leftarrow N, M \leftarrow Y \odot (e,0,e,0), \ N \leftarrow Z \odot (e,0,e,0), \\
&L(4i,\ldots,4i+3) \leftarrow (A,B)_{(0,2,4,6)} \oplus (A,B)_{(1,3,5,7)} \oplus (E,F,G)_{(2,4,6,8)} \oplus \\
&\quad (E,F,G)_{(3,5,7,9)} \oplus (K,M)_{(0,2,4,6)} \oplus (F,G)_{(1,3,5,7)} \oplus (F,G,I)_{(2,4,6,8)} \oplus \\
&\quad (B,C,D)_{(3,5,7,9)} \oplus (C,D)_{(0,2,4,6)}, \\
&P \leftarrow R, \ Q \leftarrow S, \ R \leftarrow Y \odot (p,q,p,q), \ S \leftarrow Z \odot (p,q,p,q), \\
&T \leftarrow V, \ U \leftarrow W, \ V \leftarrow Y \odot (r,s,r,s), \ \ W \leftarrow Z \odot (r,s,r,s), \\
&H(4i,\ldots,4i+3) \leftarrow (P,Q,R)_{(2,4,6,8)} \oplus (P,Q,R)_{(3,5,7,9)} \oplus (U,V)_{(0,2,4,6)} \oplus \\
&\quad (U,V)_{(1,3,5,7)} \oplus (U,V,W)_{(2,4,6,8)} \oplus (Q,R,S)_{(3,5,7,9)} \oplus (R,S)_{(0,2,4,6)}
\end{aligned}
\tag{13.23}
$$

Figure 13.4 depicts the algorithm as a data-flow diagram. After multiplying the two new source vectors by vectors of appropriate filter coefficients, they are rearranged by shuffle operations (thin arrows) so that the sum of the resulting vectors is

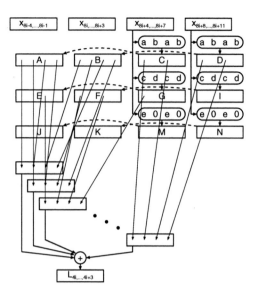

Fig. 13.4 Variant 1 of SIMD-parallel algorithm. Vectors are indicated by boxes, multiplication by boxes with rounded edges, addition by a circle with a +, shuffle operations by *thin arrows*, and the passing of values between iterations by *dashed arrows*. Only the low-pass calculations are shown, high-pass operations are similar.

the desired destination vector containing four low-pass filtered samples. Note that the intermediate vectors (after multiplication) are passed from the previous iteration (dashed arrows). In this way one can avoid half of the multiplication operations.

Only the low-pass calculations are shown. The operations for high-pass filtering are similar. A big disadvantage of this variant is that no intermediate results can be shared between the low- and high-pass part. Moreover, many shuffle operations have to be composed by two or more instructions. One reason for this is that some such operations require three source vectors. Another reason is that the Intel processor's instruction set does not allow arbitrary shuffles. Altogether this algorithm can be implemented by 10 multiplies, 14 adds, and 26 shuffles.

13.3.3.3 SIMD Parallelization – Variant 2

A major disadvantage of the first variant is that values that have to be collected in a single vector are spread over several intermediate vectors, requiring more shuffle operations. The reason for this is that downsampling causes every second value to belong together. Therefore, the second variant inserts a single step of shuffling before the multiplication, putting even and odd samples into separate vectors. This leads to the following algorithm, which is also shown in Fig. 13.5.

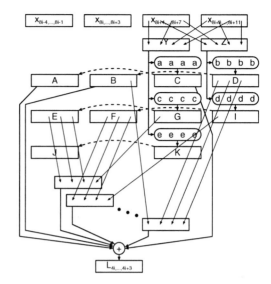

Fig. 13.5 Variant 2 of SIMD-parallel algorithm.

for all i :

$$Y \leftarrow x(8i+4, 8i+6, 8i+8, 8i+10), Z \leftarrow x(8i+5, \ldots, 8i+11),$$
$$A \leftarrow C, \; B \leftarrow D, \; C \leftarrow Y \odot (a,a,a,a), \; D \leftarrow Z \odot (b,b,b,b),$$
$$E \leftarrow G, \; F \leftarrow I, \; G \leftarrow Y \odot (c,c,c,c), \; I \leftarrow Z \odot (d,d,d,d),$$
$$J \leftarrow K, \qquad K \leftarrow Y \odot (e,e,e,e),$$
$$L(4i, \ldots, 4i+3) \leftarrow A \oplus B \oplus (E,G)_{(1,2,3,4)} \oplus (F,I)_{(1,2,3,4)} \oplus \tag{13.24}$$
$$(J,K)_{(2,3,4,5)} \oplus (F,I)_{(2,3,4,5)} \oplus (E,G)_{(3,4,5,6)} \oplus (B,D)_{(3,4,5,6)} \oplus C$$
$$P \leftarrow R, \; Q \leftarrow S, \; R \leftarrow Y \odot (p,p,p,p), \; S \leftarrow Z \odot (q,q,q,q),$$
$$T \leftarrow V, \; U \leftarrow W, V \leftarrow Y \odot (r,r,r,r), \quad W \leftarrow Z \odot (s,s,s,s),$$
$$H(4i, \ldots, 4i+3) \leftarrow (P,R)_{(1,2,3,4)} \oplus (Q,S)_{(1,2,3,4)} \oplus (T,V)_{(2,3,4,5)} \oplus$$
$$(U,W)_{(2,3,4,5)} \oplus (T,V)_{(3,4,5,6)} \oplus (Q,S)_{(3,4,5,6)} \oplus R$$

This has two advantages. First, there is one multiplication less for the e-coefficient. Second, no shuffle requires more than two source vectors. Moreover, the two results of the first shuffling step can be reused in the high-pass part. Thus, this algorithm is implemented by only 9 multiplies, 14 adds, and 20 shuffles.

13.3.3.4 SIMD Parallelization – Variant 3

The third variant adopts the scheme of the improved sequential algorithm. First, the input vectors are shuffled so that the remaining operations can be performed as in the sequential case. This reverses the order of phases completely. Then, vectors that have to be multiplied by the same filter coefficients are added, followed by multiplication and the final sum. The following algorithm is also shown in Fig. 13.6.

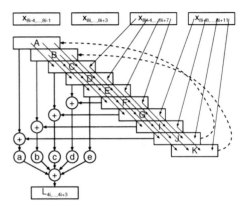

Fig. 13.6 Variant 3 of SIMD-parallel algorithm. Multiplication by a vector of equal coefficients is depicted by a single circle.

for all i :

$$Y \leftarrow x(8i+4,\ldots,8i+7), Z \leftarrow x(8i+8,\ldots,8i+11),$$
$$A \leftarrow J, B \leftarrow K, \quad C \leftarrow (A,Y)_{(1,2,3,4)}, \ D \leftarrow (B,Y)_{(1,2,3,5)},$$
$$E \leftarrow (C,Y)_{(1,2,3,6)}, \ F \leftarrow (D,Y)_{(1,2,3,7)}, \ G \leftarrow (E,Z)_{(1,2,3,4)},$$
$$I \leftarrow (F,Z)_{(1,2,3,5)}, \ J \leftarrow (G,Z)_{(1,2,3,6)}, \ K \leftarrow (I,Z)_{(1,2,3,7)}, \quad (13.25)$$
$$L(4i,\ldots,4i+3) \leftarrow (A \oplus J) \odot (a,a,a,a) \oplus (B \oplus I) \odot (b,b,b,b) \oplus$$
$$(C \oplus G) \odot (c,c,c,c) \oplus (D \oplus F) \odot (d,d,d,d) \oplus E \odot (e,e,e,e)$$
$$H(4i,\ldots,4i+3) \leftarrow (C \oplus J) \odot (p,p,p,p) \oplus (D \oplus I) \odot (q,q,q,q) \oplus$$
$$(E \oplus G) \odot (r,r,r,r) \oplus F \odot (s,s,s,s)$$

Note that only two vectors have to be passed to the next iteration. This reduces the stress on register allocation significantly. The biggest advantage of this algorithm is that all results of the shuffle phase can be reused in the high-pass part. Unfortunately, none of the shuffles, as depicted in Fig. 13.6, can be implemented as a single instruction. However, through appropriate rearrangements some of the additional instructions can be avoided. Altogether, this variant requires 9 multiplies, 14 adds, and 12 shuffles.

13.3.3.5 Experimental Results

As variants 2 and 3 of the SIMD algorithms have the same number of multiplies and adds as the improved sequential algorithm, only with vectors instead of single numbers, there is a potential speedup of 4. However, due to massive shuffle operations this speedup cannot be reached, as one can see in Fig. 13.7. According to expectations variant 3 is the best, giving speedups of 1.8.

Again, accessing cached data has only a minor influence on performance. The decay of speedup for small data sizes is due to complex startup and close-off operations, e.g., for initializing registers, which become more dominant for small data sizes. The slight decay for large data sizes is probably due to cache effects.

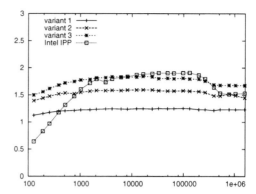

Fig. 13.7 Speedups of the SIMD parallelization variants against the improved sequential algorithm. The *horizontal axis* again shows the size of the repeatedly transformed data set.

The hand-optimized Intel IPP library has slightly better speedups for medium data sizes. However, it seems to be more dependent on cache since its performance decreases noticeably for large data sizes. Also, it seems to have even more problems with startup operations for small data sizes, although filter allocation is performed only once for all repeated calls in the experiment. Note that `ippsWTFwd_32f` is used here which does not apply lifting and where filters are not fixed, i.e., defined at runtime.

13.3.3.6 Applicability to Arbitrary Filter Banks

The approaches presented here can all be applied to other filters as well. It is not apparent, however, which one would be the best for a given filter, or if some modification of a variant can do even better. Let us, therefore, look at how the features of the presented variants behave on other kinds of filters.

Variants 1 and 2 rely on the fact that a single filter coefficient has to be applied to either even or odd samples, but not both. However, this is only true for uneven symmetrical filters, or filters without any symmetry. This means that variant 3 has even more advantages for even symmetrical filters. On the other hand, variant 3 might imply redundant multiplications for non-symmetrical filters if some low- and high-pass coefficients are equal. This happens mostly for orthogonal wavelets. In this case, however, filters have even length and, as a consequence, a low-pass coefficient for even samples always corresponds to an equal high-pass coefficient for uneven samples, or vice versa. Therefore, variant 3 does not produce redundant multiplications for orthogonal wavelets, since multiplied even samples can never be reused for the high-pass filtering.

Important questions arise for particularly long filters. Variants 2 and 3 need to store at least one vector for each filter tap to pass it to the next iteration. This requires the allocation of many CPU registers and leads to additional memory accesses when the compiler runs out of available registers. On the other hand, variant 3 has to keep all shuffled vectors in registers, whereas variants 1 and 2 can drop shuffled vectors (and even some other intermediate vectors) after having added them to the final sum. However, variant 3 can also drop these if the filter is non-symmetrical.

All these remarks are only hints, of course. Filters reveal surprisingly diverse features with respect to SIMD parallelization. Each particular filter should be examined thoroughly, based on the approaches presented above.

13.3.4 Biorthogonal 7/9 With Lifting

As most wavelet filters, the biorthogonal 7/9 filter can also be implemented by applying the lifting scheme [18]. It is a method to implement wavelet filter pairs in a joint pass. In this way it is possible to reduce the total number of operations.

13.3.4.1 Sequential Algorithm

The lifting approach factors the filter pair into several predict and update steps, where odd values (values at odd position) are predicted from even values and replaced by the difference between prediction and actual value, and even values are updated to represent a local average. This method significantly reduces the number of multiplies in the sequential algorithm. In this specific case the sequential biorthogonal 7/9 without lifting uses 9 multiplies for every two samples (improved version), whereas biorthogonal 7/9 with lifting as shown here requires only 6 multiplies.

$$
\begin{aligned}
&\text{for all } i : x(2i+1) \leftarrow x(2i+1) + a(x(2i) + x(2i+2)), \\
&\text{for all } i : x(2i) \leftarrow x(2i) + b(x(2i-1) + x(2i+1)), \\
&\text{for all } i : x(2i+1) \leftarrow x(2i+1) + c(x(2i) + x(2i+2)), \\
&\text{for all } i : x(2i) \leftarrow x(2i) + d(x(2i-1) + x(2i+1)), \\
&\text{for all } i : x(2i+1) \leftarrow -ex(2i+1), \\
&\text{for all } i : x(2i) \leftarrow \tfrac{1}{e}x(2i)
\end{aligned}
\tag{13.26}
$$

The low-pass and high-pass subbands are then found interleaved in even and odd positions, respectively. Note that the coefficients $a,\dots,e$ are not the same as in the sequential algorithm, but are the result of the factorization process on which the lifting scheme is based. Note also that each of these assignments has to be executed for all i before proceeding with the next assignment.

The lifting scheme can also be implemented in a single-loop manner in the sense that each input value is read from memory only once and each output value is written to memory once without subsequent updates. While this is an improvement in itself, since it minimizes memory access, it turns out to be the only reasonable way to go for the SIMD parallelization. To see why, let us examine the number of operations in a single lifting pass $x_{2n} \leftarrow x_{2n} + \alpha(x_{2n-1} + x_{2n+1})$. There are 2 adds and 1 multiply for every second sample, which makes 1 add and $\frac{1}{2}$ multiply per sample. We can vectorize these operations by

$$x(2n, \ldots, 2n+3) \leftarrow x(2n, \ldots, 2n+3)+$$
$$(\alpha, 0, \alpha, 0) \odot (x(2n-1, \ldots, 2n+2) + x(2n+1, \ldots, 2n+4)). \quad (13.27)$$

Since $x(2n-1, \ldots, 2n+2)$ and $x(2n+1, \ldots, 2n+4)$ require shuffle operations, we need 2 shuffles, 2 adds, and 1 multiply for every four samples, giving $\frac{1}{2}$ shuffle, $\frac{1}{2}$ add, and $\frac{1}{4}$ multiply per sample or – taken together – 1.25 operations instead of 1.5 in the non-SIMD case. This is, obviously, not a satisfying speedup, given the theoretical maximum speedup of 4.

Therefore, we develop a new algorithm with a single outer loop. To do so, we have to rewrite it by applying the well-known loop fusion technique (see Sect. 13.2.3). Immediately after iteration (i, j) of loop i, iteration $(i+1, k)$ of the subsequent loop $i+1$ is executed that depends on iteration (i, j) and does not depend on an iteration (i, l) in loop i occurring later in that loop $(l > j)$. The process begins with the first loop. After one iteration of each loop has been executed, one iteration of the fused loop is completed and the process starts over with a subsequent iteration. As iteration (i, j) also depends on iteration $(i, j-1)$, values have to be passed between iterations. For every two input values, two output values can be calculated, one low-pass and one high-pass coefficient. This leads to the following algorithm:

$$
\begin{aligned}
&\text{for all } i: \\
&\quad o \leftarrow q, \; p \leftarrow x(2i+3), \; q \leftarrow x(2i+4), \\
&\quad r \leftarrow s, \; s \leftarrow p + a(o+q), \\
&\quad t \leftarrow u, \; u \leftarrow o + b(r+s), \\
&\quad v \leftarrow w, \; w \leftarrow r + c(t+u), \\
&\quad L(i) \leftarrow t + d(v+w) \cdot \tfrac{1}{e}, \; H(i) \leftarrow w \cdot (-e).
\end{aligned}
\quad (13.28)
$$

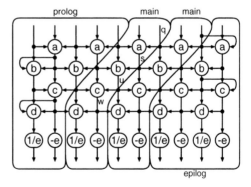

Fig. 13.8 Sequential single-loop algorithm for the biorthogonal 7/9 filter with lifting. Circles with three inputs (l left, r right, u upper) denote basic lifting operations $y = u + \alpha(l + r)$. Rounded frames indicate single iterations.

This algorithm is also shown in Fig. 13.8 for a very short data length of 10. Iterations, as described above, are denoted "main." Longer data would, of course, require more "main" iterations. Note that intermediate values q, s, u, w are passed from iteration to iteration, indicated by arrows that cross iteration borders in Fig. 13.8. These four values have to be set properly at the beginning of the loop. Also, the end of the loop needs special treatment. Figure 13.8 shows how this must be done in the case of mirroring border handling in the phases denoted by "prolog" and "epilog."

13.3.4.2 SIMD Parallel Algorithm

To be able to obtain speedup using SIMD operations, again full vectors have to be read. Like in variant 2 of the biorthogonal filter without lifting, data is shuffled after being read from memory. Then SIMD operations are applied. This leads to intermediate results which have to be shuffled again before proceeding. These results can be reused in the next iteration step, much like in the sequential algorithm, which leads to the following algorithm:

for all i :
$$
\begin{aligned}
&h \leftarrow x_2, && x_1 \leftarrow x(8i+4,\ldots,8i+7), && x_2 \leftarrow x(8i+8,\ldots), \\
&q \leftarrow (h,x_1)_{(0,2,4,6)}, \; p \leftarrow (h,x_1,x_2)_{(3,5,7,9)}, && && o \leftarrow (h,x_1)_{(2,4,6,8)}, \\
&r \leftarrow s, && s \leftarrow (a,a,a,a)\odot(o\oplus q)\oplus p, \; r \leftarrow (r,s)_{(3,5,6,7)}, \\
&t \leftarrow u, && u \leftarrow (b,b,b,b)\odot(r\oplus s)\oplus o, \; t \leftarrow (t,u)_{(3,5,6,7)}, \\
&v \leftarrow w, && w \leftarrow (c,c,c,c)\odot(t\oplus u)\oplus r, \; v \leftarrow (v,w)_{(3,5,6,7)}, \\
&L(4i,\ldots,4i+3) \leftarrow ((d,d,d,d)\odot(v\oplus w)\oplus t)\odot(\tfrac{1}{e},\tfrac{1}{e},\tfrac{1}{e},\tfrac{1}{e}), \\
&H(4i,\ldots,4i+3) \leftarrow (-e,-e,-e,-e)\odot w.
\end{aligned}
$$
(13.29)

See also Fig. 13.9 for a data-flow diagram of the algorithm.

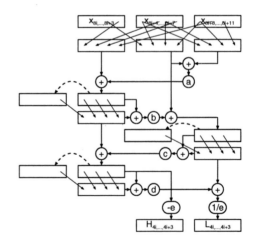

Fig. 13.9 SIMD-algorithm of biorthogonal 7/9 filter with lifting. Heavy use of shuffle-operations may cause non-optimal speedups. Like in the sequential case, intermediate values are passed between iterations (*dashed lines*).

The algorithm can also be interpreted as being equivalent to variant 3 of the non-lifting algorithm, applied to each of the four stages for coefficients a,b,c,d. To see this, consider each stage as the application of the short filters $(a,1,a),\ldots,(d,1,d)$. Then each stage consists of the steps shuffle, add, multiply, and sum, just like variant 3 in Sect. 13.3.3.2. Variants 1 and 2 could also be used here. However, considerations show that these would immediately imply unreasonable slow-downs. For other filters given in lifting scheme, a similar approach can be applied, interpreting the lifting steps as short filters.

Again, it is not possible to implement the algorithm in a straight forward way because SIMD extensions (e.g., Intel SSE instruction set) do not support shuffling from three sources into a single destination in a single instruction. However, the algorithm can be implemented with 6 multiplies, 8 adds, and 11 shuffles.

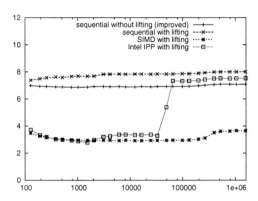

Fig. 13.10 Execution times in ns/sample of sequential and SIMD implementations with and without lifting over the size of the repeatedly transformed data set (number of floats).

13.3.4.3 Experimental Results

Figure 13.10 shows execution times of the sequential and SIMD implementations of the lifting algorithm in comparison to the non-lifting algorithm. Interestingly, the sequential implementation is slower with lifting than without, despite the reduced number of multiplies and adds. Theoretical considerations [18] would imply a speedup of 1.64. An investigation of the assembler code showed no obvious reason, the faster code being significantly longer. A guess is that there is a peculiar problem in scheduling the instructions optimally which can be resolved more easily in the longer code.

However, the SIMD implementation is able to reduce the execution times significantly. Again, cached values do not seem to play an important role. Figure 13.11 shows the speedup of the SIMD implementation compared to versions without lifting or SIMD. While, compared to the sequential lifting algorithm, we get a speedup of up to 2.66 (of a theoretical maximum of 4), the speedup is only 2.36 (of theoretical $1.64 \times 4 = 6.56$) compared to the sequential algorithm without lifting since the latter is faster, as mentioned above. However, the SIMD algorithm with lifting is faster than that without lifting. There is a speedup of about 1.3 (of theoretical 1.64). The speedup decay for large data sizes is again probably due to cache problems.

Again, the Intel IPP library is not able to outperform our SIMD implementation of wavelet lifting, as can be seen in Fig. 13.10. It shows equal performance for small

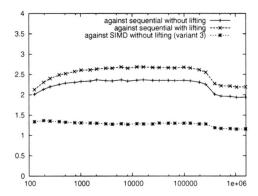

Fig. 13.11 Speedup of the SIMD implementation with lifting against implementations without lifting or SIMD.

and slightly worse performance for medium data sizes. For large data sizes there seems to be a major cache problem, since its performance even drops below that of the sequential non-lifting algorithm. Note that `ippiWTFwdRow_D97_JPEG2K_32f_C1R` is used where lifting is applied and the filter is fixed, as in our implementation.

13.3.5 Conclusion

The efficiency of the parallelization depends largely on the filter lengths, their alignments, and even on the coefficients of the filters. If some of the coefficients are equal, as there are for symmetrical filters, the sequential algorithm can be optimized by reusing computed values. To do the same in the SIMD parallelized algorithm often implies complicated shuffle operations.

Generally, the need for many shuffle operations reduces the speedup most. Memory access as a bottleneck could also limit speedups. However, investigations show that the execution times are almost invariant to whether source data is in cache or not. This means that the speedups shown above represent purely algorithmic improvements.

Apart from speedup issues, algorithms have to be found to derive optimal solutions. This is important because each parallelization presented here is one of many possible solutions and it is still possible that the shown solutions can be improved. Since in practice it would be an almost unaccomplishable amount of work to hand-code a variety of solutions to find the best, automatic optimization techniques as in [19] are required.

13.4 Recursive Algorithms

Algorithms of the convolution type are non-recursive, which means that output values are independent of each other. Whenever previous output values are reused in the computation of new values, the algorithm is called recursive. The IIR filter technique is the most important example of such an algorithm. Therefore, we shall investigate it and examine vectorization strategies.

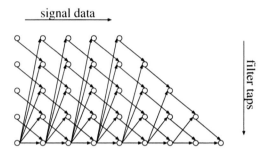

Fig. 13.12 Loop dependencies in IIR filtering.

From a computational point of view, the difference between FIR and IIR filters lies in the dependencies between loop iterations. Again, there are two loops, one over signal data and the other over filter taps. In the FIR case, iterations of the outer loop, i.e., entire inner loops, are independent of each other, leading to a rather straight-forward SIMD parallelization where the two loops (inner and outer) are transposed for a number of outer iterations equal to the SIMD vector size p, as shown in Sect. 13.3.1. In the IIR case, the dependencies are more complicated since all previous output values are required to calculate a new one. See Fig. 13.12 and compare to Fig. 13.1. Therefore, SIMD parallelization is more difficult.

In this section we will first apply usual rescheduling techniques and then show how algebraic transforms of the algorithm can improve the vectorization significantly, which is verified by experimental results. These are conducted on an Intel Pentium 4 CPU with 3.2 GHz and 2 MB cache size using the SSE extension with vectors of 4 single precision numbers. All implementations use the same amount of code optimization, i.e., memory access through incremented pointers instead of indexed arrays, and compilation with gcc 4.1.2 with the -O3 option. SIMD operations are implemented using gcc's built-in intrinsics for vector extensions and the -msse option. Note that in order to have full control over generated code, no automatic vectorization is applied. The results are compared to the hand-optimized Intel Integrated Performance Primitives (IPP) v5.3. Note that the IPP library also uses SIMD operations, but the applied methods are not known to the author.

13.4.1 Sequential IIR Algorithm

The goal of IIR filtering is to calculate the signal y from the signal x by

$$y(n) = \sum_{i=1}^{N-1} a(i)y(n-i) + \sum_{i=0}^{M-1} b(i)x(n-i), \qquad (13.30)$$

where the second term is an FIR part with coefficients $b(i)$ and the first term is the IIR part with coefficients $a(i)$. M is the number of FIR filter taps and N is the number of IIR filter taps. The formula reveals the outer loop over n and two inner loops over i.

The sequential implementation is optimized for performance to have a reasonable comparison for the SIMD parallelized version. It turns out that maintaining a pointer for $y(n)$ and $x(n)$ and addressing $x(n-i)$ and $y(n-i)$ via relative addressing is fastest. Using extra buffers or local register variables for reused values does *not* improve the performance. Therefore, a similar implementation style is adopted for the SIMD parallelization.

13.4.2 Scheduling Approach

Rescheduling approaches only change the order in which iterations and operations are executed. They have therefore limited power if there are too many data dependencies, as there are in IIR filtering. Examples can be found in [20,21]. We will use a rather straight forward approach that will be improved by algebraic transforms in the next section.

The FIR part is vectorized simply as in Sect. 13.3.1 (method B), with the result given in u. The IIR part can be parallelized in just the same way for those iterations where $i \geq p$, i.e., where the source vector $y(n-i,\ldots,n-i+p-1)$ does not overlap with the destination vector $y(n,\ldots,n+p-1)$ that is being calculated. The iterations $i = 0,\ldots,p-1$ might be implemented sequentially after computing the others in a vectorized way first by

$$v = u \oplus \sum_{i=p}^{N-1} y(n-i,\ldots,n-i+p-1) \odot (a_i,\ldots,a_i), \qquad (13.31)$$

followed by

$$y(n+k) = v_k + \sum_{i=1}^{p-1} a(i)y(n+k-i) \quad \text{for} \quad k = 0,\ldots,p-1. \qquad (13.32)$$

A first attempt to parallelize the latter part is to split it into two phases. The first phase treats those terms that reference $y(n+k-i)$ where $n+k-i < n$, i.e., already available values.

$$\text{for } i = 1, \ldots, p - 1:$$
$$v \leftarrow v \oplus (y(n - p + i), \ldots, y(n - 1), 0, \ldots) \odot \qquad (13.33)$$
$$(a(p - i), \ldots, a(p - i), 0, \ldots)$$

The second phase uses those elements of v that already represent $y(n + k)$ values. At the beginning, only $v_0 = y(n)$. Using this value, v_1 can be calculated to hold $y(n + 1)$, and so on. This leads to the following algorithm:

$$\text{for } k = 0, \ldots, p - 2:$$
$$v \leftarrow v \oplus (\ldots, 0, v_k, \ldots, v_k) \odot (\ldots, 0, a_1, \ldots, a_{p-1-k}) \qquad (13.34)$$
$$y(n, \ldots, n + p - 1) \leftarrow v$$

This first approach yields an overhead of $p - 1$ multiply-accumulate vector operations, since each phase has $p - 1$ iterations, resulting in $2(p - 1)$ operations, where only $p - 1$ would be necessary if there were no problems with data dependencies.

13.4.3 Algebraic Transforms

Algebraic transforms of the algorithm can be used to eliminate troubling data dependencies [22]. Here, we will follow an approach that fuses filter taps together to resolve data dependencies [23]. Let us look at the second iteration ($k = 1$) of the last algorithm. Here, $v_1 = y(n + 1) = v_1' + v_0 a(1)$, where v' comes from the preceding iteration. Now, we calculate the new v_2 as $v_2 + v_1 a(1)$, which can consequently be expressed as $v_2 + v_1' a(1) + v_0 a(1)^2$. Moreover, $v_2 = v_2' + v_0 a(2)$, as calculated in the first iteration. Together, we get $v_1' a(1) + v_0 (a(1)^2 + a(2))$. The term $v_1' a(1)$ could be calculated in the last iteration of the first phase, and the term $v_0 (a(1)^2 + a(2))$ can be calculated in the first iteration of the second phase because we have eliminated v_1 from the term.

Following this approach even further recursively, we get the following algorithm that substitutes both phases:

$$\text{for } i = 1, \ldots, p:$$
$$v \leftarrow v \oplus (y(n - p + i), \ldots, y(n - 1), 0, v_i, \ldots v_i) \odot s(i) \qquad (13.35)$$
$$y(n, \ldots, n + p - 1) \leftarrow v$$

$s(i)$ holds the fused filter tap coefficients and has the following form:

$$s(1) = (a(p - 1), \ldots, a(p - 1), 0)$$
$$s(2) = (a(p - 2), \ldots, a(p - 2), 0, c(1))$$
$$\cdots \qquad (13.36)$$
$$s(p - 1) = (a(1), 0, c(1), c(2), \ldots, c(p - 2))$$
$$s(p) = (0, c(1), c(2), \ldots, c(p - 1)),$$

where

$$c(k) = \sum_{i=1}^{k} a(k)c(k-i), \qquad c(0) = 1. \qquad (13.37)$$

This approach finally has only an overhead of one multiply-accumulate vector operation, since it has p iterations. For better comprehensibility, let us write the algorithm for the case $p = 4$ as in the Intel SSE architecture:

$$v \leftarrow v \oplus (y(n-3), y(n-2), y(n-1), 0) \odot (a(3), a(3), a(3), 0)$$
$$v \leftarrow v \oplus (y(n-2), y(n-1), 0, v_2) \odot (a(2), a(2), 0, a(1))$$
$$v \leftarrow v \oplus (y_{n-1}, 0, v_1, v_1) \odot (a(1), 0, a(1), a(1)^2 + a(2))$$
$$v \leftarrow v \oplus (0, v_0, v_0, v_0) \odot (0, a(1), a(1)^2 + a(2), a(1)^3 + 2a(1)a(2) + a(3))$$
$$y(n, \ldots, n+3) \leftarrow v$$

Of course, each operation requires at least one shuffle operation, maybe two on the Intel SSE architecture.

If the number of IIR-taps N is smaller than the vector size p, the above approach unfortunately only reduces to $p - 1$ operations. In this case, some divide-and-conquer algorithm might further reduce the overhead. However, $\lceil \log_2(p+1) \rceil$ seems to be the lower bound, since $y(n + p - 1)$ depends on the $p + 1$ values $u_0, \ldots, u_{p-1}, y(n-1)$ if N takes the minimal value 2.

13.4.4 Experimental Results

In Sect. 13.3 we have seen that the performance of an implementation of a filtering algorithm possibly depends on whether the signal data is in the cache or not. Therefore, we will adopt the method of varying data size to examine the cache behavior.

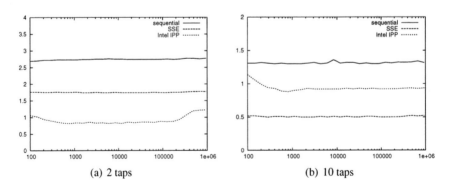

(a) 2 taps   (b) 10 taps

Fig. 13.13 Execution time in ns per sample point and filter tap depending on the data length for repeated filtering, showing the cache dependency of the algorithms.

The calculation time is expected to depend linearly on the data size and on the number of filter taps $N + M$. Therefore, we calculate the execution time per sample point and filter tap from the total execution time of the algorithm by $t_{total}/S/(N + M)$, where S is the data size.

Figure 13.13 shows the results for $N = M = 2$ and $N = M = 10$. It also includes performance measures of the Intel IPP library. While the IPP library code seems to depend a little on the data size, the major reason for this seems to be startup-overhead when filling the delay-lines, which is significant only for small data sizes. The sequential algorithm and the SIMD algorithm are completely independent of the cache state.

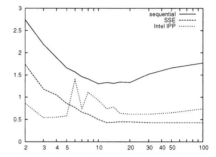

Fig. 13.14 Execution time in ns per sample point and filter tap depending on the number of filter taps.

For small numbers of taps, the IPP library code seems to be faster. This is also shown in Fig. 13.14. For $N = M \leq 5$, the SIMD algorithm cannot compete with the IPP code. The reason is probably that hand-optimized assembler code, as in the IPP library, is more important for short loops. For $N > 5$, however, our SIMD approach outperforms the IPP library by a speedup of about 1.7 and also shows more regular behavior. Compared to the sequential algorithm, speedups from 1.5 for small N to 4.5 for large N are obtained.

13.5 Block Algorithms

Algorithms that operate on blocks of signal data usually have a more irregular structure than streaming algorithms such as filtering. The most prominent example is, of course, the FFT as defined in Sect. 13.1.1. Almost all other blocked transforms are variants of the FFT and have very similar structure. As a consequence, vectorization strategies are basically the same. Therefore, we will concentrate on the FFT.

13.5.1 Data Layout

The FFT operates on complex data, which raises the question where real and imaginary parts of complex numbers are stored. The most common is an alternating scheme to keep real and imaginary parts closely together. The other possibility is to store them in separate arrays. What does that mean for vectorization efficiency? In the alternating scheme, $\frac{p}{2} = 2$ complex numbers are kept in a vector. Simultaneous addition of $2 + 2$ complex numbers simply takes the form of a vector addition. However, vectorized multiplication is more complicated. The point-wise complex product of arrays $z(n) = x(n)y(n)$ can be implemented by

$$
\begin{aligned}
&\text{for all } n : \\
&\quad a \leftarrow (\Re x(n), \Im x(n), \Re x(n+1), \Im x(n+1)) \\
&\quad b \leftarrow (\Re y(n), \Im y(n), \Re y(n+1), \Im y(n+1)) \\
&\quad c \leftarrow a \odot b, \quad d \leftarrow a \odot b_{(1,0,3,2)} \\
&\quad e \leftarrow (c,d)_{(0,4,2,6)}, \quad f \leftarrow (c,d)_{(1,5,3,7)} \odot (-1,1,-1,1) \\
&\quad (\Re z(n), \Im z(n), \Re z(n+1), \Im z(n+1)) \leftarrow e \oplus f
\end{aligned}
\tag{13.38}
$$

This scheme in principle needs two vector multiplications and one vector addition for $2 + 2$ complex numbers, whereas the sequential version needs four multiplications and two additions, or, more precisely, one addition and one subtraction for $1 + 1$ complex numbers, which seems perfect. However, there is an additional multiplication with $(-1, 1, -1, 1)$ that is necessary for the sign change in the vectorized addition, and there are 3 shuffle operations. Moreover, the two shuffles in line 4 need two instructions on Intel SSE, which makes a total of five shuffles. As a consequence, the speedup we get if we implement a sequence of complex multiplications in this way is actually a slowdown of about 0.7. This is a bad thing to start with when trying to vectorize an algorithm that is based on complex numbers.

On the other hand, the data layout with separate arrays for real and imaginary parts implies a vectorized algorithm that is equivalent to the sequential algorithm:

$$
\begin{aligned}
\Re z(n,\dots,n+3) &= \Re x(n,\dots) \odot \Re y(n,\dots) \ominus \Im x(n,\dots) \odot \Im y(n,\dots) \\
\Im z(n,\dots,n+3) &= \Re x(n,\dots) \odot \Im y(n,\dots) \oplus \Im x(n,\dots) \odot \Re y(n,\dots)
\end{aligned}
\tag{13.39}
$$

It uses 4 vector multiplications and 2 vector additions for $4 + 4$ complex numbers, which is perfect, and there are no shuffle operations at all. As a consequence, we get a speedup of about 3.7 for a sequence of multiplications.

However, the data layout might be predetermined by existing software or interface definitions. In this case, data could be rearranged after reading from memory and before writing to memory. This can be done by one shuffle operation per input and output vector. Intermediate stages of the algorithm can keep the separated data organization, though.

This rearrangement can be incorporated into the bit-reverse sorting pass that is part of the beginning or end of the FFT algorithm. Bit-reverse sorting moves $x(m)$ to $y(n)$, where the binary representations of m and n satisfy

$$m = m_0 2^0 + \cdots + m_{B-1} 2^{B-1} = BR(n) := n_{B-1} 2^0 + \cdots + n_0 2^{B-1}, \qquad (13.40)$$

hence the name. If we combine these movements with the separation of real and imaginary parts, the sorting algorithm almost does not change. Suppose the array $\tilde{x}$ holds the alternated parts of the complex x, i.e., $\tilde{x}(2n, 2n+1) = (\Re x(n), \Im x(n))$. If the data block size is at least 8, i.e., $0 \leq n < N \leq 8$, or, equivalently, $B \geq 3$, then the sorting plus separation can be vectorized by

$$
\begin{aligned}
&\text{for all } n : \\
&\quad a \leftarrow \tilde{x}(BR(n), \ldots, BR(n)+3), \\
&\quad b \leftarrow \tilde{x}(BR(n+1), \ldots, BR(n+1)+3), \\
&\quad c \leftarrow \tilde{x}(BR(n+2), \ldots, BR(n+2)+3), \\
&\quad d \leftarrow \tilde{x}(BR(n+3), \ldots, BR(n+3)+3), \\
&\quad e \leftarrow (a,b)_{(0,2,4,6)}, \quad f \leftarrow (a,b)_{(1,3,5,7)}, \\
&\quad g \leftarrow (c,d)_{(0,2,4,6)}, \quad h \leftarrow (c,d)_{(1,3,5,7)}, \\
&\quad \Re y(n, \ldots, n+3) \leftarrow (e,g)_{(0,4,1,5)}, \\
&\quad \Im y(n, \ldots, n+3) \leftarrow (f,h)_{(0,4,1,5)}, \\
&\quad \Re y(n+4, \ldots, n+7) \leftarrow (e,g)_{(2,6,3,7)}, \\
&\quad \Im y(n+4, \ldots, n+7) \leftarrow (f,h)_{(2,6,3,7)},
\end{aligned}
\qquad (13.41)
$$

where n is a multiple of $2p = 8$. This requires eight shuffles for four input vectors.

13.5.2 Basic FFT-Blocks

After bit-reverse sorting, the actual algorithm ensues with recursions such as that in Eq. (13.5). If the data size N in a recursion iteration is greater than 4, then the iteration consists of point-wise multiplication of half of the complex data by complex factors of the form $e^{-i\frac{2\pi}{N}n}$, followed by addition and subtraction with the other half of the data. Due to our data layout, this can be done easily by vectorized multiplications as in Eq. (13.39).

If the data consists of four complex values, then vector-local computations are necessary. The FFT of size $N = 4$, i.e., $y = \mathscr{F}_N x$ is written out sequentially as

$$
\begin{aligned}
b(0) &\leftarrow x(0) + x(1), & b(1) &\leftarrow x(0) - x(1), \\
b(2) &\leftarrow x(2) + x(3), & b(3) &\leftarrow x(2) - x(3), \\
y(0) &\leftarrow b(0) + b(2), & y(1) &\leftarrow b(1) - ib(3), \\
y(2) &\leftarrow b(0) - b(2), & y(3) &\leftarrow b(1) + ib(3),
\end{aligned}
\qquad (13.42)
$$

where x is assumed to be already bit-reverse sorted, i.e., $x(1)$ and $x(2)$ are swapped. This algorithm looks quite regular, but the imaginary factor $-i$ that accompanies $b(3)$ disturbs the regularity significantly. Nevertheless, a straight forward vectorization can be given by

$$
\begin{aligned}
\Re b &\leftarrow \Re x \odot (1,-1,1,-1) \oplus \Re x_{(1,0,3,2)}, \\
\Im b &\leftarrow \Im x \odot (1,-1,1,-1) \oplus \Im x_{(1,0,3,2)}, \\
\Re y &\leftarrow \Re b_{(0,1,0,1)} \oplus (\Re b, \Im b)_{(2,7,2,7)} \odot (1,1,-1,-1), \\
\Im y &\leftarrow \Im b_{(0,1,0,1)} \oplus (\Im b, \Re b)_{(2,7,2,7)} \odot (1,-1,-1,1).
\end{aligned}
\tag{13.43}
$$

We see that there are again vector multiplications for sign change. Note that the algorithm itself does not include any multiplications at all. There are six shuffle operations, whereof two require two instructions on Intel SSE. To get rid of the multiplications, we reschedule the operations so that additions and subtractions are separated, which is possible because there is always an equal number of positive and negative signs. This leads to the following algorithm:

$$
\begin{aligned}
a &\leftarrow (\Re x, \Im x)_{(0,2,4,6)}, \quad b \leftarrow (\Re x, \Im x)_{(1,3,5,7)}, \quad c \leftarrow a \oplus b, \quad d \leftarrow a \ominus b, \\
e &\leftarrow (c,d)_{(0,2,4,6)}, \quad f \leftarrow (c,d)_{(1,3,7,5)}, \quad g \leftarrow e \oplus f, \quad h \leftarrow e \ominus f, \\
\Re y &\leftarrow (g,h)_{(0,2,4,6)}, \quad \Im y \leftarrow (g,h)_{(1,7,5,3)}.
\end{aligned}
\tag{13.44}
$$

There are still six shuffle operations, only one of which needs two instructions on Intel SSE. Surprisingly, this algorithm is about 20% slower than that in Eq. (13.43). The reason is probably increased dependency of vector instructions and, thus, worse schedulability. All this shows that code optimization is difficult due to architecture dependencies, but necessary nevertheless. This problem is addressed in the next section.

13.5.3 Automatic Tuning and Signal Processing Languages (SPL)

Because implementations of algorithms show different performance characteristics on different architectures, optimal implementations have to be found on each architecture separately. This not only requires implementation efforts on each architecture, but many implementations have to be tested on each architecture. As this is rarely done manually, implementations are likely to be suboptimal.

To solve this problem, automatic tuning systems have been developed [4, 5], an approach that is well known in matrix algebra [24–26]. The idea behind these systems is that the transform is represented by a matrix M, i.e., $y = Mx$, and this matrix can be factored into sparse matrices M_k as

$$
M = M_1 M_2 \cdots M_m.
\tag{13.45}
$$

These matrices can be built from the following primitive matrices:

- the identity matrix $I_n = \mathrm{diag}(1,\ldots,1)$,
- the stride permutation matrix $L_r^{rs} = \delta(js+k, j+kr)$ of size $rs \times rs$, where $0 \le j < r$ and $0 \le k < s$, and
- the "twiddle"-matrix $T_r^{rs} = \mathrm{diag}(w^{0\cdot 0}, \ldots, w^{0\cdot(r-1)}, w^{1\cdot 0}, \ldots, w^{(s-1)(r-1)})$, where $w = e^{-i\frac{2\pi}{rs}}$.

The primitive matrices can be combined by the following operations:

- matrix multiplication,
- direct sum $A \oplus B = \begin{pmatrix} A & \\ & B \end{pmatrix}$,
- Kronecker product $A \otimes B = \begin{pmatrix} A_{0,0}B & \cdots & A_{0,s-1}B \\ \vdots & \ddots & \vdots \\ A_{r-1,0}B & \cdots & A_{r-1,s-1}B \end{pmatrix}$, and
- recursion, i.e., the use of smaller matrices with the same definition.

Together, these matrices and operations form a framework of a SPL [6]. As an example, it is possible to define the Fourier transform of size 4 (DFT$_4$) in this language through the formula

$$\text{DFT}_4 = (\text{DFT}_2 \otimes I_2)T_2^4(I_2 \otimes \text{DFT}_2)L_2^4. \tag{13.46}$$

Such a formula does not only represent a way to construct the matrix of the transform, it also defines an algorithm by which the transform can be implemented. A recursively expanded formula can automatically be converted into an actual algorithm in some programming language by substituting the primitive matrices or simple combinations M_j of them by appropriate loops of arithmetic operations. Because the matrices M_j are supposed to be sparse, the resulting algorithm usually reduces the computational complexity. For the Fourier transform, the complexity reduction is from $O(N^2)$ to $O(N \log N)$.

If a formula such as Eq. (13.46) is defined with symbolic indices (e.g., DFT$_{rs}$ = ...), then the formula constitutes a rule that can be applied in the recursive expansion of formulas. Usually, the parameters of a rule allow for several possible instantiations (e.g., $rs = 2 \cdot 4$ or $4 \cdot 2$). Moreover, there can be several applicable rules. Thus, a vast space of algorithmic implementations of a certain transform can be generated automatically.

The goal of the automatic tuning system is to traverse this space, to measure the implementations' performances, and to choose the one implementation with the best performance. However, some heuristics are necessary since it is usually too expensive to include the entire space of implementations.

There are two vectorization approaches that can be derived from this automatic tuning technique. The first one is simply to generate blocks of straight line code (i.e., code without loops) out of formulas and rules, to vectorize these "codelets" as described in Sect. 13.2.2. This is the approach taken in [7–9].

Another approach is to use the rules to generate vectorized code. If the expanded formulas contain right-sided Kronecker products with I_p, where p is the vector size, then the algorithm is directly vectorizable. This is the approach taken in [27, 28]. Special care has to be taken about shuffle operations. The formulas should be chosen so that the permutation matrices produce only permutations that are implementable as single shuffle instructions at a given architecture [29].

The question arises whether the SPL approach can also be used for convolution type streaming algorithms. A problem here is that the data size is unbounded, which would imply matrices of infinite size in the SPL formulation. To work around this problem, one could select a small number of consecutive iterations of the outer loop and apply the SPL approach to this block. To choose the vector size as the block size might be a good choice. The block algorithm is then iterated for consecutive blocks. This approach is taken in [30] for the LMS algorithm. A disadvantage is that the technique cannot automatically choose how the block iterations interact, i.e., what data is passed between iterations. An extension of SPL to infinite cyclic matrices would certainly be a general solution, but this is future work.

13.6 Mixed Algorithms

There are algorithms in signal processing that cannot be classified as either convolution or Fourier oriented. Frequently, Fourier transforms are used on blocks of streaming data. This is mostly combined with overlapped windowed blocks, i.e., window functions applied to blocks before the transform to reduce artifacts due to the lack of periodicity. The well-known short-time Fourier transform (STFT), including the Gabor transform, is the most prominent kind of such a transform in time-frequency analysis. Vectorization strategies here are basically the same as for Fourier-type transforms, as those are the main part of a STFT.

On the other hand, filter operations can be applied on blocks of data, where the handling of block borders is either zero-padded, periodic, or mirrored. Moreover, filters can be applied in several phases, which includes recursive splitting of frequency bands, as in the wavelet transform, or multi-dimensional filtering. In these cases, the passing of vector data between phases might be optimized for overall performance. Therefore, we will examine a representative example more closely.

13.6.1 Recursive Convolution – Wavelet Transforms

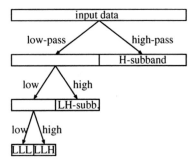

Fig. 13.15 Wavelet transform.

The wavelet transform is implemented by filter pairs such as those in Sects. 13.3.2, 13.3.3, and 13.3.4. We get a low-pass and a high-pass subband with half the size of the original data each. The low-pass subband is then filtered further to be substituted by two subbands of a quarter of the size of the original data, and so on. See Fig. 13.15.

Note that the original definition of the lifting scheme in Eq. (13.26) yields an interleaved data layout of the output data. This means that the input data of further passes is non-contiguous, which is very bad for vectorization. Fortunately, the approach with fused loops in Eq. (13.28) can separate the subbands easily, which is also true for the vectorized algorithm in Eq. (13.29).

Thus, the whole algorithm consists of several passes, where each one reads the output of the preceding pass. This is subject to cache issues, even more so with SIMD acceleration because the cache is more likely to be a bottleneck in faster algorithms. Therefore, the loop fusion technique can also be applied to all passes of the wavelet transform.

Note that special care has to be taken of block borders. See Fig. 13.8 for the case of mirrored border handling. The prolog and epilog phases in this algorithm appear in every pass of the wavelet transform. Therefore, the loop fusion has to incorporate these phases plus a certain number of main-phase iterations into big prolog and epilog phases, which can be arduous to hand-code.

13.6.2 Multi-dimensional Algorithms

The multi-dimensional Fourier transform is implemented in separate passes for each dimension. If the dimension of a certain pass accesses non-contiguous data, i.e., all passes but the first, then there is an easy method for vectorization. One simply has to perform the sequential algorithm while operating on vectors of several neighboring data values, thus transforming several columns at once. This approach can also be applied in the first dimension by transposing $p \times p$ blocks of input and output data after reading and before writing to memory, respectively, thus transforming p rows of data at once. See Eq. (13.9) for the vectorized transposition of such blocks.

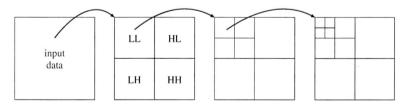

Fig. 13.16 2-D wavelet transform.

The same is true for the wavelet transform [10, 11]. Let us examine the 2-D wavelet transform. Here, each line is filtered by this scheme followed by columns being processed in the same way, giving four subbands denoted by LL, LH, HL, HH. See Fig. 13.16. As explained before, we choose a data layout with separated subbands. This has the advantage that further passes can access the subbands in the same way and the same algorithm can be used. Otherwise, methods for the transform as a whole would have to be developed [31].

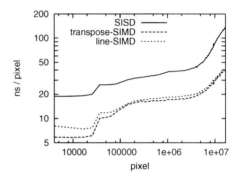

Fig. 13.17 Execution times per sample point (pixel) for one separate horizontal and vertical wavelet filtering pass, with and without SIMD.

See Fig. 13.17 for the execution times of a 2-D filtering pass. There is one horizontal and one vertical filtering step. The two vectorization approaches "line-SIMD," i.e., using the algorithm of Sect. 13.3.4.2 for horizontal filtering, and "transpose-SIMD," i.e., using the above transposition approach, are compared to the sequential "SISD" algorithm. We see that there is a performance gain by a factor of about 2.8 over the whole range of data sizes. The transposition-based parallelization is slightly better than the pure horizontal approach, mainly due to the lesser total number of shuffle operations.

We also see that there is a dependency on cached data and the algorithm does not scale linearly with the data size. To reduce cache dependencies, we will now fuse the horizontal and vertical pass [32]. In the 1-D case, we pass four values from one iteration to the other. To do a similar thing in the second dimension, we apply an approach that is known as pipeline or line-based computation [33]. If we imagine a whole row as a single value (as in the easy vertical SIMD algorithm, only with vectors of the size of a whole row), we must pass four such rows from one iteration to the other. This amounts to a buffer of four rows. In the 1-D case, we read two values from memory in a single iteration. In our row-wise approach this means that we need two new rows to start an iteration.

Since the source data for this row-wise vertical filtering is the output of the horizontal filtering, we try to use the output of the horizontal filtering in the vertical transform immediately after it is available. Thus, we have to perform two horizontal filterings (on two consecutive rows) at once. For each row we get a low-pass and a high-pass coefficient, which makes four values in total. The two low-pass values are fed into an iteration of the vertical type which produces an LL- and an LH-type coefficient, followed by the same operation on the two high-pass coefficients which produces an HL- and an HH-type coefficient. In each iteration the vertical part updates four values in the four-row buffer, which are reused when the next two rows are processed.

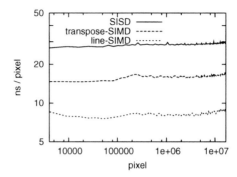

Fig. 13.18 Execution times per sample point (pixel) for the single-loop implementation with and without SIMD.

This algorithm can be vectorized without major problems, so we get a SIMD implementation of a 2-D wavelet filtering step in a single loop. The execution times are shown in Fig. 13.18. There is no cache dependency any more. This time the transposition based algorithm is significantly worse than the pure line-SIMD approach. The reason for this is increased buffer size destroying data locality, and an increased number of concurrently processed intermediate vectors per iteration making register allocation more difficult. The line-SIMD algorithm, however, performs about 3.7 times faster than the non-parallelized, which is very close to the theoretical maximum of 4.

13.7 Conclusion

Short-vector single-instruction-multiple-data (SIMD) processing is an interesting choice for parallel signal processing. The regularity of the data flow of algorithms used in signal processing enables manual and automatic vectorization techniques to efficiently exploit fine-grained parallelity for code acceleration.

The task of vectorization, however, is difficult. The reason is that there is no serve-all approach, but each algorithm has to be treated separately. This is even true if only characteristics like filter length or symmetry are changed for an otherwise simple filtering algorithm. However, most successful vectorization attempts are based on well-known strategies such as loop unrolling, loop fusion, loop transposition, and algebraic transforms. Even hard cases such as recursive filters can be parallelized efficiently in this way.

Whereas there are no general automatic vectorization systems for convolution type filtering algorithms, and manual strategies seem to be the only way to go, the space of possible implementations for Fourier-type algorithms is so large that automatic performance tuning systems that traverse this space to find the fastest implementation cannot be beat by manual implementations, at least not in the general case.

However, the approaches presented in this chapter together with automatic performance tuning techniques may spawn efficient automatic vectorization systems for a broader range of signal processing algorithms in the future. A promising way to go might be the extension of SPL, as used in block transforms, to streaming data, as processed in filter banks.

References

1. J. W. Cooley, J. W. Tukey, An algorithm for the machine calculation of complex Fourier series, Mathematics of Computation 19 (1965) 297–301.
2. P. Duhamel, M. Vetterli, Fast Fourier transforms: A tutorial review and a state of the art, Signal Processing 19 (4) (1990) 259–299.
3. C. M. Rader, Discrete Fourier transforms when the number of data samples is prime, in: Proc. of the IEEE, Vol. 56 (1968), pp. 1107–1108.
4. M. Frigo, S. G. Johnson, FFTW: An adaptive software architecture for the FFT, in: Proc. International Conference on Acoustics, Speech and Signal Processing (ICASSP), Vol. 3 (1998), pp. 1381–1384.
5. M. Püschel, B. Singer, J. Xiong, J. M. F. Moura, J. Johnson, D. Padua, M. Veloso, R. W. Johnson, SPIRAL: A generator for platform-adapted libraries of signal processing algorithms, High Performance Computing and Applications (2004) 21–45.
6. J. Xiong, J. Johnson, R. Johnson, D. Padua, SPL: A language and compiler for DSP algorithms, in: Proc. Programming Language Design and Implementation (PLDI), ACM (2001), pp. 298–308.
7. S. Kral, F. Franchetti, J. Lorenz, C. W. Überhuber, SIMD vectorization techniques for straight line code, Tech. Rep. TR2003-02, Institute of Applied Mathematics and Numerical Analysis, Vienna University of Technology (2003).
8. S. Kral, F. Franchetti, J. Lorenz, C. W. Überhuber, SIMD vectorization of straight line FFT code, in: Proc. Euro-Par (2003), pp. 251–260.
9. M. Frigo, S. G. Johnson, The design and implementation of FFTW3, in: Proc. IEEE, Vol. 93 (2005), pp. 216–231.
10. C. Tenllado, D. Chaver, L. Piñuel, M. Prieto, F. Tirado, Vectorization of the 2D wavelet lifting transform using SIMD extensions, in: Workshop on Parallel and Distributed Image Processing, Video Processing, and Multimedia, PDIVM '03, Nice, France (2003).
11. D. Chaver, C. Tenllado, L. Piñuel, M. Prieto, F. Tirado, 2-D wavelet transform enhancement on general-purpose microprocessors: Memory hierarchy and SIMD parallelism exploitation, in: Proceedings of the 2000 International Conference on High Performance Computing, Bangalore, India (2002).
12. M. Pic, H. Essafi, D. Juvin, Wavelet transform on parallel SIMD architectures, in: F. Huck, R. Juday (Eds.), Visual Information Processing II, Vol. 1961 of SPIE Proceedings, SPIE (1993) pp. 316–323.
13. C. Chakrabarti, M. Vishvanath, Efficient realizations of the discrete and continuous wavelet transforms: From single chip implementations to mappings on SIMD array computers, IEEE Transactions on Signal Processing 3 (43) (1995) 759–771.
14. M. Feil, A. Uhl, Wavelet packet decomposition and best basis selection on massively parallel SIMD arrays, in: Proceedings of the International Conference "Wavelets and Multiscale Methods" (IWC'98), Tangier, 1998, INRIA, Rocquencourt (1998), 4 pages.
15. R. Kutil, P. Eder, M. Watzl, SIMD parallelization of common wavelet filters, in: Parallel Numerics '05, Portorož, Slovenia (2005), pp. 141–149.

16. R. Kutil, P. Eder, Parallelization of wavelet filters using SIMD extensions, Parallel Processing Letters 16 (3) (2006) 335–349.
17. ISO/IEC 15444-1, Information technology – JPEG2000 image coding system, Part 1: Core coding system (Dec. 2000).
18. I. Daubechies, W. Sweldens, Factoring wavelet transforms into lifting steps, Journal of Fourier Analysis Applications 4 (3) (1998) 245–267.
19. M. Püschel, J. M. F. Moura, J. Johnson, D. Padua, M. Veloso, B. W. Singer, J. Xiong, F. Franchetti, A. Gačić, Y. Voronenko, K. Chen, R. W. Johnson, N. Rizzolo, SPIRAL: Code generation for DSP transforms, Proceedings of the IEEE, special issue on "Program Generation, Optimization, and Adaptation" 93 (2) (2005) 232–275.
20. R. Schaffer, M. Hosemann, R. Merker, G. Fettweis, Recursive filtering on SIMD architectures, in: Proc. IEEE Workshop on Signal Processing Systems (SIPS), 2003, pp. 263–268.
21. M. Hosemann, G. Fettweis, On enhancing SIMD-controlled dsps for performing recursive filtering, Journal of VLSI signal processing 43 (2–3) (2006) 125–142.
22. J. Robelly, G. Cichon, H. Seidel, G. Fettweis, Implementation of recursive digital filters into vector SIMD DSParchitectures, in: Proc. IEEE International Conference on Acoustics, Speech, and Signal Processing (ICASSP), Vol. 5 (2004), pp. 165–168.
23. R. Kutil, Parallelization of IIR filters using SIMD extensions, in: Proceedings of the 15th International Conference on Systems, Signals and Image Processing (IWSSIP), Bratislava, Slovak Republic (2008), pp. 65–68.
24. R. C. Whaley, J. Dongarra, Automatically tuned linear algebra software (ATLAS), in: Proc. Supercomputing (1998).
25. J. Bilmes, K. Asanović, C. W. Chin, J. Demmel, Optimizing matrix multiply using PHiPAC: A portable, high-performance, ANSI C coding methodoly, in: Proc. Int. Conf. Supercomputing (ICS) (1997), pp. 340–347.
26. E.-J. Im, K. Yelick, Optimizing sparse matrix computations for register reuse in SPARSITY, in: Proc. Int. Conf. Computational Sciences (ICCS) (2001), pp. 127–136.
27. F. Franchetti, M. Püschel, Short vector code generation for the discrete Fourier transform, in: Proc. IEEE International Parallel and Distributed Processing Symposium (IPDPS) (2003), pp. 58–67.
28. F. Franchetti, M. Püschel, Short vector code generation and adaption for DSP algorithms, in: Proc. International Conference on Acoutstics, Speech and Signal Processing (ICASSP), Vol. 2 (2003), pp. 537–540.
29. F. Franchetti, M. Püschel, Generating SIMD vectorized permutations, in: Proc. Compiler Construction (CC) (2008), pp. 116–131.
30. J. Robelly, G. Cichon, H. Seidel, G. Fettweis, Design and automatic code generation of the LMS algorithm for SIMD signal processors, in: Proc. IEEE International Conference on Acoustics, Speech, and Signal Processing (ICASSP), Vol. 5 (2005), pp. 81–84.
31. G. Lafruit, B. Vanhoof, L. Nachtergaele, F. Catthoor, J. Bormans, The local wavelet transform: a memory-efficient, high-speed architecture optimized to a region-oriented zero-tree coder, Integrated Computer-Aided Engineering 7 (2) (2000) 89–103.
32. R. Kutil, A single-loop approach to SIMD parallelization of 2-D wavelet lifting, in: Proceedings of the 14th Euromicro Conference on Parallel, Distributed and Network-based Processing (PDP), Montbeliard-Sochaux, France (2006), pp. 413–420.
33. C. Chrysafis, A. Ortega, Line based, reduced memory, wavelet image compression, IEEE Transactions on Image Processing 9 (3) (2000) 378–389.

Chapter 14
Financial Applications: Parallel Portfolio Optimization

Andreas Grothey

Abstract

Portfolio optimization is an area of tremendous importance for long-term investors. It is concerned with the problem of how to best diversify investment into different classes of assets (such as stock, bonds, real estate, and options) in order to meet liabilities and to maximize the expected surplus, while hedging against unacceptable risk.

Different formulations of the problem have been suggested over time, starting from Harry Markowitz' seminal mean-variance model. Practical and theoretical limitations of the mean-variance model have led to the proposal of different utility functions, risk measures, and dynamic multi-period models that allow rebalancing of the portfolio to hedge against adverse market conditions. Furthermore new legislation has often resulted in the necessity to introduce new classes of constraints on the portfolio composition.

Multi-period portfolio optimization problems are usually treated as stochastic programming problems, that is, they involve optimization over a selection of future scenarios. The desirability of having future scenarios match static and dynamic correlations between assets for all future time periods leads to problems of truly enormous sizes (often reaching millions of unknowns or more). Clearly parallel processing becomes mandatory to deal with such problems.

The most popular solution techniques for stochastic programming problems are decomposition methods and interior point methods (IPMs). Both approaches lend themselves to parallel implementations and impressive results have been achieved here in the past few years. We will review different implementations on a variety of computing platforms ranging from dedicated parallel machines to PC clusters all the way to grid environments. Typically the solution and parallelization techniques have

Andreas Grothey
School of Mathematics, University of Edinburgh, Edinburgh, UK, EH9 3JZ,
e-mail: A.Grothey@ed.ac.uk

R. Trobec et al. (eds.), *Parallel Computing*, DOI 10.1007/978-1-84882-409-6_14,

to be adapted to both the specific model formulation and the available computing platform.

14.1 Introduction

Portfolio selection is one of the most relevant and most studied topics in finance. The problem, in its basic formulation, is concerned with balancing the twin contradictory objectives of maximizing return of investment while minimizing the associated risk. Early models arising from Markowitz' seminal work [1] are static and deterministic. They consider a set $\mathscr{A}$ of possible investments and assume that joint distribution of asset returns is multivariate normal $N(\mu, \Sigma)$ with known means μ and covariance matrix Σ. The objective is to maximize expected single period return, while bounding the variance of the portfolio return as a measure of risk exposure.

There are a number of perceived weaknesses with this approach that have emerged over the years: the assumption of normal asset returns (neglecting observed "fat tails"), the assumption of known fixed means and covariances (which have to be estimated from historical data, and are clearly neither known exactly nor constant over time), and not least the inability to capture dynamic effects such as transaction costs and the possibility to hedge risk through rebalancing of the portfolio at future time stages. In the past 20 years emphasis has shifted toward stochastic dynamic models that allow the adequate representation of non-normal joint return distribution and the effects of portfolio rebalancing. An overview of these issues is given in the review paper [2].

Realistic models need to account for long planning horizons and adequate capturing of the joint distributions of all future events that can influence the return of the portfolio over the whole planning horizon. These requirements quickly result in astronomical problem sizes. While general advances in the power of desktop computers have made larger problem formulations tractable, the area is a prime candidate for the successful use of parallel algorithms. This is expected to be the case even more in future as the trend for desktop and laptop computers is increasingly to multicore architectures.

The major applications for dynamic portfolio optimization are Asset and Liability Management (ALM) models in which the investor seeks an optimal long-term investment policy that meets anticipated (but unknown) liabilities and maximizes the expected surplus return, while minimizing the risk of defaulting on the liability payments. This is a model of prime importance to long-term investors such as insurances and pension funds.

In the following section we will review various formulations of the ALM model that have been proposed in the literature and discuss their properties, in particular in view of parallel solution approaches. Sections 14.3, 14.4, 14.5, and 14.6 describe popular parallelizable solution approaches to the models such as decomposition methods (Sect. 14.3), (Sect. 14.4), IPMs and evolutionary algorithms (Sect. 14.5). In the second part (Sect. 14.7) we describe and compare in more detail some of the interior point based nested algorithms which are amongst the most promising general methods for the parallel solution of multi-period portfolio planning problems.

14.2 Asset and Liability Management by Stochastic Programming

In the Markowitz model the investor has a choice between different investments or assets $j \in \mathscr{A}$. The return of each investment is given by a random variable R_j. It is assumed that the expected return $\mu_j = I\!E[R_j]$ of each investment and their joint covariance structure $Q = \text{Cov}[R], Q_{ij} = \text{Cov}(R_i, R_j)$ are known. In the Markowitz model the twin contradictory goals of maximizing expected portfolio return $I\!E[R(x)]$ over all feasible portfolios x, while minimizing the risk measured by $\text{Var}[R(x)]$, are combined into a single combined objective[1]

$$\max_x I\!E[R(x)] - \lambda \text{Var}[R(x)]. \tag{14.1}$$

Here $\lambda > 0$ is a *risk-aversion parameter*: large values express a risk aversion, whereas small values express a (relative) risk indifference. There are various reformulations of the problem that are equivalent to (14.1), such as

$$\max_x I\!E[R(x)] \text{ s.t. } \text{Var}[R(x)] \leq \rho. \tag{14.2}$$

If we denote the relative proportion of our wealth to be invested in asset $j \in \mathscr{A}$ by x_j with $x_j \geq 0, \sum_j x_j = 1$, we can express the portfolio return $R(x)$ as

$$R(x) = \sum_{j \in \mathscr{A}} x_j R_j$$

and

$$I\!E[R] = I\!E[\sum_j x_j R_j] = \mu^T x, \qquad \text{Var}[R] = \text{Var}[\sum_j x_j R_j] = x^T Q x. \tag{14.3}$$

Model formulation (14.1) can then be expressed as

$$\max_x \mu^T x - \lambda x^T Q x \quad \text{s.t. } \sum_{j \in \mathscr{A}} x_j = 1, x_j \geq 0. \tag{14.4}$$

While the Markowitz model has become an industry standard, it suffers from important shortcomings, relating to the lack of a dynamic structure. Being a one-period

---

[1] Usually financial optimization models are expressed as maximization whereas in the discussion of solution algorithms problems are described in terms of minimization. We will keep to this convention throughout this chapter.

model, it does not allow the realistic modeling of inherently multi-period concepts, such as the inclusion of transaction costs. Further in the one-period model the investor is concerned only with the current risk and ignores the possibility to hedging via rebalancing of the portfolio. For these reasons the attention has shifted to multi-period dynamic stochastic models that are better suited to realistically reflect the financial planning problem associated with tactical asset decisions or longer-term risk management.

14.2.1 Stochastic Programming

Stochastic programming provides an appropriate framework for the mathematical description of ALM models. Stochastic programming [3,4] deals with the situation in which some of the data describing an optimization model is uncertain. At the time that the *first stage decisions* are taken, this data is unknown, but corrective actions (*recourse decisions*) can be taken after the uncertain values have become available. The classical (linear, 2-stage) stochastic programming problem can be written as

$$
\min_x c^T x + E_\xi[V(x,\xi)] \qquad V(x,\xi) = \min_y q(\xi)^T y(\xi)
$$
$$
\text{s.t. } Ax = b \qquad\qquad\qquad \text{s.t. } T(\xi)x + W(\xi)y(\xi) = h(\xi) \quad (14.5)
$$
$$
x \geq 0 \qquad\qquad\qquad\qquad\qquad y(\xi) \geq 0
$$

where the uncertainty is described by a random variable ξ, and $W(\xi), T(\xi), h(\xi)$, and $q(\xi)$ are the values of the uncertain data for a given realization of ξ. The optimal choice of the first stage decisions, x, takes into account the expected cost $E_\xi[V(x,\xi)]$ incurred by taking the corrective action $y(\xi)$ in order to satisfy the constraints. Under suitable conditions [5] problem (14.5) can be written as

$$
\min_{x,y(\xi)} c^T x + E_\xi[q(\xi)^T y(\xi)]
$$
$$
\text{s.t. } Ax = b
$$
$$
T(\xi)x + W(\xi)y(\xi) = h(\xi) \qquad\qquad (14.6)
$$
$$
x \geq 0, y(\xi) \geq 0
$$

where the constraints involving $y(\xi)$ are understood in an *almost surely* sense. Note that both (14.5) and (14.6) are linear models to keep the notation simple. A generalization of the methodology to nonlinear problems is straightforward.

In order to obtain a computationally tractable problem, the (continuous) random variable ξ is approximated by a random variable with discrete support $\tilde{\xi} \in \{\tilde{\xi}_1, \ldots, \tilde{\xi}_s\}$, $P(\tilde{\xi} = \tilde{\xi}_i) = p_i$, where the realizations $\tilde{\xi}_i$ correspond to future *scenarios*. Using the notation $T_i = T(\tilde{\xi}_i)$ (and analogously for h_i, W_i, y_i, q_i), problem (14.6) can thus be written as the *deterministic equivalent formulation*

$$\min_{x, y_i} c^T x + \sum_{i=1}^{s} p_i q_i^T y_i$$

$$
\begin{array}{llll}
\text{s.t.} & Ax & & = b \\
& T_1 x + W_1 y_1 & & = h_1 \\
& \vdots & \ddots & \vdots \\
& T_s x & + W_s y_s & = h_s
\end{array}
\tag{14.7}
$$

Problem (14.7) is an approximation of (14.6). The issue of scenario generation, that is, how to best construct an approximating discrete distribution $\tilde{\xi}$, is an active research area, but beyond the scope of this chapter. It should be noted, however, that usually a large number of scenarios are needed to adequately capture the characteristics of the underlying (continuous) distribution.

The methodology can be generalized to a multistage model in which the evolution of uncertainties can be described as an alternating sequence of decisions and random realizations that occur at different points in time (stages). Mathematically, the uncertainty is described by a stochastic process $\xi = \xi^T = (\xi_1, \ldots, \xi_T)$, where $\xi^t = (\xi_1, \ldots, \xi_t)$ represents the information available at time t. The decisions $x = (x_1, \ldots, x_T)$ are likewise described by a stochastic process, where the decision $x_t = x_t(\xi^t)$ are *non-anticipative*, that is, they depend only on information available at time t and not on future events:

$$x_1 \to \xi_2 \to x_2(\xi_2) \to \xi_3 \to x_3(\xi_2, \xi_3) \to \cdots \xi_T \to x_T(\xi_1, \ldots \xi_T).$$

In the discrete case the stochastic process can be represented as a scenario tree (Fig. 14.1):

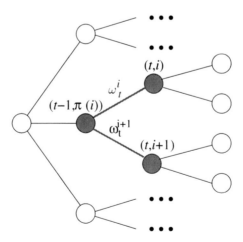

Fig. 14.1 Scenario tree.

Each level t of the tree corresponds to a point in time when a realization of the random process becomes known and a subsequent decision is taken. Each node i corresponds to a particular series of events to this point in time. The branches from a particular node represent the (discrete set) of possible future outcomes of the random variables ξ_{t+1} in the next period. The root node of the tree represents the current time $t = 0$, and the leaf nodes represent the possible states of the system at the end of the planning horizon $t = T$. We denote by $\mathcal{V}_t$ the set of nodes at level t in the tree and $\mathcal{V} = \bigcup_t \mathcal{V}_t$ the complete node set. For every node $i \in \mathcal{V}_t$, we will denote by $\pi(i) \in \mathcal{V}_{t-1}$ its immediate ancestor and by $\mathcal{C}(i) \subset \mathcal{V}_{t+1}$ its set of child nodes. For every node there is a transition probability ω_t^i of reaching this node given that its parent $\pi(i)$ has been reached. The total probability p_i of reaching node i is obtained by the product of all transition probabilities on the path from the root to node i.

Every *scenario*, that is a path through the tree from the root to a leaf, represents a particular sequence of realizations $(\hat{\xi}_1, \ldots, \hat{\xi}_T)$ of the random process ξ, that is one particular outcome of the random data in the problem. Every node further carries its own version x_i of the decision variable x_t of the appropriate stage. The multistage version of the deterministic equivalent (14.7) can thus be stated as

$$\min_x \quad \sum_{i \in \mathcal{V}} p_i d_i^T x_i \quad \text{s.t.} \quad T_i x_{\pi(i)} + W x_i = b_i \quad \forall i \in \mathcal{V} \setminus \{0\} \qquad (14.8)$$
$$W x_0 = b_0.$$

For a realistic model description the size of the scenario tree quickly reaches astronomical sizes. The number of nodes is exponential in the number of time stages considered. For a tree with $T = 5$ stages and a branching factor of 30 at each node (barely enough to capture the correlation between, say, 60 considered random variables describing the evolution of investments and liabilities), the resulting tree has 24 million scenarios.

14.2.2 Asset and Liability Management Models

An ALM model can be seen as a dynamic multistage version of the Markowitz model. Faced with future liabilities L_t of a priori uncertain amounts, the investor seeks a strategy that meets the liabilities, while maximizing the expected surplus at the end of the planning horizon. Notable models are the Frank Russel/Yassuda Kasai model [6] and the Towers Perrin model [7]. Standard references for ALM is the research work by Kusy and Ziemba [8], Mulvey and Vladimirou [9], Zenios [10], and the book by Mulvey and Ziemba [11].

Multistage stochastic programming provides a general framework for modeling financial planning problems. It provides the tools to model a wide spectrum of realistic issues in financial planning. In this section we will present a basic prototype ALM model before discussing various extensions to the model and their impact on the problem structure with particular regard to the consequences for possible parallelization. We do not aim to give a comprehensive overview of ALM models that have been suggested in the literature, nor do we attempt to evaluate their usefulness in practical terms. For these issues we refer the reader to the references given above.

An ALM model is concerned with finding the optimal way of investing into J assets $j \in \mathscr{A}$ over multiple time periods $t = 0, \ldots, T$. The returns r^t of the assets in each time period are assumed to be uncertain, but based on some (known) random distribution. An initial amount of cash b_0 is invested at $t = 0$ and the portfolio may be rebalanced at discrete times $t = 1, \ldots, T$, incurring transaction costs. At every time t a liability payment L_t of uncertain amount is due. The objective is to maximize the expectation of the final value of the portfolio at time T and to minimize the associated risk measured, for example, with the variance of the final wealth. In the stochastic programming formulation the evolution of the uncertain process driving the asset returns $r^t = (r^t_j)_{j \in \mathscr{A}}$ is described by a scenario tree (Fig. 14.1). Let v_j be the value of asset j, and c_t the transaction cost (expressed as a percentage of transaction volume). It is assumed that the value of the assets will not change throughout time and a unit of asset j can always be bought for $(1 + c_t)v_j$ or sold for $(1 - c_t)v_j$. Instead a unit of asset j held in node i (coming from node $\pi(i)$) will generate extra return $r_{i,j}$.

Model Variables

We denote by $x^h_{i,j}$ the units of asset j held at node i and by $x^b_{i,j}, x^s_{i,j}$ the transaction volume (buying, selling) of this asset at this node. We assume that we start with zero holding of all assets but with funds b_0 to invest. Further we assume that one of the assets represents cash, i.e., the available funds are always fully invested.

Model Constraints

The standard constraints on the investment policy can be expressed as follows: cash balance constraints describe possible buying and selling actions within a scenario while taking transaction costs into account. The net cash flow in each node originating from selling and buying assets must be equal to the liability payments L_i in this node

$$\sum_{j \in \mathscr{A}} (1 + c_t)v_j x^b_{i,j} - \sum_{j \in \mathscr{A}} (1 - c_t)v_j x^s_{i,j} = L_i \quad \forall i \in \mathscr{V} \setminus \{0\} \tag{14.9a}$$

$$\sum_{j \in \mathscr{A}} (1 + c_t)v_j x^b_{0,j} = b_0. \tag{14.9b}$$

Each scenario is linked to its parent through inventory constraints; these are balance constraints on asset holdings (taking into account the random return on asset):

$$(1 + r_{i,j})x^h_{\pi(i),j} = x^h_{i,j} - x^b_{i,j} + x^s_{i,j}, \quad \forall i \in \mathscr{V} \setminus \{0\}, j \in \mathscr{A}. \tag{14.10}$$

Model Objective

In the simplest case we consider an objective function that maximizes the expected portfolio surplus return over all scenarios. The wealth of the portfolio in node $i \in \mathcal{V}_T$ at final time T is given by

$$w_{T,i} = \sum_{j \in \mathcal{A}} (1 - c_t) v_j x_{i,j}^h.$$

The expected value of the portfolio at $t = T$ is thus

$$W_T = \mathbb{E}[w_T] = \sum_{i \in \mathcal{V}_T} p_i w_{T,i} = (1 - c_t) \sum_{i \in \mathcal{V}_T} p_i \sum_{j \in \mathcal{A}} v_j x_{i,j}^h. \tag{14.11}$$

The Complete Model

The prototype ALM problem can be expressed as

$$\max_x \ (1 - c_t) \sum_{i \in \mathcal{V}_T} p_i \sum_{j \in \mathcal{A}} v_j x_{i,j}^h$$

$$\text{s.t.} \quad (1 + r_{i,j}) x_{\pi(i),j}^h = x_{i,j}^h - x_{i,j}^b + x_{i,j}^s, \qquad \forall i \in \mathcal{V} \setminus \{0\}, j \in \mathcal{A} \tag{14.12}$$

$$\sum_{j \in \mathcal{A}} (1 + c_t) v_j x_{i,j}^b - \sum_{j \in \mathcal{A}} (1 - c_t) v_j x_{i,j}^s = L_i, \ \forall i \in \mathcal{V} \setminus \{0\}$$

$$\sum_{j \in \mathcal{A}} (1 + c_t) v_j x_{0,j}^b = b_0.$$

To illustrate the structure of this problem we gather decision vector components x_i for each node as $x_i = (x_{i,1}^s, x_{i,1}^b, x_{i,1}^h, \dots, x_{i,J}^s, x_{i,J}^b, x_{i,J}^h)$, and define matrices

$$W = \begin{pmatrix} 1 & -1 & 1 & & & & \\ & & & \ddots & & & \\ & & & & 1 & -1 & 1 \\ -c_1^s & c_1^b & 0 & \cdots & -c_J^s & c_J^b & 0 \end{pmatrix}, \quad T_i = \begin{pmatrix} 0 & 0 & 1 + r_{i,1} & & & \\ & & & \ddots & & \\ & & & & 0 & 0 & 1 + r_{i,J} \\ 0 & 0 & 0 & \cdots & 0 & 0 & 0 \end{pmatrix} \tag{14.13}$$

and

$$d_i \in \mathbb{R}^{1 \times 3|\mathcal{A}|} : (d_i)_{3j} = (1 - c_t) p_i v_j,$$

where $c_j^b = (1 + c_t) v_j, c_j^s = (1 - c_t) v_j$. We can now rewrite problem (14.12) as

$$\max_x \ \sum_{i \in \mathcal{V}_T} d_i^T x_i \quad \text{s.t.} \ T_i x_{\pi(i)} + W x_i = L_i e_{(J+1)} \ \forall i \in \mathcal{V} \setminus \{0\} \tag{14.14}$$

$$W x_0 = b_0 e_{(J+1)},$$

where $e_{(J+1)}$ is the $(J + 1)$-th unit vector $e_{(J+1)} = (0, \dots, 0, 1)^T$. Problem (14.14) is in deterministic equivalent multistage stochastic programming form (14.8). While the representation (14.14) is very compact, it should be kept in mind that ALM problems can grow to enormous sizes. For the example given earlier with 60 asset types,

5 times stages, and 30 branches at every stage to capture the correlations between the asset types, the resulting problem would have 24 million scenarios and 4.5×10^9 decision variables. Clearly the solution of these problems can only be attempted on massively parallel hardware. Even for more modest problem dimensions parallel solution approaches are of tremendous value.

Model Structure

If we further assemble the node-wise decision vectors $x_i, i \in \mathcal{V}$ into a global vector $x = (x_{\sigma(0)}, x_{\sigma(1)}, \ldots, x_{\sigma(|\mathcal{V}|-1)})$ where σ is a permutation of the nodes $i \in \mathcal{V}$ in reverse depth-first order, the constraint matrix of (14.14) takes the form:

$$
A = \begin{pmatrix}
\begin{array}{|ccc|} \hline W & & T_i \\ & \ddots & \vdots \\ & W & T_i \\ & & W \\ \hline \end{array} & & & & \begin{array}{c} 0 \\ \vdots \\ 0 \\ T_i \end{array} \\
& \ddots & & & \vdots \\
& & \begin{array}{|ccc|c} \hline W & & T_i & 0 \\ & \ddots & \vdots & \vdots \\ & W & T_i & 0 \\ & & W & T_i \\ & & & W \\ \hline \end{array}
\end{pmatrix}. \tag{14.15}
$$

We will later show how this rather special nested structure can be exploited in the parallel solution of ALM problems by IPMs. For decomposition approaches on the other hand it is convenient to represent problem (14.14) in recourse form:

$$
\eta_i(x_{\pi(i)}) = \max_{x_i} \sum_{j \in \mathscr{C}(i)} \eta_j(x_i), \text{ s.t. } T_i x_{\pi(i)} + W x_i = L_i e_{(J+1)}, i \in \bigcup_{t=1}^{T-1} \mathcal{V}_t \tag{14.16a}
$$

$$
\eta_i(x_{\pi(i)}) = \max_{x_i} d_i^T x_i, \text{ s.t. } T_i x_{\pi(i)} + W x_i = L_i e_{(J+1)}, \qquad i \in \mathcal{V}_T \tag{14.16b}
$$

and note that (14.14) can be expressed as

$$
\max_{x_0} \sum_{i \in \mathcal{V}_1} \eta_i(x_0), \text{ s.t. } W x_0 = b_0 e_{(J+1)}. \tag{14.17}
$$

Formulations (14.16) and (14.17) can be conveniently exploited in parallel solution approaches based on nested Benders decomposition [12–14].

14.2.3 Model Extensions

The aim of this section is to give an overview of the various extensions of the prototype ALM model presented in the previous section. We will place particular emphasis on the impact of these extensions to the problem structure from the point of view of possible parallel solution approaches.

As currently presented the ALM model is a tree-structured linear optimization problem. Various types of linear constraints can be added to the ALM model without changing the problem structure. These include short-selling constraints, limits on investment in certain types of assets (some regulations cap the amount that can be invested in risky asset classes), and composition constraints: investment in an asset class (e.g., stocks) needs to be less than (a multiple) of investments in a different asset class (e.g., bonds). Models can be expanded to include the borrowing of money, modeling of exchange rate risk in foreign currency transactions [15], or even elaborate models that predict liabilities through the modeling of the members' age structure of a pension fund, all while keeping a linear model that fits into the tree structure (14.14).

More problematic are constraints that lead to an integer programming formulation such as cardinality constraints (i.e., limiting the number of assets in which to invest at any one time) and minimum lot size constraints, i.e., investments can be only an integer multiple of a given lot size or any nonzero investment has to be above a certain threshold, after which continuous investments are allowed (Mansini and Speranza [16]). These constraints lead to very difficult (NP-hard) integer programming formulations. The review paper [17] gives an overview of different constraints and model formulations.

Many popular extensions concern the choice of objective function. The main problem with the linear objective function (14.11) is that it fails to take into account the risk of the investment. In the first instance it would be desirable to use an objective that is a multistage version of the Markowitz mean-variance objective. Following from (14.11) and using the identity $\mathrm{Var}[X] = I\!E[X^2] - I\!E[X]^2$, the variance of the final wealth can be expressed as

$$\mathrm{Var}[W_T] = (1-c_t)^2 \sum_{i\in\mathscr{V}_T} p_i (\sum_{j\in\mathscr{A}} v_j x_{i,j}^h)^2 - (1-c_t)^2 \left(\sum_{i\in\mathscr{V}_T} p_i \sum_{j\in\mathscr{A}} v_j x_{i,j}^h\right)^2 \quad (14.18)$$

so that in (14.12) we could use a combined mean-variance objective

$$\max_x I\!E[W_T] - \lambda \mathrm{Var}[W_T] \quad (14.19)$$

as in the Markowitz model (14.1). There are however two major problems with this formulation. First the use of the variance results in a quadratic model objective. More importantly, however, the squaring operation in the final term of (14.18) results in cross products between different final stage nodes $i \in \mathscr{V}_T$, resulting in a dense model Hessian and destroying the node separability of the model and hence the recourse structure (14.16) and (14.17). Both these issues are causing problems for some popular parallelizable solution approaches (notably Benders decomposition and its variants such as the L-shaped method [18]).

For these reasons there have been various suggestions to replace the $\mathrm{Var}[W_T]$ by a risk measure that leads to a linear and node-separable model formulation. The use of mean absolute deviation, leading to the objective function

$$\mathbb{E}[W_T] - \lambda \mathbb{E}\left[|W_T - \mathbb{E}[W_T]|\right], \tag{14.20}$$

is a popular choice, argued for by [19, 20] used, for example, by [21, 22]. Lucka et al. [23] use just $\mathbb{E}[W_T]$ as objective function while restricting risk through requiring that $w_{T,i} \geq C$ for all nodes $i \in \mathscr{V}_T$, again yielding a linear model. Another popular approach is to use a piecewise linear approximation of a convex risk measure such as variance or semivariance. Reference [24] uses a piecewise linear upper approximation of the semivariance. For a discussion of different linearizations of the Markowitz model and their relative performance see the review of Mansini et al. [25].

A different approach is taken by Gondzio and Grothey [26]. They introduce an explicit variable $y = \mathbb{E}[W_T]$ together with the constraint

$$y = (1 - c_t) \sum_{i \in \mathscr{V}_T} p_i \sum_{j \in \mathscr{A}} v_j x_{i,j}^h, \tag{14.21}$$

resulting in the model formulation (cf. (14.12)):

$$
\begin{aligned}
\max_{x, y \geq 0} \quad & y - \lambda \left(\sum_{i \in \mathscr{V}_T} p_i (1 - c_t)^2 [\sum_{j \in \mathscr{A}} v_j x_{i,j}^h]^2 - y^2 \right) \\
\text{s.t.} \quad & (1 - c_t) \sum_{i \in \mathscr{V}_T} p_i \sum_{j \in \mathscr{A}} v_j x_{i,j}^h = y \\
& (1 + r_{i,j}) x_{\pi(i),j}^h = x_{i,j}^h - x_{i,j}^b + x_{i,j}^s, \qquad \forall i \in \mathscr{V} \setminus \{0\}, j \in \mathscr{A} \\
& \sum_{j \in \mathscr{A}} (1 + c_t) v_j x_{i,j}^b - \sum_{j \in \mathscr{A}} (1 - c_t) v_j x_{i,j}^s = L_i, \forall i \in \mathscr{V} \setminus \{0\} \\
& \sum_{j \in \mathscr{A}} (1 + c_t) v_j x_{0,j}^b = b_0.
\end{aligned} \tag{14.22}
$$

Defining the lattice matrices Q_i,

$$Q_i \in \mathbb{R}^{3J \times 3J} : \begin{cases} (Q_i)_{3j,3k} = p_i (1 - c_t)^2 v_j v_k, \ j,k \in \mathscr{A}, & i \in \mathscr{V}_T \\ Q_i = 0, & i \notin \mathscr{V}_T \end{cases}$$

that have entries only in elements with a row and column index divisible by 3 (corresponding to the x^h variables), problem (14.22) can be written more compactly as (cf. (14.14))

$$
\max_{x,y} \quad y - \lambda [\sum_{i \in \mathscr{V}_T} x_i^T Q_i x_i - y^2] \quad \text{s.t.} \quad
\begin{aligned}
& \sum_{i \in \mathscr{V}_T} d_i^T x_i = y \\
& T_i x_{\pi(i)} + W x_i = L_i e_{(J+1)} \quad \forall i \in \mathscr{V} \setminus \{0\} \\
& W x_0 = b_0 e_{(J+1)},
\end{aligned}
\tag{14.23}
$$

restoring the node-separability of the objective function and the sparsity of the model Hessian, albeit at the expense of introducing links between the final stage nodes $i \in \mathcal{V}_T$ through constraint (14.21). While this is still causing problems for solution approaches based on Benders' decomposition, the structure in (14.23) can be efficiently exploited in IPMs.

On the other hand the use of the variance as a risk measure has been criticized in various places for its simplicity. Variance fails to properly take into account the observed "fat tails" of the portfolio return distribution as well as undesirably penalizing for over-performance of the portfolio as much as for under-performance. Practitioners often recommend the use of a von Neumann–Morgenstern type [27] nonlinear utility formulation

$$\mathbb{E}[U(W_T)], \tag{14.24}$$

where $U : \mathbb{R} \to \mathbb{R}$ is a (usually convex) *utility function*. A popular choice for $U(x)$ is $U(x) = -\log x$ as suggested by Kelly [28] and used by [29,30]. Konno et al. [31,32] suggest the use of skewness (third moment of W_T) in the objective to adequately cover nonsymmetrical distribution of return. Pflug [33,34] suggests the use of lower semivariance

$$\mathbb{E}[W_T] - \lambda \sqrt{\mathbb{E}[([W_T - \mathbb{E}[W_T]]^-)^2]}$$

for the same reason. As shown in [30] these formulations can be incorporated into the model (14.23), at the expense of introducing nonlinear constraint and objective terms. More recently other risk measures such as VaR [35] or CVaR [36] have been suggested. It is shown in [36] that the use of CVaR results in a linear model, whereas VaR leads to difficult nonlinear nonconvex models [37]. Other approaches for ALM that have been suggested involve robust optimization [38,39] and optimization under stochastic dominance constraints [40]. However, as far as we are aware there are so far no parallel implementations of any of these models.

14.3 Parallel Solution Approaches: Decomposition

Decomposition approaches such as nested Benders decomposition [12–14,41,42], the L-shaped method [18], and scenario decomposition [43] have for a long time been the methods of choice for the solution of stochastic programming problems. Benders decomposition and related approaches work directly with the recourse formulations (14.16) and (14.17): for the concave recourse functions $\eta_i(x)$ in (14.16), piecewise linear overestimating approximations $\tilde{\eta}_i(x)$ are built up from cuts received from its children. Scenario decomposition on the other hand uses the scenario formulation of the deterministic equivalent (14.8), in which every scenario $i = 1, \ldots, s$ has its own copy x_i of the first stage decision variables. Their equality is imposed by explicit *non-anticipativity* constraints $x_i = x$.

$$\min_{x,y_j} c^T x + \sum_{j=1}^{s} p_j q_j^T y_j, \quad \text{s.t.} \qquad Ax = b$$

$$T_i x_i + W_i y_i = h_i, \qquad i \in \mathscr{V}$$
$$x_i = x \qquad i \in \mathscr{V}$$

(14.25)

A multistage version of the scenario formulation in which there are separate versions $x_{t,i}$ of the decision variables x_t in all stages t for all scenarios $i \in \mathscr{V}_T$ with appropriate non-anticipativity constraints can be derived accordingly (see Fig. 14.2). Here variables in all but the last stage are replicated and independent copies, one per scenario, are created. Vertical lines on the right-hand side of Fig. (14.2) indicate duplicated nodes; variables associated with these nodes are forced to be identical by non-anticipativity constraints.

Fig. 14.2 Multistage scenario formulation with explicit non-anticipativity constraints.

In a scenario decomposition scheme the non-anticipativity constraints are relaxed using (augmented) Lagrangians, yielding a formulation that decomposes into single scenarios.

Decomposition methods are naturally suited for parallelization, with each node approximation $\tilde{\eta}_i(x)$, for each scenario subproblem, being assigned to one processor. There are many parallel implementations for the solution of ALM problems that use variants of this idea. See [12, 44] for a review of some of these methods. More recently Edirisinghe and Patterson [24] use a variant of the L-shaped method for problems with block-separable recourse to solve a portfolio optimization problem with a piecewise linear mean-semivariance approximation as objective function. Though not implemented in parallel the authors point that a parallel version of their algorithm would be possible. Nevertheless they are able to solve a problem with 5 periods, 100,000 scenarios, and 26 million variables in 1334 seconds on a 2.5 GHz Pentium IV system. Pflug and coauthors in [34] describe a parallel decomposition algorithm based on Diagonal Quadratic Approximation (DQA) [43] within a scenario decomposition method based on augmented Lagrangians, which is used as part of the AURORA Financial Management System developed at the University of Vienna (see also Sect. 14.6).

Traditionally decomposition algorithms are synchronous: there is a tight coupling between parent and child nodes. Each parent sends its current solution to its children and waits until it has received a new cut from each child before continuing. In an asynchronous version, local problems at each node are solved continuously, each with the best information available at that point in time. A node does not have to wait for all children to respond or even for its own parent to send a new trial point. Communication between parent and child nodes is organized through buffers. At each new iteration, a node scans the buffers for new information from either its children or its parent and solves its corresponding local problem based on that information.

An interesting application of this idea is by Linderoth and Wright. In [45] they solve a linear stochastic programming problem with 10^7 scenarios and 1.26×10^{10} variables on a heterogeneous computational grid with 1024 nodes in 32 hours, using an asynchronous version of the L-shaped method. Although the application in question was not a portfolio optimization problem, there is no doubt that the methodology can be applied to linear variants of the ALM problem. Laure and Moritsch [46] on the other hand report on a parallel asynchronous Benders decomposition algorithm specifically targeted at ALM problems and also used as part of the AURORA Financial Management System [34]. Their implementation is in the OpusJava framework for maximal portability. They consider problems using binary scenario trees with five to eight stages, resulting in 63–511 nodes. They report a speed-up of 3.48 on a cluster of 4 Sun Ultra 10 workstations and a superlinear speedup of 18.51 on a 16-node Beowulf Linux cluster.

A disadvantage of decomposition approaches are that their applicability is largely limited to linear problem formulations. While generalizations to at least convex nonlinear problems exist, they cannot usually match the efficiency displayed for linear problems. However, Parpas and Rustem [47] review the relative performance of a regularized Benders decomposition methods and a scenario decomposition employing augmented Lagrangians for the mean-variance formulation of an ALM on problems with up to 6250 scenarios and 590,000 decision variables and find these to be competitive. Also, Mulvey and Shetty in [48] investigate the parallelization of a scenario decomposition method for an ALM model using the expectation of a nonlinear (convex) utility function as objective function. They solve multistage problems with upto 3072 scenarios and 480,000 variables and report a speedup of 2.7 when going from 32 to 128 processors (a parallel efficiency of 69%).

14.4 Parallel Solution Approaches: Interior Point Based Algorithms

One of the most successful methods for the parallel solution of stochastic programming problems are IPMs [49] and multitude of applications to ALM exist [26, 30, 50, 51]. There are various reasons for this popularity: their applicability to a wide range of formulations spanning linear, quadratic, and nonlinear models, their comparative nonsensitivity to large problem sizes (IPMs are in practice observed to converge in $\mathcal{O}(\log N)$ iterations, where N is the problem size), and not least the amenability of the linear algebra operations to parallelization. For these reasons we will devote a major part of this chapter to the discussion of IPMs in the context of portfolio optimization. In this section we will give an overview of IPMs and the linear algebra issues that make them so well suited to the solution of ALM problems on parallel hardware. In the later part of Sect. 14.7 we give a detailed analysis and comparison of some notable implementations.

There are various different variants of IPMs, the most popular being primal–dual [49] and primal IPM. They differ in details of the algorithm logic. Up to small differences, the linear algebra computations (and therefore the main computational effort) of these variants are identical. In what follows we restrict our attention to the primal–dual IPM and point out differences to primal IPMs where necessary. Further IPMs can be applied with minor modifications to linear, quadratic, and nonlinear optimization problems. We will derive IPMs for the quadratic case, again pointing out differences for linear and nonlinear problem formulations when appropriate. Consider the quadratic programming problem:

$$
\begin{aligned}
\min \quad & c^T x + \tfrac{1}{2} x^T Q x \\
\text{s.t.} \quad & A x = b, \\
& x \geq 0,
\end{aligned}
\tag{14.26}
$$

where $Q \in \mathscr{R}^{n \times n}$, $A \in \mathscr{R}^{m \times n}$ is a full rank matrix of linear constraints and Q is positive semidefinite on the null-space of the constraints $Ax = b$. Vectors x, c, and b are assumed to have appropriate dimensions. IPMs can be motivated by replacing the inequality constraints $x \geq 0$ with logarithmic barrier terms to get

$$
\begin{aligned}
\min \quad & c^T x + \frac{1}{2} x^T Q x - \mu \sum_{j=1}^{n} \ln x_j \\
\text{s.t.} \quad & A x = b,
\end{aligned}
$$

where $\mu \geq 0$ is a barrier parameter. The Lagrangian associated with this problem has the form:

$$
L(x, y, \mu) = c^T x + \frac{1}{2} x^T Q x - y^T (A x - b) - \mu \sum_{j=1}^{n} \ln x_j
$$

and the conditions for a stationary point are thus

$$
\begin{aligned}
\nabla_x L(x, y, \mu) &= c - A^T y - \mu X^{-1} e + Q x = 0 \\
\nabla_y L(x, y, \mu) &= \qquad\qquad\quad A x - b = 0,
\end{aligned}
$$

where $X^{-1} = \mathrm{diag}\{x_1^{-1}, x_2^{-1}, \ldots, x_n^{-1}\}$. Denoting

$$
s = \mu X^{-1} e, \qquad \text{i.e.,} \qquad X S e = \mu e,
$$

where $S = \mathrm{diag}\{s_1, s_2, \ldots, s_n\}$ and $e = (1, 1, \ldots, 1)^T$. The first-order optimality conditions (for the barrier problem) are

$$\begin{aligned}
Ax &= b, \\
A^T y + s - Qx &= c, \\
XSe &= \mu e \\
(x, s) &\geq 0.
\end{aligned} \tag{14.27}$$

Under appropriate conditions [49] the nonlinear system of equations (14.27) has a unique solution (x_μ, s_μ, y_μ) for every $\mu > 0$, and as $\mu \to 0$, this point converges to the solution (x^*, s^*, y^*) of the original problem (14.26). IPMs employ a homotopy approach: at every iteration a Newton step toward the solution of (14.27) is performed, followed by a gradual reduction of the barrier parameter μ. Appropriate safeguards guarantee the convergence to the optimal solution of the original problem. The Newton direction is obtained by solving the system of linear equations:

$$\begin{bmatrix} A & 0 & 0 \\ -Q & A^T & I \\ S & 0 & X \end{bmatrix} \begin{bmatrix} \Delta x \\ \Delta y \\ \Delta s \end{bmatrix} = \begin{bmatrix} \xi_p \\ \xi_d \\ \xi_\mu \end{bmatrix}, \tag{14.28}$$

where

$$\xi_p = b - Ax, \qquad \xi_d = c - A^T y - s + Qx, \qquad \xi_\mu = \mu e - XSe.$$

By elimination of

$$\Delta s = X^{-1}(\xi_\mu - S\Delta x) = -X^{-1}S\Delta x + X^{-1}\xi_\mu$$

we get from (14.28) the symmetric indefinite augmented system of linear equations

$$\begin{bmatrix} -Q - \Theta^{-1} & A^T \\ A & 0 \end{bmatrix} \begin{bmatrix} \Delta x \\ \Delta y \end{bmatrix} = \begin{bmatrix} \xi_d - X^{-1}\xi_\mu \\ \xi_p \end{bmatrix}, \tag{14.29}$$

where $\Theta = XS^{-1}$ is a diagonal scaling matrix. The main difference in primal IPMs is that $\Theta = X^2$ is used. By eliminating Δx from the first equation we can reduce (14.29) further to normal equations form:

$$(A(Q + \Theta^{-1})^{-1}A^T)\Delta y = r, \tag{14.30}$$

for appropriate r. For an IPM applied to a linear problem, $Q = 0$ can be substituted throughout. For an IPM applied to the nonlinear problem

$$\min_x f(x) \text{ s.t. } g(x) = 0, x \geq 0,$$

we need to use $A = \nabla g(x), Q = \nabla^2 f(x) + \sum_{i=1}^m y_i \nabla^2 g_i(x)$ in (14.28)–(14.30). Systems (14.29) and (14.30) are equivalent and it depends on the problem in question which one is preferable. A relatively dense matrix Q can result in a dense inverse $(Q + \Theta^{-1})^{-1}$ and hence a dense system (14.30). For this reason (14.29) would be preferred despite its larger size. For linear problems or problems with a sparse (or even diagonal) Q, either system can be solved efficiently.

Since the main computation work of an IPM is concentrated on the solution of system (14.29) or (14.30), they are the primary target of parallelization efforts.

14.4.1 IPM Applied to Stochastic Programming Problems

For an IPM applied to the multistage stochastic programming problem formulation (14.23) constituent matrices of the augmented system (14.29) take the form:

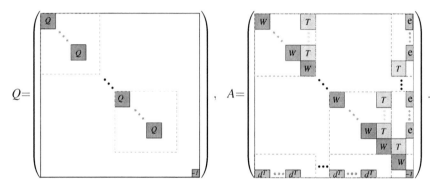

The resulting augmented system matrix is hence a structured matrix of the form displayed in Fig. 14.3. By a symmetric block row and column reordering, corresponding to gathering together node contributions $(\Delta x_i, \Delta y_i)$, the matrix (14.3) can be transformed into the form displayed in Fig. 14.4 which is of nested bordered block-diagonal form (14.31)

$$
\Psi =
\begin{pmatrix}
\begin{array}{|ccc|} \hline
\Phi & & \Gamma^T \\
& \ddots & \vdots \\
& \Phi & \Gamma^T \\
\Gamma & \cdots \Gamma & \Phi \\ \hline
\end{array}
& & & & \Gamma^T \\
& & & & \vdots \\
& & & & \Gamma^T \\
& & & & \Gamma^T \\
& \ddots & & & \vdots \\
& & \begin{array}{|ccc|c|} \hline
\Phi & & \Gamma^T & \Gamma^T \\
& \ddots & \vdots & \vdots \\
& \Phi & \Gamma^T & \Gamma^T \\
\Gamma & \cdots \Gamma & \Phi & \Gamma^T \\ \hline
\end{array} \\
\Gamma & \cdots \Gamma \; \Gamma & \cdots & \Gamma \cdots \Gamma \; \Gamma & \Phi
\end{pmatrix}
. \qquad (14.31)
$$

The main computational work in an IPM applied to stochastic programming problems, such as ALM, is in the solution of linear systems with the system matrix (14.31). A common feature of many approaches is that they use the nested struc-

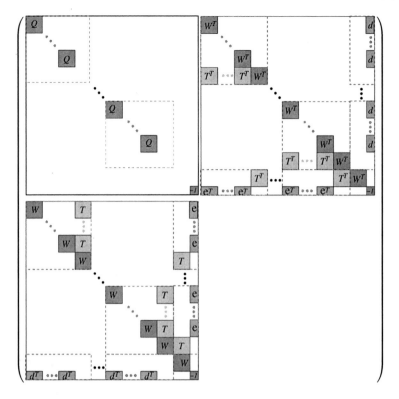

Fig. 14.3 Augmented system matrix for IPM applied to model (14.23).

ture of (14.31) to obtain a compact implicit representation of the Cholesky factors of this matrix. The underlying methodology is a nested Schur complement factorization. While this mechanism is not always made explicit, it provides a unifying framework for the description of these approaches. The basic idea is that a bordered block-diagonal matrix of the form

$$\Psi = \begin{pmatrix} \Phi_1 & & & \Gamma_1^T \\ & \Phi_2 & & \Gamma_2^T \\ & & \ddots & \vdots \\ & & & \Phi_s & \Gamma_s^T \\ \Gamma_1 & \Gamma_2 & \cdots & \Gamma_s & \Phi_0 \end{pmatrix}, \tag{14.32}$$

where $\Phi_i \in I\!\!R^{n_i \times n_i}, \Gamma \in I\!\!R^{n_0 \times n_i}$ allows for a (conceptual) block Cholesky factorization

$$\Psi = LDL^T$$

of the form

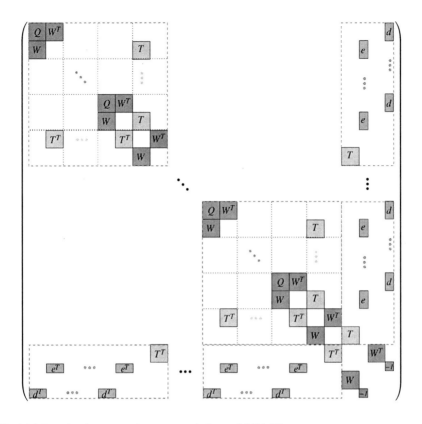

Fig. 14.4 Reordered augmented system matrix for model (14.23).

$$
L = \begin{pmatrix} L_1 & & & & \\ & L_2 & & & \\ & & \ddots & & \\ & & & L_s & \\ L_{0,1} & L_{0,2} & \cdots & L_{0,s} & L_c \end{pmatrix}, \quad
D = \begin{pmatrix} D_1 & & & & \\ & D_2 & & & \\ & & \ddots & & \\ & & & D_s & \\ & & & & D_c \end{pmatrix}, \tag{14.33}
$$

where

$$
L_i D_i L_i^T = \Phi_i, \qquad i = 1, \dots, s \tag{14.34a}
$$

$$
L_{0,i} = \Gamma_i L_i^{-T} D_i^{-1}, \quad i = 1, \dots, s \tag{14.34b}
$$

$$
C = \Phi_0 - \sum_{i=1}^{s} \Gamma_i \Phi_i^{-1} \Gamma_i^T \tag{14.34c}
$$

$$
L_c D_c L_c^T = C. \tag{14.34d}
$$

The representation (14.34) can be used to compute the solution to the system

$$\Psi x = b,$$

where $x = (x_1, \ldots, x_s, x_0)^T$, $b = (b_1, \ldots, b_s, b_0)^T$ as follows:

$$z_i = L_i^{-1} b_i, \quad i = 1, \ldots, s \tag{14.35a}$$

$$z_0 = L_c^{-1} \left(b_0 - \sum_{i=1}^{s} L_{0,i} z_i \right) \tag{14.35b}$$

$$y_i = D_i^{-1} z_i, \quad i = 0, \ldots, s \tag{14.35c}$$

$$x_0 = L_c^{-T} y_0 \tag{14.35d}$$

$$x_i = L_i^{-T} (y_i - L_{0,i}^T x_0), \quad i = 1, \ldots, s. \tag{14.35e}$$

For a multistage stochastic programming problem the augmented system matrix in (14.29) is a nested chain of matrices of type (14.32), as depicted in (14.31). Every diagonal block Φ_i in (14.32), (14.33), (14.34), and (14.35) itself is of bordered block-diagonal structure. This nesting is as deep as the number of stages in the multistage problem. In order to factorize a nested system matrix (14.31), computations (14.34) can also be nested. Whenever (14.34a) asks for the factorization of Φ_i, the entire process (14.34) can be applied at the next lower level to obtain the required factorization. Similarly, if for the backsolves (14.35), the matrices L_i are themselves of form (14.33), the process (14.35) can be employed at the next lower level to compute the result of the operations in (14.35a) to (14.35e).

The principal advantage of this nested structure exploitation is a gain in complexity. Processes (14.34) and (14.35) need to be performed on every node of the scenario tree; any actual explicit factorization and backsolves are only performed for matrices of size n (the size of the model on each node). An upper bound for the total complexity of factorizing the matrix (14.31) occurring in stochastic programming is therefore $\mathcal{O}(|\mathcal{V}|n^3)$, compared with up to $\mathcal{O}(N^3)$ in the unstructured case. Of course the complexity of the linear algebra can be significantly less than $\mathcal{O}(N^3)$ if sparsity can be exploited, but is unlikely to be able to match the linear complexity $\mathcal{O}(|\mathcal{V}|)$ of the nested approach. An additional advantage is a reduction in memory use, which again is reduced to $\mathcal{O}(|\mathcal{V}|n^2)$.

It is important to realize that these operations are amenable to parallel computations. Calculations (14.34a), (14.34b) and (14.35a) to (14.35e) are independent for each i and can be performed simultaneously on different processors. Further (14.34c) and (14.35b) require communications and gather-type operations, whereas only (14.34d) and (14.35b) to (14.35d) require global operations. All the global calculations and communications involve entities of size n_0 (the size of Φ_0). Since in multistage stochastic programming n_0 is small compared to the sizes n_i of the diagonal blocks Φ_i (which are themselves block-structured), the potential for efficient parallelization is huge. The parallel distribution of calculations is summarized in Fig. 14.5.

This basic principle of nested Schur complement computations is employed by many parallel IPM approaches to the ALM model.

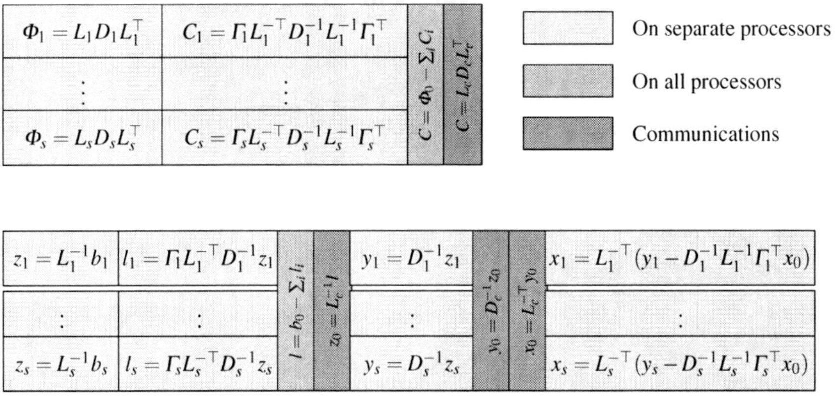

$\Phi_1 = L_1 D_1 L_1^\top$	$C_1 = \Gamma_1 L_1^{-\top} D_1^{-1} L_1^{-1} \Gamma_1^\top$		
$\vdots$	$\vdots$	$C = \Phi_0 - \Sigma_i C_i$	$C = L_c D_c L_c^\top$
$\Phi_s = L_s D_s L_s^\top$	$C_s = \Gamma_s L_s^{-\top} D_s^{-1} L_s^{-1} \Gamma_s^\top$		

On separate processors
On all processors
Communications

$z_1 = L_1^{-1} b_1$	$l_1 = \Gamma_1 L_1^{-\top} D_1^{-1} z_1$		$y_1 = D_1^{-1} z_1$		$x_1 = L_1^{-\top}(y_1 - D_1^{-1} L_1^{-1} \Gamma_1^\top x_0)$
$\vdots$	$\vdots$	$l = b_0 - \Sigma_i l_i$ / $l = L_c$	$\vdots$	$y_0 = D_c^{-1} z_0$ / $x_0 = L_c^{-\top} y_0$	$\vdots$
$z_s = L_s^{-1} b_s$	$l_s = \Gamma_s L_s^{-\top} D_s^{-1} z_s$		$y_s = D_s^{-1} z_s$		$x_s = L_s^{-\top}(y_s - D_s^{-1} L_s^{-1} \Gamma_s^\top x_0)$

Fig. 14.5 Split of computations (14.34) and (14.35) between processors.

Nested Birge–Qi Factorization

The Schur complement approach presented above exploits the nested structure in the augmented system (14.29) for multistage stochastic programming. As an alternative one can consider applying a similar approach to the normal equation form (14.30). This idea goes back to Birge and Qi [52]. For ease of presentation we consider a 2-stage linear stochastic programming problem with $Q = 0$ and a system matrix

$$A = \begin{pmatrix} W_1 & & T_1 \\ & \ddots & \vdots \\ & & W_s\ T_s \\ & & W_0 \end{pmatrix}, \tag{14.36}$$

where W_0 can be rectangular, but is assumed to have full row rank. In this case the normal equations (14.30) reduce to

$$(A\Theta A^T)\Delta y = b, \tag{14.37}$$

where $\Theta = \mathrm{diag}(\Theta_1, \ldots, \Theta_s, \Theta_0)$, yielding

$$A\Theta A^T = \begin{pmatrix} W_1\Theta_1 W_1^T & & \\ & \ddots & \\ & & W_s\Theta_s W_s^T \\ & & & I \end{pmatrix} + \begin{pmatrix} T_1 & 0 \\ \vdots & \vdots \\ T_s & 0 \\ W_0 & I \end{pmatrix} \begin{pmatrix} T_1\Theta_0 & 0 \\ \vdots & \vdots \\ T_s\Theta_0 & 0 \\ W_0\Theta_0 & -I \end{pmatrix}^T. \tag{14.38}$$

The inverse of $A\Theta A^T$ can be obtained from the Sherman–Morrison–Woodbury formula

$$(W + UV^T)^{-1} = W^{-1} - W^{-1}UG^{-1}V^T W^{-1}, \qquad G = I + V^T W^{-1} U.$$

Using $W = \text{diag}(W_1 \Theta_1 W_1^T, \ldots, W_s \Theta_s W_s^T, I)$ and

$$U = \begin{pmatrix} T_1 & 0 \\ \vdots & \vdots \\ T_s & 0 \\ W_0 & I \end{pmatrix}, \qquad V = \begin{pmatrix} T_1 \Theta_0 & 0 \\ \vdots & \vdots \\ T_s \Theta_0 & 0 \\ W_0 \Theta_0 & -I \end{pmatrix},$$

we can compute

$$G = \begin{pmatrix} \Theta_0 & \\ & I \end{pmatrix} \begin{pmatrix} \hat{G} & W_0^T \\ -W_0 & 0 \end{pmatrix}, \qquad \hat{G} = \Theta_0^{-1} + W_0^T W_0 + \sum_{i=1}^{s} T_i^T (W_i \Theta_i W_i^T)^{-1} T_i.$$

Hence (14.37) can be solved by the sequence

$$z_i = (W_i \Theta_i W_i)^{-1} b_i, \qquad i = 1, \ldots, s \tag{14.39a}$$

$$v = \sum_{i=1}^{s} T_1^T z_i + W_0^T b_0 \tag{14.39b}$$

$$\begin{pmatrix} u_1 \\ u_2 \end{pmatrix} = \begin{pmatrix} \hat{G} & W_0^T \\ -W_0 & \end{pmatrix}^{-1} \begin{pmatrix} v \\ -b_0 \end{pmatrix} \tag{14.39c}$$

$$w_i = (W_i \Theta_i W_i^T)^{-1} T_i u_1, \qquad i = 1, \ldots, s \tag{14.39d}$$

$$w_0 = W_0 u_1 + u_2 \tag{14.39e}$$

$$\Delta y_i = z_i - w_i, \qquad i = 1, \ldots, s \tag{14.39f}$$

$$\Delta y_0 = z_0 - w_0, \tag{14.39g}$$

where (14.39c) can be solved by

$$[W_0 \hat{G}^{-1} W_0^T] u_2 = W_0 \hat{G}^{-1} v - b_0 \tag{14.40a}$$

$$\hat{G} u_1 = v - W_0^T u_2. \tag{14.40b}$$

As before we can consider a nested version of the factorization, by realizing that for a nested system matrix, matrices W_i are again of the form (14.36) and hence (14.39a) to (14.39d) can be solved by employing the process (14.39) on the next lower level. As with the Schur complement approach, the nested Birge–Qi factorization can be parallelized by distributing computations (14.39a), (14.39d), and (14.39f) among the available processors. The Birge–Qi factorization has been applied to 2-stage stochastic programming in parallel by Jessup et al. [53]. A nested version for multi-stage stochastic programming has been derived by Pflug and Halada [54].

14.4.2 Results

In this section we give an overview of the results that have been reported in the literature for the parallel solution of portfolio optimization problems by interior point approaches.

Blomvall and Lindberg in [29] solve a 7-stage ALM with 1.0 million scenarios, 15.7 million constraints, and 16.9 million variables in 20 hours on a serial 200 MHz Sparc station with 2 GB of memory. In [55] they report on the solution of a 10-stage ALM problem with 1.9 million scenarios, 67 million variables, and 118 million constraints in 97 minutes on 32 processors in a Beowulf cluster of 900 MHz Athlon CPUs with 512 MB memory per node. Further Blomvall [56] solves a series of ALM problems with the largest having 24 stages, 8.3 million scenarios, 50 million variables, and 92 million constraints in 1136 seconds on 32 processors on the same Beowulf cluster. He also reports consistent parallel efficiencies of 93–97% on 1–32 processors on a smaller problem with 2.8 million variables. All these results use a primal IPM implementations employing a specialized version of the nested Schur complement factorization.

Grothey and Gondzio also employ a nested Schur complement factorization but within a primal–dual IPM framework. In [26] they apply sequential quadratic programming with the interior point code OOPS as the QP solver to solve various nonlinear variations of the ALM model, including logarithmic utility functions, skewness objectives, and constraints on variance and semivariance. They use a set of randomly generated test problems with the largest one optimizing 20 assets over 4 periods with a branching factor of 55 leading to 170,000 scenarios and 10.5 million constraints. They parallelize the computations on 8 processors of a SunFire 15 K, 48-processor machine with 900 MHz UltraSparc-II processors and 48 GB of shared memory. They can solve the largest problem in under 2500 seconds on 8 processors and report a speedup of 7.28–7.64 over the formulations considered. They report that a serial version of their algorithm achieves considerable gains in terms of both solution speed and memory usage compared to commercial solver CPLEX 7.0. In [26] they solve a mean-variance formulation with 4 stages, 1.7 million scenarios, 10 assets, and over 52 million variables in 8500 seconds on 16 processors of the above machine. Further on smaller versions of the problem ranging between 5–9 million variables they report a parallel efficiency of 87.7–94.0% on 8 processors. Finally in [57] they report on the solution of an ALM problem with 6 stages, 12.8 million scenarios, and 1.02×10^9 variables on 1280 processors of the 1600 1.7 GHz-processor machine HPCx in 3600 seconds. On a smaller problem with 10 million variables they report of a parallel efficiency of 86% when comparing 512 processors with 16 processors on a BlueGene/L machine.

Lucka, Melichercik, and Halada in [23] use a primal–dual IPM employing a nested Birge–Qi factorization to solve 2- and 3-stage linear ALM models. Their largest model has 3 stages, 90,000 scenarios, and 450,000 unknowns. They report serial solution time of 203 seconds on a 3.6 GHz Pentium IV processor and 23.4 seconds on a cluster of 16 such machines, resulting in a parallel efficiency of 54%. Pflug and Halada in [54] with the same algorithm report a reduction in computation time from 20.7 seconds on 2 processors to 5.8 seconds on 32 processors resulting in a parallel efficiency of 22% for a 3-stage model with 640 scenarios and 25,600 variables. These results were however obtained on a heterogeneous cluster incorporating 400 MHz Pentium-II and 700 MHz Pentium-III processors.

A different approach is taken by Durazzi, Ruggiero, and Zanghirati [58]. They use a primal–dual IPM in which the Newton system is solved by an iterative method with a structured block-preconditioner. They solve 2-stage linear stochastic programming problems (although not specifically ALM models) on a Cray T3E using up to 120 processors. On their largest problem with 480 scenarios and 46,000 variables they report near perfect speedup (and in some cases even superlinear speedup).

14.5 Parallel Solution Approaches: Heuristics and Evolutionary Algorithms

As pointed out in Sect. 14.2.3 many realistic formulations of the ALM model lead to difficult integer programming or global optimization problems that are beyond the reach of the solution approaches reviewed so far. However there is a keen interest in applying evolutionary algorithms such as Genetic Algorithms (GA) to portfolio optimization problems. These algorithms can be employed in two different lines. First, as single-objective evolutionary algorithms (SOEA) that are used to an otherwise intractable integer or non-convex formulation (arising from cardinality constraints or minimum lot size constraints) of a (usually single period) portfolio selection problem (see for example [59, 60] for a survey). The other line are multi-objective evolutionary algorithms (MOEA) that are employed to directly calculate the Pareto efficient set in a multi-objective version of the portfolio selection problem (using expected return and a risk measure as the objective functions). See [61] for a survey. While most of these only deal with sequential computing, evolutionary algorithms in general are easily parallelized [62]. Loraschi and coauthors [63] present a distributed parallel GA for the computation of the efficient frontier in a portfolio selection problem consisting of the expected return and semivariance as risk measure on a cluster of 10 Sun Sparc machines.

14.6 Other Approaches

Moreno-Vozmediano and coauthors [64] use grid computing in the context of VaR-constrained portfolio optimization. They use Monte Carlo simulation to evaluate the VaR of a given portfolio by distributing the considered scenarios among the available grid computation nodes. They report a speedup of 2.0 for 4 grid nodes compared to the serial version. Similarly Zanghirati and coauthors [65] use Monte Carlo simulation to compute the final value of an ALM portfolio with 200 different products. They report near linear speedup on a Cray T3E machine with up to 128 processors, simulating 10,240 scenarios.

A notable development is the AURORA Financial Management System [34] at the University of Vienna. It aims to provide a decision support system in a parallel framework for the whole process required in portfolio optimization starting from data analysis to create a scenario tree for the underlying asset return and liability distributions, the actual optimization of the ALM problem, through to the post-optimization analysis. Various parallelizable optimization methods have been described as part of the AURORA framework, notably asynchronous nested Benders decomposition [46] and scenario decomposition using augmented Lagrangians [34]. An overview is given in [22]. Efforts are underway for a grid-enabled DSS where not only the optimization, but every component of the whole process (possibly for several models at the same time) can be distributed over available network resources [66, 67].

14.7 Detailed Comparison of Parallel Interior Point Approaches

In this last section we will give detailed comparison of some of the parallel interior point implementations that have been applied to portfolio optimization problems.

14.7.1 OOPS (Gondzio and Grothey)

The object-oriented parallel solver (OOPS) of Gondzio and Grothey [26, 30, 57, 68] explicitly uses the Schur complement mechanism (14.34) and (14.35) in an object-oriented framework. OOPS represents a multistage stochastic programming as a tree of matrix objects. Each node of the tree, at any level, represents a matrix of type (14.32) together with the appropriate linear algebra implementations (14.34) and (14.35) to support factorize and back-solve operations involving this matrix. The matrix object also includes obvious secondary routines such as matrix-vector products. The object-oriented layout of OOPS hides the linear algebra implementations of a particular node in the scenario tree (or the matrix tree) from its parent and child nodes. Indeed OOPS does not require the subblocks Φ_i and Γ_i to be of bordered block-diagonal type themselves; they can represent *any* exploitable block matrix structure. This layout makes OOPS applicable to a wide range of structured problems beyond ALM or even stochastic programming.

On the other hand this generality prevents OOPS from exploiting the fine-level structure of the matrices Γ_i, Φ_i, which is apparent from Fig. 14.4 and (14.13), as is done in competing implementations. In OOPS the Schur complement matrix C and its Cholesky factor L_C in operations (14.34d), (14.35b), and (14.35d) are treated as dense matrices (as they will be in general), although, as we will see when discussing alternative IPM approaches, they have further exploitable structure. The main computational effort in OOPS is in the assembling of the Schur complement matrix C in (14.34c):

$$C = \Phi_0 - \sum_{i=1}^{s} \Gamma_i \Phi_i^{-1} \Gamma_i^T.$$

Since the decomposition $\Phi_i = L_i D_i L_i^T$ is known, (14.34c) can be computed as

$$C = \Phi_0 - \sum_{i=1}^{n} V_i^T D_i^{-1} V_i,$$

where

$$V_i = L_i^{-1} \Gamma_i^T.$$

Generally Γ_i^T will be a tall, thin, sparse matrix. If the factors L_i are sparse as well, we can expect V_i to display some sparsity. The local contribution $V_i^T D_i^{-1} V_i$ to the Schur complement is therefore most efficiently obtained as

$$V_i^T D_i^{-1} V_i = \sum_k v_{ik} D_i^{-1} v_{ik}^T, \qquad (14.41)$$

where v_{ik} is the sparse k-th row of V_i (see Fig. 14.6). This uses the observation that

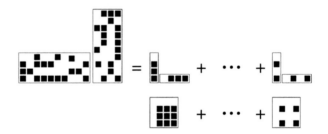

Fig. 14.6 Calculation of $V_i^T V_i = \sum_k v_{ik} v_{ik}^T$.

outer products of sparse vectors can be efficiently computed without the need to scan for matching sparse elements (as would need to be done for the alternative of computing pairwise inner product of columns of V_i). However, it requires changing the representation of V_i from column-wise to row-wise sparse format.

It is paramount for efficiency in Eq. (14.41) that V_i is as sparse as possible, hence fill-in in the solution step $V_i = L_i^{-1} \Gamma_i^T$ needs to be avoided at all cost. This is however not easily achievable. For the ALM problem in Fig. 14.4, the Γ_i^T matrices are of the form $(0, \ldots, 0, T)^T$, so fill-in in V_i should be limited to the bottom block corresponding to T. However usually the factorization of $\Phi_i = L_i D_i L_i$ involves pivoting for sparsity, resulting in an unavoidable reordering of $\Gamma_i = (0, \ldots, 0, T)^T$ and subsequent fill-in (see Fig. 14.7). Note however that this problem is partially alleviated by the use of an implicit Schur complement factorization for Φ; this effectively restricts the pivot choices to within each current block (Fig. 14.8). While not completely avoiding unnecessary fill-in, any occurring fill-in in V_i is restricted to the Φ_0 block.

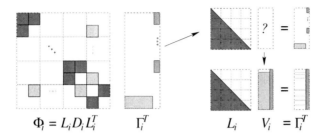

Fig. 14.7 Sparse calculation of $V_i = L_i^{-1}\Gamma_i^T$: Fill-in for sparse factors of L_i.

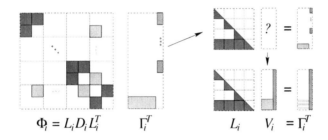

Fig. 14.8 Sparse calculation of $V_i = L_i^{-1}\Gamma_i^T$: Restricted fill-in for block factors of L_i.

14.7.2 Riccati-Based IPM Solver (Blomvall and Lindberg)

A related approach is used by Blomvall and Lindberg [29, 55] and implemented in parallel in [56]. Like OOPS they use a nested implicit factorization of the system (14.31). Differently, however they use the fact that ALM problems (as indeed most stochastic programming problems) can be written in dynamic stochastic programming form. This enables them to use a more compact nested implicit inverse factorization. The price they pay, however, is a loss of generality. Their solution approach is only applicable to dynamic stochastic programming; it cannot be applied to general nested structures in the way that OOPS can.

In detail [29] uses the control formulation:

$$\min_{x_i,u_i,i\in\mathcal{V}} \sum_{i\in\mathcal{V}} U_i(x_i,u_i)$$

$$\begin{aligned} \text{s.t.}\quad x_i - A_i x_{\pi(i)} - B_i u_{\pi(i)} &= b_i, & i &\in \mathcal{V}\setminus\{0\} \\ E_i x_i + F_i u_i &\geq e_i, & i &\in \mathcal{V} \\ x_i &\in [x_i^l, x_i^u], & i &\in \mathcal{V} \\ u_i &\in [u_i^l, u_i^u], & i &\in \mathcal{V}, \end{aligned}\qquad(14.42)$$

where $x_i \in \mathbb{R}^{n_x}, u_i \in \mathbb{R}^{n_u}, e_i \in \mathbb{R}^{m_e}$ and the other matrices and vectors have corresponding dimensions. The ALM formulation (14.12) naturally fits into this framework with the identification $x_i = (x_i^h)$, $u_i = (x_i^b, x_i^s)$. Blomvall and Lindberg proceed by applying a primal interior point solver to this problem and derive a nested elimination scheme based on the use of discrete time Riccati equations from control theory. Details are given in [29, 55]. In order to compare the approach with that taken by [26, 30, 51] we present the Riccati scheme of Blomvall et al. in the framework of Sect. 14.4. A primal IPM applied to system (14.42) needs to solve at each iteration an augmented system matrix of form (14.31) with the specific choices

$$\Phi_i = \begin{bmatrix} \nabla_{xx}^2 U_i + \Theta_i^x & \nabla_{xu}^2 U_i & & E_i^T & -I \\ \nabla_{ux}^2 U_i & \nabla_{uu}^2 U_i + \Theta_i^u & & F_i^T & \\ & & \Theta_i^s & -I & \\ E_i & F_i & -I & & \\ -I & & & & \end{bmatrix} \qquad \Gamma_i^T = \begin{bmatrix} 0 & 0 & 0 & 0 & 0 \\ \vdots & \vdots & \vdots & \vdots & \vdots \\ 0 & 0 & 0 & 0 & 0 \\ 0 & 0 & 0 & A_i & B_i \end{bmatrix},$$

$$(14.43)$$

where we have introduced slack variables s_i for second constraint in (14.42), Lagrange multipliers y and w for the two constraints in (14.42), and the columns correspond to components $(x_i, u_i, s_i, w_i, y_i)$ and $(x_{\pi(i)}, u_{\pi(i)}, s_{\pi(i)}, w_{\pi(i)}, y_{\pi(i)})$, respectively. Rather than obtaining a sparse LDL^T decomposition of each local node block Φ_i as in OOPS, Blomvall and Lindberg use an implicit factorization of Φ_i. They realize that through elimination of the Δs and Δw components, the system

$$\begin{bmatrix} \nabla_{xx}^2 U_i + \Theta_i^x & \nabla_{xu}^2 U_i & & E_i^T & -I \\ \nabla_{ux}^2 U_i & \nabla_{uu}^2 U_i + \Theta_i^u & & F_i^T & \\ & & \Theta_i^s & -I & \\ E_i & F_i & -I & & \\ -I & & & & \end{bmatrix} \begin{bmatrix} \Delta x \\ \Delta u \\ \Delta s \\ \Delta w \\ \Delta y \end{bmatrix} = \begin{bmatrix} \xi_x \\ \xi_u \\ \xi_s \\ \xi_c \\ \xi_b \end{bmatrix} \qquad (14.44)$$

can be reduced to

$$\begin{bmatrix} Q_i & P_i & -I \\ P_i^T & R_i & \\ -I & & \end{bmatrix} \begin{bmatrix} \Delta x \\ \Delta u \\ \Delta y \end{bmatrix} = \begin{bmatrix} \xi_x + E_i^T \Theta_i^s \xi_c + E_i^T \xi_s \\ \xi_u + F_i^T \Theta_i^s \xi_c + F_i^T \xi_s \\ \xi_b \end{bmatrix}, \qquad (14.45)$$

where

$$Q_i = \nabla^2_{xx}U_i + \Theta^x_i + E^T_i \Theta^s_i E_i \tag{14.46a}$$

$$R_i = \nabla^2_{uu}U_i + \Theta^u_i + F^T_i \Theta^s_i F_i \tag{14.46b}$$

$$P_i = \nabla^2_{ux}U_i + E^T_i \Theta^s_i F_i \tag{14.46c}$$

and hence further to

$$\Delta x = -\xi_b \tag{14.47a}$$

$$\Delta u = R^{-1}_i(-P^T_i \Delta x + F^T_i \Theta^s_i \xi_c + F^T_i \xi_s + \xi_u) \tag{14.47b}$$

$$\Delta y = Q_i \Delta x + P_i \Delta u - E^T_i \Theta^s_i \xi_c - E^T_i \xi_s - \xi_x \tag{14.47c}$$

$$\Delta w = \Theta^s_i E_i \Delta x + \Theta^s_i F_i \Delta u - \xi_s \tag{14.47d}$$

$$\Delta s = (\Theta^s_i)^{-1}(\Delta w + \xi_s). \tag{14.47e}$$

Note that Blomvall and Lindberg use a logarithmic barrier term directly for the slack in the $E_i x_i + F_i u_i \geq e_i$ constraint, thus the $\Delta s, \Delta w$ (slack and multiplier for this inequality) never explicitly appear in their presentation. Reductions (14.44), (14.45), (14.46), and (14.47), together with the sparsity of Γ^T_i in (14.43), means that the Schur complement computations (cf. (14.34c))

$$\Phi^+_0 = \Phi_0 - \sum \Gamma_i \Phi^{-1}_i \Gamma^T_i \tag{14.48}$$

affect only the Q, R, and P blocks in Φ_0, hence preserves the sparsity, so that the same implicit factorization can be used for Φ^+_0 after the update

$$\tilde{Q}_i = Q_i + \sum_{j \in \mathscr{C}(i)} A^T_j W_j A_j \tag{14.49a}$$

$$\tilde{R}_i = R_i + \sum_{j \in \mathscr{C}(i)} B^T_j W_j B_j \tag{14.49b}$$

$$\tilde{P}_i = P_i + \sum_{j \in \mathscr{C}(i)} A^T_j W_j B_j, \tag{14.49c}$$

where

$$W_i = Q_i - P_i R^{-1}_i P^T_i . \tag{14.50}$$

Blomvall and Lindberg use these updates together with the Schur complement calculations (14.34) and (14.35) in a nested tree elimination scheme. Equation (14.46) is used to calculate Q_i, R_i, and P_i on every node. Then, starting at the leaf nodes they factorize R_i to get a representation for R^{-1}_i and calculate W_i from (14.50). Once calculations on the leaf nodes are finished they are aggregated onto the next level by the updates (14.49). After this the process continues further up the tree using $\tilde{Q}_i, \tilde{R}_i, \tilde{P}_i$ instead of Q_i, R_i, P_i for all remaining nodes. The work involved in building the Schur complement (14.34c) thus reduces to the recursions (14.49) and the only explicit factorization necessary is that of $\tilde{R}_i$ on every node, which is of dimension n_u of the number of control variables u. OOPS in comparison needs to factorize Φ_i which is of dimension $2n_x + n_u + 2m_e$. Through this approach Blomvall and Lindberg further manage to reduce the work involved in forming the Schur complement (14.34c). Whereas in OOPS the calculation of $L^{-1}_i \Gamma^T_i$ fills in over the complete Φ_i

block (see Fig. 14.8) even for very sparse Γ_i, the fill-in in Blomvall and Lindberg is limited to the (already dense) Q_i, R_i, P_i blocks.

As pointed out, the drawback is a loss in generality, in that the approach is only applicable to problems that can be stated in nested control formulation. This, however, is generally the case for ALM models. Another drawback is a possible loss of stability, since the block-elimination scheme (14.47) effectively restricts the choice of pivots as compared to a general sparse factorization $\Phi_i = L_i D_i L_i^T$.

14.7.3 Tree Sparse IPM Solver (Steinbach)

Steinbach in [51,69,70] presents a tree-sparse IPM approach which is very similar to that of Blomvall and Lindberg. Steinbach, however, works within the primal–dual interior point framework. He starts from the same control formulation (14.42) as Blomvall and Lindberg. However, rather than dealing with a general local constraint

$$E_i x_i + F_i u_i \geq e_i$$

as Blomvall and Lindberg, Steinbach deals with a finer local constraint structure:

$$E_i^r x_i + F_i^r u_i \in [r_{lj}, r_{ul}] \tag{14.51a}$$
$$E_i^c x_i + F_i^c u_i = e_i^c \tag{14.51b}$$
$$E_i^x x_i \qquad = e_i^x \tag{14.51c}$$
$$\qquad F_i^u u_i = e_i^u \tag{14.51d}$$

distinguishing explicitly local state, control, and mixed constraints, with the assumption that for every local node $i \in \mathcal{V}$, the matrix

$$F_i^l = \begin{pmatrix} E_i^c & F_i^c \\ E_i^x & \\ & F_i^u \end{pmatrix}$$

has full rank. He uses these constraints to eliminate as many variables as possible from the local nodes in a first step, to further reduce the dimension of the recursions (14.49). In detail, through an incomplete LU-decomposition with row pivoting of F_i^c (since F_i^c does not need to have full rank), appropriate partitions of E_i^c and e_i^c

$$\Pi_i F_i^c = \begin{pmatrix} L_{Fi} & 0 \\ Z_i & 0 \end{pmatrix} U_{Fi}, \qquad \Pi_i E_i^c = \begin{pmatrix} E_i^{c,1} \\ E_i^{c,2} \end{pmatrix}, \qquad \Pi_i e_i^c = \begin{pmatrix} e_i^{c,1} \\ e_i^{c,2} \end{pmatrix},$$

and a change of variables $\hat{u}_i = U_{Fi} u_i$, the mixed constraint $E_i^c x_i + F_i^c u_i = e_i^c$ can be expressed as

$$L_{Fi}\hat{u}_i^1 + E_i^{c,1}x_i = e_i^{c,1} \tag{14.52}$$

$$Z_i\hat{u}_i^1 + E_i^{c,2}x_i = e_i^{c,2} \tag{14.53}$$

and thus

$$\hat{u}_i^1 = L_{Fi}^{-1}(e_i^{c,1} - E_i^{c,1}x_i) \tag{14.54a}$$

$$(E_i^{c,2} - Z_i L_{Fi}^{-1} E_i^{c,1})x_i = e_i^{c,2} - Z_i L_{Fi}^{-1} e_i^{c,1}. \tag{14.54b}$$

By substituting $u_i = U_{Fi}^{-1}\hat{u}_i$ into every equation of (14.42) that contains u_i, variables $\hat{u}_i^1$ are eliminated from each local node, and constraint (14.51b) is transformed into (14.54a) which is a reduced dimensional constraint of mixed type and (14.54b), an additional constraint of local-state type (14.51c). In the next step through an LU-decomposition

$$E_i^x = (L_{Ei}\ 0)U_{Ei},$$

(E_i^x being of full rank) and a change of variables $\hat{x}_i = U_{Ei}x_i$, the local constraint $E_i^x x_i = e_i^x$ can be expressed as

$$e_i^x = (L_{Ei}\ 0)\begin{pmatrix} U_{Ei}^1 x_i \\ U_{Ei}^2 x_i \end{pmatrix} = L_{Ei}\hat{x}_i^1$$

and hence as

$$\hat{x}_i^1 = L_{Ei}^{-1} e_i^x, \qquad \hat{x}_i^2 \text{ free .}$$

By substituting $x_i = U_{Ei}^{-1}\hat{x}_i$ into every equation of (14.42) that contains x_i, variables $\hat{x}_i^1$ are eliminated from the system. Finally the same step is applied to the local constraint $F_i^u u_i = e_i^u$ to eliminate a subset of the control variables u_i at every local node. These node eliminations are applied as a preprocessing step, effectively reducing the dimension of the state and control variables and the mixed constraint $E_i x_i + F_i u_i \geq e_i$ at every node. The recursions (14.49) are thus performed in a reduced space, yielding further efficiency gains. As with the approach of Blomvall and Lindberg the price to pay for the efficiency gains is a lack of generality and a possible instability resulting from the restriction of pivot choices.

Although the scheme of Steinbach could clearly be parallelized along the lines of [56], unfortunately no such implementation seems to exist. It has, however, been applied to ALM problems in serial [51]. For a problem with 8 assets and 6 stages, leading to 218,000 scenarios and over 2 million decision variables one complete factorization could be performed in 33 seconds on a 175 MHz R10000 processor.

14.8 Conclusions

We have presented recent approaches to the parallel solution of multistage portfolio optimization problems. Traditionally decomposition has dominated this field, but recently interior point based approaches have proved to be a serious competitor.

Problems of many millions of variables can now be routinely solved on moderate parallel hardware, while the use of dedicated massively parallel machines makes the solution of problems with 10^9 variables and more feasible.

High-performance parallel computing is expected to continue to make an impact on solution approaches to financial planning problems, especially as new, inexpensive parallel hardware becomes commonplace. For many popular solution approaches such as decomposition, IPMs, and evolutionary algorithms promising parallel implementations exist; research is underway to adapt these approaches to distributed parallel systems such as in grid computing.

New computing hardware will require new efficient implementations and modifications of existing algorithm, while on the other hand, progress in the study of appropriate model formulations including risk measures results in new mathematical challenges for successful parallelism exploitation. The field continues to be exciting.

References

1. H. M. Markowitz, Portfolio selection, J Financ (1952) 77–91.
2. M. Steinbach, Markowitz revisited: Mean variance models in financial portfolio analysis, SIAM Rev 43 (1) (2001) 31–85.
3. J. R. Birge, F. Louveaux, Introduction to Stochastic Programming, Springer-Verlag, New York (1997).
4. P. Kall, S. W. Wallace, Stochastic Programming, John Wiley & Sons, Chichester (1994).
5. S. E. Wright, Primal-dual aggregation and disaggregation for stochastic linear programs, Math Oper Res 19 (4) (1994) 893–908.
6. D. Cariño, T. Kent, D. Myers, C. Stacy, M. Sylvanus, A. Turner, K. Watanabe, W. Ziemba, The Russel-Yasuda Kasai model: An asset/liability model for Japanese insurance company using multistage stochastic programming, Interfaces 24 (1) (1994) 29–49.
7. J. M. Mulvey, A. E. Thorlacius, The Towers Perrin global capital market scenario generation system, in: W. T. Ziemba, J. M. Mulvey (Eds.), World Wide Asset and Liability Management, Cambridge University Press, Cambridge (1998) pp. 286–312.
8. M. Kusy, W. Ziemba, A bank asset and liability model, Oper Res 34 (1986) 356–376.
9. J. Mulvey, H. Vladimirou, Stochastic network programming for financial planning problems, Manage Sci 38 (1992) 1643–1664.
10. S. Zenios, Asset/liability management under uncertainty for fixed-income securities, Ann Oper Res 59 (1995) 77–97.
11. W. T. Ziemba, J. M. Mulvey, Worldwide Asset and Liability Modeling, Publications of the Newton Institute, Cambridge University Press, Cambridge (1998).
12. G. Consigli, M. Dempster, Dynamic stochastic programming for asset–liability management, Ann Oper Res 81 (1998) 131–162.
13. M. A. H. Dempster, R. T. Thompson, Parallelization and aggregation of nested Benders decomposition, Ann Oper Res 81 (1998) 163–188.
14. A. Ruszczyński, Parallel decomposition of multistage stochastic programs, Math Program 58 (1993) 201–228.
15. A. Consiglio, S. A. Zenios, Integrated simulation and optimization models for tracking international fixed income indices, Technical Report, Department of Public and Business Administration, University of Cyprus, Nikosia, Cyprus (1998).
16. R. Mansini, M. G. Speranza, Heuristic algorithms for the portfolio selection problem with minimum transaction lots, Eur J Oper Res 114 (1999) 219–233.

17. G. Mitra, T. Kyriakis, C. Lucas, M. Pirbhai, A review of portfolio planning: Models and systems, in: S. Satchell, A. E. Scowcroft (Eds.), Advances in Portfolio Construction and Implementation, Butterworth and Heinemann, Oxford (2003) pp. 1–39.
18. R. Van Slyke, R. J.-B. Wets, L-shaped linear programs with applications to optimal control and stochastic programming, SIAM J Appl Math 17 (1969) 638–663.
19. P. Artzner, F. Daelben, J.-M. Eber, D. Heath, Coherent measures of risk, Math Financ 9 (1999) 203–228.
20. W. Ogryczak, A. Ruszczynski, From stochastic dominance to mean-risk models: Semideviations as risk measures, Eur J Oper Res 116 (1999) 35–50.
21. H. Konno, H. Shirakawa, H. Yamazaki, Mean absolute deviation portfolio optimization model and its applications to Tokyo stock market, Manage Sci 37 (1991) 519–531.
22. G. C. Pflug, A. Świętanowski, Selected parallel optimization methods for financial management under uncertainty, Parallel Comput 26 (2000) 3–25.
23. M. Lucka, I. Melichercik, L. Halada, Application of multistage stochastic programs solved in parallel in portfolio management, Parallel Comput 34 (2008) 469–485.
24. N. C. P. Edirisinghe, E. I. Patterson, Multi-period stochastic portfolio optimization: Block-separable decomposition, Ann Oper Res 157 (2007) 367–394.
25. R. Mansini, W. Ogryczak, M. G. Speranza, LP solvable models for portfolio optimization: A classification and computational comparison, IMA J Manag Math 14 (2003) 187–220.
26. J. Gondzio, A. Grothey, Parallel interior point solver for structured quadratic programs: Application to financial planning problems, Ann Oper Res 152 (1) (2007) 319–339.
27. J. von Neumann, O. Morgenstern, Theory of Games and Economic Behaviour, Princeton University Press, Princeton (1953).
28. J. L. Kelly, A new interpretation of information rate, AT&T Tech J 35 (1956) 917–926.
29. J. Blomvall, P. O. Lindberg, A Riccati-based primal interior point solver for multistage stochastic programming, Eur J Oper Res 143 (2002) 452–461.
30. J. Gondzio, A. Grothey, Solving nonlinear portfolio optimization problems with the primal-dual interior point method, Eur J Oper Res 181 (2007) 1019–1029.
31. H. Konno, H. Shirakawa, H. Yamazaki, A mean-absolute deviation-skewness portfolio optimization model, Ann of Oper Res 45 (1993) 205–220.
32. H. Konno, K.-I. Suzuki, A mean-variance-skewness portfolio optimization model, J Oper Res Soc Jpn 38 (1995) 173–187.
33. G. C. Pflug, How to measure risk, in: Modelling and Decisions in Economics, Physica-Verlag (1999) pp. 39–59.
34. G. C. Pflug, A. Świętanowski, E. Dockner, H. Moritsch, The AURORA financial management system: Model and parallel implementation design, Ann Oper Res 99 (2000) 189–206.
35. P. Jorion, Value at Risk: The New Benchmark for Controlling Market Risk, McGraw-Hill (2000).
36. R. T. Rockafellar, S. Uryasev, Optimization of conditional value-at-risk, J Risk 2 (3) (2000) 21–41.
37. A. A. Gaivoronski, G. C. Pflug, Value-at-risk in portfolio optimization: Properties and computational approach, J Risk 7 (2) (2005) 1–31.
38. A. Ben-Tal, A. Nemirovski, Robust convex optimization, Math Oper Res 23 (1998) 769–805.
39. D. Goldfarb, G. Iyengar, Robust portfolio selection problems, Math Oper Res 28 (2003) 1–38.
40. D. Dentcheva, A. Ruszczyński, Portfolio optimization with stochastic dominance constraints, J Bank Financ 30 (2006) 433–451.
41. J. R. Birge, Decomposition and partitioning methods for multistage stochastic linear programs, Oper Res 33 (1985) 989–1007.
42. H. I. Gassmann, MSLiP: A computer code for the multistage stochastic linear programming problems, Math Program 47 (1990) 407–423.
43. J. Mulvey, A. Ruszczyński, A new scenario decomposition method for large scale stochastic optimization, Oper Res 43 (1995) 477–490.
44. S. Zenios, High-performance computing in finance: The last 10 years and the next, Parallel Comput 25 (1999) 2149–2175.

45. J. Linderoth, S. J. Wright, Decomposition algorithms for stochastic programming on a computational grid, Comput Optim Appl 24 (2/3) (2003) 207–250.
46. E. Laure, H. Moritsch, Portable parallel portfolio optimization in the AURORA financial management system, in: Commercial Applications of High-Performance Computing, Vol. 4528 of Proceedings of SPIE, Society of Photo-Optical Instrumentation Engineers (2001) pp. 193–204.
47. P. Parpas, B. Rustem, Computational assessment of nested Benders and augmented Lagrangian decomposition for mean-variance multistage stochastic problems, INFORMS J Comput 19 (2) (2007) 239–247.
48. J. Mulvey, B. Shetty, Financial planning via multi-stage stochastic optimization, Comput Oper Res 31 (2004) 1–20.
49. S. J. Wright, Primal-Dual Interior-Point Methods, SIAM, Philadelphia (1997).
50. J. Blomvall, P. O. Lindberg, Backtesting the performance of an actively managed option portfolio at the Swedish stock market, 1990–1999, J Econ Dyn Control 27 (2003) 1099–1112.
51. M. Steinbach, Recursive direct algorithms for multistage stochastic programs in financial engineering, in: P. Kall, H.-J. Lüthi (Eds.), Operations Research Proceedings, Selected Papers of the International Conference on Operations Research Zürich 1998, Springer-Verlag (1999) pp. 241–250.
52. J. R. Birge, L. Qi, Computing block-angular Karmarkar projections with applications to stochastic programming, Manage Sci 34 (12) (1988) 1472–1479.
53. E. R. Jessup, D. Yang, S. A. Zenios, Parallel factorization of structured matrices arising in stochastic programming, SIAM J Opimiz 4 (4) (1994) 833–846.
54. G. C. Pflug, L. Halada, A note on the recursive and parallel structure of the Birge and Qi factorization for tree structured linear programs, Comput Optim Appl 24 (2003) 251–265.
55. J. Blomvall, P. O. Lindberg, A Riccati-based primal interior point solver for multistage stochastic programming – Extensions, Optim Method Softw 17 (3) (2002) 383–407.
56. J. Blomvall, A multistage stochastic programming algorithm suitable for parallel computing, Parallel Comput 29 (2003) 431–445.
57. J. Gondzio, A. Grothey, Direct solution of linear systems of size 10^9 arising in optimization with interior point methods, in: R. Wyrzykowski (Ed.), Parallel Processing and Applied Mathematics, Vol. 3911 of Lect Notes Comput Sc, Springer-Verlag, Berlin (2006) pp. 513–525.
58. C. Durazzi, V. Ruggiero, G. Zanghirati, Parallel interior-point method for linear and quadratic programs with special structure, J Optimiz Theory App 110 (2001) 289–313.
59. S.-H. Chen, Evolutionary Computation in Economics and Finance, Physica-Verlag, Heidelberg (2002).
60. F. Schlottmann, D. Seese, Modern heuristics for finance problems: A survey of selected methods and applications, in: S. Rachev (Ed.), Handbook of Computational and Numerical Methods in Finance, Birkhäuser, Berlin (2004) pp. 331–360.
61. M. G. C. Tapia, C. A. C. Coello, Application of multi-objective evolutionary algorithms in economics and finance: A survey, in: Proceedings of the Conference on Evolutionary Computation (2007) pp. 532–539.
62. Z. Konfrst, Parallel genetic algorithms: Advances, computing trends, applications and perspectives, in: 18th International Parallel and Distributed Processing Symposium (2004) pp. 162–169.
63. A. Loraschi, A. Tettamanzi, M. Tomassini, C. Svizzero, C. Scientifico, P. Verda, Distributed genetic algorithms with an application to portfolio selection problems, in: Artificial Neural Nets and Genetic Algorithms, Springer-Verlag (1995) pp. 384–387.
64. R. Moreno-Vozmediano, K. Nadiminti, S. Venugopal, A. B. Alonso-Conde, H. Gibbins, R. Buyya, Portfolio and investment risk analysis on global grids, J Comput Syst Sci 73 (2007) 1164–1175.
65. G. Zanghirati, F. Cocco, G. Paruolo, F. Taddei, A Cray T3E implementation of a parallel stochastic dynamic asset and liabilities management model, Parallel Comput 26 (2000) 539–567.

66. C. Wiesinger, D. Giczi, R. Hochreiter, An open grid service environment for large-scale computational finance modelling systems, in: M. Bubak, G. Albada, P. Sloot, J. Dongarra (Eds.), International Conference on Computational Science, Vol. 3036 of Lect Notes Comput Sc, Springer-Verlag, Berlin (2004) pp. 83–90.

67. R. Hochreiter, C. Wiesinger, D. Wozabal, Large-scale computational finance applications on the open grid service environment, in: P. Sloot, A. G. Hoekstra, T. Priol, A. Reinefeld, M. Bubak (Eds.), Advances in Grid Computing: EGC 2005, Vol. 3470 of Lect Notes Comput Sc, Springer-Verlag, Berlin (2005) pp. 891–899.

68. J. Gondzio, A. Grothey, Exploiting structure in parallel implementation of interior point methods for optimization, Technical Report MS-04-004, School of Mathematics, University of Edinburgh, Edinburgh EH9 3JZ, Scotland, UK (December 2004).

69. M. Steinbach, Hierarchical sparsity in multistage convex stochastic programs, in: S. Uryasev, P. M. Pardalos (Eds.), Stochastic Optimization: Algorithms and Applications, Kluwer Academic Publishers (2000) pp. 363–388.

70. M. Steinbach, Tree-sparse convex programs, Math Method Oper Res 56 (3) (2002) 347–376.

Chapter 15
The Future of Parallel Computation

Selim G. Akl and Marius Nagy

Abstract

As any other scientific discipline, computing science is undergoing a continuous process of transformations and innovations driven by theoretical research and technological advancements. Inspired by physical and biological phenomena occurring in nature, new computational models are proposed, with the potential to greatly increase the efficiency of computational processes. Another direction of development pertains to the characteristics of the problems tackled by computing science. With the increasingly ubiquitous and pervasive nature of computers in the modern society, the class of problems and applications computing science has to address is continuously expanding.

The importance played by parallelism in each of these two major development trends confirms the fundamental role parallel processing continues to occupy in the theory of computing. The idea of massive parallelism permeates virtually all unconventional models of computation proposed to date and this is shown here through examples such as DNA computing, quantum computing or reaction–diffusion computers. Even a model that is mainly of theoretical interest, like the accelerating machine, can be thought of as deriving its power from doubling the number of processing units (operating in parallel) at each step.

The scope of computing science has expanded enormously from its modest boundaries formulated at the inception of the field and many of the unconventional problems we encounter today in this area are inherently parallel. We illustrate this by presenting five examples of tasks in quantum information processing that can only be carried out successfully through a parallel approach. It is one more testimony to

Selim G. Akl
School of Computing, Queen's University, Kingston, Ontario, Canada,
e-mail: akl@cs.queensu.ca

Marius Nagy
School of Computing, Queen's University, Kingston, Ontario, Canada,
e-mail: marius@cs.queensu.ca

R. Trobec et al. (eds.), *Parallel Computing*, DOI 10.1007/978-1-84882-409-6_15,

the fact that parallelism is universally applicable and that the future of computing cannot be conceived without parallel processing.

15.1 Introduction

The purpose of this final chapter is to glance into the future and sketch the most probable forms parallel computing may take, having as a starting point the trends we can observe today. When it comes to computing in parallel, we can distinguish two major directions heading into the future. The first is strongly related to the continuous development and expansion of the Internet and the improvement in network management schemes. Having better means available for a group of computers to cooperate in solving a computational problem – whether they are homogeneous or heterogeneous, geographically close to each other or distributed over the Internet, small or large in number – will inevitably translate into a higher profile of clusters and grids in the landscape of parallel and distributed computing. We are not going to insist further here on the increasing role played by clusters and grids, especially since we have already discussed them in Chapter 2 in the context of models of parallel computing.

Another, more revolutionary, direction taken by parallel computation challenges the very physical level at which information is stored and manipulated in a computing machine. The electronic computers now in use are based on large-scale integration of transistors onto silicon chips such that a logical bit is physically realized as a voltage level in an electronic circuit. Although this technology was able to sustain a steady increase in the speed of processors over the past few decades, its limits are well in sight by now. Consequently, researchers focused on finding alternate ways of encoding and processing information that have the potential to speed up computation beyond what is possible using an electronic computer.

Proposals for an unconventional computing device include, but are not limited to, computing with DNA strands, quantum information processing, membrane computing (P systems) or computations in reaction – diffusion systems. All these alternatives are inspired from natural phenomena and each advances a fundamentally new physical support for information.

DNA Computing

In DNA computing, the computation is performed by synthetically obtained DNA strands. Performing an algorithm in this context amounts to applying some standard lab manipulation techniques (annealing and denaturation, polymerase chain reaction, gel electrophoresis, etc.) to the DNA in a test tube [1]. The strands act as both processors and memory units. Using this "bioware," NP-complete problems can be solved in linear time by covering an exponential search space in parallel [2–4]. The Watson–Crick complementarity, responsible for the formation of the hydrogen bonds that allow two strands of DNA to anneal together, is the key mechanism used to explore all possible computational paths simultaneously. Unfortunately, the scalability of this technique is severely restricted by the amount of DNA required to ensure an exhaustive search and by the error rate of the molecular operations involved.

Membrane Computing

Another biologically inspired computational model bears the name of membrane computing or P systems, in honor of its founder Gheorghe Păun [5]. The model employs a hierarchy of membranes, with each membrane separating a region, just as cell components (the nucleus, the Golgi Apparatus, mitochondria and various vesicles) as well as the whole cell itself are identified by a separating membrane. The mathematical equivalent of molecules (chemicals) inside a cell component are symbols belonging to a certain region. The analogy continues by considering chemical reactions as production rules.

All membrane regions evolve simultaneously, according to a global clock. During each time unit in each region, all applicable rules are applied nondeterministically, in a maximally parallel manner. The computation stops when no further rules can be applied and the result (output) is read either from the environment (outside the skin membrane) or as the content of some nondestructible membrane.

Since the inception of the field, a plethora of variants of membrane systems have been defined and their computational power studied. In particular, algorithms have been designed to solve NP-complete problems in polynomial time, with the trade-off of exponential space [6–8]. Through membrane division, however, this exponential space can be created in polynomial (linear) time.

Quantum Information Processing

A strong candidate for tomorrow's computing paradigm is manipulating information at the quantum level. The idea of harnessing quantum mechanical effects in order to improve the efficiency of computation naturally follows the miniaturization trend witnessed in the computer industry for so many years now. According to this trend, we will soon reach the atomic and subatomic level for the embodiment of a logical bit and, inevitably, the laws of quantum mechanics will have to be taken into consideration.

Similar to DNA computing, quantum algorithms attempt to find a solution to a problem through an exhaustive search. The efficiency of the procedure comes again from the fact that an exponential number of computational paths can be pursued in parallel, by preparing a quantum register in a superposition state comprising an exponential number of classical states. Thus, for applications like integer factorization and finding discrete logarithms, a quantum computer offers an exponential speedup over a conventional one [9]. Quantum algorithms acting on small inputs have been successfully implemented in practical experiments [10–12], but the main difficulty to overcome remains the scalability of the various techniques proposed to build a quantum computer.

The Reaction–Diffusion Computer

Proposals to improve the efficiency of computation can come from any branch of science, not only biology or physics. Our last example draws its inspiration from chemistry: the reaction–diffusion computer. In this truly novel paradigm, both the data and the results of the computation are encoded as concentration profiles of the reagents. The computation itself is performed via the spreading and interaction of wave fronts. Because molecules diffuse and react in parallel, a computer based on reaction–diffusion is endowed with a natural parallelism that allows it to efficiently solve combinatorial problems [13].

Probably the most evident and, at the same time, fundamental observation about the unconventional computing paradigms enumerated above is that they owe their computational power to some form of *massive parallelism*. In a full test tube acting as a DNA computer we may have 10^{15}–10^{17} operations performed in parallel, while a small chemical reactor may host millions of elementary (2–4 bit) processors operating in parallel through reaction–diffusion means. Similar characteristics empower the other two paradigms mentioned. It is therefore justified to affirm, without the risk of making an overstatement, that parallel processing lies at the heart of the quest for efficiency in computation. As various computing devices are infiltrating every aspect of human life and the pervasive nature of computers is on the rise, the need for a parallel approach comes also from an increasing number of applications dealing with real-time requirements and inherently parallel problems.

Parallel processing may even be the power behind some hypercomputational models credited with capabilities that go beyond those of a Turing machine. Thus, the accelerating machine, a computational model of mainly theoretical interest, can double its speed at each step [14]. More precisely, the time required by an operation at any given step of a computation is only half (or some other constant fraction) of that required to perform the same operation in the previous step. This property allows the accelerating machine to perform any number of iterations of a computational step in a finite amount of time. As a consequence, solving the Halting Problem is within the reach of the accelerating machine. This result is entirely due to the accelerating feature of the model and one way the speed can be doubled each step is by doubling the number of processors operating in parallel at each step.

But the particular form that parallelism will take in the operation of tomorrow's computing machines remains for the future to decide. The few paradigms briefly discussed in this section are representative for the efforts made nowadays toward a radically new computing technology, with important advantages over the electronic computer. As things stand today, we credit quantum information processing with the highest chances of playing an important role in the way computations are going to be performed in a few decades time. This attitude is encouraged by the tremendous effort put today into finding a viable design for a practical quantum computer and the impressive achievements, already commercially available, in quantum cryptography. Consequently, the remainder of this chapter will focus on uncovering the "secrets" responsible for the potential quantum computation has to offer and the different ways parallelism is encountered in this novel paradigm of computation.

15.2 Quantum Computing

The field of quantum information processing is based on the postulates governing quantum mechanics. The aim of this section is to familiarize the reader with these postulates and the mathematical formalisms required to work with them to the extent needed for quantum computing. Good introductions to quantum mechanics for computing scientists can be found in [15–19], but for a detailed exposition of the field one should see [20].

15.2.1 Quantum Mechanics

We begin our presentation by describing a few experiments that, in our opinion, best illustrate those features of quantum mechanics that are at the heart of quantum information processing, namely, *superposition*, *measurement* and *interference*.

15.2.1.1 Double-Slit Experiment

This experiment was first conducted by Thomas Young in 1801, and it demonstrated that light behaves like waves. Young projected light onto a screen through a barrier pierced with two closely spaced slits (see Fig. 15.1). What he observed on the screen was an *interference* pattern, the hallmark of waves. The importance of modern-day versions of Young's experiment is best illustrated by Richard Feynman in his *Lectures* [20]. He believed that the result of the double-slit experiment was the fundamental mystery of quantum mechanics.

If Young performed his experiment using simple screens and candlelight, the tremendous advances in technology allow us today to repeat the experiment with very weak light, that is, light produced as one photon at a time. Thus, it is very unlikely that several photons would be found within the experimental apparatus at the same time. Surprisingly (and against our intuitions), given that enough time elapses as to allow the photons, arriving one at a time, to accumulate on the screen, the same interference pattern will appear. The obvious question is: what was each photon interfering with, if it was alone in the experimental apparatus?

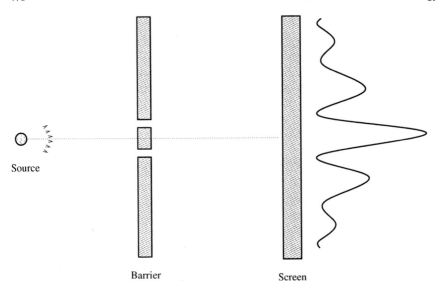

Fig. 15.1 Young's double-slit experiment. Light projected onto a screen through a barrier pierced with two closely spaced slits creates an *interference pattern*.

According to the Copenhagen interpretation (the standard view among many physicists), the only possible answer can be: with itself. In the absence of any observations, it does not make sense to associate a specific route to the photon in its way from the light source to the screen. In a sense, each particle went not through one slit, but rather through both slits, and, as it appeared on the other side, it interfered with itself. This behavior is a manifestation of the quantum principle of superposition of states, a principle without which quantum computation and quantum information would be inconceivable.

If we choose to observe the particle as it goes through the experimental apparatus (that is, to measure its state), the wave function describing it will collapse into one of the two possible outcomes and the particle will be detected passing through one of the two slits with equal probability. In either case, the superposition is destroyed and with it any chance of interference. But if the particle is not observed until the end, as it collects on the screen, then the superposition holds through to the end, enabling the interference phenomenon witnessed on the screen. The duality between particles and waves has also been demonstrated for other quanta that can be localized (electrons, neutrons, atoms) and even for larger entities, like complex molecules composed of tens of atoms.

15.2.1.2 Single Photon Interferometry

The Mach–Zehnder interferometer (depicted in Fig. 15.2) is an optical device composed of beam splitters, mirrors and photon detectors carefully placed to bring about

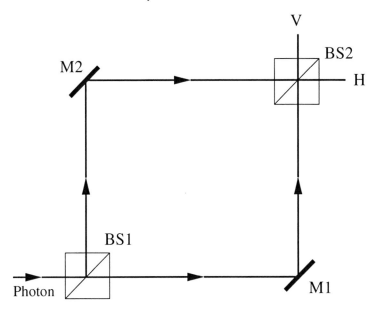

Fig. 15.2 A Mach–Zehnder interferometer (BS = beam splitter; M = mirror). A photon entering the first beam splitter horizontally will always emerge from the horizontal port of the second beam splitter due to *self-interference*.

quantum interference when a photon travels through the apparatus. A beam splitter is a half-silvered mirror that will let half of the incident beam pass through and reflect the other half. But when a single photon is confronted with a beam splitter, its state becomes a superposition of being reflected and going through at the same time. Thus, a photon entering the first beam splitter horizontally will always emerge from the horizontal port of the second beam splitter, provided the two arms of the interferometer have equal lengths. As in the case of Young's two-slit experiment, the reason is *self-interference*.

The probabilities of leaving the interferometer horizontally in the two possible histories (traveling the upper arm and lower arm, respectively) reinforce each other during the interference process that takes place in the second beam splitter. At the same time, the probabilities of leaving the experimental apparatus vertically cancel each other out. Any attempt to find out which way the photon took through the experimental device will collapse the superposition and ruin the interference. In such a case, there will be an equal probability of detecting the photon exiting horizontally or vertically, regardless of the path the photon was observed to take between the beam splitters.

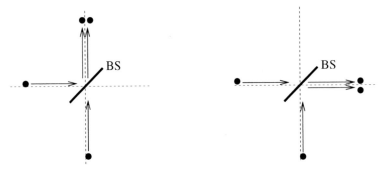

Fig. 15.3 The two photons always emerge from the beam splitter (BS) along the same output due to a quantum interference effect.

15.2.1.3 Two-Photon Interferometry

The quantum interference effect witnessed at the second beam splitter in the previous experiment can also occur if two single-mode, but otherwise independent, photons enter a $50-50$ beam splitter, as shown in Fig. 15.3. The "mode" of a photon refers to the physical properties, like frequency and polarization, that together define the electromagnetic field with which a photon is associated. When the two photons are in the same mode, all the properties of the two photons are identical at the beam splitter output, so they become essentially indistinguishable.

As a consequence of this "bosonic" character of photons, the probabilities that both photons will be transmitted or both reflected interfere destructively, canceling each other. As a result, the two photons will always be seen emerging from the beam splitter along the same output, either both horizontally or both vertically. This surprising quantum interference effect was demonstrated for independent photons, emitted from a single-photon source [21]. Such an experiment is also important from the practical viewpoint of building quantum logic gates for photon-based quantum computing [22].

15.2.2 Mathematical Framework

Quantum mechanics takes place in the framework provided by linear algebra. We can associate to any isolated physical system a complex vector space with an inner product defined on it, known as the state space of the system. Mathematically, such a vector space with an inner product is called a Hilbert space. At any given point in time, the system is completely described by its state vector, which must be a unit vector in the system's state space.

Quantum state spaces and the transformations acting on them are traditionally described in terms of vectors and matrices using the compact *bra/ket* notation introduced by Dirac [23]. According to his conventional notation, for states that correspond to discrete values of an observable, *kets* like $|x\rangle$ are simply column vectors, typically used to describe quantum states. Similarly, the matching *bra* $\langle x|$ is a row vector denoting the conjugate transpose of $|x\rangle$.

15.2.2.1 The Qubit

At an abstract level, the simplest quantum mechanical system is the quantum bit, or *qubit*. A qubit is a unit vector in a two-dimensional state space for which a particular orthonormal basis, denoted by $\{|0\rangle, |1\rangle\}$, has been fixed. The two basis vectors $|0\rangle$ and $|1\rangle$ correspond to the possible values a classical bit can take. However, unlike classical bits, a qubit can also take many other values. In general, an arbitrary qubit $|\Psi\rangle$ can be written as a linear combination of the computational basis states:

$$|\Psi\rangle = \alpha|0\rangle + \beta|1\rangle, \tag{15.1}$$

where α and β are complex numbers such that $|\alpha|^2 + |\beta|^2 = 1$. This is the fundamental difference distinguishing quantum bits from classical ones and is a direct application of the quantum principle of superposition of states. The qubit $|\Psi\rangle$ in Eq. (15.1) is in a superposition of $|0\rangle$ and $|1\rangle$, a state in which it is not possible to say that the qubit is definitely in the state $|0\rangle$ or definitely in the state $|1\rangle$. After all, what better intuition about the superposition principle than the idea (quite old and widely accepted now) that each particle is also a wave.

For a single qubit, there is a very intuitive geometric representation of its state as a point on a sphere. Taking $\alpha = e^{i\gamma}\cos(\theta/2)$ and $\beta = e^{i\gamma}e^{i\varphi}\sin(\theta/2)$ in Eq. (15.1), we can rewrite the state of qubit $|\psi\rangle$ as

$$|\psi\rangle = e^{i\gamma}(\cos\frac{\theta}{2}|0\rangle + e^{i\varphi}\sin\frac{\theta}{2}|1\rangle), \tag{15.2}$$

where θ, φ and γ are real numbers. Note that this is always possible since $|\alpha|^2 + |\beta|^2 = 1$. Also, because a global phase factor like $e^{i\gamma}$ has no observable effects (i.e., it does not influence the statistics of measurement predicted for qubit $|\psi\rangle$), we can effectively ignore it. Consequently, the pair (θ, φ) uniquely identifies a point $(\cos\varphi\sin\theta, \sin\varphi\sin\theta, \cos\theta)$ on a unit three-dimensional sphere called the *Bloch sphere* [17, 24].

Figure 15.4 depicts four possible states of a qubit using the Bloch sphere representation. Note that the states corresponding to the points on the equatorial circle have all equal contributions of 0-ness and 1-ness. What distinguishes them is the *phase*. For example, the two states displayed above, $1/\sqrt{2}(|0\rangle + |1\rangle)$ and $1/\sqrt{2}(|0\rangle - |1\rangle)$ are the same up to a relative phase shift of π, because the $|0\rangle$ amplitudes are identical and the $|1\rangle$ amplitudes differ only by a relative phase factor of $e^{i\pi} = -1$.

We have described qubits as mathematical objects, but there are real physical systems which may be described in terms of qubits. Possible physical realizations of a qubit include two different polarizations of a photon, the alignment of a nuclear spin in a uniform magnetic field or two electronic levels in an atom. In the

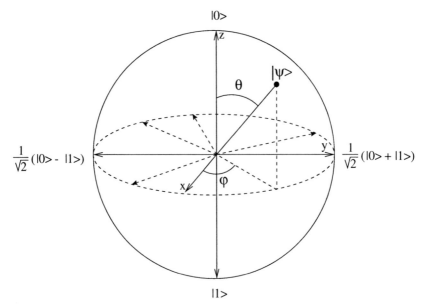

Fig. 15.4 The Bloch sphere representation of a qubit.

experiments presented at the beginning of the section, the state of a photon is described in terms of the two possible routes that can be used when traversing the experimental apparatus.

15.2.2.2 Measurements

We now turn our attention on the amount of information that can be stored in a qubit and, subsequently, retrieved from a qubit. Since any point on the Bloch sphere can be characterized by a pair of real-valued parameters taking continuous values, it follows that, theoretically, a qubit could hold an infinite amount of information. As it turns out, however, we cannot extract more information from such a qubit than we are able to do it from a classical bit.

The reason is that we have to *measure* the qubit in order to determine which state it is in. And another of the fundamental postulates of quantum mechanics, the one regarding measurements (Postulate 3 in [17]), restricts us in the amount of information that can be gained about a quantum state through measurement. According to this postulate, when we measure a qubit $|\Psi\rangle = \alpha|0\rangle + \beta|1\rangle$ with respect to the standard basis for quantum computation $\{|0\rangle, |1\rangle\}$, we get either the result 0 with probability $|\alpha|^2$ or the result 1 with probability $|\beta|^2$. The condition that the probabilities must sum to one corresponds geometrically to the requirement that the qubit state be normalized to length 1, that is the inner product $\langle\Psi|\Psi\rangle$ equals 1.

Furthermore, measurement alters the state of a qubit, collapsing it from its superposition of $|0\rangle$ and $|1\rangle$ to the specific state consistent with the measurement result.

For example, if we observe $|\Psi\rangle$ to be in state $|0\rangle$ through measurement, then the post-measurement state of the qubit will be $|0\rangle$, and any subsequent measurements (in the same basis) will yield 0 with probability 1. In general, measurement of a state transforms the state into one of the eigenvectors of the observable being measured. The probability that the state is measured as basis vector $|u\rangle$ is the square of the norm of the amplitude of the component of the original state in the direction of the basis vector $|u\rangle$. The implicit assumption we adopt herein is that a measurement is performed in the standard basis for quantum computation, whenever the basis vectors associated with the measurement operation are not stated explicitly.

15.2.2.3 No-Clonability

Naturally, measurements in bases other than the computational basis are always possible, but this will not help us in determining α and β from a single measurement. One might think of solving this problem by making multiple copies of the initial qubit $|\Psi\rangle$ and then measuring each of the copies in order to obtain an estimation of α and β. In fact, it turns out to be impossible to make a copy of an unknown quantum state. The *no-cloning* theorem, one of the earliest results of quantum computation and quantum information [25], states that quantum mechanics prevents us from building a quantum cloning device capable of copying nonorthogonal quantum states. The ability to clone orthogonal quantum states translates into the ability to copy classical information, since the different states of classical information can be thought of merely as orthogonal quantum states. So it seems that quantum mechanics places severe limitations on the accessibility of quantum information, but in some circumstances (like devising secure quantum cryptographic protocols, for instance) this can be turned into an advantage.

15.2.2.4 Quantum Registers

Let us examine now more complex quantum systems composed of multiple qubits. In classical physics, individual two-dimensional state spaces of n particles combine through the Cartesian product to form a vector space of 2^n dimensions, representing the state space of the ensemble of n particles. However, this is not how a quantum system can be described in terms of its components. Quantum states combine through the tensor product to give a resulting state space of 2^n dimensions, for a system of n qubits. It is this exponential growth of the state space with the number of particles that quantum computers try to exploit in their attempt to achieve exponential speedup of computation over classical computers.

For a system of two qubits, each with basis $\{|0\rangle, |1\rangle\}$, the resulting state space is the set of normalized vectors in the four-dimensional space spanned by basis vectors $\{|0\rangle \otimes |0\rangle, |0\rangle \otimes |1\rangle, |1\rangle \otimes |0\rangle, |1\rangle \otimes |1\rangle\}$, where $|x\rangle \otimes |y\rangle$ denotes the tensor product between column vectors $|x\rangle$ and $|y\rangle$. It is customary to write the basis in the more compact notation $\{|00\rangle, |01\rangle, |10\rangle, |11\rangle\}$. This generalizes in the obvious way to an n-qubit system with 2^n basis vectors.

15.2.2.5 Quantum Evolution

The next step after laying the mathematical foundation for describing quantum registers is to focus on the "circuits" composing a hypothetical quantum computer. Operating a quantum gate is strongly related to the way an isolated quantum system evolves over time. We already saw what happens when we try to measure such a quantum system. If, for example, we are trying to read the content of a quantum memory register, the system will undergo a sudden, unpredictable jump into one of the eigenvectors associated with the measurement operator. In other words, there will be a discontinuity in the evolution of the quantum memory register. But, if we leave the register unobserved, the system will undergo a smooth, continuous evolution governed by Schrödinger's equation, a deterministic differential equation which enables us to predict the future or uncover the past evolution of the memory register. Consequently, any quantum computation is reversible and therefore quantum gates (the quantum analog of classical gates) must always have as many outputs as they have inputs, in order to avoid any loss of information that would prevent the computation from being undone.

15.2.2.6 Quantum Gates

A quantum NOT gate acting on a single qubit will evolve the initial state $\alpha|0\rangle + \beta|1\rangle$ into the final state $\alpha|1\rangle + \beta|0\rangle$, in which the roles of $|0\rangle$ and $|1\rangle$ have been interchanged. Because every quantum gate acts linearly, the transformation is fully specified by its effect on the basis vectors. Hence, there is a very convenient representation of a quantum gate in matrix form. Starting from the expressions of the two basis vectors in column form:

$$|0\rangle = \begin{bmatrix} 1 \\ 0 \end{bmatrix}, \ |1\rangle = \begin{bmatrix} 0 \\ 1 \end{bmatrix}, \tag{15.3}$$

the matrix X representing the quantum NOT gate is then defined as follows:

$$X = \begin{bmatrix} 0 & 1 \\ 1 & 0 \end{bmatrix}. \tag{15.4}$$

The first column represents the effect of applying the NOT gate to state $|0\rangle$, while the second column is the result of applying the NOT gate to state $|1\rangle$. We can now describe the operation of the quantum NOT gate, acting on an arbitrary qubit state, through the following equation:

$$X \cdot \begin{bmatrix} \alpha \\ \beta \end{bmatrix} = \begin{bmatrix} \beta \\ \alpha \end{bmatrix}. \tag{15.5}$$

Other examples of single-qubit gates are the Z gate:

$$Z = \begin{bmatrix} 1 & 0 \\ 0 & -1 \end{bmatrix},$$ (15.6)

which leaves $|0\rangle$ unchanged, but introduces a phase shift by flipping the sign of $|1\rangle$, and the Hadamard gate:

$$H = \frac{1}{\sqrt{2}} \begin{bmatrix} 1 & 1 \\ 1 & -1 \end{bmatrix},$$ (15.7)

which is one of the most useful quantum gates, because it creates superpositions of $|0\rangle$ and $|1\rangle$.

Although there are an infinite number of single-qubit gates, not any two-by-two matrix is a legitimate representation of a quantum gate. Schrödinger's equation states that the dynamics of a quantum system must take states to states in a way that preserves orthogonality. In other words, the normalization condition $|\alpha|^2 + |\beta|^2 = 1$ for the initial state $\alpha|0\rangle + \beta|1\rangle$ must also be true for the quantum state after the gate has acted. This translates into the requirement that the matrix U describing the single-qubit gate be *unitary*, that is, $U^* \cdot U = I$, where U^* is the conjugate transpose of U. Single-qubit gates can be conveniently visualized as rotations of the arrow representing the qubit state on the surface of the Bloch sphere.

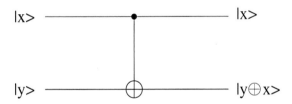

Fig. 15.5 Controlled-NOT quantum gate.

Quantum gates on multiple qubits can also be defined. Figure 15.5 depicts a controlled-NOT gate, an instance of the more abstract controlled-U gate, where $U = X$. The target bit $|y\rangle$ is flipped if and only if the control bit $|x\rangle$ is set to 1. The matrix describing the operation of the controlled-NOT gate is

$$CNOT = \begin{bmatrix} 1 & 0 & 0 & 0 \\ 0 & 1 & 0 & 0 \\ 0 & 0 & 0 & 1 \\ 0 & 0 & 1 & 0 \end{bmatrix}.$$ (15.8)

Multiple qubit gates must also satisfy the requirement that probability be conserved, so they too must be unitary transformations. Since any unitary matrix is invertible and the inverse is also a unitary matrix, it follows that a quantum gate can always be inverted by another quantum gate. The set of all one-qubit rotations (gates) together with the controlled-NOT gate is universal for quantum computation. But finite universal sets of gates exist as well. Two researchers working independently have shown that any imaginable quantum computation can be performed by connecting together multiple copies of a certain two-qubit gate [26, 27]. Such universal quantum gates are analogous to the NAND gate in classical computation.

15.2.3 Entanglement

Entanglement is probably the strangest and most controversial aspect of quantum mechanics, but at the same time it is credited with the most surprising applications. This section contains a brief discussion of this unusual phenomenon.

Similar to single qubits, multiple-qubit systems can also be in a superposition state. The vector

$$|\Psi\rangle = \frac{1}{2}(|00\rangle + |01\rangle + |10\rangle + |11\rangle) \tag{15.9}$$

describes a superposition state of a two-qubit system in which all four components (corresponding to the four basis vectors) have equal amplitudes. What about the two qubits composing the system? Can we characterize their states individually? If we rewrite Eq. (15.9) in order to express $|\Psi\rangle$ as the tensor product

$$|\Psi\rangle = (\frac{1}{\sqrt{2}}|0\rangle + \frac{1}{\sqrt{2}}|1\rangle) \otimes (\frac{1}{\sqrt{2}}|0\rangle + \frac{1}{\sqrt{2}}|1\rangle), \tag{15.10}$$

then we can legitimately assert that each of the component qubits is also in a superposition state, perfectly balanced between $|0\rangle$ and $|1\rangle$. Now let us drop the two middle terms in Eq. (15.9) and consider the superposition state described by

$$|\Phi\rangle = \frac{1}{\sqrt{2}}|00\rangle + \frac{1}{\sqrt{2}}|11\rangle. \tag{15.11}$$

In this case it is no longer possible to find complex numbers α, β, γ and δ such that

$$(\alpha|0\rangle + \beta|1\rangle) \otimes (\gamma|0\rangle + \delta|1\rangle) = \frac{1}{\sqrt{2}}|00\rangle + \frac{1}{\sqrt{2}}|11\rangle. \tag{15.12}$$

The state of the system cannot be decomposed into a product of the states of the constituents. Even though the state of the system is well defined (through the state vector $|\Phi\rangle$), neither of the two component qubits is in a well-defined state. This is again in contrast to classical systems, whose states can always be broken down into the individual states of their components. Furthermore, if we try to measure the two qubits, the superposition will collapse into one of the two basis vectors contributing to the superposition and the outcomes of the two measurements will always coincide. In other words, if one of the qubits is found to be in state $|0\rangle$, then the second one will necessarily be in the same state, while a state $|1\rangle$ assumed after measurement will be shared by both qubits. Therefore, we say that the two qubits are entangled and $|\Phi\rangle$ describes an entangled state of the system. It was Schrödinger who actually named the phenomenon *entanglement* in 1935 [28].

Entanglement defines the strong correlations exhibited by two or more particles when they are measured and which cannot be explained by classical means. This does not imply that entangled particles will always be observed in the same state, as entangled states like

$$\frac{1}{\sqrt{2}}|01\rangle \pm \frac{1}{\sqrt{2}}|10\rangle \tag{15.13}$$

prove it. States like these or the one in Eq. (15.11) are known as Bell states or EPR pairs, named after the people who pointed out their strange properties [29].

In some sense, we can say that superposition encompasses entanglement, since entanglement can be viewed as a special case of superposition. It is also interesting to make an analogy between entanglement and the concept of primality from number theory. Indeed, an entangled state of the system corresponds to a prime number, since it cannot be factored or decomposed as a product of subsystem states.

15.3 Parallelism in Quantum Computing

We now have the necessary tools to discuss the role of parallelism in quantum computing. When talking about parallelism in the context of quantum computation, the immediate understanding given to the term refers to the ability of a quantum computer to simultaneously evolve (transform) a potentially large number of classical states, by preparing a quantum register in a superposition of all those states and then applying the desired transformation on the quantum register. This form of parallelism is specific to quantum computing because it exploits the quantum mechanical principle of superposition of states and, hence, it is termed *quantum parallelism*. We describe in detail the mechanism of quantum parallelism in the following section and show that it is the key ingredient in obtaining an exponential speedup over a conventional computer for some applications.

15.3.1 Quantum Parallelism

Suppose we want to evaluate an arbitrary function $f : N \rightarrow N$ for various inputs x. Then we can define a unitary quantum gate U_f, whose action on the inputs x and y is shown in Fig. 15.6. Since U_f must be reversible by definition, we need input y in order to ensure that x is "remembered" at the output (no loss of information). The image of x through f XOR-ed with y is obtained on the bottom output line ($\oplus$ denotes an Exclusive OR operation or, equivalently, addition modulo 2). In general, the input and corresponding output lines depicted in Fig. 15.6 may represent an arbitrary number of qubits, such that $\oplus$ is applied bitwise. This construction is possible for any function f.

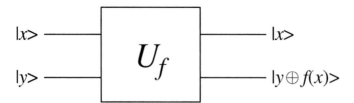

Fig. 15.6 A generic quantum gate designed to compute the values of a function f.

In order to compute $f(x)$, for some input x, we set y to zero and then $f(x)$ can be read from the bottom ouput line(s):

$$U_f(|x\rangle \otimes |0\rangle) = |x\rangle \otimes |f(x)\rangle. \tag{15.14}$$

The advantage of the quantum paradigm of computation now becomes apparent. If we want to compute $f(x)$ for an arbitrary number of inputs x, all we have to do is to prepare the x part of the quantum register as a superposition of all inputs that we want to be evaluated and then apply the gate U_f. The y part of the register, which was initially 0, now stores a superposition of all images $f(x)$ that we sought to compute.

In particular, if we start with n qubits, each in the state $|0\rangle$, and apply a Hadamard gate to each of them, then what we get is a superposition of all inputs from 0 to $2^n - 1$:

$$\frac{1}{2^n} \sum_{i=0}^{2^n-1} |i\rangle. \tag{15.15}$$

Now, with a single application of the gate U_f we obtain all 2^n corresponding images in a superposition:

$$U_f((\frac{1}{2^n} \sum_{i=0}^{2^n-1} |i\rangle) \otimes |0\rangle) = (\frac{1}{2^n} \sum_{i=0}^{2^n-1} |i\rangle) \otimes (\frac{1}{2^n} \sum_{i=0}^{2^n-1} |f(i)\rangle). \tag{15.16}$$

In this way, a quantum computer can evaluate an exponential number of inputs in the time it takes a conventional electronic computer to evaluate just one input. This type of transformation operating in parallel on all inputs is known as *quantum parallelism*. The enormous potential of a quantum computer to outperform a classical machine lies precisely in the massive parallelism it offers "within a single piece of hardware" [30].

This form of parallelism, however, does not automatically translate into an exponential speedup for any computational problem. The difficulty resides in extracting the information we have computed in quantum parallel. In order to see what are the values $f(x)$ obtained, we must read (that is, measure) the quantum register. And we have already seen in Sect. 15.2.2.2 that measuring is a disruptive process implying a

loss of information by collapsing the superposition state of the quantum register to a
state compatible with the outcome obtained through measurement. This means that
from the 2^n values encoded in the state of the quantum register before measurement,
we can only read out one and, worse still, we do not even have control over which
one we get, since the measurement process is a probabilistic one.

Nevertheless, an exponential speedup can still be obtained if the information
sought through measurement is a global property of all terms in the superposition
and not just one particular term. It is the case, for example, of the quantum algorithm devised by Shor to factorize an integer in polynomial time [9]. Knowing that
factoring n is as hard as computing orders modulo n, Shor set out to find the period
of a function $f_{x,n}(a) = x^a \bmod n$, for some x chosen to be coprime with n. Once the
period is found, the divisors of n can easily be inferred using standard techniques
from number theory.

Classically, in order to find the period of a function, we need to evaluate that
function over and over again for many different inputs. But using quantum parallelism, we need only one evaluation. Furthermore, what we need afterward is a
global property of all images through f (the period) and not a particular image of a
particular input. In Shor's algorithm, the quantum Fourier transform (QFT) is used
in order to interfere the computational paths and bring out the period. Some kind of
Fourier transform is usually employed in quantum algorithms to constructively re-
combine different alternatives in a superposition such that the amplitude of solutions
is strengthened, while nonsolutions interfere destructively, canceling each other.

Thus, factoring integers (and the related problem of finding discrete logarithms)
can be solved in quantum polynomial time, while the best known classical technique
for factorization (the number field sieve) is super-polynomial or sub-exponential in
the size of the integer to be decomposed. On the other hand, for the vast majority of
problems in computer science (including NP-complete ones) quantum parallelism
is not expected to bring more than a quadratic speedup [31].

But *quantum parallelism* is not the only form of parallelism encountered in the
context of quantum information processing. This syntagm is used to denote the ability to perform a certain computation simultaneously on all terms of a quantum superposition, regardless of the number of qubits composing the quantum register whose
state is described by that superposition. A different interpretation refers to parallelism as the ability to act simultaneously on a certain number of qubits, whether for
the purpose of measuring them or evolving their quantum state.

In the following section, we illustrate this second meaning of the term *parallelism* in quantum computation by presenting five examples in which a parallel
computing approach is most appropriate, if not vital, for the success of the computation. The common theme of all these examples, apart from the fact that they are
all drawn from the field of quantum information processing, is their *evolving* nature,
in the sense that their characteristics vary during the computational process itself.
Because of their dynamic nature, these computations may be labeled as *unconventional*, as opposed to the computation performed by a Turing machine, for example. The problems we are about to describe may also be interpreted as quantum

mechanical instances of the unconventional computing paradigms introduced in Chapter 2 as computations that cannot be simulated sequentially.

15.4 Examples

In each of the five cases enumerated below, we describe the problem as it is formulated in quantum information processing and emphasize the importance of a parallel approach in order to reach a solution. Furthermore, we identify the characteristics that make it belong to a certain class of unconventional (evolving) computations.

15.4.1 Parallelizing the Quantum Fourier Transform

The Fourier transform is a very useful tool in computer science and it proved of crucial importance for quantum computation as well. Since it can be computed much faster on a quantum computer than on a classical one, the discrete Fourier transform allows for the construction of a whole class of fast quantum algorithms. Shor's quantum algorithms for factoring integers and computing discrete logarithms [9] are the most famous examples in this category.

The QFT is a linear operator whose action on any of the computational basis vectors $|0\rangle, |1\rangle, \cdots, |2^n - 1\rangle$ associated with an n-qubit register is described by the following transformation:

$$|j\rangle \longrightarrow \frac{1}{\sqrt{2^n}} \sum_{k=0}^{2^n-1} e^{2\pi i jk/2^n} |k\rangle, \ 0 \le j \le 2^n - 1. \tag{15.17}$$

However, the essential advantage of quantum computation over classical computation is that the quantum mechanical principle of superposition of states allows all possible inputs to be processed at the same time. Consequently, if the quantum register is in an arbitrary superposition of the basis vectors $\sum_{j=0}^{2^n-1} x_j |j\rangle$, then the QFT will rotate this state into another superposition of the basis vectors $\sum_{k=0}^{2^n-1} y_k |k\rangle$, in which the output amplitudes y_k are the classical discrete Fourier transform of the input amplitudes x_j. Classically, we can compute the numbers y_k from x_j using $\Theta(2^{2n})$ elementary arithmetic operations in a straightforward manner and in $\Theta(n2^n)$ operations by using the Fast Fourier Transform algorithm [32].

In contrast, a circuit implementing the QFT requires only $O(n^2)$ elementary quantum gates, as proved by Coppersmith [33]. Such a circuit can be easily derived if Eq. (15.17) is rewritten as a tensor product of the n qubits involved:

$$|j_1 \cdots j_n\rangle \longrightarrow \frac{(|0\rangle + e^{2\pi i 0.j_n}|1\rangle) \otimes (|0\rangle + e^{2\pi i 0.j_{n-1}j_n}|1\rangle) \otimes \cdots \otimes (|0\rangle + e^{2\pi i 0.j_1 \cdots j_n}|1\rangle)}{2^{n/2}}, \tag{15.18}$$

using the binary representation $j_1 j_2 \cdots j_n$ of j and binary fractions in the exponents (for full details see [17]).

Note that each Fourier transformed qubit is in a balanced superposition of $|0\rangle$ and $|1\rangle$. These qubits differ from one another only in the relative phase between the $|0\rangle$ and the $|1\rangle$ components. For the first qubit in the tensor product, j_n will introduce a phase shift of 0 or π, depending on whether its value is 0 or 1, respectively. The phase of the second qubit is determined (controlled) by both j_n and j_{n-1}. It can amount to $\pi + \pi/2$, provided j_{n-1} and j_n are both 1. This dependency on the values of all the previous qubits continues up to (and including) the last term in the tensor product. When $|j_1\rangle$ gets Fourier transformed, the coefficient of $|1\rangle$ in the superposition involves all the digits in the binary expansion of j.

In the case of each qubit, the 0 or π phase induced by its own binary value is implemented through a Hadamard gate. The dependency on the previous qubits is reflected in the use of controlled phase shifts, as depicted in Fig. 15.7. In the figure, H denotes the Hadamard transformation

$$H \equiv \frac{1}{\sqrt{2}} \begin{bmatrix} 1 & 1 \\ 1 & -1 \end{bmatrix}, \tag{15.19}$$

while the gate R_k implements a $\pi/2^{k-1}$ phase shift of the $|1\rangle$ component, according to the unitary transformation

$$R_k \equiv \begin{bmatrix} 1 & 0 \\ 0 & e^{2\pi i/2^k} \end{bmatrix}. \tag{15.20}$$

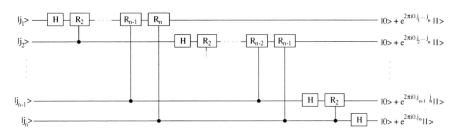

Fig. 15.7 Quantum circuit performing the discrete Fourier transform. The final swapping of qubits was omitted for simplicity.

15.4.1.1 Rank-Varying Complexity

Computing the QFT and its inverse can be viewed as examples of algorithms with rank-varying complexity. According to the quantum circuit above, we need n Hadamard gates and $n - 1 + n - 2 + \cdots + 1$ conditional rotations, for a total of $n(n + 1)/2$ gates required to compute the Fourier transform on n qubits. But this total amount of work is not evenly distributed over the n qubits. The number of gates a qubit needs to be passed through is in inverse relation with its *rank*. $|j_1\rangle$ is subjected to n elementary quantum gates, $n-1$ elementary unitary transformations are applied to $|j_2\rangle$, and so on, until $|j_n\rangle$, which needs only one basic operation.

If we break down the QFT algorithm into n steps (one for each qubit involved), then its complexity varies with each step. Starting with $|j_1\rangle$, the time needed to complete each step decreases over time. Since the rank of each step dictates its complexity, the circuit implementing the QFT is an example of a rank-varying complexity algorithm.

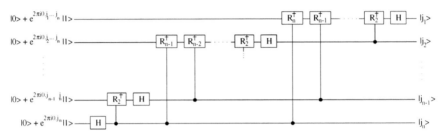

Fig. 15.8 Quantum circuit performing the inverse Fourier transform.

Naturally, the computation of the inverse QFT can also be decomposed into steps of varying complexity. Reversing each gate in Fig. 15.7 gives us an efficient quantum circuit (depicted in Fig. 15.8) for performing the inverse Fourier transform. Note that the Hadamard gate is its own inverse and $R_k^\dagger$ denotes the conjugate transpose of R_k:

$$R_k^\dagger \equiv \begin{bmatrix} 1 & 0 \\ 0 & e^{-2\pi i/2^k} \end{bmatrix}. \tag{15.21}$$

Getting back to the original $|j_1 j_2 \cdots j_n\rangle$ from its Fourier transformed expression has a certain particularity however. Because of the interdependencies introduced by the controlled rotations, the procedure must start by computing $|j_n\rangle$ and then work its way up to $|j_1\rangle$. The value of $|j_n\rangle$ is needed in the computation of $|j_{n-1}\rangle$. Both $|j_n\rangle$ and $|j_{n-1}\rangle$ are required in order to obtain $|j_{n-2}\rangle$. Finally, the values of all the higher rank bits are used to determine $|j_1\rangle$ precisely. Thus, computing the inverse Fourier transform by the quantum circuit illustrated in Fig. 15.8 is a procedure the complexity of whose steps increases with their rank.

Certainly, the fact that the total amount of operations (work) is not evenly distributed over the steps composing a certain algorithm does not change the overall complexity of the algorithm in any way. But the study of computations that can be characterized as having rank-varying complexity is important especially in the field of parallel computing. Operations pertaining to the same step or belonging to distinct steps may be executed in parallel, leading to an important reduction in the overall running time of the respective algorithm. In the particular case of the QFT, the transformation of the first qubit has the highest computational complexity.

However, the use of an appropriate parallel architecture allows us to complete the entire computation during the n time units required just for the first qubit. Since the solution we describe can be characterized as a parallelization of the semiclassical solution due to Griffiths and Niu [34], we analyze the advantages offered by the former with respect to the performance of the latter.

15.4.1.2 Semiclassical (Sequential) Solution

Although the circuits for computing the QFT and its inverse are efficient in terms of the total number of gates employed, the majority of these gates operate on two qubits. This makes a practical implementation difficult, since arranging for one qubit to influence another in a desired way is far greater a challenge than evolving a single-qubit closed quantum system in accordance with any unitary transformation.

A method to replace all the two-qubit gates in the circuit performing the QFT by a smaller number of one-qubit gates controlled by classical signals has been developed by Griffiths and Niu [34] under the assumption that a measurement of the quantum register follows the application of the QFT, as it is usually the case, including in Shor's factoring quantum algorithm. Their approach takes advantage of the fact that the roles of the control and target qubits in any of the two-qubit gates required to carry on the computation of the QFT are interchangeable. Consequently, the quantum circuit in Fig. 15.7 is equivalent to the one depicted in Fig. 15.9 (for inputs restricted to four qubits).

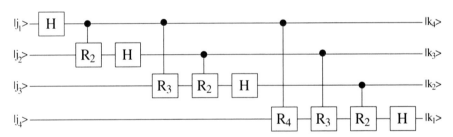

Fig. 15.9 Alternative arrangement of gates in the circuit performing the quantum Fourier transform. The roles of the control and target qubits in the controlled phase shift gates can be switched.

Note that, from this new perspective, the computation of the QFT appears to be a procedure whose steps are of increasing complexity. However, under the assumption that the Fourier transform is immediately followed by a quantum measurement, the complexity of each step in the computation can be made constant. Since a control qubit enters and leaves a two-qubit gate unchanged, it follows that the top qubit in Fig. 15.9 yields the same result regardless of whether it is measured as it exits the circuit or immediately after undergoing the Hadamard transform. In the latter case, the result of the measurement can be used to determine the phase shift that needs to be applied on the second qubit, before it too is subjected to a Hadamard transform and then measured. The phase computed for the second qubit together with the result of the second measurement is passed down as classical inputs for the rotation applied to the third qubit.

The computation proceeds in this manner all the way down to the last qubit, with a phase rotation, a Hadamard gate and a measurement being performed at each step. The process is illustrated in Fig. 15.10, where double lines have been used to denote a classical signal, according to the usual convention. Although the phase shift applied to each qubit is considered a single operation, conceptually it is a combination of the gates depicted in the corresponding box, with each component being applied only if the controlling qubit was measured as 1.

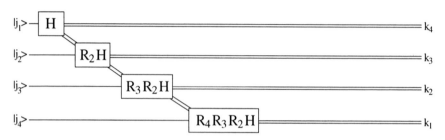

Fig. 15.10 Semiclassical circuit for computing the quantum Fourier transform. *Single lines* convey quantum information, while *double lines* carry classical information.

Example

Here is an example of how the outcome of measurements determines the phase rotation that will be applied to subsequent qubits. If the top qubit in Fig. 15.10 yields a 1 ($k_4 = 1$) when measured, then the second qubit undergoes a $\pi/2$ phase shift before the Hadamard gate and then it is measured. Suppose now that the outcome of this measurement is a 0 ($k_3 = 0$). Then the third qubit is phase shifted by

$$k_4\frac{\pi}{4} + k_3\frac{\pi}{2} = \frac{\pi}{4} \tag{15.22}$$

and then the Hadamard gate is applied. Again, without loss of generality, let the measurement yield a 1 ($k_2 = 1$). Then the phase shift applied to the bottom qubit is

$$k_4\frac{\pi}{8} + k_3\frac{\pi}{4} + k_2\frac{\pi}{2} = \frac{5\pi}{8}. \tag{15.23}$$

This semiclassical approach to computing the QFT achieves optimality in terms of the number of elementary unitary transformations that have to be applied. It also has the important advantage of employing only quantum transformations acting on a single qubit at a time. However, there is still room for improvement, as the total time needed to complete the computation can be further squeezed down if parallelism is brought into play. In the next section we show how a quantum pipeline architecture is able to speed up the computation of the Fourier transform.

15.4.1.3 Parallel Approach

The solution developed in [34] to reduce the complexity of the QFT envisages a purely sequential approach, which is motivated by the same data dependency that causes the complexity of a step to vary with its rank. Nevertheless, there is a certain degree of parallelism that is worth exploiting in the computation of the QFT (or its inverse) in order to minimize the overall running time.

Our parallel approach is based on the observation that once a qubit has been measured, all phase shift gates classically controlled by the outcome of that measurement can be applied in parallel. The arrangement, again for just four qubits, is shown in Fig. 15.11. The one-qubit gates are ordered into a linear array having a Hadamard transform at the top and followed by a $\pi/2$ phase shift gate. The phase shift induced by any other gate down the array is just half the rotation performed by the immediately preceding gate.

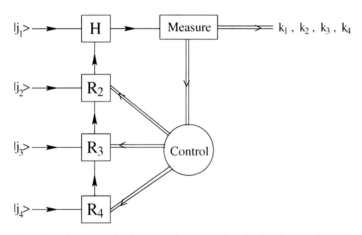

Fig. 15.11 Quantum pipeline array for computing the Fourier transform. The input is quantum, but the output is classical. At each step, qubits move one position up in the array.

This architecture allows R_2, R_3 and R_4 to be performed in parallel during the first cycle. Since each phase shift gate acts on a different qubit, they can all be applied simultaneously, if the top qubit yielded a 1 upon measurement. In the second cycle, each qubit in the array travels up one position, except of course for the top one, which has already been measured. Now, depending on the outcome of the second measurement, R_2 and R_3 can be simultaneously effected on the corresponding qubits. In the third cycle, only R_2 is needed and only if the control is 1. The computation ends with the last qubit reaching the Hadamard gate and being measured afterward. A formal description of the procedure, in the general case, is given as in Algorithm 15.1.

Algorithm 15.1 Parallel_Quantum_Fourier_Transform

1: *Input:* $|j_1 j_2 \cdots j_n\rangle$
2: *Output:* $k_1 k_2 \cdots k_n$
3:
4: **for** $i = 1$ to n **do**
5:     $|j_i\rangle \longleftarrow H|j_i\rangle$;
6:     Measure $|j_i\rangle$ as k_{n-i+1};
7:     **if** $k_{n-i+1} = 1$ **then**
8:         **for** $l = 2$ to $n - i + 1$ in parallel **do**
9:             $|j_{i+l-1}\rangle \longleftarrow R_l|j_{i+l-1}\rangle$;
10:            $|j_{i+l-1}\rangle$ moves one position up in the array
11:         **end for**
12:     **end if**
13: **end for**

In the worst case, when all qubits are measured as 1, there is no difference between the parallel algorithm outlined above and the sequential solution envisaged by Griffiths and Niu [34] with respect to the overall running time. Assuming, for analysis purposes, that measuring a qubit, applying a phase shift and performing a Hadamard transformation, each takes one time unit, then the total time necessary to complete the Fourier transform on a quantum register with n qubits is $3n - 1$, as the top qubit in both the sequential circuit of Fig. 15.10 and the parallel circuit of Fig. 15.11 does not require a phase shift.

However, in the average case, some of the classical signals controlling the array of phase shift gates in Fig. 15.11 will have been observed as 0, meaning that no phase shifts have to be performed during those respective cycles. In contrast, the sequential solution depicted in Fig. 15.10 requires the application of a phase shift at every step following the first measurement with outcome 1. If the expected probability of a measurement yielding 0 equals the expected probability to observe a 1 following a measurement, then the running time of the parallel solution is shorter than the sequential running time by a difference proportional to the time it takes to effect a number of $O(n)$ phase shift gates, where n is the size of the input register.

The difference between the sequential running time and the parallel running time is maximum when $|j_1\rangle$ is measured as 1 and all the other qubits are observed in the state 0. In this case, the circuit in Fig. 15.10 still performs $n - 1$ phase shifts for a total running time of $3n - 1$ time units, while the circuit in Fig. 15.11 executes all $n - 1$ phase shifts in parallel during the first cycle, thus completing the computation in $2n + 1$ time units.

The second advantage of the parallel approach is that the phase shift gates that need to be applied during the computation are known at the outset, making it easy to set them up beforehand in order to form the required linear array architecture. In other words, regardless of the initial quantum state of the register on which the

QFT is to be performed, the first gate in the linear array (top gate in Fig. 15.11) will always perform a Hadamard gate, the second gate always performs a $\pi/2$ phase shift, the third gate is "hardwired" to effect a $\pi/4$ phase shift and so on. The systolic mode of operation of the quantum array compensates for the fixed characteristics of each gate, the qubits traversing the array to undergo a specific quantum evolution at each node. In the current context, the attribute "systolic" describes the rhythmic mode in which data travel through the array of gates, much like blood does through the circulatory system.

In the sequential approach, on the other hand, the phase shift applied to each qubit is not known at the outset, as it is computed on the fly based on the information about the measurements performed so far and transmitted as classical signals. This means that the gates effecting the necessary phase shifts in the semiclassical approach of Griffiths and Niu [34] have to be "programmed" or adjusted during the computation, in order to accommodate a discrete set of possible values for the phase shift.

In the example given at the end of previous section, the phase shift applied to the bottom qubit is $5\pi/8$ because the previous measurements yielded $k_4 = 1$, $k_3 = 0$ and $k_2 = 1$. But the phase shift could have been $7\pi/8$ if all the measurements yielded a 1 or just $\pi/8$ if $k_4 = 1$ and $k_3 = k_2 = 0$. Therefore, we do not know at the outset how to "set" the quantum gates responsible with the phase shift performed on each qubit, as this information becomes available only during the computation, depending on the probabilistic results of the measurements. Technologically, this is more difficult to implement than a linear array of gates whose characteristics are fixed for any possible course of the computation.

The semiclassical Fourier transform and its parallelization are applicable to those quantum computations in which the Fourier transform immediately precedes a measurement of the qubits involved in the computation, like in Shor's algorithms for factoring integers and computing discrete logarithms [9]. Furthermore, the quantum systolic array architecture works equally fine if the input is already classical, in which case the restriction to measure the qubits after applying the Fourier transform can be lifted altogether.

When $j_1, j_2, \ldots, j_n$ are classical bits, the topology of the circuit in Fig. 15.11 remains unchanged, except that no measurements are performed and the flow of data through the linear array is reversed, as shown in Fig. 15.12. As more data are fed into the linear array through the Hadamard gate, after having "controlled" the parallel execution of a set of phase shifts, the computational complexity of each step increases with its rank. When j_1 enters the array, only the Hadamard gate is active, but with each consecutive step, a new gate down the array joins the ones above it to operate on the qubits traversing the array. Because these gates operate in parallel, the execution time of each step is maintained constant. Also note that, in this case, all outputs are simultaneously obtained during the last step of the computation.

The overall parallel running time, in the worst case, is therefore $2n - 1$ time units, as there are no measurements to perform. The worst case occurs when $j_2, j_3, \ldots, j_n$ all have value 1. For all other inputs (that is, when at least one of $j_2, j_3, \ldots, j_n$ is 0), the parallel running time is smaller than the time needed to complete the computation in a purely sequential manner, where each qubit is dealt with one after the other, in decreasing order of their ranks.

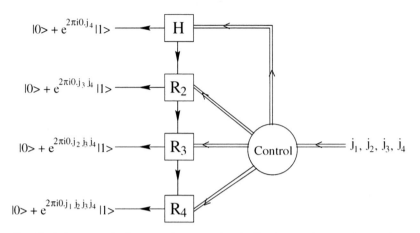

Fig. 15.12 Quantum pipeline array for computing the Fourier transform on classical inputs. The output is now quantum and the flow of qubits in the array is downward.

Quantum algorithms employ the Fourier transform in order to create an interference among the terms in a superposition. From this point of view, the QFT offers little advantage, if any, when applied to a classical input. However, the situation is different for quantum cryptography. Distributing classical keys through quantum means is a procedure that may use the QFT and its inverse as encoding and decoding algorithms to protect vital information while in transit [35].

Naturally, the parallel approach detailed in this section for the computation of the direct Fourier transform is also applicable, with the same results, to the circuit in Fig. 15.8, performing the inverse Fourier transform. The difference in time complexity between the sequential approach and the parallel one, although seemingly insignificant from a theoretical perspective, may prove essential under practical considerations, as we show in our next example.

15.4.2 Quantum Decoherence

Qubits are fragile entities and one of the major challenges in building a practical quantum computer is to find a physical realization that would allow us to complete a computation before the quantum states we are working with become seriously affected by quantum errors. In an ideal setting, we evolve our qubits in perfect isolation from the outside world. But any practical implementation of a quantum computation will be affected by the interactions taking place between our system and the environment. These interactions cause quantum information to leak out into the environment, leading to errors in our qubits. Different types of errors may affect an ongoing computation in different ways, but *quantum decoherence*, as defined below, usually occurs extremely rapidly and can seriously interfere with computing the QFT and its inverse.

In the context of a quantum key distribution protocol [35], consider the task of recovering the original (classical) bit string $j = j_1 j_2 \cdots j_n$ from its QFT form. The circuit performing this computation (see Fig. 15.8) takes as input n qubits. The state of each qubit can be described by the following general equation:

$$|\psi_k\rangle = \frac{1}{\sqrt{2}}|0\rangle + \frac{e^{i\theta_k}}{\sqrt{2}}|1\rangle, \ 1 \leq k \leq n, \tag{15.24}$$

where the relative phase θ_k, characterizing the qubit of rank k, depends on the values of bits $j_k, j_{k+1}, \cdots, j_n$. The corresponding density operator is given by

$$\rho_k = |\psi_k\rangle\langle\psi_k| = \frac{1}{2}|0\rangle\langle0| + \frac{e^{-i\theta_k}}{2}|0\rangle\langle1| + \frac{e^{i\theta_k}}{2}|1\rangle\langle0| + \frac{1}{2}|1\rangle\langle1|, \tag{15.25}$$

or in matrix form

$$\rho_k = \frac{1}{2}\begin{bmatrix} 1 & e^{-i\theta_k} \\ e^{i\theta_k} & 1 \end{bmatrix}. \tag{15.26}$$

The diagonal elements (or the *populations*) measure the probabilities that the qubit is in state $|0\rangle$ or $|1\rangle$, while the off-diagonal components (the *coherences*) measure the amount of interference between $|0\rangle$ and $|1\rangle$ [36]. Decoherence then, resulting from interactions with the environment, causes the off-diagonal elements to disappear. Since that is where the whole information carried by a qubit is stored, the input qubits for computing the inverse Fourier transform are very sensitive to decoherence. When they become entangled with the environment, the interference brought about by the Hadamard gate is no longer possible, as the system becomes effectively a statistical mixture. In other words, decoherence makes a quantum system behave like a classical one.

Naturally, this process is not instantaneous, but it usually occurs extremely rapidly, subject to how well a qubit can be isolated from its environment in a particular physical realization. Because of decoherence, we must obtain the values of $j_1, j_2, \cdots, j_n$ before time limit δ, after which the errors introduced by the coupling with the environment are too serious to still allow the recovery of the binary digits of j.

The precise value of δ will certainly depend on the particular way chosen to embody quantum information. If the qubits are implemented as trapped ions, then usually such a physical system is relatively well isolated and decoherence is not a major concern. Nevertheless, other impediments make the design of a scalable quantum architecture a very challenging task. As a consequence, current experiments are only able to manipulate a handful of qubits. An illustrative example is a result from

2005 reporting the implementation of the semiclassical QFT on a "quantum register" composed of three beryllium ion qubits [12].

At the other end of the spectrum, we have attempts to implement the QFT, and quantum algorithms in general, using the well-established technology of nuclear magnetic resonance (NMR). In this case, decoherence plays a much more important role, directly affecting the accuracy of the results and placing a serious limitation on the scalability of this type of quantum computing architecture.

Experimental arrangements to compute the QFT on a 3-qubit NMR quantum information processor are reported by Weinstein et al. [11, 37]. Also, a 7-qubit experiment to implement the simplest meaningful instance of Shor's algorithm for factoring the number 15 uses the QFT as an important step of the computation [10]. Again we can see that scalability is the main obstacle toward building a practical quantum computer.

Of course, one of the possibilities to cope with the errors introduced by quantum decoherence is to use quantum codes to correct them. But here too, there are limitations. The more serious the errors are, the more ancillary qubits are required to correct them and consequently, the higher the probability of an error occurring in the correcting circuit itself. Therefore, we can only use that many auxiliary qubits to correct quantum errors before no advantage whatsoever in the accuracy of the solution is gained. From this point of view, parallelism offers a "clean" solution, avoiding the errors caused by quantum decoherence altogether and completing the computation before the entanglement with the surrounding environment seriously affects the ongoing quantum transformations.

The point we wish to make here is that when all other means have been used, a parallel approach may be the only way to further improve scalability by reducing the running time of the quantum algorithm and keep it below the decoherence threshold. In the particular case of computing the QFT and its inverse for cryptographic purposes, when δ lies between the parallel completion time and the sequential completion time, the quantum pipeline array may be the only architecture capable to precisely recover all digits in the binary expansion of j. From a different perspective, the parallel solution allows for longer bit strings to be transmitted between the communicating parties, thus achieving better scalability over the purely sequential approach. With respect to scalability, it is also important to note that the parallel solution scales up linearly in the number of quantum gates employed, when the number of qubits on which the QFT is performed increases.

15.4.2.1 Time-Varying Variables

We have already seen that the computation of the Fourier transform by quantum means belongs to the class of computations in which the complexity of each step depends on its rank. In addition, if we also take into consideration the properties of the computational environment, we are faced with the negative effects caused by quantum decoherence. Formally, the data stored in the quantum register before time limit δ is significantly different from what the same qubits encode after the decoherence threshold δ. The coupling between our qubits and their surrounding environment effectively places a hard deadline on the computation. After this deadline, the input data (variables) will have changed and if the computation is not yet complete, it has inevitably failed. From this perspective, the computation of the QFT (whether direct or inverse) in the presence of decoherence is an example of the paradigm dealing with *time-varying variables*.

As we have demonstrated above, parallelism can help us cope with variables whose values change over time. The use of a parallel approach becomes critical when the solution to a certain problem must accommodate a deadline. In our case, quantum decoherence places an upper bound on the scalability of computing the QFT or its inverse, and the only chance to reach beyond that limit is through a parallel solution.

15.4.3 Quantum Error-Correction

Parallel processing is often the best alternative to avoid quantum errors in general and not just decoherence. The following examples on correcting quantum errors using specialized *quantum codes* or via *symmetrization* clearly show this.

In the computation of the QFT and its inverse the complexity of each step evolves with its rank. The more steps are executed before the current one, the higher the computational resources required to complete it. In this section, we still focus on steps of variable complexity, but in this case the variation is *time driven* rather than *rank driven*. In other words, we can have a high computational complexity even for the first step, if we allow some time to pass before starting the computation. The amount of computational resources required to successfully carry out a certain step is directly proportional to the amount of time elapsed since the beginning of the computation. We illustrate this paradigm through the use of error-correcting codes employed to maintain a quantum computation error-free.

The laws of quantum mechanics prevent, in general, a direct application of the classical error-correction techniques. We cannot inspect (measure) at leisure the state of a quantum memory register to check whether an ongoing computation is not off track without the risk of altering the intended course of the computation. Moreover, because of the no-cloning theorem, quantum information cannot be amplified in the same way that digital signals can. Correcting quantum errors certainly requires much more ingenuity than fixing classical bits, but the basic idea of using redundancy is still useful.

Like in the classical case, the information contained in a qubit is spread out over several qubits so that damage to any one of them will not influence the outcome of the computation. In the quantum case, though, the encoding of the logical qubit is achieved through the use of specific resources, by entangling the logical qubit with several ancilla qubits. In this way, the information in the state of the qubit to be protected is spread out among the correlations characterizing an entangled state.

Paradoxically enough, entanglement with the environment can be fought back using quantum error-correcting codes based on entanglement [38].

15.4.3.1 Quantum Codes

The construction of all quantum error-correcting codes is based on the surprising, yet beautiful idea of *digitizing the errors*. Any possible error affecting a single qubit can be expressed as a linear combination of no errors (I), bit flip errors (X), phase errors (Z) and bit flip phase errors (Y), where I, X, Z and Y are the Pauli operators describing the effect of the respective errors. Generalizing to the case of a quantum register, an error can be written as $\sum_i e_i E_i$ for some error operators E_i and coefficients e_i. The error operators can be tensor products of the single-bit error transformations or more general multibit transformations. An error-correcting code that can undo the effect of any error belonging to a set of correctable errors E_i will embed n data qubits (logical qubits) in $n + k$ code qubits (physical qubits). The joint state of the ensemble of code qubits is subject to an arbitrary error, mathematically expressed as a linear combination of the correctable error operators E_i.

To recover the original encoded state, a syndrome extraction operator has to be applied that uses some ancilla qubits to create a superposition of the error indices i corresponding to those correctable error operators E_i that have transformed the encoded state. Measuring only the ancilla qubits will collapse the superposition of errors, yielding only one index k. But because the ancilla qubits were entangled with the code qubits through the application of the syndrome extraction operator, the side effect of the measurement is that the corruption caused by all error transformations will be undone, save for the one corresponding to index k. Consequently, only one inverse error transformation is required in order to complete the recovery process. In essence, knowing how to deal with a set of fundamental error transformations allows us to tackle any linear combination of them by projecting it to one of the basis components. This process is referred to as *digitizing* or *discretizing* the errors.

Peter Shor's second major contribution to the advancement of quantum computation was the creation in 1995 of an algorithm that could correct any kind of error (amplitude and/or phase errors) affecting a single qubit in a 9-qubit code [39]. In a different approach, Steane studied the interference properties of multiple particle entangled states and managed to devise a shorter 7-qubit code [40]. The number of qubits necessary for a perfect recovery from a single error was later squeezed down to a minimum of five [41, 42].

Naturally, in order to cope with more than one error at a time, it is necessary to use larger and more elaborate codes. The book of Nielsen and Chuang [17] offers a detailed treatment of quantum codes, explaining how ideas from classical linear codes can be used to construct large classes of quantum codes, such as the Calderbank-Shor–Steane (CSS) codes [43, 44] or the stabilizer codes (also known as additive quantum codes), which are even more general than the CSS codes and are based on the stabilizer formalism developed by Gottesman [45].

The major drawback in using large and intricate quantum codes is that the corrective circuit itself is as much prone to errors as the quantum circuit responsible for the main computation. The more errors we are attempting to rectify, the more the complexity and length of the recovery procedure will increase (see [46] for some theoretical bounds on the relationship between the number of data qubits, the total number of entangled qubits and the maximal number of errors that can be tolerated). Thus, we can only increase the size of the error correction codes up to a certain cut-off point, past which no further gains in accuracy can be made.

One attempt to overcome this limitation are the *concatenated* codes. If a certain code uses n physical qubits to encode one logical qubit, a concatenated version of that code is obtained by further encoding each of the n qubits in another block of n. This hierarchical structure (tree) can be further expanded to accommodate as many levels as desired. By adding more levels of concatenation, the overall chance for an error can be made arbitrarily small, provided that the probability of an individual error is kept below a certain critical threshold [47]. Of course, the high cost of using concatenated codes lies in the exponential increase in the number of qubits with the number of levels added.

15.4.3.2 Time-Varying Complexity

This short exposition of the various quantum error-correcting codes devised to maintain the coherence of fragile quantum states and to protect them from dissipative errors caused by spontaneous emissions, for example, clearly shows one thing. The more time it takes to complete a quantum computation, the more errors are introduced in the process. Quantum error-correcting schemes can be employed to deal with these errors, but the running time, number of ancilla qubits and complexity of the correcting algorithm are directly proportional to the number and seriousness of the errors introduced into the computation. Correcting quantum errors is an important task executed alongside the mainstream computation and its complexity is heavily dependent on time. Steps executed soon after the initialization of the quantum register will require none or low complexity recovery techniques, while steps executed long after the initialization time may require complicated schemes and heavy resources allocated to deal with quantum errors.

Again, parallelism can help avoid this increase in the complexity of the recovery procedure and ultimately ensure the success of the computation. If the steps of the algorithm are independent of one another and can be executed in any order, then the most straightforward application of parallelism is to execute all steps simultaneously and thus complete the computation before any serious errors can accumulate over time. In this way we try to avoid or elude quantum errors rather than deal with them. But parallelism, in the form of redundancy, can also be used to correct quantum errors.

15.4.3.3 Error Correction via Symmetrization

The technique called *error correction via symmetrization* [48, 49] is yet another example of how the duality of quantum mechanical laws can be exploited for the benefit of quantum computation. Although the measurement postulate severely restricts us in recycling techniques from classical error correction, it can still offer conceptually new ways of achieving error correction that are simply unavailable to classical computers. Error correction via symmetrization relies on the projective effect of measurements to do the job. The technique uses n quantum computers, each performing the same computation. Provided no errors occur, the joint state of the n computers is a symmetric one, lying somewhere in the small symmetric subspace of the entire possible Hilbert space. Devising a clever measurement that projects the joint state back into the symmetric subspace should be able to undo possible errors, without even knowing what the error is.

To achieve this, the n quantum computers need to be carefully entangled with a set of ancilla qubits placed in a superposition representing all possible permutations of n objects. In this way, the computation can be performed over all permutations of the computers simultaneously. Then, by measuring the ancilla qubits, the joint state of the n computers can be projected back into just the symmetric computational subspace, without the errors being measured explicitly. Peres has shown that this technique is most appropriate for correcting several qubits that are slightly wrong, rather than correcting a single qubit that is terribly wrong [50]. Error correction via symmetrization can be applied repeatedly, at regular time intervals, to avoid the accumulation of large errors and continually project the computation back into its symmetric subspace.

No matter which parallel approach is employed, if the required number of quantum processing units is provided, then the algorithm is successful. Simulating the same solution on an insufficient number of quantum computers will lead to a gradual accumulation of the quantum errors up to the point where the results of the computation are compromised.

15.4.4 Quantum Distinguishability

The problem of distinguishing among entangled quantum states is a quantum mechanical instance of the interacting variables paradigm (Sect. 2.7.2.4). Suppose we have a fixed set of quantum states described using the usual Dirac notation $|\Psi_i\rangle$ ($1 \leq i \leq n$) known to both Alice and Bob. Alice randomly chooses a state from the set and prepares a qubit (or set of qubits) in that particular state. She then gives the qubit(s) to Bob who is free to measure them in any way he likes, without applying any quantum operation on them (Bob lacks the power of a quantum computer). To be more specific, Bob can apply any kind of measurement on the qubit(s) and possibly process and/or interpret the classical information acquired through measurement, but he cannot manipulate quantum information using unitary evolution.

In the end, his task is to identify the index i of the state characterizing the qubit(s) Alice has given him. The only case in which a set of quantum states can be reliably (that is, 100% of the time) distinguished from one another is if they are pairwise orthogonal.

Now consider the case in which we try to distinguish among the four Bell states

$$\frac{1}{\sqrt{2}}|00\rangle + \frac{1}{\sqrt{2}}|11\rangle, \frac{1}{\sqrt{2}}|00\rangle - \frac{1}{\sqrt{2}}|11\rangle, \frac{1}{\sqrt{2}}|01\rangle + \frac{1}{\sqrt{2}}|10\rangle, \frac{1}{\sqrt{2}}|01\rangle - \frac{1}{\sqrt{2}}|10\rangle$$

by resorting only to direct quantum measurements (in other words, no quantum transformations are possible before a measurement). In these circumstances, any sequential approach (that is, measuring the qubits one after the other) will be of no help here, regardless of the basis in which the measurements are performed. By measuring the two qubits in sequence, in the computational basis, Bob can distinguish the states $\frac{1}{\sqrt{2}}(|00\rangle \pm |11\rangle)$ from $\frac{1}{\sqrt{2}}(|01\rangle \pm |10\rangle)$. He does this by checking if the outcomes of the two measurements are the same or not. But this kind of measurement makes it impossible to differentiate between $\frac{1}{\sqrt{2}}(|00\rangle + |11\rangle)$ and $\frac{1}{\sqrt{2}}(|00\rangle - |11\rangle)$, or between $\frac{1}{\sqrt{2}}(|01\rangle + |10\rangle)$ and $\frac{1}{\sqrt{2}}(|01\rangle - |10\rangle)$.

Alternatively, Bob can decide to perform his measurements in a different basis, like $(|+\rangle, |-\rangle)$, where the basis vectors are

$$|+\rangle = \frac{1}{\sqrt{2}}|0\rangle + \frac{1}{\sqrt{2}}|1\rangle \tag{15.27}$$

and

$$|-\rangle = \frac{1}{\sqrt{2}}|0\rangle - \frac{1}{\sqrt{2}}|1\rangle. \tag{15.28}$$

Due to the fact that

$$\frac{|00\rangle + |11\rangle}{\sqrt{2}} = \frac{|++\rangle + |--\rangle}{\sqrt{2}} \tag{15.29}$$

and

$$\frac{|00\rangle - |11\rangle}{\sqrt{2}} = \frac{|+-\rangle + |-+\rangle}{\sqrt{2}}, \tag{15.30}$$

Bob can now reliably distinguish the quantum state $\frac{1}{\sqrt{2}}(|00\rangle + |11\rangle)$ from $\frac{1}{\sqrt{2}}(|00\rangle - |11\rangle)$. Indeed, if the two qubits yield identical outcomes when measured in this new basis, then we can assert with certainty that the state was not $\frac{1}{\sqrt{2}}(|00\rangle - |11\rangle)$. Similarly, if the measurement outcomes for the qubits are different, the original state could not have been $\frac{1}{\sqrt{2}}(|00\rangle + |11\rangle)$. Unfortunately, in this new setup, the quantum states $\frac{1}{\sqrt{2}}(|00\rangle + |11\rangle)$ and $\frac{1}{\sqrt{2}}(|01\rangle + |10\rangle)$ become indistinguishable and the same is true for $\frac{1}{\sqrt{2}}(|00\rangle - |11\rangle)$ and $\frac{1}{\sqrt{2}}(|01\rangle - |10\rangle)$.

The computational bases $(|0\rangle, |1\rangle)$ and $(|+\rangle, |-\rangle)$ are, respectively, the two extremities of an (theoretically) infinite number of choices for the basis relative to which the quantum measurements are to be performed. But even though the separation line between the four Bell states will drift with the choice of the basis vectors, the two extreme cases discussed above offer the best possible distinguishability.

Intuitively, this is due to the entanglement exhibited between the two qubits in all four states. As soon as the first qubit is measured (regardless of the basis), the superposition describing the entangled state collapses to the specific state consistent with the measurement result. In this process, some of the information originally encapsulated in the entangled state is irremediably lost. Consequently, measuring the second qubit cannot give a complete separation of the four EPR states. But the Bell states do form an orthonormal basis, which means that (at least theoretically) they can be distinguished by an appropriate quantum measurement. However, this measurement must be a *joint* measurement of both qubits simultaneously, in order to achieve the desired distinguishability.

15.4.4.1 Generalization

A more compact representation of the Bell basis is through a square matrix where each column is a vector describing one of the Bell states:

$$\frac{1}{\sqrt{2}} \begin{pmatrix} 1 & 0 & 0 & 1 \\ 0 & 1 & 1 & 0 \\ 0 & 1 & -1 & 0 \\ 1 & 0 & 0 & -1 \end{pmatrix}. \tag{15.31}$$

The elements of each column are the amplitudes or proportions in which the computational basis states $|00\rangle$, $|01\rangle$, $|10\rangle$ and $|11\rangle$ are present in the respective EPR state.

This scenario can be extended to ensembles of more than two qubits. The following matrix describes eight different entangled states that cannot be reliably distinguished unless a joint measurement of all three qubits involved is performed:

$$\frac{1}{\sqrt{2}} \begin{pmatrix} 1 & 0 & 0 & 0 & 0 & 0 & 0 & 1 \\ 0 & 1 & 0 & 0 & 0 & 0 & 1 & 0 \\ 0 & 0 & 1 & 0 & 0 & 1 & 0 & 0 \\ 0 & 0 & 0 & 1 & 1 & 0 & 0 & 0 \\ 0 & 0 & 0 & 1 & -1 & 0 & 0 & 0 \\ 0 & 0 & 1 & 0 & 0 & -1 & 0 & 0 \\ 0 & 1 & 0 & 0 & 0 & 0 & -1 & 0 \\ 1 & 0 & 0 & 0 & 0 & 0 & 0 & -1 \end{pmatrix}. \tag{15.32}$$

In general, for a quantum system composed of n qubits, one can define the following 2^n entangled states of the system:

$$\frac{1}{\sqrt{2}}(|000\cdots0\rangle \pm |111\cdots1\rangle)$$

$$\frac{1}{\sqrt{2}}(|000\cdots1\rangle \pm |111\cdots0\rangle)$$

$$\vdots$$ (15.33)

$$\frac{1}{\sqrt{2}}(|011\cdots1\rangle \pm |100\cdots0\rangle).$$

These vectors form an orthonormal basis for the state space corresponding to the n-qubit system. The only chance to differentiate among these 2^n states using quantum measurement(s) is to observe the n qubits simultaneously, that is, perform a single joint measurement of the entire system. In the given context, *joint* is really just a synonym for *parallel*. Indeed, the device in charge of performing the joint measurement must possess the ability to "read" the information stored in each qubit, in parallel, in a perfectly synchronized manner. In this sense, at an abstract level, and just for the sake of offering a more intuitive understanding of the process, the measuring apparatus can be viewed as having n probes. With all probes operating in parallel, each probe can "peek" inside the state of one qubit, in a perfectly synchronous operation. The information gathered by the n probes is seen by the measuring device as a single, indivisible chunk of data, which is then interpreted to give one the 2^n entangled states as the measurement outcome.

From a mathematical (theoretical) point of view, such a measurement operator can be easily constructed by defining each of the 2^n states that are to be distinguished to be a projector associated with the measurement operation. We are well aware though that a physical realization of this mathematical construction is extremely difficult, if not impossible to achieve in practice, with today's technology. Yet, if there is any hope to see a joint measurement performed in the future, then only a device operating in a parallel synchronous fashion on all n qubits (as explained above) would succeed.

It is perhaps worth emphasizing that if such a measurement cannot be applied then the desired distinguishability can no longer be achieved regardless of how many other measuring operations we are allowed to perform. In other words, even an infinite sequence of measurements touching at most $n-1$ qubits at the same time cannot equal a single joint measurement involving all n qubits.

Furthermore, with respect to the particular distinguishability problem that we have to solve, a single joint measurement capable of observing $n-1$ qubits simultaneously offers no advantage whatsoever over a sequence of $n-1$ consecutive *single* qubit measurements. This is due to the fact that an entangled state like

$$\frac{1}{\sqrt{2}}(|000\cdots0\rangle + |111\cdots1\rangle)$$ (15.34)

cannot be decomposed neither as a product of $n - 1$ individual states nor as a product of two states (one describing a single qubit and the other describing the subsystem composed of the remaining $n - 1$ qubits). Any other intermediate decomposition is also impossible.

Overall, our distinguishability problem can only be tackled successfully within a parallel approach, where we can measure all qubits simultaneously. Conceptually, distinguishing among entangled quantum states is a quantum example of measuring interdependent variables. In this particular quantum instance, the interdependence between variables takes the form of entanglement between qubits, the phenomenon ultimately responsible for making a parallel approach imperative. But not only measuring entangled states requires a parallel solution, quantum evolutions that have to maintain a certain entangled state may also resort to parallelism in order to achieve their goal. In our last example, we investigate entanglement as a global mathematical constraint that has to be satisfied throughout a quantum computation.

15.4.5 Transformations Obeying a Global Condition

Some computational problems require the transformation of a mathematical object in such a way that a property characterizing the original object is to be maintained at all times throughout the computation. This property is a global condition on the variables describing the input state and it must be obeyed at every intermediate step in the computation, including for the final state. Geometric flips, map recoloring and rewriting systems are three examples of transformations that can be constrained by a global mathematical condition [51].

Here, we show that some quantum transformations acting on entangled states may also be perceived as computations obeying a global mathematical constraint. Consider, for example, an ensemble of n qubits sharing the following entangled state:

$$\frac{1}{\sqrt{2}}|000\cdots 0\rangle + \frac{1}{\sqrt{2}}|111\cdots 1\rangle. \qquad (15.35)$$

The entanglement characterizing the above state determines a strict correlation between the values observed in case of a measurement: either all qubits are detected in the state 0 or they are all seen as 1. Suppose that this correlation has to be maintained unaltered, regardless of the local transformations each of the qubits may undergo. Such a transformation may be the application of a *NOT* quantum gate to any of the qubits forming the ensemble. After such an event, the particular entangled state given in Eq. (15.35) is no longer preserved and as a consequence the correlation between the qubits will be altered. The qubit whose state was "flipped" will be observed in the complementary state, with respect to the other qubits. The global mathematical constraint is no longer satisfied.

Parallelism can once again make the difference and help maintain the required entangled state. If, at the same time one or more of the qubits are "flipped", we also apply a *NOT* gate to all remaining qubits simultaneously, then the final state coincides with the initial one. In this way, although the value of each qubit has been switched, the correlation we were interested to maintain remains the same. Also note that any attempt to act on less than n qubits simultaneously is doomed to failure.

The state given in Eq. (15.35) is not the only one with this property. Any entangled state from the orthonormal basis set (15.33) could have been used in the example presented above. The correlation among the qubits would have been different, but the fact that applying a *NOT* gate, in parallel, to all qubits does not change the quantum state of the ensemble is true for each entangled state appearing in system (15.33).

Perhaps the scenario described above can be extended to other quantum transformations besides the *NOT* gate. Another, perhaps more interesting generalization would be a quantum computation that has to maintain entanglement as a generic, global mathematical constraint and not a specific type of entanglement with a particular correlation among the qubits involved. Such a computation would allow entanglement to change form, but the mathematical definition of entanglement would still have to be obeyed at each step with each transformation.

15.5 Looking Ahead

In this final chapter, we have reviewed some of the most promising computing paradigms that have emerged from our relentless quest to make computation more efficient in terms of speed or accuracy of the result obtained. A special attention was given to the quantum computing paradigm, which still has the potential to radically transform the field of computer science, provided that experimentalists will eventually find a viable design for a practical quantum computer. However, the essential observation that can be formulated seeing all these efforts is that parallel processing and the future of computation go hand in hand. Whether we are discussing conventional computing architectures (like clusters and grids) or more exotic proposals (DNA computing, quantum computing, etc.), they all draw their power from some form of parallelism and they all can be considered as massively parallel computing devices.

Moreover, the polymorphic nature of parallelism becomes evident by surveying all these different ways to envisage computation. In most paradigms, parallelism refers to a large (and sometimes huge) number of processing elements operating simultaneously, whether these are conventional electronic processors or DNA molecules. But in quantum information processing, for instance, massive parallelism is an attribute characterizing the "software" rather than the "hardware", since it refers to how a huge computational space can be explored in parallel by manipulating only a relatively small number of qubits. Furthermore, quantum computing is also the perfect example of how different instances of parallelism can be encountered within the same computational paradigm, as we showed in this chapter.

It is difficult to foretell, at this point, what will be the dominant computing technology in a few decades' time. It could be one of the alternatives described at the beginning of the chapter or a hybrid solution involving a combination of two or more paradigms. Yet another possibility would be the emergence of a totally new and revolutionary way to perform computations. What we can say for sure, though, is that parallel processing has reaffirmed its importance with every novel model of computation proposed over time and it will continue to do so. Its capital role in the theory of computing will not change, regardless of the physical layer used to represent and manipulate information. In this respect, the universal attribute of parallelism becomes apparent, ensuring its perennity.

References

1. G. Păun, G. Rozenberg, A. Salomaa, DNA Computing – New Computing Paradigms, Springer (1998).
2. L. Adleman, Molecular computation of solutions to combinatorial problems, Science 266 (1994) 1021–1024.
3. R. J. Lipton, DNA solution of hard computational problems, Science 268 (5210) (1995) 542–545.
4. W.-L. Chang, M. Guo, J. Wu, Solving the independent-set problem in a DNA-based supercomputer model, Parallel Processing Letters 15 (4) (2005) 469–479.
5. G. Păun, Computing with membranes, Journal of Computer and System Sciences 61 (1) (2000) 108–143.
6. M. Pérez-Jiménez, A. Riscos-Núñez, A linear solution for the knapsack problem using active membranes, in: Membrane Computing. Lecture Notes in Computer Science, Vol. 2933, Springer (2004) pp. 250–268.
7. G. Păun, P systems with active membranes: Attacking NP-complete problems, Journal of Automata, Languages, Combinatorics 6 (1) (2001) 5–90.
8. C. Zandron, C. Ferretti, G. Mauri, Solving NP-complete problems using P systems with active membranes, in: I. Antoniou, C. Calude, M. Dinneen (Eds.), Unconventional Models of Computation, Springer, London (2000) pp. 289–301, dISCO – Universita di Milano-Bicocca, Italy.
9. P. W. Shor, Polynomial-time algorithms for prime factorization and discrete logarithms on a quantum computer, Special issue on Quantum Computation of the SIAM Journal on Computing 26 (5) (1997) 1484–1509.
10. L. M. K. Vandersypen, M. Steffen, G. Breyta, C. S. Yannoni, M. H. Sherwood, I. L. Chuang, Experimental realization of Shor's quantum factoring algorithm using nuclear magnetic resonance, Nature 414 (2001) 829–938.
11. Y. S. Weinstein, et al., Quantum process tomography of the quantum fourier transform, Journal of Chemical Physics 121 (13) (2004) 6117–6133, http://arxiv.org/abs/quant-ph/0406239v1.
12. J. Chiaverini, et al., Implementation of the semiclassical quantum fourier transform in a scalable system, Science 308 (5724) (2005) 997–1000.
13. A. Adamatzky, B. D. L. Costello, T. Asai, Reaction-Diffusion Computers, Elsevier, 2005.
14. R. Fraser, S. G. Akl, Accelerating machines: A review, International Journal of Parallel, Emergent and Distributed Systems 23 (1) (2008) 81–104.
15. N. D. Mermin, From Cbits to Qbits: Teaching Computer Scientists Quantum Mechanics, http://arxiv.org/abs/quant-ph/0207118 (July 2002).
16. E. Rieffel, W. Polak, An introduction to quantum computing for non-physicists, ACM Computing Surveys 32 (3) (2000) 300–335.

17. M. A. Nielsen, I. L. Chuang, Quantum Computation and Quantum Information, Cambridge University Press (2000).
18. M. Hirvensalo, Quantum Computing, Springer-Verlag (2001).
19. A. Berthiaume, Quantum computation, in: L. A. Hemaspaandra, A. L. Selman (Eds.), Complexity Theory Retrospective II, Springer-Verlag, New York (1997) pp. 23–51.
20. R. Feynman, R. B. Leighton, M. Sands, The Feynman Lectures on Physics, Vol. III, Addison-Wesley, Reading, Mass. (1965).
21. C. Santori, et al., Indistinguishable photons from a single-photon device, Nature 419 (2002) 594–597.
22. E. H. Knill, R. Laflamme, G. J. Milburn, A scheme for efficient quantum computation with linear optics, Nature 409 (2001) 46–52.
23. P. Dirac, The Principles of Quantum Mechanics, 4th Edition, Oxford University Press, 1958.
24. E. W. Weisstein, et al., Bloch sphere, From *MathWorld* – A Wolfram Web Resource, http://mathworld.wolfram.com/BlochSphere.html.
25. W. K. Wootters, W. H. Zurek, A single quantum cannot be cloned, Nature 299 (1982) 802–803.
26. A. Barenco, A universal two-bit gate for quantum computation, Proceedings of the Royal Society of London A 449 (1995) 679–683.
27. D. DiVincenzo, Two-bit gates are universal for quantum computation, Physical Review A 51 (1995) 1015–1022.
28. E. Schrödinger, Discussion of probability relations between separated systems, Proceedings of the Cambridge Philosophical Society 31 (1935) 555–563.
29. A. Einstein, B. Podolsky, N. Rosen, Can quantum-mechanical description of physical reality be considered complete?, Physical Review 47 (1935) 777–780.
30. A. Berthiaume, G. Brassard, Oracle quantum computing, Journal of Modern Optics 41 (12) (1994) 2521–2535.
31. S. Robinson, Emerging insights on limitations of quantum computing shape quest for fast algorithms, SIAM News 36 (1) (2003).
32. J. W. Cooley, J. Tukey, An algorithm for the machine calculation of complex fourier series, Mathematics of Computation 19 (1965) 297–301.
33. D. Coppersmith, An approximate fourier transform useful in quantum factoring, Technical Report RC19642, IBM (1994).
34. R. Griffiths, C.-S. Niu, Semiclassical Fourier transform for quantum computation, Physical Review Letters 76 (1996) 3228–3231.
35. M. Nagy, S. G. Akl, S. Kershaw, Key distribution based on the quantum Fourier transform, in: Proceedings of the International Conference on Security and Cryptography (SECRYPT 2008), Porto, Portugal (2008) pp. 263–269.
36. C. Cohen-Tannoudji, B. Diu, F. Laloe, Quantum Mechanics, Vols. 1 and 2, Wiley, New York (1977).
37. Y. S. Weinstein, et al., Implementation of the quantum fourier transform, Physical Review Letters 86 (9) (2001) 1889–1891.
38. J. Preskill, Fault-tolerant quantum computation, in: H.-K. Lo, S. Popescu, T. Spiller (Eds.), Introduction to Quantum Computation and Information, World Scientific (1998) pp. 213–269, http://xxx.lanl.gov/abs/quant-ph/9712048.
39. P. W. Shor, Scheme for reducing decoherence in quantum computer memory, Physical Review A 52 (1995) 2493–2496.
40. A. M. Steane, Error correcting codes in quantum theory, Physical Review Letters 77 (5) (1996) 793–797.
41. C. H. Bennett, D. P. DiVincenzo, J. A. Smolin, W. K. Wootters, Mixed state entanglement and quantum error correction, Physical Review A 54 (1996) 3824–3851, http://arxiv.org/abs/quant-ph/9604024.
42. R. Laflamme, C. Miquel, J. P. Paz, W. H. Zurek, Perfect Quantum Error Correction Code, http://arxiv.org/abs/quant-ph/9602019 (February 1996).
43. A. R. Calderbank, P. W. Shor, Good quantum error-correcting codes exist, Physical Review A 54 (2) (1996) 1098–1106, http://arxiv.org/abs/quant-ph/9512032.

44. A. M. Steane, Multiple particle interference and quantum error correction, Proceedings of the Royal Society of London A 452 (1996) 2551–2576.
45. D. Gottesman, Class of quantum error-correcting codes saturating the quantum hamming bound, Physical Review A 54 (1996) 1862–1868, http://arxiv.org/abs/quant-ph/9604038.
46. A. Ekert, C. Macchiavello, Quantum error correction for communication, Physical Review Letters 77 (1996) 2585–2588.
47. J. Preskill, Reliable quantum computers, Proceedings of the Royal Society of London A 454 (1998) 385–410, http://xxx.lanl.gov/abs/quant-ph/9705031.
48. A. Berthiaume, D. Deutsch, R. Jozsa, The stabilization of quantum computation, in: Proceedings of the Workshop on Physics and Computation: PhysComp '94, IEEE Computer Society Press, Los Alamitos, CA (1994) pp. 60–62.
49. A. Barenco, A. Berthiaume, D. Deutsch, A. Ekert, R. Jozsa, C. Macchiavello, Stabilization of Quantum Computations by Symmetrization, http://xxx.lanl.gov/abs/quant-ph/9604028 (April 1996).
50. A. Peres, Error Symmetrization in Quantum Computers, http://xxx.lanl.gov/abs/quant-ph/9605009 (May 1996).
51. S. G. Akl, Evolving computational systems, in: S. Rajasekaran, J. H. Reif (Eds.), Parallel Computing: Models, Algorithms, and Applications, CRC Press (2007) a modified version is available as Technical Report No. 2006-526, School of Computing, Queen's University, Kingston, Ontario, Canada.

Index

Breinigsville, PA USA
29 December 2009
229761BV00007B/60/P